Sex Ratios
Concepts and Research Methods

Covering sex allocation, sex determination and operational sex ratios, this multi-author volume provides both a conceptual context and an instruction in methods for many aspects of sex ratio research. Theory, statistical analysis and genetics are each explained and discussed in the first three sections. The remaining chapters each focus on research in one of a wide spectrum of animal, plant and microbial taxa, including sex-ratio-distorting bacteria in invertebrates, malaria parasites, birds, humans and other mammals, giving critical appraisals of such research. *Sex Ratios: Concepts and Research Methods* is primarily intended for graduate and professional behavioural and evolutionary ecologists in this field, but it will also be useful to biologists building evolutionary models, and researchers analysing data involving proportions or comparisons across phylogenetically related species.

IAN HARDY is Lecturer in Animal Population Biology at the School of Biosciences, University of Nottingham. He has published more than 50 articles, about half of which focus on sex ratios.

Sex Ratios

Concepts and Research Methods

Edited by

Ian C.W. Hardy

School of Biosciences, University of Nottingham, United Kingdom

CAMBRIDGE
UNIVERSITY PRESS

University Printing House, Cambridge CB2 8BS, United Kingdom

One Liberty Plaza, 20th Floor, New York, NY 10006, USA

477 Williamstown Road, Port Melbourne, VIC 3207, Australia

4843/24, 2nd Floor, Ansari Road, Daryaganj, Delhi - 110002, India

79 Anson Road, #06-04/06, Singapore 079906

Ruiz de Alarcón 13, 28014 Madrid, Spain

Dock House, The Waterfront, Cape Town 8001, South Africa

Cambridge University Press is part of the University of Cambridge.

It furthers the University's mission by disseminating knowledge in the pursuit of education, learning and research at the highest international levels of excellence.

www.cambridge.org
Information on this title: www.cambridge.org/9780521665780

First published 2002

A catalogue record for this publication is available from the British Library

Library of Congress Cataloging in Publication data
Sex ratios: concepts and research methods / edited by Ian C.W. Hardy.
    p.   cm.
Includes bibliographical references and index.
ISBN 0 521 66578 7
1. Sex ratio. I. Hardy, Ian.
QH481 .S47 2002
576.8′55–dc21   2001037938

ISBN 978-0-521-66578-0 Paperback

To my daughters Fenna and Zenta, who were conceived after this book, but came out long before, and to my wife Lidia who was present at the conception of all three.

Contents

Contributors

Ingrid Ahnesjö
Department of Animal Ecology
Evolutionary Biology Centre
Uppsala University
Norbyvägen 18D
752 36 Uppsala
Sweden
Email: Ingrid.Ahnesjo@zoologi.uu.se

Jacobus J. Boomsma
Department of Population Ecology
Zoological Institute
University of Copenhagen
Universitetsparken 15
2100 Copenhagen East
Denmark
Email: JJBoomsma@zi.ku.dk

Johannes A.J. Breeuwer
Institute for Biodiversity and Ecosystem
   Dynamics
University of Amsterdam
P.O. Box 94062
1090 GB Amsterdam
The Netherlands
Email: breeuwer@science.uva.nl

Andrew Cockburn
Evolutionary Ecology Group
School of Botany and Zoology
Australian National University
Canberra ACT 0200
Australia
Email: Andrew.Cockburn@anu.edu.au

James M. Cook
Department of Biology
Imperial College at Silwood Park
Ascot
Berkshire SL5 7PY
United Kingdom
Email: j.cook@ic.ac.uk

Tom J. de Jong
Institute of Evolutionary and Ecological Sciences
University of Leiden
P.O. Box 9516
2300 RA Leiden
The Netherlands
Email: DeJong@rulsfb.LeidenUniv.nl

Michael C. Double
Evolutionary Ecology Group
School of Botany and Zoology
Australian National University
Canberra ACT 0200
Australia
Email: Mike.Double@anu.edu.au

William A. Foster
Department of Zoology
University of Cambridge
Downing Street
Cambridge CB2 3EJ
United Kingdom
Email: waf1@cam.ac.uk

Ian C.W. Hardy
School of Biosciences
University of Nottingham
Sutton Bonington Campus
Loughborough, Leicestershire LE12 5RD
United Kingdom
Email: ian.hardy@nottingham.ac.uk

Edward Allen Herre
Smithsonian Tropical Research Institute
Unit 0948
APO AA 34002-0948
USA
and
Smithsonian Tropical Research Institute
Apartado 2072
Balboa
Republic of Panama
Email: HERREA@gamboa.si.edu

Martha S. Hunter
Dept. of Entomology
410 Forbes Building
University of Arizona
Tucson, AZ 85721
USA
Email: mhunter@ag.arizona.edu

Gregory D.D. Hurst
Department of Biology
University College London
4 Stephenson Way
London NW1 2HE
United Kingdom
Email: g.hurst@ucl.ac.uk

Peter G.L. Klinkhamer
Institute of Evolutionary and Ecological
  Sciences
University of Leiden
P.O. Box 9516
2300 RA Leiden
The Netherlands
Email: KLINKHAMER@rulsfb.LeidenUniv.nl

Sarah B.M. Kraak
Netherlands Institute for Fisheries
  Research
PO Box 68
1970 AB IJmuiden
The Netherlands
Email: sarahbmkraak@yahoo.co.uk

Sven Krackow
Animal Behaviour
Institute of Zoology
University of Zürich-Irchel
Winterthurerstrasse 190
8057 Zürich
Switzerland
Email: skrackow@zool.unizh.ch

Charlotta Kvarnemo
Department of Zoology
Stockholm University
106 91 Stockholm
Sweden
Email: Lotta.Kvarnemo@zoologi.su.se

John Lazarus
Evolution and Behaviour Research Group
Department of Psychology
University of Newcastle
Newcastle Upon Tyne
NE1 7RU
United Kingdom
Email: j.lazarus@newcastle.ac.uk

Sarah Legge
Evolutionary Ecology Group
School of Botany and Zoology
Australian National University
Canberra ACT 0200
Australia
Email: Sarah.Legge@anu.edu.au

Peter J. Mayhew
Department of Biology
University of York
PO Box 373
York YO10 5YW
United Kingdom
Email: pjm19@york.ac.uk

Evert Meelis
Institute of Evolutionary and Ecological Sciences
University of Leiden
P.O. Box 9516
2300 RA Leiden
The Netherlands
Email: meelis@rulsfb.leidenuniv.nl

Gösta Nachman
Zoological Institute
Department of Population Ecology
Copenhagen University
Universitetsparken 15
2100 Copenhagen East
Denmark
Email: GNachman@zi.ku.dk

Cornelis J. Nagelkerke
Institute of Systematics and Population Biology
Section Population Biology
University of Amsterdam, Kruislaan 320
1098 SM Amsterdam
The Netherlands
Email: nagelkerke@science.uva.nl

Sean Nee
Institute of Cell, Animal and Population Biology
University of Edinburgh
Edinburgh EH9 3JT
United Kingdom
Email: sean.nee@ed.ac.uk

Paul J. Ode
USDA-ARS Beneficial Insects Introduction
   Research
University of Delaware
501 S. Chapel St.
Newark, DE 19713 USA
Email: paulode@UDel.Edu

Steven Hecht Orzack
The Fresh Pond Research Institute
64 Fairfield Street
Cambridge, MA 02140
USA
orzack@freshpond.org

Ido Pen
Zoological Laboratory
University of Groningen
P.O. Box 14
9750 AA Haren
The Netherlands
Email: i.r.pen@biol.rug.nl

Andrew F. Read
Institute of Cell, Animal and Population Biology
University of Edinburgh
Edinburgh EH9 3JT
United Kingdom
Email: aread@holyrood.ed.ac.uk

Maurice W. Sabelis
Institute for Systematics and Population Biology
University of Amsterdam
Kruislaan 320
1098 SM Amsterdam
The Netherlands
Email: sabelis@science.uva.nl

Jon Seger
Department of Biology
University of Utah
257 South 1400 East
Salt Lake City
UT 84112 0840
USA
Email: seger@bionix.biology.utah.edu

Todd G. Smith
Department of Medicine
University of Toronto
Medical Sciences Building
1 King's College Circle
Toronto M5S 1A5
Ontario, Canada
Email: toddia@zoo.utoronto.ca

Richard Stouthamer
Department of Entomology
University of California
Riverside CA 92521
USA
Email: richard.stouthamer@ucr.edu

J. William Stubblefield
Cambridge Energy Research Associates
20 University Road
Cambridge MA 02138
USA
Email: bstubblefield@cera.com

Franz J. Weissing
Department of Genetics
University of Groningen
Kerklaan 30
9751 NN Haren
The Netherlands
Email: weissing@biol.rug.nl

Stuart A. West
Institute of Cell, Animal and Population Biology
University of Edinburgh
Edinburgh EH9 3JT
United Kingdom
Email: Stu.West@ed.ac.uk

Kenneth Wilson
Institute of Biological Sciences
University of Stirling
Stirling FK9 4LA
United Kingdom
Email: ken.wilson@stirling.ac.uk

Preface and acknowledgements

Like so many other things, this book was conceived while lying in bed. I had just finished writing a statistically oriented commentary on opossum sex ratios and it wasn't the first time I had felt the need for such methodological comments, but usually these had been confined to referee's reports and unpublished correspondence. It would be useful, I thought, if a collection of such comments, and much more, was compiled. The book that sprang to mind would cover not only analytical methods, but also modelling and empirical techniques for sex ratio research, and its applications, and would cover all taxa from micro-organisms to mammoths (actually, mammoths are only mentioned here but see Chapters 9 and 15 for micro-organisms). Further, methodology would be presented very much in a theoretical context, there being no reason to break with the long and excellent tradition of close exchange between theoretical and empirical sex ratio research. Inevitably, discussion of some issues could be quite practical ('hands on') while other issues would require a greater emphasis on concepts ('minds on'). Realizing what was needed, I immediately discounted the idea of writing the whole book myself. No individual could possibly have both the depth of knowledge and breadth of experience required to cover all of this adequately: it would have to be a multi-author edited volume.

The common belief that editing a multi-authored book takes just as much time and effort as simply writing the whole oneself (if that were possible) is probably an underestimate. Nonetheless, this book could not possibly have been completed without the other 34 authors, and I thank them all for their collaborations and contributions. Each chapter was formally peer reviewed by both internal referees (authors of other chapters) and

external referees: these deserve many thanks for their excellent comments and suggestions, especially those* who refereed more than one chapter. The internal referees were: Andrew Cockburn, James Cook*, William Foster, Molly Hunter, Sarah Kraak, Sven Krackow*, John Lazarus, Sarah Legge*, Peter Mayhew*, Evert Meelis*, Gösta Nachman, Kees Nagelkerke, Paul Ode, Andrew Read, Jon Seger*, Richard Stouthamer, Franjo Weissing, Stu West and Ken Wilson. The external referees were: Monique Borgerhoff Mulder, Andrew Bourke, Graeme Buchanan, Austin Burt, Conrad Cloutier, Mick Crawley, Martin Daly, Tony Dixon, Alison Dunn, Mercy Ebbert, Mark Fellowes, Scott Field, Steve Frank*, Simon Gates, Richard Green, Darryl Gwynne, Mel Hatcher, Kate Lessells, Mike Mesterton-Gibbons*, Nancy Moran, Pekka Pamilo, Dave Parker*, Ric Paul, Jay Rosenheim, Ben Sheldon*, Mike Strand, Ian Swingland, Doug Taylor* and two who wish to remain anonymous. I thank Tracey Sanderson, Sarah Jeffery and Sarah Price at Cambridge University Press for help and support and Rob Heinsohn and Jean-Yues Rasplus for the cover photographs. I take this opportunity to thank Charles Godfray (Silwood Park), Jacques van Alphen (Leiden), Malcolm Young (Newcastle), Jan Kozłowski (Kraków), Dave Parker (Århus), Alberto Tinaut (Granada), Wallace Arthur (Sunderland) and Don Grierson (Nottingham) for employing or hosting me in the above places over the years, and all those who have ever collaborated with me, in this book and elsewhere, on sex ratio research and related issues. Finally, I thank the late WD Hamilton for his extraordinary insights into sex ratio evolution.

I.C.W.H.
Sutton Bonington, December 2001

Part I

Sex ratio theory

Chapter 1

Models of sex ratio evolution

Jon Seger & J. William Stubblefield

1.1 | Summary

Our understanding of sex ratio evolution depends strongly on models that identify: (1) constraints on the production of male and female offspring, and (2) fitness consequences entailed by the production of different attainable brood sex ratios. Verbal and mathematical arguments by, among others, Darwin, Düsing, Fisher, and Shaw and Mohler established the fundamental principle that members of the minority sex tend to have higher fitness than members of the majority sex. They also outlined how various ecological, demographic and genetic variables might affect the details of sex-allocation strategies by modifying both the constraints and the fitness functions. Modern sex-allocation research is devoted largely to the exploration of such effects, which connect sex ratios to many other aspects of the biologies of many species. The models used in this work are of two general kinds: (1) expected-future-fitness or tracer-gene models that ask how a given sex allocation will affect the future frequencies of neutral genes carried by the allocating parent, and (2) explicit population-genetic models that consider the dynamics of alleles that determine alternative parental sex allocation phenotypes. Each kind of model has different strengths and weaknesses, and both are often essential to the full elucidation of a given problem.

1.2 | Introduction

Males and females are produced in approximately equal numbers in most species with separate sexes, regardless of the mechanism of sex determination, and in most hermaphroditic species individuals expend approximately equal effort on male and female reproductive functions. Why should this be so? Sex allocation is a frequency-dependent evolutionary game (Charnov 1982, Maynard Smith 1982, Bulmer 1994). The basic principle that explains why balanced sex ratios evolve so often was described in a limited and tentative way by Darwin (1871), further developed by Karl Düsing (1883, 1884) and several early twentieth century authors, and then summarized concisely by RA Fisher in *The Genetical Theory of Natural Selection* (1930) (Edwards 1998, 2000). Subsequent work has generalized the principle and extended it to cover a great variety of special circumstances to which Fisher's elegant but elementary account does not apply.

Sex allocation is now remarkably well understood, and this understanding is often hailed as a triumph of evolutionary theory. However, to say that the fundamentals may be well understood is not to say that all of the interesting and important discoveries have been made. Despite its focus on a seemingly simple and singular phenomenon, sex-allocation research has become a rich and diverse enterprise that makes contact with many aspects of biology in a wide range of taxa. The field has continued to yield surprising phenomena and novel insights, and the pace of discovery shows no signs of slowing. The book you are now holding illustrates the field's amazing richness and describes many of the current research frontiers. But because the field is so large, not even a multi-authored volume can cover it all. Reviews of varying emphasis and scope have been provided by Williams (1979), Charnov (1982, 1993), Trivers (1985), Clutton-Brock (1986, 1991), Clutton-Brock and Iason (1986), Karlin and Lessard (1986), Nonacs (1986), Bull and Charnov (1988), Frank (1990, 1998), Wrensch and Ebbert (1993), Bulmer (1994), Godfray (1994), Bourke and Franks (1995), Crozier and Pamilo (1996), Herre *et al.* (1997), Klinkhamer *et al.* (1997), Hewison and Gaillard (1999) and West *et al.* (2000), among many other authors.

In this chapter we introduce the central principle of sex ratio evolution and some of the techniques used to model it. We emphasize basic concepts and issues that appear (at least implicitly) in all models, and we attempt to place these ideas in their historical context.

1.3 | Models have always been central

Mathematical models are, and always have been, central to the study of sex ratios. Indeed, it is hard to think of any biological field, associated with specific phenotypes, that is more thoroughly model-driven. Population genetics is also model-driven in this sense, of course, but its models concern genes *in general*; the genes of population genetics are abstracted, intentionally, from any particular class of phenotypic effects.

The sex ratio, by contrast, could hardly be more concrete. This is sometimes forgotten, because every sexual species has a sex ratio (or at least allocation to male and female functions). But in fact the phenotype at issue (the relative numbers of two reproductive morphs) is in many ways an extremely particular and mundane fact

of life. Even so, biologists from Darwin to the present have sensed an underlying generality of principle. They have spoken of 'the' sex ratio (singular), as if to understand the sex ratio of any *one* species would be (obviously) to understand the sex ratios of many *others*. Today we have good reasons to view 'the' sex ratio in this way, but most of these were unknown to Darwin. Nonetheless, he initiated the modern discussion of sex ratios, in *The Descent of Man and Selection in Relation to Sex* (1871), by describing the outlines of a quantitative, dynamical model that includes most of the essential features of everything that would follow. Formal mathematical analysis came later, as did direct connections to genetics, and these developments gave rise to a richness that Darwin could not have anticipated. Even so, he saw that there must be a simple underlying principle to be elucidated and then (by implication) applied to a broad diversity of special cases. We still see the subject in this way.

The principle emerges from an analysis of the reproductive consequences of an elementary but generic model of reproduction. The principle is then applied and extended by specifying details that may be left vague in the generic model, which is to say by modifying various implicit and explicit assumptions of the model.

Sex ratio modelling has been an extremely successful enterprise. This success can be attributed to three features of the relationship between the models and reality. First, the relevant biological factors can be specified and represented appropriately in simple mathematical expressions. Second, these factors can be observed and measured in nature, and many of them vary both within and among species in ways that are predicted to change the sex ratios produced by different individuals or species. Third, the fitness differences arising from sex ratio behaviours are often large, so real organisms are expected to show sex ratio modifications at least qualitatively like those predicted by theory, and in fact they often do. In this chapter we focus mainly on the first of these three features of sex ratio research: how biology is represented in models, and how the models are then analysed to uncover predictions that might (at least in principle) be tested in nature. Other chapters more thoroughly explore the variations that have been incorporated into sex ratio models, and the ways in which experimental and observational data have been used to test these models.

1.4 | Darwin's argument

As its title implies, *The Descent of Man and Selection in Relation to Sex* (1871) is really two books merged into one. The book on human origins begins with Chapter I, 'The Evidence of the Descent of Man from some Lower Form', and the book on sexual selection begins with Chapter VIII, 'Principles of Sexual Selection'. Darwin opens the chapter by explaining that sexual selection is 'that kind of selection' that 'depends on the advantage which certain individuals have over other individuals of the same sex and species, in exclusive relation to reproduction' (page 256). Sexual selection is about *relative* advantage in the competition for mates, not about survival or absolute competence to reproduce.

> When the two sexes follow exactly the same habits of life, and the male has more highly developed sense or locomotive organs than the female, it may be that these in their perfected state are indispensable to the male for finding the female; but in the vast majority of cases, they serve only to give one male an advantage over another, for the less well-endowed males, if time were allowed them, would succeed in pairing with the females; and they would in all other respects, judging from the structure of the female, be equally well adapted for their ordinary habits of life.
>
> *(page 257)*

Darwin then describes several kinds of sex differences that seem to make sense on this principle; for example, the generally earlier emergence of male insects. He notes that the intensity of the competition for mates will be a function of the sex ratio and then opens a section titled 'Numerical Proportion of the Two Sexes' (page 263).

> I have remarked that sexual selection would be a simple affair if the males considerably exceeded

in number the females. Hence I was led to investigate, as far as I could, the proportions between the two sexes of as many animals as possible; but the materials are scanty. I will here give only a brief abstract of the results, retaining the details for a supplementary discussion, so as not to interfere with the course of my argument. Domesticated animals alone afford the opportunity of ascertaining the proportional numbers at birth; but no records have been specially kept for this purpose. By indirect means, however, I have collected a considerable body of statistical data, from which it appears that with most of our domestic animals the sexes are nearly equal at birth.

Darwin's numbers show rough equality or modest male excesses at birth for various domestic species and for humans. He then points out that 'we are concerned with the proportion of the sexes, not at birth, but at maturity,' because that is when the competition for mates will occur. His data here are less definite, but they suggest greater male mortality and thus a relative *deficit* of males at maturity. However, 'The practice of polygamy leads to the same results as . . . an actual inequality . . . for if each male secures two or more females, many males will not be able to pair; and the latter assuredly will be the weaker or less attractive individuals.' Pages 266–279 then review patterns of polygamy and sexual dimorphism, and pages 279–300 discuss the 'laws of inheritance' of secondary sexual characters.

The chapter then returns to the problem of the sex ratio. Pages 300–315 present a detailed 'Supplement on the proportional numbers of the two sexes in animals belonging to various classes' (humans, horses, sheep, birds, fish and insects). A final short section 'On the Power of Natural Selection to regulate the proportional Numbers of the Sexes, and General Fertility' (pages 315–320) lays out the evolutionary argument. Its second paragraph (page 316) begins as follows:

> Let us now take the case of a species producing . . . an excess of one sex—we will say of males—these being superfluous and useless, or nearly useless. Could the sexes be equalized through natural selection? We may feel sure, from all characters being variable, that certain pairs

would produce a somewhat less excess of males over females than other pairs. The former, supposing the actual number of the offspring to remain constant, would necessarily produce more females, and would therefore be more productive. On the doctrine of chances a greater number of the offspring of the more productive pairs would survive; and these would inherit a tendency to procreate fewer males and more females. Thus a tendency towards the equalization of the sexes would be brought about.... The same train of reasoning is applicable... if we assume that females instead of males are produced in excess, for such females from not uniting with males would be superfluous and useless.

Parents that produce an excess of the minority sex will be 'more productive' because fewer of their offspring will be 'superfluous'. The paragraph says more of these offspring will 'survive', but this is illogical. Perhaps Darwin meant 'reproduce', or perhaps he was confused about the cause of the differential productivity. The paragraph asserts that parents of the minority sex will enjoy a productivity advantage, no matter which sex is 'produced in excess', and it indicates that the sex in excess will suffer increased failure to mate ('not uniting'). But does the paragraph show *how* these effects modulate parental fitness? It certainly contains all the elements and reaches the right conclusion, but it does not clearly explain why, or in what sense, parents of the minority sex are 'more productive'. In retrospect it comes extremely close (see Sober 1984, Bulmer 1994, Edwards 1998), but it does not explain what will happen in the generation of the parents' grandprogeny.

The next paragraph (pp. 317–318) presents both an advance and a retreat. The advance is an overt anticipation of the concept of parental expenditure or investment (as 'force'). In the previous paragraph, Darwin had explicitly noted the trade-off between numbers of male and numbers of female offspring; in this paragraph he explicitly notes the trade-off between offspring number and offspring quality. Parents that produce fewer 'superfluous males' but 'an equal number of productive females' would probably benefit, as a consequence, from 'larger and finer' ova or embryos, and 'their young [would be] better

nurtured in the womb and afterwards.' In support of this idea, Darwin notes that inverse relationships between seed number and seed size can be seen both among and within species of plants. 'Hence the offspring of the parents which had wasted least force in producing superfluous males would be the most likely to survive, and would inherit the same tendency not to produce superfluous males, whilst retaining their full fertility in the production of females. So it would be with the converse case of [an excess of] the female sex.'

The retreat is a muddled explanation of the disadvantages experienced by 'superfluous' offspring. For purposes of argument, Darwin had begun the paragraph assuming that there was an excess of males, and that some parents produced fewer of them but a typical number of females. 'When the offspring from the more and the less male-productive parents were all mingled together, none would have any direct advantage over the others.' This is not true in the sense that he seems to intend. The offspring might be equivalent *individually* (leaving aside the 'indirect' benefits noted above), but not *collectively*; parents that produced more males would have more descendants *through males* than those that produced fewer males, given that the offspring 'were all mingled together'. In this sense sons are not 'superfluous' even when produced in excess. Darwin seems to be imagining that parents that contribute to the male excess will have no more grandoffspring through their sons (collectively) than those parents that refrain from producing excess males.

In the second edition of the *Descent* (1874), most of Chapter VIII is similar to that of the first edition, but the final section is completely different. It is renamed 'The proportion of the sexes in relation to natural selection', and it consists mainly of an inconclusive discussion of the relationship between sex-biased infanticide and the primary sex ratio. It concludes:

> In no case, as far as we can see, would an inherited tendency to produce both sexes in equal numbers or to produce one sex in excess, be a direct advantage or disadvantage to certain individuals more than to others; for instance, an individual with a tendency to produce more

males than females would not succeed better in the battle for life than an individual with an opposite tendency; and therefore a tendency of this kind could not be gained through natural selection.... I formerly thought that when a tendency to produce the two sexes in equal numbers was advantageous to the species, it would follow from natural selection, but I now see that the whole problem is so intricate that it is safer to leave its solution for the future.

Why did Darwin abandon his own previous argument which was close to the 'solution' and clearly moving in the right direction? On one reading of the 1874 retraction, he considers the 1871 argument to be flawed by a reliance on species-benefit reasoning. Consistent with such an interpretation, the paragraph laying out the evolutionary argument (1871, p 316) includes an extraneous and confused aside on the adjustment of fertility, which we deleted from our earlier quotation.

> ...But our supposed species would by this process be rendered, as just remarked, more productive; and this would in many cases be far from an advantage; for whenever the limit to the numbers which exist, depends, not on destruction by enemies, but on the amount of food, increased fertility will lead to severer competition and to most of the survivors being badly fed. In this case, if the sexes were equalized by an increase in the number of the females, a simultaneous decrease in the total number of the offspring would be beneficial, or even necessary, for the existence of the species; and this, I believe, could be effected through natural selection in the manner hereafter to be described.

Why Darwin should invoke, here, the concept of species' benefit (or need!) seems baffling. Two pages later, as promised, he describes in two paragraphs how reduced fertility (offspring number) could evolve by ordinary natural selection, given trade-offs between maintenance and reproduction, and between offspring number and quality. These two paragraphs end the chapter and brilliantly anticipate late-twentieth-century developments in life-history theory. They contain no species-benefit reasoning that we can detect. Darwin credits Herbert Spencer's *Principles of Biology* (1867) for inspiration on this subject.

It seems odd that Darwin should have lost his nerve and failed to correct confusions that were probably no worse than hundreds that he must have surmounted in other contexts. His decision to remove the entire argument from the second edition of the *Descent* (1874) can be taken to support the view that he never really understood the principle as well as a generous reading might suggest he did at the time he wrote it. He sees a close connection between sexual selection and the sex ratio: as the number of males competing for each productive mating increases, their average reproductive success must decrease. But he does not seem to recognize that he should directly compare the average fitnesses of males and females, and that he should evaluate the fitnesses of parents by counting their grandprogeny. In any case, his decision to remove the evolutionary argument from the second edition undoubtedly changed the history of behavioural ecology. The second edition was reprinted far more extensively than the first and became the edition read by almost everyone, including RA Fisher (Edwards 1998).

The recognition that sex ratios evolve through negatively frequency-dependent selection on the relative reproductive success of male and female offspring is traditionally attributed to Fisher (1930). His two-page verbal argument is informed by a knowledge of genes and it is far more lucid than Darwin's, but otherwise it is very similar in spirit. Why does Fisher not credit Darwin? One explanation is that, like most of his contemporaries, Fisher had read the second edition of the *Descent* and understandably failed to see any reason to persue Darwin's hint about what he 'formerly thought'. Edwards (1998) has shown that Darwin's initial lead was picked up by Düsing and several early twentieth century authors who further clarified the argument, and that Fisher was almost certainly aware of at least some of these later works. Why does Fisher not cite them either? Edwards suggests that Fisher understood his own account of the principle to be derived from these sources, that he assumed his interested contemporaries also would have been aware of them, and that standards of scholarly attribution were not as strict in 1930 as they are today. These factors could explain why Fisher (1930) presents the principle so casually.

1.5 | The elements of a sex ratio model

A fully specified model of evolutionary adaptation can be viewed as a proposal showing how certain biological circumstances will give rise to a *fitness function* and a set of *constraints*. These relations are typically referred to as the *assumptions*, because the modeller is free, in principle, to change them in arbitrary ways. Models based on relatively 'realistic' assumptions are often considered more scientifically 'interesting' than those based on unrealistic assumptions, but, as Fisher himself points out in the preface to *The Genetical Theory of Natural Selection* (1930), models cannot really illuminate the natural world without also illuminating unnatural worlds. 'No practical biologist interested in sexual reproduction would be led to work out the detailed consequences experienced by organisms having three or more sexes; yet what else should he do if he wishes to understand why the sexes are, in fact, always two?' A model becomes explicitly mathematical when it embodies its assumptions ('three sexes', for example) in a set of formal quantitative relations that can be evaluated to reveal expected evolutionary outcomes. These deductions, following from the assumptions, can be interpreted as predictions about what would be expected to happen if the world actually worked as the assumptions propose it does. Such a derivation of expected consequences of the assumptions is often referred to as an *analysis* of the model.

There are two distinctive but complementary approaches to setting up and analysing explicit sex ratio models. The older, more intuitive and more expressive approach employs 'expected-future-fitness' calculations similar in spirit to those used in many inclusive-fitness and quantitative-genetic models. In this approach, sex-allocation strategies are evaluated with respect to the expected future frequencies of selectively neutral genes (tracers of descent) carried by an individual parent that exhibits a given sex ratio phenotype. The younger, more rigorous but less transparent approach employs dynamical population-genetic models to ask under what circumstances an allele that determines a specific parental sex ratio phenotype can invade (or fix) against an allele that determines a different phenotype. Both kinds of models can vary widely in sophistication and complexity. Neither is inherently 'better'; the choice of approach is largely a matter of taste and the nature of the problem being considered (see Bulmer 1994). We will illustrate both approaches.

Even in its original verbal form, the Darwin–Fisher argument is a legitimate (if primitive) sex ratio model. It is only marginally mathematical, but that does not disqualify it as a model. The relevant assumptions are clearly identified, most importantly: (1) that sex-specific fitness differences arise from an inevitable 'competition' for mates, which implies a *fitness function*, and (2) that parents that produce more sons (or daughters) must necessarily produce fewer daughters (or sons), which implies a *constraint*. The implicitly quantitative analysis proceeds as follows. Parents that overproduce the minority sex will have offspring that enjoy greater than average reproductive success, on average. Therefore, any heritable variants that tend to cause overproduction of the minority sex will increase in frequency, and as they do so the sex ratio imbalance will decrease. Because this is true no matter which sex is currently under-represented, there must be a stable evolutionary equilibrium at which male and female offspring are produced in approximately equal numbers. If all parents were to produce equal numbers of females and males, then no other sex ratio phenotype could increase under selection. Today we would call this unbeatable phenotype an 'evolutionarily stable strategy' or ESS (Maynard Smith & Price 1973).

1.6 | Düsing's model

The first general mathematical treatment of sex ratio evolution has long been attributed to Shaw and Mohler (1953), who derived an elegant formalization of Fisher's argument. However,

Edwards (1998), in reconstructing Fisher's sources, discovered that a similar mathematical treatment had been published almost 70 years earlier by Karl Düsing. His Ph.D. dissertation (Düsing 1883, expanded to book length in 1884) is mainly a study of factors associated with variation in progeny sex ratios in 'man, animals and plants' (e.g. maternal age and parity). In the early pages of this work, Düsing poses and answers a question that leads him to construct what is undoubtedly the first formal sex ratio model and perhaps the first mathematical model in evolutionary biology.

Given that animals vary their sex ratios in response to particular conditions of life, why do we not see large overall sex ratio imbalances? The reason, Düsing says, is that deviations from a balanced sex ratio will tend to be self-correcting: an excess of one sex provides a reason to produce more of the other. To make the argument concrete, he assumes a population in which there is a lack of females, and points out that *all the males together have the same number of offspring as all the females*. Because the latter are (by assumption) in the minority, each will have on average more offspring. For example, if there are x females and nx males, and if they produce z offspring in all, then each female will produce z/x offspring and each male will produce z/nx (Düsing 1884 p. 10, see Edwards 2000 for a full translation of the argument). He points out that if a female produced more female offspring, these daughters would produce, collectively, a larger than average number of offspring. Suppose a female produces A sons and a daughters, and another produces the converse (A daughters and a sons). The first will have

$$A\frac{z}{nx} + a\frac{z}{x} \qquad (1.1)$$

grandchildren and the second will have

$$a\frac{z}{nx} + A\frac{z}{x}. \qquad (1.2)$$

If we assume that $A > a$, such that $A = ba$ (with $b > 1$), then the first female will contribute

$$\frac{az}{x}\left[\frac{b}{n} + 1\right] \qquad (1.3)$$

individuals to the second generation, while the second female will contribute

$$\frac{az}{x}\left[\frac{1}{n} + b\right] \qquad (1.4)$$

which is

$$\frac{1 + bn}{b + n} \qquad (1.5)$$

times as many. Düsing notes that if the population sex ratio is balanced ($n = 1$), then this expression evaluates to 1 for any sex ratio. *No matter what progeny sex ratio a female produces, she will have the same number of descendants in the second generation.* But not so if the sex ratio is unbalanced. For example, if there are twice as many males as females, then the ratio of grandchildren will be

$$\frac{1 + 2b}{b + 2} \qquad (1.6)$$

as a function of the difference in progeny sex ratios (b). Düsing contrasts the fitnesses of two females for which $b = 3$; the one producing a threefold female excess has 7/5 as many grandchildren as the one producing a threefold male excess.

In less than two pages, Düsing both clarifies Darwin's argument and quantifies it. He identifies the key underlying fact that total male and female fitnesses must be equal; he identifies relative numbers of grandchildren as the appropriate measure of fitness; he writes a general expression for fitness as a function of the parent's progeny sex ratio b given the population sex ratio n; and he discovers that fitness is unaffected by progeny sex ratios if and only if the population sex ratio is balanced (in effect, the ESS argument). Having given a general theoretical reason why progeny sex ratio adjustment might be advantageous to individuals, he then embarks on a massive empirical review of such adjustments and their correlates in many species. Apparently this subject was as interesting and controversial in Düsing's time as it is today; his analyses of the patterns were much discussed, and his evolutionary model was forgotten (Edwards 1998, 2000, SH Orzack, pers. comm.).

1.7 Fisher's equal-investment principle

Fisher's (1930) explanation of the sex ratio principle is as brief as Düsing's, but purely verbal and very well known, so we will not dwell on it here except to note that it presents a very important generalization of the earlier arguments. Fisher carefully considers the nature of the constraint on male and female offspring production, and discovers that the sex ratio equilibrium concerns the distribution of parental effort or 'expenditure' (later generalized by Trivers 1972 as 'investment'), not numbers *per se*. For example, suppose daughters are twice as costly to produce as sons. Then a parent with the resources to produce 12 sons might instead produce six daughters. What is the evolutionary equilibrium in this case? At a numerical sex ratio of 1:1 a typical parent could have four sons and four daughters. Males and females will have equal average reproductive success, so a rare male-specialist parent (with 12 sons) would have many more grand-offspring than an average parent (with eight offspring in all), and this advantage would increase the proportion of male-specialist parents (and males) in future generations. Only when males became twice as numerous as females (six sons and three daughters in a typical brood) would parents become evolutionarily indifferent to the sexes of their offspring. Sons would be only half as successful as daughters, but also only half as expensive. Thus, over the population as a whole, we expect to find equal expenditure or investment in the two sexes, not necessarily equal numbers.

This generalization leads immediately to testable predictions. In species where one sex is more costly (to parents) than the other, that sex should tend to be produced in correspondingly smaller numbers. This prediction has held up well in many recent studies of sexually dimorphic social and solitary Hymenoptera. Fisher was aware that human males suffer higher mortality rates in childhood than do females, rendering them less costly per infant born. He argues that the slight but conspicuous male excess at birth is plausibly an adjustment to equalize overall investment in the sexes (at least under patterns of mortality that would have existed in early human societies). This example illustrates the logic that has been used many times since then to connect sex allocation with other aspects of biology.

1.8 Genetic models I: tracer genes and the Shaw–Mohler equation

Shaw and Mohler (1953) set out to formalize Fisher's argument and connect it more closely to genetics. Their model is extremely simple and transparent, and it forms the basis of most subsequent sex ratio models. The key idea is to calculate the contribution that a parent in one generation (P) makes to the gene pool in the second descending generation (that of its offspring's offspring, G_2), if the parent produces a sex ratio x (proportion males) in the G_1 (offspring) generation where the average sex ratio is X. The focal parent produces n offspring in all, and the population at large produces N. In G_1 there will be NX males and 'all together they will supply half the genes which are transmitted from G_1 to G_2', so each male's share will be $1/2NX$. The focal parent's sons therefore contribute $nx/2NX$ of the genes in G_2, and its daughters contribute $n(1 - x)/2N(1 - X)$, for a total of

$$\frac{1}{2}\left[\frac{nx}{NX} + \frac{n(1 - x)}{N(1 - X)} \right].$$

(1.7)

The parent contributes half of the genes carried by each of its nx sons and $n(1 - x)$ daughters in G_1, so its net genetic contribution to G_2 is

$$C = \frac{1}{4}\left[\frac{nx}{NX} + \frac{n(1 - x)}{N(1 - X)} \right],$$

(1.8)

or

$$C = \frac{1}{4}\left(\frac{m}{M} + \frac{f}{F} \right),$$

(1.9)

'where m and f are the numbers of male and female zygotes in the [focal] progeny while M and F stand for the corresponding numbers in the entire G_1'. C is a measure of genetic fitness because it can be interpreted as the expected frequency in future generations of a selectively neutral tracer allele that in generation P was carried only by the focal parent. If the population sex ratio is balanced ($X = 1/2$), then $C = n/2N$ independent of the focal parent's sex ratio (as long as n does not depend on x). But if X is any other value, then some sex ratios x will give rise to larger contributions than others. 'The gene or genes favored are always those whose increase will shift the population sex ratio (X) toward 0.5.'

The paper goes on to show that the equilibrium progeny sex ratio is not affected if male and female offspring, once produced, survive to adulthood with different probabilities; these probabilities cancel out of the expressions for m/M and f/F. Curiously, the paper does *not* extend the analysis to include sexually dimorphic mortality rates *during* the period of parental care, or other sources of differential offspring *costs*, even though Fisher considered this extension verbally and noted the implication that m, M, f and F can be interpreted more generally as net parental expenditures on behalf of male and female offspring. Bodmer and Edwards (1960) modelled Fisher's argument by writing an expression for the reproductive value produced by a unit of parental expenditure, given sex-specific intrinsic costs and probabilities of surviving the period of parental care. This *rate of return* measures 'the selective advantage attached to reproduction with particular sex and parental expenditure ratios'; it is independent of the focal parent's progeny sex ratio when 'the total parental expenditure incurred in respect of children of each sex is equal', confirming 'Fisher's Law'.

Because the *total* (population-wide) male and female investments M and F are directly proportional to the *average* (individual) investments, we are free to normalize the Shaw–Mohler equation to give an average fitness of 1, in keeping with modern conventions in other areas of population genetics

$$W = \frac{1}{2}\left(\frac{f}{F} + \frac{m}{M}\right),\qquad(1.10)$$

where F is the average value of f in the population and M is the average value of m. A simple analysis that explicitly incorporates differential costs can then be carried out as follows. The constraint on allowable combinations of female and male offspring can be represented by an equation that specifies the number of daughters that a parent will produce if it also produces m sons. For example, assume the simplest kind of linear trade-off between male and female production, and let each daughter cost c times as much as a son. Then the constraint is $cf + m = r$, or

$$f = \frac{r - m}{c},\qquad(1.11)$$

where r is the total resource available for offspring production (in units of the cost of a son). Substituting the constraint (eq. 1.11) into the fitness function (eq. 1.10) we get

$$W = \frac{1}{2}\left(\frac{\frac{r-m}{c}}{\frac{R-M}{c}} + \frac{m}{M}\right).\qquad(1.12)$$

Note that the relative cost of a female (c) cancels out, and that without any loss of generality we are free to set the average resource (R) equal to 1. Thus the constrained fitness function simplifies to

$$W = \frac{1}{2}\left(\frac{r - m}{1 - m} + \frac{m}{M}\right).\qquad(1.13)$$

What sex ratio (m) will maximize our parent's fitness, given that it has r units of resource? Of course we already know what the answer is supposed to be: parents should expend equal amounts of resource on each sex. If that's what *typical* parents are doing, then $M = 1/2$, and the fitness function further simplifies to

$$W = [(r - m) + (m)] = r.\qquad(1.14)$$

Now the parent's sex allocation (m) also cancels out, and its fitness is simply equal to its resource pool. We have explicitly derived the result that each parent is indifferent to the sexes of its own offspring, even where male and female costs differ, as long as there is equal overall *investment* in the population at large ($M = 1/2$).

Of course parents are far from indifferent when overall investment is not equal, and the Shaw–Mohler equation quantifies the fitness differences associated with atypical ('mutant') progeny sex ratios in populations that are away from the evolutionary equilibrium (Figure 1.1). Not only do parents that invest equally in sons and daughters do better than average when most other parents are investing unequally, but parents that over-compensate do even better, and

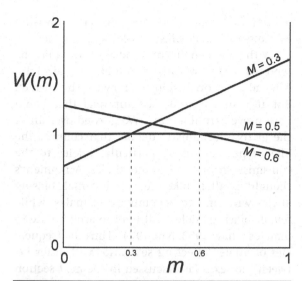

Fig 1.1 Fitness as a function of brood sex ratio in populations with different average sex ratios, as described by the Shaw–Mohler equation. $W(m)$ is the parent's expected genetic representation in the generation of its grandprogeny, given that it has one unit of resource ($r = 1$) which it uses to produce m male and $1 - m$ female offspring in a population where the average proportional allocation to males is M (0.3, 0.5, or 0.6).

those that produce the under-represented sex exclusively do best of all. As we mentioned earlier, the selection coefficients associated with sex ratio differences can be huge relative to those believed to account for much real adaptive evolution, and they can be large relative to those needed theoretically to overpower drift even in very small populations. For example, consider a population out of equilibrium by only 1% ($M = 0.49$). Then a parent with one unit of resource will have a fitness of

$$W = \frac{1}{2}\left(\frac{1-m}{0.51} + \frac{m}{0.49}\right). \qquad (1.15)$$

A typical parent ($m = 0.49$) has a fitness of 1.0, while a nearly identical 'Fisherian' parent ($m = 0.5$) has a fitness of 1.0004 (0.04% above average) and a fully overcompensating parent ($m = 1$) has $W = 1.02$ (2% above average). The Fisherian ($m = \frac{1}{2}$) parent's advantage increases rapidly with the size of the population's deviation from equilibrium, reaching 1% when the deviation reaches 5% ($M = 0.45$ or $M = 0.55$).

In deriving this model we made some simplifying assumptions that do not always hold

true. For example, we assumed that the population is effectively infinite and randomly mating, that generations are discrete and nonoverlapping, that the constraint on male and female offspring numbers is linear ($cf + m = r$) and that the fitnesses of individual female and male offspring are independent of brood sex ratios. With respect both to their production and to their reproductive values, positively and negatively synergistic interactions between sons and daughters can be imagined and, for certain taxa, documented. The Shaw–Mohler framework can be extended to allow for such nonlinearities, and two classic examples (local mate competition and hermaphroditic plants) are considered in section 1.10. The fitness function can be expanded to account for differences in ploidy (e.g. haplodiploidy), to account for differences in the focal parent's (or other caregiver's) relatedness to male and female offspring (e.g. workers in social Hymenoptera), to account for differences in situation-specific male and female fitnesses (e.g. offspring of high- and low-ranking mothers in some ungulate and primate species), and to account for overlapping generations (see Chapter 2).

Although straightforward in principle, these and other extensions may greatly complicate the analysis of the resulting model. Since both the evolutionary and analytical objectives are to maximize $W(m, f)$ subject to constraints, techniques from optimal control theory and other branches of applied mathematics are sometimes used to find the evolutionarily stable allocation strategies (e.g. Macevicz & Oster 1976, Oster & Wilson 1978). Probabilistic approaches may also be necessary, as in the small-population case where stochastic fluctuations of the sex ratio will be large and the total allocation to males and females (M, F) will not be effectively independent of the focal parent's allocation (m, f). Here parents are not indifferent to their own progeny sex ratios even when the population is at its evolutionary equilibrium, as first noted by Verner (1965).

MacArthur (1965) identified an interesting corollary of the Shaw–Mohler formulation that holds in many but not all models with nonoverlapping generations: at equilibrium the *product* of the *numbers* of females and males ($N_f N_m$) is maximized, even where individuals of one sex cost more than the other (see Maynard Smith 1978, 1982, Charnov 1982, Karlin & Lessard 1986).

At this point it may be useful to review the core assumptions and logic of this historic neutral-gene-transmission model that still strongly influences the way we conceive and analyse selection on sex ratios. The evolutionary pay-off associated with a given sex ratio phenotype is assumed to be proportional to the reproduction of the parent's offspring. Shaw and Mohler explicitly invoke the transmission of genes, and even refer to their paper as a discussion of the 'population genetics of autosomal genes affecting the primary sex ratio', although no alleles or gene frequencies appear anywhere in it. (Düsing knew nothing of genes, of course, but intuitively knew that maternal and paternal contributions would be of equal evolutionary importance.) Thus the sex ratio differences among parents are assumed to be caused at least in part by genetic variants that the parents will transmit to their offspring. Then, as Darwin almost argued, the offspring of parents that over-produce the under-represented sex will have more offspring, and their offspring will (as he did argue, but then doubted) 'inherit a tendency to procreate fewer' of the over-represented sex and more of the under-represented sex, so that 'a tendency towards the equalization of the sexes [will] be brought about'.

In the Shaw–Mohler formulation, this reasoning is embodied in an equation that expresses the total expected reproduction by offspring of parents that produce different sex ratios in a population with a given overall sex ratio. This measure of fitness is explicitly constructed to reflect the transmission to distant generations of selectively neutral genes carried by the parents. We find that overproducing the minority sex (more generally, the under-invested sex) always yields greater than average fitness. We interpret fitness (defined in this way) as a metric indicating the expected evolutionary fates of alleles that incline their bearers to produce different progeny sex ratios; such alleles are implicitly ones of small individual effect, possibly occurring at many different genetic loci scattered throughout the genome. We conclude that a population fixed for a 'Fisherian' genotype

$(m = M = \frac{1}{2})$ should not be subject to invasion by male- or female-biasing alleles at any loci.

With the benefit of hindsight we may be tempted to view this argument as air-tight. After all, its conclusion is known to be correct. But it rests on some assumptions that could have proved troublesome. We glossed over all of the gritty mechanistic details that connect the phenotypes caused by particular alleles to the transmission of those same alleles. A moment's thought is all it takes to see that male-biasing alleles will tend to accumulate in males, while female-biasing alleles will tend to accumulate in females (Shaw 1958, Nur 1974). Thus the frequencies of alleles affecting sex ratios will differ between the sexes (as discussed in the next section and illustrated in Figure 1.3). Might this affect the evolutionary outcome in a species where sex ratios are determined by just one parent (say, the mother)? The answer is not obvious, so we need also to construct and analyse dynamic population-genetic models in which these potentially critical connections are represented explicitly.

1.9 | Genetic models II: alleles that determine parental sex ratios

We were able to develop the logic of expected-future-fitness models along historical lines, because the history begins simply and then adds layers of complexity. By contrast, the history of models with alleles that specifically affect parental sex ratios is not so straightforward. The first models were, for the most part, relatively complex, opaque and lacking in generality, so they do not provide good examples with which to introduce the subject as presently understood. For this reason we will first describe a more highly derived but simple and generic model, and then look back briefly at some pioneering models from the literature.

Genetic evolution will not occur unless the genome includes at least one locus with two or more different genotypes that tend to produce different phenotypes. Often we can reasonably assume that what's true for one locus will be true (qualitatively) for others, in which case the

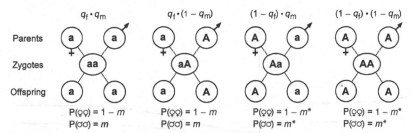

Fig 1.2 Mating types and offspring productions in a haploid genetic model. Alleles **a** and **A** occur at frequencies q_f and $1 - q_f$ in females and q_m and $1 - q_m$ in males. Male and female parents mate at random, so the four mating types occur in the proportions indicated above the pedigrees. Diploid zygotes form briefly and then undergo meiosis to form haploid spores that develop into the next generation of adults. Brood sex ratios are determined by the mother's genotype (m for mothers of genotype **a**, and m^* for mothers of genotype **A**).

problem can be represented adequately by just one locus with two alleles. Often it is also reasonable to let the species be haploid, so as to reduce the number of distinct genotypes to an absolute minimum. If we assume, in addition, that progeny sex ratios are determined by the mother's phenotype, and that parents mate randomly with respect to their sex ratio genotypes, then we have defined the very simple model that is shown in Figure 1.2 and summarized algebraically in Table 1.1. Each row in the table represents one of the four mating types illustrated in the figure. Alleles **a** and **A** act in the mother to determine her expected progeny sex ratio, m or m^*, respectively. The frequency of **a** is q_f in females and q_m in males. The entries under 'daughters' and 'sons' are the expected proportions of each progeny resulting from matings of a given type (row) that will be females or males of genotypes **a** or **A** (columns).

The *state variables* in this model are the genotype frequencies q_f and q_m. We want to know whether these gene frequencies will change and, if so, in which direction. To do this we need to write equations for q_f' and q_m' (the allele frequencies *next* generation) as functions of q_f and q_m (the allele frequencies *this* generation). This may sound difficult, but in fact it is easy given the preliminary calculations we have already placed in the table.

By definition, q_f' is the *proportion* of all daughters that will be of genotype **a**. The *total* production of daughters is $q_f(1 - m) + (1 - q_f)(1 - m^*)$. This expression goes in the denominator. For the numerator we need the total production of daughters *of genotype* **a**. This can be read directly from the table, as the sum of the products formed by multiplying each term in the 'frequency' column by the term in the 'daughters/a' column of the same row. Thus the *recurrence equation* for the female genotype frequency is

$$q_f' = \frac{q_f q_m(1-m) + \frac{1}{2}q_f(1-q_m)(1-m) + \frac{1}{2}(1-q_f)q_m(1-m^*)}{q_f(1-m) + (1-q_f)(1-m^*)}.$$

(1.16)

To make the origin and meaning of each term as easy to see as possible, we have written the equation without any further algebraic simplifications. By a similar train of reasoning, we obtain

Table 1.1 Frequencies and outputs of the four mating types in a haploid model

Mating	Frequency	Daughters a	Daughters A	Sons a	Sons A
a × a	$q_f q_m$	$(1 - m)$		m	
a × A	$q_f(1 - q_m)$	$\frac{1}{2}(1 - m)$	$\frac{1}{2}(1 - m)$	$\frac{1}{2}m$	$\frac{1}{2}m$
A × a	$(1 - q_f)q_m$	$\frac{1}{2}(1 - m^*)$	$\frac{1}{2}(1 - m^*)$	$\frac{1}{2}m^*$	$\frac{1}{2}m^*$
A × A	$(1 - q_f)(1 - q_m)$		$(1 - m^*)$		m^*

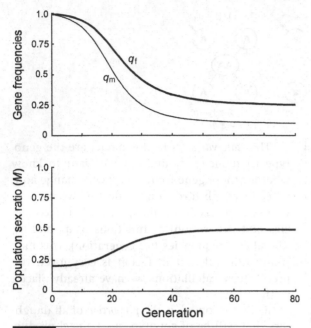

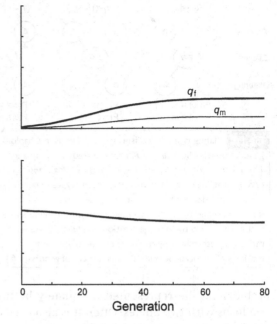

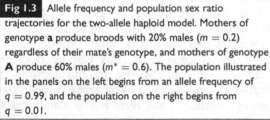

Fig 1.3 Allele frequency and population sex ratio trajectories for the two-allele haploid model. Mothers of genotype **a** produce broods with 20% males ($m = 0.2$) regardless of their mate's genotype, and mothers of genotype **A** produce 60% males ($m^* = 0.6$). The population illustrated in the panels on the left begins from an allele frequency of $q = 0.99$, and the population on the right begins from $q = 0.01$.

the corresponding recurrence equation for the male allele frequency

$$q'_m = \frac{q_f q_m(m) + \frac{1}{2}q_f(1 - q_m)(m) + \frac{1}{2}(1 - q_f)q_m(m^*)}{q_f(m) + (1 - q_f)(m^*)}.$$

(1.17)

Everything on both right-hand sides is known, so by evaluating this pair of equations we obtain the genotype frequencies for the next generation; these can then be used to obtain the genotype frequencies for the generation after that, and so on for as long as we care to *iterate* this dynamical system. Doing so by hand would be tedious, but it is easy by computer. Figure 1.3 illustrates two such calculations. Given alleles with phenotypic values flanking $^1/_2$ (in this case, $m = 0.2$ and $m^* = 0.6$), the system always converges to genotype frequencies that give $M = ^1/_2$ (in this case, $q_f = 0.25$). Males have a much lower frequency of the female-biasing allele **a** ($q_m = 0.1$

at equilibrium) because a disproportionate number of their mothers carry the male-biasing allele A. Despite this complication, our conclusion from the expected-future-fitness approach is supported, at least for this genetic system.

Any population-genetic model that can be written down in this way (as a set of recurrence equations) can be studied by iteration in the manner illustrated in Figure 1.3. For example, Shaw (1958) did this for a diploid model (without the benefit of a computer!), and Hartl and Brown (1970) did so for a haplodiploid model. But each set of parameters and initial conditions considered in this way is just an anecdote. Even in the present very simple model, there are infinitely many combinations of m and m^*. Since not all of them can be considered, how are we to be confident of any general conclusions we might want to draw about the model's behaviour throughout certain regions of its parameter space?

Fortunately, we can often obtain rigorous general results by restricting our attention to the dynamics of *invasion* and *fixation*. We consider the system's behaviour at the 'boundary' of the state space, where one allele (A) is nearly fixed and the other (a) is vanishingly rare. Here the system can be represented as a set of linear equations far simpler than the full system. In matrix form, the system can be written $\mathbf{q}' = \mathbf{Mq}$ where $\mathbf{q}'$ and $\mathbf{q}$ are vectors of the (infinitesimal) genotype frequencies (here, q_f and q_m) and $\mathbf{M}$ is a matrix of coefficients representing the system's dynamics in the immediate neighbourhood of the boundary. The elements of $\mathbf{M}$ are partial derivatives of the full recurrence equations evaluated at $\mathbf{q} = 0$, where allele $\mathbf{a}$ is imagined to have just entered the population as a very rare migrant or a new mutation. As long as $\mathbf{a}$ remains very rare, the linearized system accurately represents the behaviour of the full system. Standard techniques of linear algebra allow us to determine under what conditions a small 'disturbance' of the equilibrium (i.e. tiny positive allele frequencies) will grow. Typically this is all we really care about because we need only show that an allele with a certain value of m^* will not be invaded by any allele with an m different from m^*. This is usually a straightforward task.

In the present case, as in many, the matrix $\mathbf{M}$ takes a simple and illuminating form. Here we write it out explicitly, with its associated vectors $\mathbf{q}'$ and $\mathbf{q}$, and the result of the multiplication, so as to make the meanings of all the terms easy to grasp.

$$
\begin{bmatrix} q'_{\mathrm{f}} \\ q'_{\mathrm{m}} \end{bmatrix} = \begin{bmatrix} \dfrac{1}{2}\dfrac{1-m}{1-m^*} & \dfrac{1}{2} \\ \dfrac{1}{2}\dfrac{m}{m^*} & \dfrac{1}{2} \end{bmatrix} \begin{bmatrix} q_{\mathrm{f}} \\ q_{\mathrm{m}} \end{bmatrix}
$$

$$
= \begin{bmatrix} \dfrac{1}{2}\dfrac{1-m}{1-m^*}\, q_{\mathrm{f}} + \dfrac{1}{2}q_{\mathrm{m}} \\ \dfrac{1}{2}\dfrac{m}{1-m^*}\, q_{\mathrm{f}} + \dfrac{1}{2}q_{\mathrm{m}} \end{bmatrix} \qquad (1.18)
$$

Note that the phenotypic values (m, m^*) of the genotypes ($\mathbf{a}$, $\mathbf{A}$) appear only in the first column of the matrix, while nothing but a simple constant ($1/2$) appears in the second column. The coefficients in the first column are multiplied by q_f, the frequency of the mutant genotype in *females*. When m is smaller than m^*, the upper left-hand coefficient is greater than one-half and the lower left-hand coefficient is less than one-half. As a consequence, increases in q_f will tend to cause larger increases of q'_f than of q'_m and the relatively female-biasing mutant allele $\mathbf{a}$ will tend to concentrate itself in females. The opposite happens when m is larger than m^*. Males transmit their genotypes without bias (passively) to daughters and sons, because males do not influence the sexes of their offspring.

The vector $\mathbf{q}$ (which is near zero) will tend to grow when the largest eigenvalue of $\mathbf{M}$ is greater than 1, and it will shrink towards zero when the eigenvalue is less than 1. (The eigenvalues λ are solutions of the characteristic equation $\det(\mathbf{M} - \lambda\mathbf{I}) = 0$, where $\mathbf{I}$ is the identity matrix.) The overall magnitude of $\mathbf{q}$ will not change when the largest eigenvalue is exactly 1, and this will obviously be the case at least whenever $m = m^*$, even if m^* is far from its evolutionary equilibrium. (When $m = m^*$, all four elements of $\mathbf{M}$ are $1/2$ exactly.) The eigenvalue is easy to calculate, but it takes a rather messy and unrevealing form involving nonintegral powers of m and m^*. However, since we are really most interested to know what conditions other than $m = m^*$ will give an eigenvalue of 1, we can greatly simplify the problem by *setting* $\lambda = 1$ and expanding the resulting characteristic equation $\det(\mathbf{M} - \mathbf{I}) = 0$ which usually takes an understandable and revealing form. In the present case we get

$$
\frac{1}{2} - \frac{1}{4}\frac{1-m}{1-m^*} - \frac{1}{4}\frac{m}{m^*} = 0, \qquad (1.19)
$$

which easily rearranges to give

$$
1 = \frac{1}{2}\left[\frac{1-m}{1-m^*} + \frac{m}{m^*} \right]. \qquad (1.20)
$$

This should look familiar: it is the Shaw–Mohler equation with the parent's fitness (the left-hand side) set equal to 1. This equality will hold, for arbitrary values of m, if and only if $m^* = 1/2$, consistent with our previous analysis.

It may seem that we have worked through a great deal of 'intricate' genetics and mathematics (as Darwin predicted), only to return to a place very close to the one he reached 130 years ago without the benefit of genes or calculations. In fact, the theory has been greatly augmented and strengthened. We understand that the equilibrium entails equal net investment in the sexes, which will not mean equal numbers if female offspring cost more or less than males. We understand the basic principle much more deeply, having seen it implemented in two different genetic frameworks, and this gives us both the understanding and the confidence to construct complex and sophisticated tracer-gene models (e.g. Frank 1998). In short, we now have tools with which to attack a nearly unlimited range of problems that could not be solved securely (or in some cases, at all) without formal methods such as these. The next section considers how such methods have been used to extend the central principle to situations not encompassed by the original model.

1.10 | Themes and variations

In nature, the biological circumstances surrounding sex allocation are as variable as the ecologies and life histories of real organisms. This variation motivates a seemingly endless diversity of sex ratio models. Does this diversity of models undermine the supposed generality of sex ratio theory? Not really, because the variations play on just a few underlying themes, if often in combinations with each other.

1. *Differential costs of the sexes.* This theme was clearly identified by Fisher. It arises from differential rates of mortality during the period of parental care, from differential resource needs of male and female offspring (reflected for example in different sizes at weaning), and from other ways in which male and female offspring may differentially affect a parent's future reproduction (for example, by compromising the parent's future growth, or by exposing the parent to increased risks of mortality).

2. *Condition-dependent benefits and reproductive values of the sexes.* Both the constraints and the genetic payoffs associated with different progeny sex ratios may change with a variety of environmental factors including seasonality, resource availability, parental size (competitive ability, fecundity) and aspects of local population structure. In some social species, for example, members of the less-dispersing sex may help their parents to defend a territory or to feed subsequent offspring, thereby providing direct reproductive benefits that partially offset the costs of their own production. Alternatively, under certain circumstances the members of one sex may be more likely to reproduce or to benefit from increased parental investment than the other (as first argued by Trivers & Willard 1973). Finally, the offspring of one sex may be more likely to compete with each other for the same matings, in which case they are partially reproductively redundant from the parent's point of view (local mate competition, section 1.10.1). Some of these effects can be subtle, giving rise to selection pressures far weaker than those associated with the population's mean allocation ratio.

3. *Mode of inheritance and locus of control.* Genes with unusual patterns of inheritance, such as those on Y chromosomes and mitochondrial chromosomes, sometimes have different ESS sex ratios than do those on autosomes, giving rise to evolutionary 'conflicts' over the sex ratio (Hamilton 1967). In haplodiploid species where males transmit their genes only to daughters, mates 'disagree' profoundly as to what their progeny sex ratio should be (e.g. Brockmann & Grafen 1989), and in some social Hymenoptera, workers and queens disagree as to what sex ratio their colony should produce (section 1.10.2). In principle, embryonic offspring may disagree with their parents over what sex they themselves should become; for example, if one sex is more costly and therefore produced in smaller numbers, all offspring, given the choice, would prefer to be of that sex.

We now briefly describe three classic extensions of the basic Darwin–Fisher model to show how it has been adapted to such special circumstances.

1.10.1 Local mate competition

The current 'Golden Age' of sex ratio research could be said to have begun with W.D. Hamilton's (1967) paper 'Extraordinary sex ratios'. Hamilton reviews Fisher's argument and then relaxes two of Fisher's implicit assumptions. The paper's first section relaxes the assumption that sex allocation is controlled by autosomal genes (sex-linked meiotic drive, an instance of theme 3, above). The second section relaxes the assumption of random mating in a large population and considers 'Local mate competition' or LMC (theme 2). Hamilton considers species such as fig wasps, where mating takes place in small aggregations representing the offspring of just a few females, followed by dispersal of the mated females. Under this population structure, brothers compete relatively directly with each other for matings, but sisters do not compete directly for the resources on which their own reproductive success depends. This asymmetry requires a modified *fitness function*.

Under the LMC scenario, a mother's fitness rises linearly with the number of dispersing females she produces, as in the Darwin–Fisher model, but *not* with the number of sons. Because her sons tend to compete with each other, each *additional* son yields a *smaller increase* in the *total number of inseminations* achieved by all her sons together, which is the other source of her fitness. In other words, male production obeys a law of diminishing reproductive returns that does not apply to daughters. Hamilton writes expected-future-fitness expressions for females producing sex ratios x_A and x_B in a population where each mating aggregation contains the offspring of n randomly chosen females, and he finds that the 'unbeatable' (ESS) sex ratio is

$$x^* = \frac{n-1}{2n}. \tag{1.21}$$

In an aggregation containing the offspring of two unrelated mothers, the ESS allocation is extremely unequal: 25% effort to sons, 75% to daughters. However, as n increases beyond just a few mothers contributing offspring to each aggregation, brothers compete less directly with each other (so they are less redundant from their mother's point of view), and the optimal male investment rises toward 50%. In an aggregation founded by just one female, the theoretical optimum sex ratio is 0% males, which is interpreted as '[no] more males than are necessary to ensure the fertilization of all her daughters'. Hamilton reviews sex ratio data from wasps, beetles, mites and thrips that mate in small aggregations and that have haplodiploid genetic systems permitting females to freely control their progeny sex ratios. Broods are strongly female-biased in almost every case but tend to include at least one male.

To test the model's logic, Hamilton constructs an explicit dynamical haplodiploid genetic model and iterates it on the computer for the case $n = 2$. Surprisingly, the unbeatable sex ratio turns out to be approximately 0.21 rather than 0.25 as predicted by the general analytical model. This discrepancy was later confirmed to be a real difference between the ESSs for diploid (biparental) and haplodiploid (arrhenotokous) genetic systems, through the analysis of more sophisticated expected-future-fitness models and explicit genetic models (Hamilton 1979, Taylor & Bulmer 1980, Uyenoyama & Bengtsson 1982, Frank 1985, Herre 1985, Taylor 1988, Stubblefield & Seger 1990). The exact ESS for haplodiplody is

$$m^* = \frac{(n-1)(2n-1)}{n(4n-1)}. \tag{1.22}$$

The downward deviation of m^* (for a given n), relative to Hamilton's original solution, is a consequence of arrhenotoky (males developing from unfertilized eggs), not haplodiploidy *per se*. The same ESS (eq. 1.22) holds for hypothetical haplo-haploid and diplodiploid genetic systems under which, as in ordinary haplodiploidy, females arise from biparentally produced zygotes while males arise from unfertilized eggs (Stubblefield & Seger 1990). Under arrhenotoky, but not under biparental genetic systems (regardless of the ploidys of males and females), inbreeding has unequal effects on genetic transmission through the two sexes, and this leads to the difference between eqs. 1.21 and 1.22.

Under the assumptions of the original Darwin–Fisher model, equal investment in the sexes is a 'weak-form' ESS, not an optimum: if the population as a whole invests equally, then all individual allocations (from 100% sons through to 100% daughters) are equally fit. But under LMC, the ESS is 'strong-form' (Uyenoyama & Bengtsson 1982): even if the population is at equilibrium, individuals suffer reduced fitness if their own progeny sex allocations depart from the ESS, because fitness is determined by the sex ratios within local aggregations.

During the last two decades of the twentieth century, local mate competition became a centrepiece of sex ratio research both through experimental studies of several species of parasitoid wasps and through field studies of entire communities of fig wasps (Chapters 6, 10, 19 and 20). An important theoretical and empirical issue running through much of this work concerns sex ratio adjustments made by individual mothers in response to information about the numbers and fecundities of other females likely to have contributed offspring to the same mating aggregation. What cues might females use to gather information about other females contributing to their aggregation? What responses should they try to make? How accurate might their responses be? Both theoretically and empirically, the answers depend in interesting ways on a variety of biological details. Given his interest in sex ratio variation among families, Karl Düsing clearly would have enjoyed the current state of LMC research.

1.10.2 Sex ratio conflict in ants

The theory of inclusive fitness (Hamilton 1964) was conceived with social insects in mind, and in his 1972 paper Hamilton considers them at length. Focusing on the Hymenoptera, he points out that haplodiploidy gives rise to a peculiar pattern of relatedness among family members. Owing to their father's haploidy, outbred full sisters are related by $r = 3/4$, but a mother is related to offspring of both sexes by the usual $r = 1/2$. (Coefficients of relatedness are reviewed by Bulmer 1994.) Thus a female would transmit more of her genes to future generations by rearing a sister than by rearing a daughter. Hamilton proposes that as a consequence, a hymenopteran female will 'easily [evolve] an inclination to work in the maternal nest rather than start her own.' However, a female is related to her brothers by only half the usual amount ($r = 1/4$), so she is *not* more related to her mother's offspring *as a whole* than to her own, unless 'the sex ratio or some ability to discriminate allows the worker to work mainly in rearing sisters.' Hamilton suggests that inbreeding might lead to female-biased sex ratios and thereby to eusociality, but a female's average r to her siblings remains the same as her average r to offspring under inbreeding, if mothers control their own sex ratios (Trivers & Hare 1976), so inbreeding does not of itself favour the evolution of eusocial workers. Trivers and Hare (1976) argue that Hamilton's suggestion will work only if daughters actively promote their

own reproductive interests at the expense of their mother's.

> The asymmetrical degrees of relatedness in haplodiploid species predispose daughters to the evolution of eusocial behavior, provided that they are able to capitalize on the asymmetries, either by producing more females than the queen would prefer, or by gaining partial or complete control of the genetics of male production.
>
> *(p. 250)*

Trivers and Hare then outline several different steps that workers could take to 'capitalize' on their closer relationships to sisters, sons and nephews than to daughters and brothers. The most important for our purposes (and the most famous) is 'Skewing the colony's investment toward reproductive females and away from males.'

In a colony with just one queen who is singly mated, and who lays all the reproductive eggs, females will be three times as related to their sisters as to their brothers ($3/4 : 1/4$). If the ratio of investment were 1:1 over the population as a whole, then workers would gain three times as much fitness from rearing sisters as from rearing brothers and might therefore benefit from biasing their investment towards sisters, as pointed out by Hamilton. Trivers and Hare argue: (1) that there is little to stop the workers from doing this, counter to their mother's interests, since they do all the work, and (2) that at the resulting evolutionary equilibrium

> we expect three times as much to be invested in females as in males, for at this ratio of investment [3:1] the expected [reproductive success] of a male is three times that of a female, per unit investment, exactly canceling out the workers' greater relatedness to their sisters. Were the mother to control the ratio of investment, it would equilibrate at 1:1, so that in eusocial species in which all reproductives are produced by the queen but reared by their sisters, strong mother-daughter conflict is expected regarding the ratio of investment, and a measurement of the ratio of investment is a measure of the relative power of the two parties.

The paper presents an extensive analysis of data on investment ratios in ants, bees and wasps with different kinds of social structures, and these are broadly in agreement with the sex ratio arguments. In particular, an average allocation ratio of roughly 3:1 is found for 20 species of monogynous ants, in agreement with the model on the assumptions: (1) that the queens in most of these species are singly mated, (2) that the relative dry weights of males and females indicate their relative costs to the colony, and (3) that workers tend to control the sex ratio. In 1976 there was very little evidence about the mating frequencies of queen ants. Subsequent work has shown that mate numbers can vary within as well as between species, and has exploited this fact to produce some very clean and elegant tests of the model (see below). Subsequent work also has shown that the dry weights of females may tend to over-represent their costs relative to males, such that the average allocation ratio of Trivers and Hare's 20 monogynous ants may actually be closer to 2:1 than 3:1 (see Boomsma 1989, Bourke & Franks 1995, Crozier & Pamilo 1996), as might be expected if multiple mating is common in some of these species. Trivers and Hare estimate allocation ratios for a number of polygynous ants (those with several to many queens per colony) and 'slave makers' (in which the queen's offspring are reared by workers of another species); as predicted, the apparent allocation ratios of these species are less female biased than those of the monogynous species, on average.

Having introduced the concept of an irreducible conflict over the sex ratio, Trivers and Hare go on to dissect it in some detail. For example, if the queen lays only a fraction (p) of the male eggs, with unmated workers laying the remainder ($1 - p$), then the equilibrium ratio of investment (F/M) for workers is

$$\frac{3(3 + p)}{(3 - p)^2}. \tag{1.23}$$

This declines from 3:1 when the queen lays all the male eggs ($p = 1$) to 1:1 when workers lay all the male eggs ($p = 0$). From the queen's point of view, the corresponding ESS is

$$\frac{(3 + p)}{(3 - p)(1 + p)}, \tag{1.24}$$

which is 1:1 at both endpoints and slightly lower in the middle. The conflict over the *sex ratio* disappears when workers produce all the males, but it is replaced by a conflict over *male production*, since the queen's inclusive fitness is reduced by worker laying.

Like other models in the paper, this one is derived within an expected-future-fitness framework that takes Fisher's principle as an axiom and that extends it, using Hamilton's inclusive-fitness theory, to account for unequal coefficients of relatedness and indirect parentage. By 1976 this framework seemed so obvious and secure to the authors that they could present expressions such as (1.23) and (1.24) without derivation and with little or no comment. Other theorists, not so readily persuaded by the logic of Trivers and Hare's novel and intricate arguments, soon started testing these arguments by analysing explicit genetic models (e.g. Oster *et al.* 1977, Benford 1978, Charnov 1978, Oster & Wilson 1978, Craig 1980, Taylor 1981, Pamilo 1982, Bulmer 1983). Trivers and Hare's central conclusions were all upheld, although a number of previously unsuspected complications were uncovered by these models; for example, the way in which workers with different sex-allocation phenotypes interact behaviourally to determine the colony's sex ratio can affect the nature of the ESS (Charnov 1978, Craig 1980, Pamilo 1982, Bulmer 1983).

Social insect colonies differ in many relevant ways among and even within species. For example, as mentioned above they may have little or no worker production of males; they may have one or several queens (or no queen); and the queen or queens may mate with one or several males. Models incorporating each of these contingencies have been analysed. Variation in the queen's mate number within species is especially interesting because it affects the workers' but not the queen's equilibrium sex allocation, with potentially dramatic effects on the outcome of the worker–queen conflict (Boomsma & Grafen 1990, 1991).

Consider a species where there is always one queen but she may have mated with one or two males, and suppose that colonies with once- and twice-mated queens are equally frequent and equally productive. Workers in the once-mated colonies would be indifferent to their colony's allocation ratio if the population-wide ratio were 3:1, as in the simplest model considered above. But workers in the twice-mated colonies would be indifferent if the population ratio were 2:1. (Their average relatedness to sisters is $r = \frac{1}{2}(\frac{3}{4} + \frac{1}{4}) = \frac{1}{2}$.) Thus both colony types cannot be indifferent simultaneously. If the population-wide allocation ratio is more female-biased than 2:1, then workers in twice-mated colonies would do best to produce males (brothers) exclusively, because males would be the under-represented sex from their point of view. Similarly, if the population-wide ratio is less female-biased than 3:1, then workers in once-mated colonies should produce females (sisters) exclusively.

Suppose that workers can assess the queen's mate number (for example, by perceiving the genetic diversity of their sisters). Then, given our assumptions that the two colony types are about equally frequent and that workers can 'assume' a roughly equal mixture of colony types, the evolutionarily stable outcome should be a polymorphism among colonies, with the singly mated colonies specializing in exclusive female production and the twice-mated colonies producing (at least on average) a 1:2 *male* bias such that the combined population-wide investment ratio is 2:1. These divergent allocations by colonies with different patterns of relatedness asymmetry give rise to a 'split sex ratio' distribution over colonies. Neither type of colony can improve its fitness by making a different sex ratio, and neither actually *produces* its own ESS, although in our example the population average is the ESS for twice-mated colonies.

Colony sex ratios are distributed bimodally in many ant species, and some recent genetic studies of mate numbers in such species are qualitatively consistent with the predictions of this split sex ratio model. Colonies with once-mated queens tend to produce sex ratios that are more strongly female-biased than those of colonies with twice-mated queens (Sundström 1994, Sundström *et al.* 1996). These findings support the idea that sex ratio 'imbalances' (from any actor's point of view) create significant opportunities to increase fitness by making adjustments that exploit the imbalance, and they suggest that worker ants can in fact assess levels of relatedness within their own colonies. Other

sources of among-colony variation in relatedness asymmetry, with expected or observed effects on among-colony sex ratio variation, have been considered by Trivers and Hare (1976), Ward (1983), Yanega (1989), Boomsma (1991), Mueller (1991), Chan and Bourke (1994), Evans (1995) and others.

1.10.3 Hermaphrodites

Most plants are simultaneous hermaphrodites (Chapter 16), as are some animals. In such species, sex allocation is a matter of relative effort devoted to male and female functions (for example, to pollen and seed production). Given a linear trade-off between male and female functions, the ESS in an outbreeding population is to invest equally in each kind of function (Maynard Smith 1971). Empirically, relative investment in male and female functions is more difficult to estimate than the numbers of male and female offspring in a dioecious species, and there are additional reasons why hermaphroditism tends to strike us as complicated, even messy. But the hermaphroditic model is actually much easier to solve than the Darwin–Fisher model, because the fitness differences associated with different relative male investments appear sooner (in the first, offspring generation) rather than later (in the second, grandoffspring generation). Perhaps, if human beings were outbreeding simultaneous hermaphrodites, sex ratio evolution would not have baffled Darwin to the extent that it did.

Why are some species hermaphroditic rather than dioecious? Various reasons have been suggested, and most are plausible. For example, some hermaphrodites self-fertilize, and this permits colonization of unoccupied habitats by single immature individuals (e.g. seeds). Self-fertilization also shifts the ESS sex allocation strongly toward investment in female functions, through a principle closely related to that of local mate competition. By economizing on male function, individuals can increase their genetic contributions to future generations; in effect, they escape part of the 'cost of sex' (e.g. Maynard Smith 1978).

However, many plants are self-incompatible (outcrossing) simultaneous hermaphrodites. They pay the full cost of sex, and they need unrelated mates. What are the benefits in this case? One popular and well supported idea is that owing to their immobility and their reliance on animal vectors for pollination and/or seed dispersal, many plants may experience diminishing returns on investment in one or both sex functions. In addition, the temporal separation of male and female functions may reduce the degree to which those functions draw from the same pool of resources. Under such conditions, an individual may be able to achieve greater net reproduction by being partly male and partly female than it could by devoting all of its resources to just one sex function. In other words, the *constraint* on possible combinations of male and female reproduction may be nonlinear in a way that makes hermaphroditism more efficient than dioecy.

This idea is often modelled by representing an individual's realized or effective male and female reproductive *outputs* as arbitrary powers of its internal resource *allocations*: $m = x^a$ and $f = (1 - x)^b$. The exponent a controls the shape of the function that scales reproductive returns on investment in male function, and the exponent b scales returns on female investment. If a or b is less than 1, then the corresponding sex function shows diminishing returns to scale, but if a or b is greater than 1, then the corresponding function shows increasing returns to scale. For example, suppose $a = 0.25$ because pollinators are easily saturated, but $b = 1$ because fruits and their seeds will be eaten by birds in direct proportion to their abundance and then dispersed widely. The *fitness set* representing possible combinations of realized male and female outputs (m and f, corresponding to values of x between 0 and 1) bends outward with respect to the origin, as shown in Figure 1.4. This graph makes it easy to see that hermaphrodites (individuals with some degree of mixed sex expression) will tend to have larger total reproductive outputs than pure males or pure females, because $m + f$ is clearly greater for intermediate points on the fitness set than it is for points at the ends where m or f is zero. But where exactly is the ESS? The unbeatable allocation is not obvious, because the fitness set is not symmetrical.

We can find the ESS by writing a Shaw–Mohler equation for the fitness of a focal individual that allocates x to male function in a population

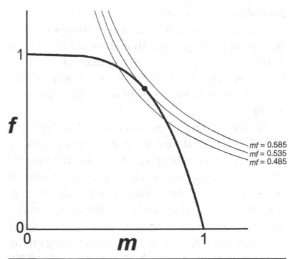

Fig 1.4 Fitness set for a simple model of hermaphroditic sex allocation. The heavy curve shows possible combinations of male (m) and female (f) reproductive outputs for a plant whose male gain exponent ($a = 0.25$) provides diminishing returns on investment in male function, while its female gain exponent ($b = 1$) provides linear returns. The equilibrium resource allocation ($X^* = 0.2$) is highly female biased, as is the realized output of male and female reproductive functions ($m^* = 0.67$, $f^* = 0.80$), indicated by the filled circle. The light curves are hyperbolas representing constant values of the product mf. They can be viewed as a contour map, and they demonstrate that MacArthur's (1965) principle applies to this model; the evolutionarily stable strategy (ESS) coincides with the highest value of mf attainable on the fitness set.

where the average allocation is X

$$W = \frac{1}{2}\left[\frac{x^a}{X^a} + \frac{(1 - x)^b}{(1 - X)^b}\right]. \tag{1.25}$$

When the population-wide average allocation X is at the evolutionary equilibrium, our focal individual should be unable to increase its fitness W by choosing an allocation x that differs from X. In other words, $W(x)$ should be maximized when $x = X$. To find this unbeatable allocation we differentiate the Shaw–Mohler equation with respect to x, set the derivative equal to zero, set $x = X$, and then solve for X. Doing so gives the solution

$$X^* = \frac{a}{a + b}. \tag{1.26}$$

On substituting the exponents discussed above ($a = 0.25, b = 1$) into this general solution we get $X^* = 0.20$, which is highly female-biased with respect to resources invested and substantially biased even with respect to the resulting reproductive outputs $(m^* = 0.67, f^* = 0.80)$ (Figure 1.4). The hermaphrodite's fitness $(W = 1)$ is substantially greater than that achieved by a male $(x = 1, W = 0.75)$ or a female $(x = 0, W = 0.63)$, so hermaphroditism is clearly stable against invasion by the pure sexes (Charnov et al. 1976).

If the scaling exponents a and b both exceed 1, then the fitness set bends inward towards the origin and dioecy (with equal numbers of males and females) is evolutionarily stable. If one exponent is less than 1 and the other is greater than 1, then either androdioecy (males and hermaphrodites) or gynodioecy (females and hermaphrodites) may be stable, depending on the exact values of a and b (Charnov et al. 1976). This simple model has been extended in many ways to reflect potentially important aspects of plant physiology, development and ecology. For example, models have been studied that incorporate vegetative growth between bouts of reproduction, with trade-offs between investment in growth (for future reproduction) and investment in reproduction (for current fitness); in some such models the regions of parameter space supporting gynodioecy and dioecy expand while those supporting androdioecy shrink, in ways that may help to explain why androdioecy is very rare in nature (Seger & Eckhart 1996, Eckhart & Seger 1999). Plant sex-allocation strategies are further complicated by several other features of plant biologies including strong spatial population structures, mating systems that involve mixed selfing and outcrossing, and cytoplasmically determined 'male sterility' (femaleness), which gives rise to an evolutionarily unstable but widespread form of gynodioecy.

In 1941, the botanist D. Lewis published an explicit population-genetic model for the relative frequencies and fecundities of females and hermaphrodites in populations where male-sterile individuals (females) are determined by genotypes at a nuclear locus. He discovered that females cannot invade an outbreeding hermaphroditic population unless they set at least twice as many seeds as a typical hermaphrodite, that the equilibrium frequency of females will approach one-half as hermaphrodites increase their male function and decrease their female function (thereby becoming male-like)

and that none of this is affected by the dominance or recessiveness of the alleles that convert hermaphrodites to females. Lewis does not cite Darwin, Düsing, Fisher or any other previous sex ratio theorist, and he does not seem to realize that his model illuminates general issues in sex allocation and represents a major methodological advance. The paper seems not to have been noticed by subsequent sex ratio theorists until much later, after explicit genetic models had been reinvented.

1.11 | Conclusion: diversity in unity

Sex ratios evolve according to simple, aesthetically beautiful principles, but they often affect and respond to many particular and even idiosyncratic aspects of a species' biology. Sex ratio theory therefore establishes concrete links between some of the most specific and some of the most general phenomena in biology, and it does so in a rich and productive way. The theory also accommodates a wide variety of styles and techniques of analysis that continue to grow in sophistication and rigour. Yet despite the field's ever-growing diversity in these respects, its theoretical structure is becoming simpler and more transparent. The field as a whole is much larger than it once was because more is known, and in more detail. But the central ideas and principles seem more coherent, more clearly articulated, and therefore easier to master than they were a few decades ago. As we stressed at the outset, there is still a great deal to be done, both empirically and theoretically, and no one can predict what will turn up next. Sex touches almost everything, so the study of its allocation will lead us to many new problems, and a large fraction of these seem certain to be interdisciplinary.

For example, as genetics becomes increasingly genomic in scale, we are encouraged to think in increasingly concrete terms about the possibilities for intragenomic conflict over sex ratios. When Hamilton introduced this subject in the first half of his 1967 paper, it was little more than an abstract possibility supported by a few observations of sex-chromosome meiotic drive. Now, only 35 years later, we know the exact chromosomal locations of all the genes in several species' genomes, and soon we will know when and where these genes are expressed and at least something about the physiological properties of their products. There are many situations in which sex-linked and autosomal genes might 'disagree' about their carrier's parental investments or other sex-biased interactions with kin. How are these conflicts 'settled'? Our growing ability to observe both the expression and the evolution of arbitrarily large sets of genes should bring new life to both the theoretical and empirical aspects of this very interesting problem.

On the theoretical side, expected-future-fitness and population-genetic models will again play complementary roles. Obviously, with specific genes in mind, it will be natural and necessary to construct explicit multi-locus dynamical models that show how the interactions of alleles with various different phenotypic effects might lead to various alternative outcomes. Perhaps less obviously, it will also be necessary to ask how neutral alleles at typical loci, elsewhere in the genome, might be affected by the possible outcomes of the conflict. If different outcomes are better and worse for large numbers of genes throughout the genome, then such genes might be recruited into the conflict as modifiers of small effect. Because fitness-based models describe how particular sex-allocation phenotypes affect the future frequencies of neutral genes associated with those phenotypes, such models will be used to study how the genetic background at large might be expected to respond to particular genetic conflicts.

Acknowledgments

We thank Steve Frank, Ian Hardy, Mike Mesterton-Gibbons, Steve Orzack and Franjo Weissing for encouragement and for many helpful criticisms and suggestions.

References

Benford FA (1978) Fisher's theory of the sex ratio applied to the social Hymenoptera. *Journal of Theoretical Biology*, **72**, 701–727.

Bodmer W & Edwards AWF (1960) Natural selection and the sex ratio. *Annals of Human Genetics*, **24**, 239–244.

Boomsma JJ (1989) Sex-investment ratios in ants: has female bias been systematically overestimated? *American Naturalist*, **133**, 517–532.

Boomsma JJ (1991) Adaptive colony sex ratios in primitively eusocial bees. *Trends in Ecology and Evolution*, **6**, 92–95.

Boomsma JJ & Grafen A (1990) Intraspecific variation in ant sex ratios and the Trivers–Hare hypothesis. *Evolution*, **44**, 1026–1034.

Boomsma JJ & Grafen A (1991) Colony-level sex-ratio selection in the eusocial Hymenoptera. *Journal of Evolutionary Biology*, **4**, 383–407.

Bourke AFG & Franks NR (1995) *Social Evolution in Ants*. Princeton: Princeton University Press.

Brockmann HJ & Grafen A (1989) Mate conflict and male behaviour in a solitary wasp, *Trypoxylon (Trypargilum) politum* (Hymenoptera: Sphecidae). *Animal Behaviour*, **37**, 232–255.

Bull JJ & Charnov EL (1988) How fundamental are Fisherian sex ratios? *Oxford Surveys in Evolutionary Biology*, **5**, 96–135.

Bulmer M (1983) Sex ratio evolution in social Hymenoptera under worker control with behavioral dominance. *American Naturalist*, **121**, 899–902.

Bulmer M (1994) *Theoretical Evolutionary Ecology*. Sunderland, MA: Sinauer Associates.

Chan GL & Bourke AFG (1994) Split sex ratios in a multiple-queen ant population. *Proceedings of the Royal Society of London*, series B, **258**, 261–266.

Charnov EL (1978) Sex-ratio selection in eusocial Hymenoptera. *American Naturalist*, **112**, 317–326.

Charnov EL (1982) *The Theory of Sex Allocation*. Princeton: Princeton University Press.

Charnov EL (1993) *Life-history Invariants*. Oxford: Oxford University Press.

Charnov EL, Maynard Smith J & Bull JJ (1976) Why be an hermaphrodite? *Nature*, **263**, 125–126.

Clutton-Brock TH (1986) Sex ratio variation in birds. *Ibis*, **128**, 317–329.

Clutton-Brock TH (1991) *The Evolution of Parental Care*. Princeton: Princeton University Press.

Clutton-Brock TH & Iason GR (1986) Sex ratio variation in mammals. *Quarterly Review of Biology*, **61**, 339–374.

Craig R (1980) Sex investment ratios in social Hymenoptera. *American Naturalist*, **116**, 311–323.

Crozier RH & Pamilo P (1996) *Evolution of Social Insect Colonies: Sex Allocation and Kin Selection*. Oxford: Oxford University Press.

Darwin C (1871) *The Descent of Man and Selection in Relation to Sex*. London: John Murray.

Darwin C (1874) *The Descent of Man and Selection in Relation to Sex*, 2nd edn. London: John Murray.

Düsing K [sic] (1883) Die Factoren, welche die Sexualität entscheiden. *Jenaische Zeitschrift für Naturwissenschaft*, **16**, 428–464.

Düsing C [sic] (1884) *Die Regulierung des Geschlechtsverhältnisses bei der Vermehrung der Menschen, Tiere und Pflanzen*. Jena: Gustav Fischer Verlag.

Eckhart VM & Seger J (1999) Phenological and developmental costs of male sex function in hermaphroditic plants. In: TO Vuorisalo & PK Mutikainen (eds), *Life History Evolution in Plants*, pp 195–213. Dordrecht: Kluwer.

Edwards AWF (1998) Natural selection and the sex ratio: Fisher's sources. *American Naturalist*, **151**, 564–569.

Edwards AWF (2000) Carl Düsing (1884) on *The Regulation of the Sex-Ratio*. *Theoretical Population Biology*, **58**, 255–257.

Evans JD (1995) Relatedness threshold for the production of female sexuals in colonies of a polygynous ant, *Myrmica tahoensis*, as revealed by microsatellite DNA analysis. *Proceedings of the National Academy of Sciences of the United States of America*, **92**, 6514–6517.

Fisher RA (1930) *The Genetical Theory of Natural Selection*. Oxford: Clarendon Press.

Frank SA (1985) Hierarchical selection theory and sex ratios. II. On applying the theory, and a test with fig wasps. *Evolution*, **39**, 949–964.

Frank SA (1990) Sex-allocation theory for birds and mammals. *Annual Review of Ecology and Systematics*, **21**, 13–55.

Frank SA (1998) *Foundations of Social Evolution*. Princeton: Princeton University Press.

Godfray HCJ (1994) *Parasitoids: behavioural and evolutionary ecology*. Princeton: Princeton University Press.

Hamilton WD (1964) The genetical evolution of social behaviour. *Journal of Theoretical Biology*, **7**, 1–52.

Hamilton WD (1967) Extraordinary sex ratios. *Science*, **156**, 477–488.

Hamilton WD (1972) Altruism and related phenomena, mainly in social insects. *Annual Review of Ecology and Systematics*, **3**, 193–232.

Hamilton WD (1979) Wingless and fighting males in fig wasps and other insects. In: MS Blum & NA Blum (eds), *Reproductive Competition and Sexual Selection in Insects*, pp 167–220. New York: Academic Press.

Hartl DL & Brown SW (1970) The origin of male haploid genetic systems and their expected sex ratio. *Theoretical Population Biology*, **1**, 165–190.

Herre EA (1985) Sex ratio adjustment in fig wasps. *Science*, **228**, 896–898.

Herre EA, West SA, Cook JM, Compton SG & Kjellberg F (1997) Fig-associated wasps: pollinators and parasites, sex-ratio adjustment and male polymorphism, population structure and its consequences. In: JC Choe & BJ Crespi (eds), *The Evolution of Mating Systems in Insects and Arachnids*, pp 226–239. Cambridge: Cambridge University Press.

Hewison AJM & Gaillard J-M (1999) Successful sons or advantaged daughters? The Trivers-Willard model and sex-biased maternal investment in ungulates. *Trends in Ecology and Evolution*, **14**, 229–234.

Karlin S & Lessard S (1986) *Theoretical Studies on Sex Ratio Evolution*. Princeton: Princeton University Press.

Klinkhamer PGL, de Jong TJ & Metz H (1997) Sex and size in cosexual plants. *Trends in Ecology and Evolution*, **12**, 260–265.

Lewis D (1941) Male sterility in natural populations of hermaphrodite plants. *New Phytologist*, **40**, 56–63.

MacArthur RH (1965) Ecological consequences of natural selection. In: TH Waterman & H Morowitz (eds), *Theoretical and Mathematical Biology*, pp 388–397. New York: Blaisdell.

Macevicz S & Oster GF (1976) Modelling social insect populations II: optimal reproductive strategies in annual eusocial insect colonies. *Behavioral Ecology and Sociobiology*, **1**, 265–282.

Maynard Smith J (1971) The origin and maintenance of sex. In: GC Williams (ed), *Group Selection*, pp 163–175. Chicago: Aldine-Atherton.

Maynard Smith J (1978) *The Evolution of Sex*. Cambridge: Cambridge University Press.

Maynard Smith J (1982) *Evolution and the Theory of Games*. Cambridge: Cambridge University Press.

Maynard Smith J & Price GR (1973) The logic of animal conflict. *Nature*, **246**, 15–18.

Mueller UG (1991) Haplodiploidy and the evolution of facultative sex ratios in a primitively eusocial bee. *Science*, **254**, 442–444.

Nonacs P (1986) Ant reproductive strategies and sex allocation theory. *Quarterly Review of Biology*, **61**, 1–21.

Nur U (1974) The expected changes in the frequency of alleles affecting the sex ratio. *Theoretical Population Biology*, **5**, 143–147.

Oster GF & Wilson EO (1978) *Caste and Ecology in the Social Insects*. Princeton: Princeton University Press.

Oster GF, Eshel I & Cohen D (1977) Worker-queen conflict and the evolution of social castes. *Theoretical Population Biology*, **12**, 49–85.

Pamilo P (1982) Genetic evolution of sex ratios in eusocial Hymenoptera: allele frequency simulations. *American Naturalist*, **119**, 638–656.

Seger J & Eckhart VM (1996) Evolution of sexual systems and sex allocation in plants when growth and reproduction overlap. *Proceedings of the Royal Society of London*, series B, **263**, 833–841.

Shaw RF (1958) The theoretical genetics of the sex ratio. *Genetics*, **43**, 149–163.

Shaw RF & Mohler JD (1953) The selective significance of the sex ratio. *American Naturalist*, **87**, 337–342.

Sober E (1984) *The Nature of Selection: Evolutionary Theory in Philosophical Focus*. Cambridge, MA: MIT Press.

Spencer H (1867) *Principles of Biology*, volume 2. London: Williams & Norgate.

Stubblefield JW & Seger J (1990) Local mate competition with variable fecundity: dependence of offspring sex ratios on information utilization and mode of male production. *Behavioral Ecology*, **1**, 68–80.

Sundström L (1994) Sex ratio bias, relatedness asymmetry and queen mating frequency in ants. *Nature*, **367**, 266–268.

Sundström L, Chapuisat M & Keller L (1996) Conditional manipulation of sex ratios by ant workers: a test of kin selection theory. *Science*, **274**, 993–995.

Taylor PD (1981) Sex ratio compensation in ant populations. *Evolution*, **35**, 1250–1251.

Taylor PD (1988) Inclusive fitness models with two sexes. *Theoretical Population Biology*, **34**, 145–168.

Taylor PD & Bulmer M (1980) Local mate competition and the sex ratio. *Journal of Theoretical Biology*, **86**, 409–419.

Trivers RL (1972) Parental investment and sexual selection. In: BG Campbell (ed), *Sexual Selection and the Descent of Man 1871–1971*, pp 136–179. Chicago: Aldine.

Trivers RL (1985) *Social Evolution*. Menlo Park: Benjamin/Cummings.

Trivers RL & Hare H (1976) Haplodiploidy and the evolution of the social insects. *Science*, **191**, 249–263.

Trivers RL & Willard DE (1973) Natural selection of parental ability to vary the sex ratio of offspring. *Science*, **179**, 90–92.

Uyenoyama MK & Bengtsson BO (1982) Towards a genetic theory for the evolution of the sex ratio III. Parental and sibling control of brood investment ratio under partial sib-mating. *Theoretical Population Biology*, **22**, 43–68.

Verner J (1965) Selection for sex ratio. *American Naturalist*, **99**, 419–421.

Ward PS (1983) Genetic relatedness and colony organization in a species complex of ponerine ants II. Patterns of sex ratio investment. *Behavioral Ecology and Sociobiology*, **12**, 301–307.

West SA, Herre EA & Sheldon BC (2000) The benefits of allocating sex. *Science*, **290**, 288–290.

Williams GC (1979) The question of adaptive sex ratio in outcrossed vertebrates. *Proceedings of the Royal Society of London, Series B*, **205**, 553–559.

Wrensch DL & Ebbert MA (eds) (1993) *Evolution and Diversity of Sex Ratio in Insects and Mites*. New York: Chapman and Hall.

Yanega D (1989) Caste determination and differential diapause within the first brood of *Halictus rubicundus* in New York (Hymenoptera: Halictidae). *Behavioral Ecology and Sociobiology*, **24**, 97–107.

Chapter 2

Optimal sex allocation: steps towards a mechanistic theory

Ido Pen & Franz J. Weissing

2.1 | Summary

Sex-allocation theory is often hailed as the most successful branch of evolutionary ecology, yet its success has been limited to a relatively small number of taxa, mostly haplodiploid insects. Sex ratio variation in vertebrates is still poorly understood. We argue that this is due to the failure of current sex-allocation models to sufficiently take into account the complexities of vertebrate sex determination and life histories. Our main purpose here is to discuss how more 'mechanistic' models might be constructed to help answer some of the many open questions regarding vertebrate sex allocation. In particular, we discuss the importance of costs of control, the multidimensional nature of allocation decisions and conflicts over allocation decisions. We give an overview of optimality or evolutionarily stable strategy (ESS) techniques that are useful in analysing sex-allocation problems, and we present a series of models to illustrate several of the concepts and techniques.

2.2 | Introduction

Sex-allocation theory (Charnov 1982) has been very successful in gaining insight into the ultimate causes of sex ratio variation, but in applications the extent of its success has proven rather taxon-specific. Especially in haplodiploid insects relatively simple models seem to be able to correctly predict qualitative features of sex ratio variation, and sometimes even quantitative predictions have been met with remarkable precision (e.g. Werren 1980, but see Hardy *et al.* 1998). In contrast, the theory has shed a rather pale light on sex ratio variation in vertebrates. This discrepancy is usually attributed to taxonomic differences in the mode of sex determination (Williams 1979, Maynard Smith 1980, Bull & Charnov 1988). Relative to the accuracy of sex ratio control in haplodiploids, a chromosomal mechanism of sex determination, which is common in vertebrates, supposedly hinders parental sex ratio manipulation. This cannot, however, be the whole story because there are several well-documented examples of adaptive sex allocation in vertebrates (Clark 1978, Conover & Voorhees 1990, Daan *et al.* 1996, Komdeur *et al.* 1997, Kruuk *et al.* 1999).

Perhaps equally important is that most models of sex allocation lack the sophistication required to tackle the complexities of vertebrate sex allocation and life histories. In addition to the mechanism of sex determination there are several other aspects of vertebrate biology that are usually ignored in sex-allocation models.

1. Vertebrates are often long-lived and face a trade-off between current and future reproduction. Strategic allocation decisions such as reproductive effort and sex allocation should reflect this trade-off. In contrast, standard sex-allocation models typically assume very simple life histories with nonoverlapping generations and reproduction at a fixed rate (for rare exceptions, see Olivieri *et al.* 1994, Zhang & Wang 1994, Lessells

1998). For a better understanding of sex allocation in long-lived species it will be crucial to study sex allocation, reproductive effort and other reproductive allocation decisions within a single framework.

2. In many vertebrates it seems likely that both parents can have some degree of influence on sex allocation. Asymmetries between the parents may cause their optimal allocation patterns to differ (Trivers 1974). Theoretical analyses tend to ignore the fact that males and females differ fundamentally in their means of exerting control of resource allocation to their offspring. For example, in birds with biparental care, females are the more likely sex to be in control of clutch size and the sex ratio, whereas both sexes might be able to practise sex-biased food allocation.

3. If the mode of sex determination prevents manipulation of the primary sex ratio, there are almost invariably costs involved in manipulating the sex ratio at some later stage. There is certainly awareness of this, but it is seldomly explicitly included in models of sex allocation (but see Maynard Smith 1980, Eshel & Sansone 1991, 1994, Leimar 1996, Pen *et al.* 1999).

In this chapter we discuss these and other biological factors that might be important for improving our understanding of sex allocation in vertebrates. Most of these aspects have been addressed before, but only some have been explicitly included in models; others have not been addressed at all, and many questions remain unanswered. Our main purpose is to discuss how models might be constructed to answer some of these questions and we provide several examples. We refer to our modelling approach as 'mechanistic' because we emphasize the evolutionary implications of specific mechanisms of control over allocation decisions. We do not advocate models that include as much realism as possible, because such models tend to obscure, rather than enhance, our understanding. Instead, we advocate models of 'intermediate complexity' that include several crucial aspects of vertebrate biology, without making the models too difficult to analyse.

This chapter is structured as follows: In section 2.3 we give a mainly verbal account of some mechanistic aspects of sex allocation. We discuss the importance of costs of control, the multidimensional nature of allocation decisions and conflicts over allocation decisions. In section 2.4 we give an overview of optimality or evolutionarily stable strategy (ESS) techniques that are useful in analysing sex-allocation problems. We show how to derive an expression for the fitness of an individual based on its life history, and we show how to analyse fitness such that meaningful conclusions can be inferred from it. Finally, in section 2.5 we present and analyse a series of models, ranging from very simple classical models to more advanced ones, in order to illustrate the concepts and techniques introduced in the previous sections.

2.3 | Mechanistic aspects

In this section we discuss in a nontechnical way some mechanistic aspects of sex allocation. In section 2.4 we show how ESS techniques can be used to formally analyse the problems addressed here.

The cornerstone of sex-allocation theory is Fisher's (1930) principle of equal allocation, which states that selection on the sex ratio comes to a halt if and only if allocation to sons equals allocation to daughters. Although it is now recognized (see Edwards 1998 and Chapter 1) that the basic idea behind the principle goes back to Darwin (1871) and the first mathematical treatment to Düsing (1884), we still call it 'Fisher's principle' here. It is well-known that Fisher's principle relies on many tacit assumptions (Bull & Charnov 1988). We will not review all these assumptions here, but discuss a few that are relevant to our mechanistic approach. One of the most important assumptions of Fisher's principle is that parental control over the sex ratio is cost-free. That is, parents do not waste resources and/or suffer a higher mortality due to sex ratio control. We discuss some biological mechanisms that may render sex ratio control costly, and we use some simple models to investigate how much parents can afford to waste in adjusting the sex ratio.

Fisher's argument assumes that parents have a fixed amount of resource to allocate to their offspring. Iteroparous organisms must decide several times how much to invest in reproduction versus other activities that may affect their future reproduction. Does this affect sex allocation? Does it matter how parents can adjust sex allocation, by adjusting the sex ratio or by investing more or less in individual sons and daughters? We discuss the consequences for optimal sex allocation when different allocation components contribute to the overall allocation pattern, not necessarily all under parental control.

Fisher's equal allocation principle (also known as the equal investment principle) assumes parental control, but if gametes or offspring could somehow affect or obstruct sex allocation, selection may have a different outcome. Gametes and offspring are more closely related to themselves than to their siblings, unlike the parents who are usually equally related to all offspring. This asymmetry typically causes gametes and offspring to favour a sex ratio less biased than under parental control. If parents employ a mechanism of selective killing of gametes or offspring, this is likely to be opposed by the potential 'victims'.

Parents may also disagree over sex allocation if the costs of parental care are unequally shared between them. We discuss how differences between parents in control mechanisms and the information parents have about their partner's decision may affect the outcome of selection.

2.3.1 Costs

If parents are unable to adjust the primary sex ratio (at fertilization), then sex ratio manipulation is likely to be costly to the parents. The nature and the magnitude of the cost depends on the mechanism of secondary sex ratio control. At least two types of cost may be important. First, a loss of invested resources. If parents cannot directly change the sex of their offspring, they must selectively kill offspring of a particular sex, thereby losing some of the resources already invested in those offspring (Myers 1978, Williams 1979, Maynard Smith 1980). This may be costly if parents are resource-limited, or if the number of foetuses that can be simultaneously implanted is limited. Even if no energy or resources are lost, selective abortion may involve a second type of cost. There are a number of rea-

sons why selective abortion may reduce the fitness prospects of the nonaborted offspring. For example, it may take time to replace a jettisoned offspring with a new one. Such delayed reproduction may be costly: it has been suggested that in birds, a female might be able to recognize the sex of a developing egg just before laying, and resorb it if it is of the undesirable sex (Emlen 1997, Oddie 1998, Pen *et al.* 1999). Since it takes time for the next egg to reach the same developmental stage, this means of manipulating the sex ratio may result in a later laying date, which in turn may be detrimental to the survival of offspring (Klomp 1970). This type of control may also lead to a more pronounced hatching asynchrony, which might also be detrimental to offspring survival.

Despite its potential importance, costs of sex ratio control are hardly ever explicitly included in evolutionary models of sex allocation (exceptions are Maynard Smith 1980, Eshel & Sansone 1991, 1994, Leimar 1996, Pen *et al.* 1999). One would like to know the cost parents can afford for sex ratio adjustment to remain adaptive, but this problem has never been analysed systematically. To give at least an indication of the magnitude of such costs, in sections 2.5.2 and 2.5.5 we analyse two models with costly sex ratio control. In the first model, parents are selected to bias the sex ratio because sons and daughters have different costs. In the second model, parents are selected to bias the sex ratio according to some environmental cue that differentially affects the survival of sons and daughters. In both models parents have a fixed amount of resource to allocate to reproduction, the primary sex ratio is fixed at parity, and the sex ratio is secondarily adjusted by selective abortion after a certain initial investment (the cost of control) in the aborted offspring.

In the first model (section 2.5.2), a son requires E_m units of resource and a daughter E_f units. All parents produce the same sex ratio. Without costs of sex ratio control equal allocation is expected: the investment per son (E_m) times the number of sons (n_m) equals the investment per daughter (E_f) times the number of daughters (n_f), $E_m \times n_m = E_f \times n_f$. Figure 2.1 shows for several values of the investment ratio $\rho = E_m/E_f$ how the sex ratio bias decreases with the

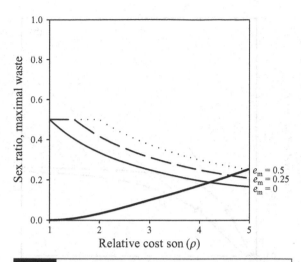

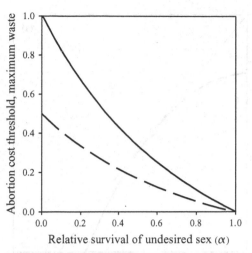

Fig 2.1 ESS sex ratio with costly selective abortion (see section 2.5.2). The decreasing solid line illustrates Fisher's classic result that – in the absence of any abortion costs ($e_m = 0$) – the ESS sex ratio should decline with the relative cost $\rho = $ cost son/cost daughter of a son according to $s^* = 1/(1 + \rho)$. A substantially smaller sex ratio bias is to be expected if sex ratio control can only be achieved by abortion involving some cost e_m per aborted offspring (relative to a cost of unity per daughter). Notice that for small values of ρ it does not pay at all to adjust the sex ratio. The increasing solid line indicates the maximal proportion of total resources available for reproduction wasted during abortion.

Fig 2.2 Selective abortion in the Trivers and Willard model. The solid line depicts the threshold cost of abortion as a function of alpha, the relative survival of the 'undesired' sex. If abortion costs are below the threshold, parents should abort all offspring of the undesired sex. For costs above the threshold, parents should not adjust the sex ratio at all. The dashed line indicates the maximal proportion of reproductive resources wasted by abortion. See section 2.5.5 for details.

cost of control e_m (relative to $E_f = 1$), up to the point where it no longer pays to bias the sex ratio at all. This point is precisely where $e_m = (E_m - E_f)/2$, half the cost difference between sons and daughters. Figure 2.1 also depicts the maximal proportion of total resources wasted on sex ratio control. The maximal waste is surprisingly small: even when a son costs twice the resource of a daughter, and a sex ratio bias of 1:2 is to be expected in the absence of costs, it never pays to waste more than about 3.5% of resources on sex ratio adjustment.

In the second model (section 2.5.5), we analyse a variant of Trivers and Willard's (1973) model of conditional sex ratio adjustment. Parents can be in two states, one favourable for sons, the other for daughters. Specifically, in each state the survival of one sex is a proportion α of the opposite sex's survival. If the costs of control (i.e. the initial investment in aborted offspring) are small, the ESS is a 'bang-bang' strategy: produce only one sex in each state. If, however, the costs

of control are higher than a certain threshold, then it does not pay to bias the sex ratio at all (Figure 2.2). Hence, it is not true that costs of control select for sex ratio trends intermediate between bang-bang strategies and no adjustment at all (Oddie 1998). We have again calculated the maximal waste of resources on sex ratio adjustment, which is plotted in Figure 2.2. In contrast to the 'Fisherian' model, we now find that waste can be much larger. For example, when the benefit ratio equals 2 ($\alpha = \frac{1}{2}$), the maximal waste is about 17% of total resources. A tentative conclusion of this analysis is therefore that under conditional sex ratio adjustment a much larger cost of sex ratio control can be sustained compared to unconditional sex ratio adjustment.

2.3.2 Components

Sex allocation is usually defined as the division of resources between male and female function (Charnov 1982). For dioecious organisms, the problem is how to split resources between sons and daughters. Suppose a parent 'wishes' to allocate A_m resources to sons and A_f to daughters. The parent can do this in several different ways.

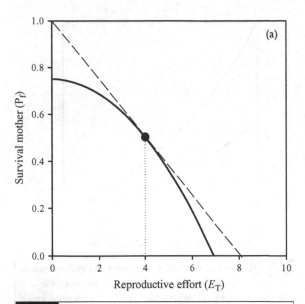

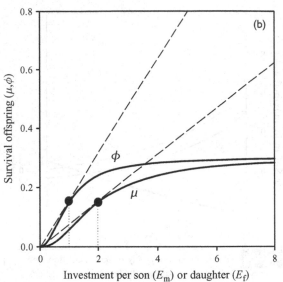

Fig 2.3 Survival of mother (a) as a function of her reproductive effort and survival of sons and daughters (b) as a function of the investment per son and daughter. Dashed lines indicate graphically the evolutionarily stable reproductive effort E_T^* of the mother and the ESS investment (E_m^* and E_f^*) per offspring for the case that the mother controls all allocation components. In the example the following functions are used: $P_f = P_{max} - \gamma E_T^2$, $\mu(E_m) = \mu_{max} E_m^2/(\alpha^2 + E_m^2)$, $\phi(E_f) = \phi_{max} E_f^2/(\beta^2 + E_f^2)$. Parameters: $P_{max} = 0.75$, $\gamma = 0.016$, $\alpha = 2$, $\beta = 1$, $\phi_{max} = \mu_{max} = 0.3$.

It can keep the allocation to individual sons and individual daughters fixed, and manipulate the numerical sex ratio. Or, conversely, the parent can keep the sex ratio fixed, but vary allocation to individual sons and daughters. Thus, the overall allocation pattern can be partitioned into at least four allocation components: the number of offspring or clutch size c, the primary sex ratio s (proportion of sons), the amount of resources E_m allocated per individual son (assuming all sons receive the same amount) and the allocation E_f per individual daughter. The total amount of resources allocated to sons is then given by

$$A_m = scE_m, \tag{2.1}$$

and the amount allocated to daughters by

$$A_f = (1 - s)cE_f. \tag{2.2}$$

Total allocation to reproduction, the reproductive effort, can now be written in terms of the different allocation components as

$$E_T = A_m + A_f = scE_m + (1 - s)cE_f. \tag{2.3}$$

In this way, both sex allocation (A_m relative to A_f) and reproductive effort (E_T) can be viewed as the outcome of selection on four allocation components (s, c, E_m, E_f). The outcome is likely to depend on the way the allocation components affect various fitness components. In the simplest scenario, a parent's survival subsequent to reproduction depends only on its reproductive effort, and the survival of an offspring depends only on the amount of resources invested in it (see Figure 2.3). However, it is possible that the survival of offspring in addition depends on clutch size and/or the sex composition of the clutch.

As the analysis in section 2.5.6 shows, the partitioning of allocation into components is helpful for several reasons. First, it gives us insight into the minimum number of allocation components that must be able to evolve in order for certain outcomes to be expected. For example, in order for Fisher's (1930) equal allocation principle ($A_m = A_f$) to be valid, selection on the sex ratio alone is not sufficient. It also necessary for selection to act on clutch size (see section 2.5.6). On the other hand, selection on the investment per individual son or daughter is not necessary for Fisher's principle to hold true (see Chapter 3 in Pen 2000). Numerical results of a specific example are shown in Table 2.1.

Table 2.1 | Evolutionarily stable allocation components, reproductive effort and sex allocation under different scenarios of maternal control. In the first scenario, which corresponds to Figure 2.3, all components are under maternal control, leading to equal allocation ($A_f/A_m = 1$). In the second scenario, the primary sex ratio is not under maternal control, but fixed at $s = 1/2$. As a consequence, at the ESS less is invested in daughters than in sons ($A_f/A_m < 1$). In the third scenario, the clutch size is fixed at $c = 5$, implying a female-biased investment ($A_f/A_m > 1$) at ESS. The functions P, ϕ and μ, and all parameters are as in Figure 2.3

Allocation component	No constraints	Sex ratio constrained	Clutch size constrained
Clutch size (c)	3.00	2.79	5.00
Sex ratio (s)	0.33	0.50	0.25
Investment per daughter (E_f)	1.00	1.19	0.72
Investment per son (E_m)	2.00	1.68	1.80
Total allocation to daughters (A_f)	2.00	1.66	2.69
Total allocation to sons (A_m)	2.00	2.34	2.27
Reproductive effort (E_T)	4.00	4.00	4.96
Allocation ratio (A_f/A_m)	1.00	0.71	1.19

Another advantage of partitioning allocation into components is that it forces us to think about what fitness components are affected by what allocation components. This may have important consequences. For example, we need to consider whether survival of offspring depends only on the amount of resources invested in them, or also on the clutch size or the sex ratio. For Fisher's equal allocation to hold, it is necessary that clutch size and sex ratio have no effect on offspring survival (Chapter 3 in Pen 2000).

For small litter sizes, the problem of simultaneous selection of multiple allocation components can be very complicated, especially if parents vary in the amount of resource available for reproduction and can adjust clutch size and the sex ratio accordingly. Without quite precise knowledge of the distribution of resource availability in the population, even qualitative predictions about the relationship between resource availability and sex allocation are difficult to make (Williams 1979, Frank 1987).

It is possible to construct models with more than just the four allocation components we have considered so far, although such a model will obviously be more difficult to analyse. For example,

one can explicitly take the time structure of a reproductive episode into account by splitting the episode into a number of time units and defining for each time unit one or more allocation components. This would raise some interesting questions. Since parents usually have to decide on clutch size and the sex ratio before making any further investment in their offspring, not all of the above four allocation components are simultaneously under control. A systematic investigation of the consequences for sex allocation of such a temporal decoupling of allocation decisions has not yet been made (Stubblefield & Seger 1990, Seger & Eckhart 1996).

The study of sex allocation and reproductive effort in a single framework raises more interesting questions. For example, the benefits of some forms of risky investment are difficult to assign to just one sex, such as nest defence behaviour. There is currently no theory on how such investments, of which the benefits are shared equally by all offspring, affect sex allocation.

2.3.3 Conflicts

The concept of a conflict of interest in the context of sex allocation was introduced by Trivers (1974).

Trivers noted that if the sexes are not equally costly to raise, then Fisher's principle of equal allocation only holds if the sex ratio of the offspring is under maternal control. If the offspring themselves were able to determine their sex, then selection would favour a different outcome. Since the optimal strategies of mother and offspring do not coincide, they have a conflict of interest. The reason for the discrepancy between mother and offspring is that the very tendency of the mother to produce fewer of the expensive sex renders this sex the one with the highest reproductive value. Offspring will therefore have a greater tendency to become the more expensive sex, even at the expense of the number of siblings their mother can raise. Because of the latter aspect of the problem, kin selection arguments play a role in the analysis. In section 2.5.3 we show how a simple kin selection model can be used to analyse the offspring's strategy.

Eshel and Sansone (1991, 1994) concluded that when both mother and offspring have some influence on the sex ratio, a compromise between the mother's and the offspring's optima is the most likely evolutionary outcome. They also investigated what happens when offspring signal their vulnerability to maternal manipulation, a stronger signal being associated with the handicap of greater vulnerability. The interesting conclusion arose that the mother may be more likely to 'win' the conflict if she has no information about the state of her offspring. However, it is not clear under what circumstances selection could favour a reduction in a mother's discriminatory power. It is also difficult to envisage mechanisms that might allow offspring to determine their own sex or influence the sex of others. Of greater practical relevance seems the possibility of offspring hiding or revealing their sex. This may affect the outcome of the conflict between parent and offspring, but it may also lead to a new conflict, between sons and daughters. This problem has yet to be analysed.

A conflict may arise at an even earlier stage, between a parent and its gametes. Reiss (1987) has shown that only when fitness differences between sons and daughters are sufficiently large, gametes of the 'wrong' sex may be selected to sacrifice themselves in favour of gametes of the 'right' sex. However, the biological interpretation of this conclusion is not clear. Does it mean that gametes actively oppose parental manipulation, or does it mean that they signal their identity under the appropriate circumstances? More work is needed to analyse specific biological scenarios of the conflict between parents and gametes to see how the outcome depends on specific mechanistic details.

Most sex-allocation models assume that just one parent, by default the mother, has control over sex-allocation decisions. In many vertebrates both parents have some share in the allocation of resources to their offspring. Asymmetries in the nature of their contribution and asymmetries in their life history may cause the parents to disagree over the preferred allocation pattern. In order to properly analyse such a scenario, several mechanistic aspects of the asymmetry between the parents have to be considered:

1. What components of allocation are influenced by each parent?

2. In what temporal order do the parents make their decisions?

3. What information about their decisions is transferred between the parents, and when?

In many cases some allocation components will be under the control of one parent, and others under the control of the other parent. For example, one parent might be able to determine the sex ratio while the other parent determines the clutch size. If the sex ratio is determined before the clutch size, then the outcome of selection on the sex ratio depends on whether the parent in control of clutch size can detect the sex ratio before determining the clutch size. This kind of situation can be analysed relatively straightforwardly with standard methods (see section 2.5.7).

Some components of allocation might be influenced by both parents. For example, it is conceivable that both parents can adjust the sex ratio to some extent. If the parents favour different sex ratios they have a conflict. This could lead to an arms race or a stable resolution of the conflict. In the case of a stable outcome, either one parent 'wins' or there will be a compromise between the parental optima. Of major importance for the outcome is the order in which the parents can make their decisions. For example, in mammals it seems likely that males, being the heterogametic sex, have the first opportunity to bias the primary sex ratio. After insemination or fertilization it is the female's turn to adjust the sex ratio. Once the offspring are born, males might again have an opportunity to alter the sex ratio, for example by differentially provisioning sons and daughters. In birds, where females are heterogametic, females are most likely to have the first option. We are not aware of any models that investigate the consequences of biparental control over the same allocation problem.

2.4 | Fitness

In this section we give a brief overview of basic evolutionarily stable strategy (ESS) techniques that are useful for analysing sex allocation problems. More elaborate treatments can be found in Taylor (1996), McNamara and Houston (1996), Taylor and Frank (1996) and Frank (1998).

2.4.1 Evolutionary stability

The recipe for an ESS analysis has three ingredients (e.g. Parker & Maynard Smith 1990): (1) define a strategy set consisting of all phenotypically feasible traits; (2) define a fitness function relating the 'adaptedness' of a trait to properties of the population and/or the environment; (3) use an ESS criterion to find those strategies that are evolutionarily stable. So far we have considered the first ingredient, describing how the strategy set may reflect mechanistic aspects of sex allocation. Here we focus on the rest of the recipe.

In an ESS analysis one seeks an expression for the fitness $W(x, x^*)$ of a rare mutant phenotype x in a resident population that is monomorphic for phenotype x^*. A necessary condition for x^* to be a ESS is that in a resident population of individuals with phenotype x^*, rare mutants with phenotype $x \neq x^*$ do not have a higher fitness than the residents. In other words

$$W(x^*, x^*) \geq W(x, x^*). \tag{2.4}$$

This implies that $W(x, x^*)$ achieves its maximal value with respect to x when $x = x^*$, giving us

the conditions

$$\left.\frac{\partial W}{\partial x}\right|_{x=x^*} = 0 \quad \text{and} \quad \left.\frac{\partial^2 W}{\partial x^2}\right|_{x=x^*} \leq 0. \tag{2.5}$$

To avoid unneccesary technical details, in the rest of the chapter we focus only on the first condition, the equilibrium condition. For completeness we note that the conditions in eq. 2.5 do not imply that in a resident population with phenotype $\hat{x}$ close to an ESS x^* selection favours mutants that are even closer to the ESS x^*. To check whether an ESS has this extra stability property (convergence stability), additional second-order conditions have to be verified (Taylor 1996, Geritz et al. 1998).

We shall see in sections 2.4.2 and 2.4.3 that in many sex-allocation models the fitness function $W(x, x^*)$ can be written in the 'Shaw–Mohler' form

$$W(x, x^*) = \frac{m(x)}{m(x^*)} + \frac{f(x)}{f(x^*)}, \tag{2.6}$$

named after Shaw and Mohler (1953) who first derived such an expression in the context of sex ratio evolution. The term $m(x)$ represents the returns on investment in sons, as determined by the mutant's behaviour x, and $f(x)$ the returns on investment in daughters. 'Returns on investment' seems rather vague, but this is because its interpretation may vary from one model to another. Often it simply means 'number of individuals that survive until adulthood'. The returns on investment are scaled by the 'reproductive values' (see section 2.4.3) $1/m(x^*)$ and $1/f(x^*)$ to take into account that an individual of one sex may have a greater expected contribution to future generations than an individual of the opposite sex. If we take m and f to mean the number of sons and daughters surviving until adulthood, the value of a son relative to that of a daughter equals the ratio $f(x^*)/m(x^*)$ of adult females to adult males in the population, because everyone has one father and one mother.

The first of the ESS conditions (eq. 2.5) applied to eq. 2.6 yields

$$\frac{m'(x^*)}{m(x^*)} + \frac{f'(x^*)}{f(x^*)} = 0, \tag{2.7}$$

and this equation is known as the Shaw–Mohler equation (Charnov 1982). Together with the second-order condition, it shows that the ESS x^* maximizes the product $m(x^*) \times f(x^*)$. Similar product-maximization theorems hold for many sex-allocation models (Lessard 1989). In such cases an ESS analysis may be skipped altogether and one may rely on the easier product-maximization technique. However, for relatively complex life histories, it is not always possible to derive a fitness function as simple as eq. 2.6, and one has to derive a 'custom-made' version. It turns out that for more complex life histories, simple product-maximization criteria are in general not valid (e.g. Charnov 1982, p 17).

2.4.2 Life history and population dynamics

It is obvious that the derivation of evolutionarily stable sex-allocation patterns crucially depends on a proper measure of the fitness $W(x, x^*)$ of a rare mutant phenotype x in a resident population with phenotype x^*. Unfortunately, many sex-allocation models are based on *ad hoc* measures of fitness. A proper measure of fitness is given by the growth rate λ of a subpopulation of mutants relative to the growth rate λ^* of the resident population (Metz *et al.* 1992). If the growth rate of the mutant subpopulation is less than that of the resident population, the mutant cannot increase in frequency, and vice versa. The growth rates are derived from population dynamical models, a model for the resident population and one for the mutant subpopulation. The structure of the models is determined by the life history of the (class of) organisms we wish to study. It is often helpful to characterize life histories by a number of distinct states or 'stages' in which the organism can exist, and the transitions between the stages (Caswell 1989). Dioecious organisms can obviously be in at least two states: female or male. One may further distinguish between different kinds of females and males; for example, according to how old they are, whether they have a territory or not, etc. States can also be continuous; typical continuous state variables are time of birth, mass, size. Here we restrict our attention to life histories with a finite number of discrete states.

To illustrate the methods discussed in this section, we analyse the simplest possible life history with two states (males and females) and overlapping generations. Consider a monomorphic resident population with allocation strategy or behavior x^*. Each adult female produces $f^* = f(x^*)$ daughters and $m^* = m(x^*)$ sons that survive until adulthood, one season later. Adult females survive with probability $P_f^* = P_f(x^*)$, adult males with probability P_m. This gives us the population dynamical model

$$n_f' = \left(P_f^* + \frac{1}{2}f^*\right)n_f + \frac{1}{2}f^*Q^*n_m \tag{2.8a}$$

$$n_m' = \frac{1}{2}m^*n_f + \left(P_m + \frac{1}{2}m^*Q^*\right)n_m, \tag{2.8b}$$

where n_f' and n_m' are the number of adult females and males in the next season as a function of the numbers n_f and n_m in the current season. Note that f^* and m^* are multiplied by $\frac{1}{2}$, to account for the genetic share of parents in their offspring, or equivalently, to prevent counting the same offspring twice. The reproductive output of males is Q^* times that of females because the per capita reproductive output of males differs from that of females whenever the adult sex ratio differs from 1:1. In fact, from the viewpoint of an autosomal gene, every season the output of all females must equal that of all males since every offspring has one mother and one father: $n_f = Q^*n_m$, or equivalently $Q^* = n_f/n_m$. Substituting this expression for Q^* in eqs. 2.8a and 2.8b gives the simplified system

$$n_f' = (P_f^* + f^*)n_f \tag{2.9a}$$
$$n_m' = m^*n_f + P_m n_m. \tag{2.9b}$$

In the example, the state of the population can be described by a vector $\mathbf{n} = (n_f, n_m)$. More generally, if we discriminate between k different states, the population can be described by a vector $\mathbf{n} = (n_1, \ldots, n_k)$. If the per capita contribution of individuals in state j to individuals of state i in the next season is given by $a_{ij} = a_{ij}(x^*)$, then the number of individuals n_i in state i changes from one season to the next according to the recurrence relations

$$n_i' = \sum_{j=1}^{k} a_{ij}n_j, \quad i = 1, \ldots, k. \tag{2.10}$$

In matrix notation, $\mathbf{n}' = \mathbf{An}$, where $\mathbf{A} = \mathbf{A}(x^*)$ is the $k \times k$ matrix with elements $a_{ij}(i, j = 1, \ldots, k)$.

The matrix corresponding to eqs. 2.8a and 2.8b is given by

$$A = \begin{pmatrix} P_f^* + \frac{1}{2}f^* & \frac{1}{2}f^*Q^* \\ \frac{1}{2}m^* & P_m + \frac{1}{2}m^*Q^* \end{pmatrix} \quad (2.11)$$

or equivalently, according to eqs. 2.9a and 2.9b, by

$$\begin{pmatrix} P_f^* + f^* & 0 \\ m^* & P_m \end{pmatrix}. \quad (2.12)$$

Under mild conditions (e.g. Caswell 1989), a population with dynamics governed by A will eventually reach demographic equilibrium and grow with constant rate $\lambda^* = \lambda(x^*)$. In demographic equilibrium the distribution of individuals over the different states is stable, given by a vector $\mathbf{u}(x^*) = (u_1, \ldots, u_k)$, hence every stage grows with rate λ^*. Technically, λ^* equals the dominant eigenvalue of A and $\mathbf{u}(x^*)$ is a dominant right eigenvector of A. There are standard methods for calculating the eigenvalues of a given matrix (e.g. Lancaster 1969). In case of a 2×2 nonnegative matrix (every element ≥ 0), the dominant eigenvalue is given by

$$\lambda = \frac{1}{2}(a_{11} + a_{22})$$
$$+ \sqrt{(a_{11} + a_{22})^2 - 4(a_{11}a_{22} - a_{12}a_{21})}. \quad (2.13)$$

In our example, a straightforward calculation shows that $\lambda(x^*) = P_f(x^*) + f(x^*)$, i.e. a population in demographic equilibrium grows by a factor $P_f^* + f^*$ per season. This is quite obvious from eq. 2.9a in our example, but in general dominant eigenvalues are often difficult to calculate.

How does selection enter the picture? After specifying a population dynamical model for the resident population with behaviour x^*, we need a model for a mutant subpopulation with behaviour x. Writing $\mathbf{m} = (m_1, \ldots, m_k)$ for the numbers of mutants in the different states we can write $\mathbf{m}' = \mathbf{B}\mathbf{m}$, where $\mathbf{B}$ is a state-transition matrix which depends on the mutant behaviour x. How does $\mathbf{B}$ look for our example? Similar to eq. 2.11, with a few modifications. We assume that females are in control of behaviour x. Then survival of mutant females and their reproductive output (corresponding to the first column of A) depend on x. However, if mutants are rare, then it is very unlikely that mutant males mate with

mutant females, hence the reproductive output of mutant males is determined by the resident behaviour x^*. $\mathbf{B}$ can therefore be written as

$$\mathbf{B} = \begin{pmatrix} P_f + \frac{1}{2}f & \frac{1}{2}f^*Q^* \\ \frac{1}{2}m & P_m + \frac{1}{2}m^*Q^* \end{pmatrix}. \quad (2.14)$$

The rareness of mutants causes the adult sex ratio $Q^* = u_f/u_m$ to be determined by the resident behaviour x^*, and can be calculated from eq. 2.9b: replacing the n's by u's, setting the left-hand side equal to $\lambda^* u_m$ and dividing both sides by u_m gives

$$Q^* = \frac{u_f}{u_m} = \frac{\lambda^* - P_m^*}{m^*} = \frac{P_f^* + f^* - P_m^*}{m^*}. \quad (2.15)$$

If adult females and males have identical survival ($P_f = P_m$), then the ratio of adult females to males is just f^*/m^*, the ratio of surviving daughters to surviving sons. Otherwise ($P_f \neq P_m$), the adult sex ratio is more biased towards the sex with the highest adult survival. Note that $\mathbf{B}$ does not only depend on the mutant behaviour x, but also on the resident behaviour x^*, so in general we have $\mathbf{B} = \mathbf{B}(x, x^*)$. Clearly, if the mutant behaviour does not differ from the resident behaviour, that is, $x = x^*$, then we should have $\mathbf{B}(x^*, x^*) = \mathbf{A}(x^*)$. This is obviously the case in our example.

Given a transition matrix $\mathbf{B}(x, x^*)$ we can calculate the growth rate $\lambda = \lambda(x, x^*)$ of the mutant subpopulation, which determines whether the mutants will increase in frequency ($\lambda > \lambda^*$) or not ($\lambda < \lambda^*$). Hence, $\lambda(x, x^*)$ is a suitable fitness measure, and by eq. 2.5 an ESS must obey

$$\left.\frac{\partial \lambda}{\partial x}\right|_{x=x^*} = 0. \quad (2.16)$$

In our example, the dominant eigenvalue of eq. 2.14 can be found by inserting the elements of eq. 2.14 into eq. 2.13. The readers may check for themselves that a rather nasty formula results. Differentiating this formula with respect to x yields, after some tedious calculations, the ESS condition

$$\left.\frac{\partial \lambda}{\partial x}\right|_{x=x^*} = \frac{f(x^*)m(x^*)}{\sqrt{z^*}} \left[\frac{P_f'(x^*)}{f(x^*)} \right.$$
$$\left. + \frac{1}{2}\left(\frac{f'(x^*)}{f(x^*)} + \frac{m'(x^*)}{m(x^*)}\right) \right] = 0, \quad (2.17)$$

where primes denote differentiation and $\sqrt{z^*}$ corresponds to the square root in eq. 2.13. Since $f^*m^*/\sqrt{z^*}$ is positive, eq. 2.17 is equivalent to

$$\frac{P_f'(x^*)}{f(x^*)} + \frac{1}{2}\left(\frac{f'(x^*)}{f(x^*)} + \frac{m'(x^*)}{m(x^*)}\right) = 0. \qquad (2.18)$$

The first term is the marginal cost of increasing x in terms of the female's own survival, the second term is the marginal benefit in terms of offspring production. This is a generalization of the Shaw–Mohler equation (eq. 2.7), as we shall discuss below in more detail.

Even for such a simple model, it is quite hard to calculate the dominant eigenvalue $\lambda(x, x^*)$ and to derive the ESS condition (eq. 2.17). For models with more than two states, it becomes very impractical. In the next section, we shall see that the calculations can be simplified considerably with the aid of reproductive values.

2.4.3 Reproductive values

The expected contribution to future generations may differ between individuals in different states. For example, if males are the rare sex, then they are more 'valuable' because they can, on average, expect more offspring than females. To quantify such differences in value in general, we need the concept of reproductive value. The reproductive value v_i of an individual in state i relative to the reproductive values of individuals in other states j can be defined such that $u_i v_i / \sum_j u_j v_j$ is the expected proportion of genes in the far future that reside in individuals of state i now. In view of this interpretation, reproductive values can be defined recursively according to

$$v_j = \frac{1}{\lambda^*}\sum_i a_{ij}v_i, \qquad (2.19)$$

where a_{ij} is the *per capita* contribution (through survival and/or reproduction) of individuals in state j to individuals in state i in the next time period. In other words, the current reproductive value of an individual is the total amount of reproductive value it contributes to the next time period. The contribution is scaled by the inverse of the growth rate λ^* of the population to account for the 'diluting' effect of a growing population. Technically, the vector of reproductive values $\mathbf{v} = (v_1, \ldots, v_k)$ is a dominant left

eigenvector of the transition matrix $\mathbf{A}$. Since only the relative reproductive values are meaningful in this context (the direction of the vector, not its length), we may arbitrarily set one of them to a fixed value, leaving $k - 1$ equations in $k - 1$ unknowns. For our example, we have

$$\lambda^* v_f = \left(P_f^* + \frac{1}{2}f^*\right)v_f + \frac{1}{2}m^* v_m. \qquad (2.20)$$

Substituting $\lambda^* = P_f^* + f^*$ yields $v_f = (m^*/f^*)v_m$. We may set $v_m = 1/m^*$, giving

$$v_m = \frac{1}{m^*} \quad \text{and} \quad v_f = \frac{1}{f^*}. \qquad (2.21)$$

The usefulness of the reproductive values arises from the fact that instead of $\lambda(x, x^*)$, which is hard to calculate, we can use as a fitness function

$$W(x, x^*) = \sum_{i,j} v_i(x^*)b_{ij}(x, x^*)u_j(x^*). \qquad (2.22)$$

This fitness function is the total amount of reproductive value contributed by mutants to the next time period. Note that the v_i and the u_j do not depend on the mutant behaviour x, but only on the resident behaviour x^*. In other words, W can be interpreted as the fitness of mutants that display mutant behaviour during a single time period, after which they revert to the resident behaviour.

The use of W as a fitness function instead of λ is justified by the fact that $\partial\lambda/\partial x$ and $\partial W/\partial x$ have the same sign. In fact, it follows from simple algebraic considerations (e.g. Taylor 1996) that

$$\left.\frac{\partial W}{\partial x}\right|_{x=x^*} = \sum_{i,j} v_i u_j \left.\frac{\partial b_{ij}}{\partial x}\right|_{x=x^*} = \sum_i v_i u_i \left.\frac{\partial\lambda}{\partial x}\right|_{x=x^*}. \qquad (2.23)$$

Hence $\partial W/\partial x$ is identical to $\partial\lambda/\partial x$, up to the positive factor $\sum_i v_i u_i$.

Notice that all terms b_{ij} that do not depend explicitly on the mutant behaviour x can be omitted from the sum (eq. 2.22) since they would vanish after differentiation anyhow. This is particularly convenient when the mutant behaviour is restricted to individuals of only one state, say state j. We can then skip all terms in eq. 2.22 that do not explicitly depend on x and obtain (up to the positive constant $u_j(x^*)$) the simplified fitness function

$$W(x, x^*) = \sum_i v_i(x^*) b_{ij}(x, x^*).$$ (2.24)

This applies to our example, because we assumed that only females are in control. Applying eq. 2.24 yields

$$W(x, x^*) = v_f(x^*) \left(P_f(x) + \frac{1}{2} f(x) \right)$$

$$+ v_m(x^*) \frac{1}{2} m(x)$$ (2.25a)

$$= \frac{P_f(x)}{f(x^*)} + \frac{1}{2} \left(\frac{f(x)}{f(x^*)} + \frac{m(x)}{m(x^*)} \right),$$ (2.25b)

and we get the same ESS condition as before (see eq. 2.18)

$$\frac{P'_f(x^*)}{f(x^*)} + \frac{1}{2} \left(\frac{f'(x^*)}{f(x^*)} + \frac{m'(x^*)}{m(x^*)} \right) = 0.$$ (2.26)

This condition is a generalization to overlapping generations of the Shaw–Mohler equation (eq. 2.7). It is a special case of the more general equation

$$P'_f(x^*) v_P(x^*) + \frac{1}{2} \left(\frac{f'(x^*)}{f(x^*)} + \frac{m'(x^*)}{m(x^*)} \right) = 0,$$ (2.27)

where $v_P(x^*)$ is the reproductive value of a surviving female. This relatively simple and transparent condition does not just hold for our simple example, but it also holds for a fairly general class of life histories, as we show in the Appendix in section 2.6. The only part that depends on life history details is the reproductive value v_P^* of a surviving female.

2.4.4 Density dependence

In section 2.4.3 we showed that calculation of the growth rate $\lambda(x, x^*)$ of the mutant subpopulation can be avoided with the aid of reproductive values, and we only need to compute the dominant eigenvalue and eigenvectors of the resident population. However, one might question whether a value $\lambda^* \neq 1$ can be considered reasonable, since no biological population can grow or decline forever. In reality, density-dependent factors will ensure that a resident population will, in the long run, be more or less stationary, corresponding to $\lambda^* = 1$. To ensure that $\lambda^* = 1$ in the model, it is necessary to specify a particular mechanism of density dependence. For example, at high population densities juvenile sur-

vival might be compromised or the age of first breeding might be delayed. It is then assumed that the density-dependent life history parameters take on such values in equilibrium that a stationary population results. For such a population with $\lambda^* = 1$, it is then relatively easy to calculate the left and right eigenvectors and to derive the fitness function of eq. 2.22.

The choice of a paticular mechanism of density dependence can have important evolutionary implications (see Mylius & Diekman 1995, Pen & Weissing 2000). Those life history parameters that are supposed to be affected by density dependence can no longer be considered as independent entities. Instead, the constraint $\lambda^* = 1$ can be used to express the density-dependent life history parameters in terms of the other parameters. As a consequence, different mechanisms of density dependence may lead to rather different fitness functions and, hence, to rather different evolutionary predictions.

2.4.5 Direct fitness and relatedness

Sometimes the fitness of an individual with behaviour x does not just depend on its own behaviour and the behaviour x^* of the entire population, but also on the behaviour $\bar{x}$ in a local environment of the individual. That is, fitness can be written as $W(x, \bar{x}, x^*)$. The effect on fitness of a small change in x can then be written as

$$\frac{dW}{dx} = \frac{\partial W}{\partial x} R_0 + \frac{\partial W}{\partial \bar{x}} R_1 + \frac{\partial W}{\partial x^*} R_\infty,$$ (2.28)

where the R_i's can be interpreted as coefficients of relatedness (Michod & Hamilton 1980) between the individual doing x and itself ($R_0 = 1$), the local individuals (R_1) and the entire population ($R_\infty = 0$). Hence, if we write $R_1 = R$, the ESS condition (eq. 2.5) can be generalized to

$$\left(\frac{\partial W}{\partial x} + R \frac{\partial W}{\partial \bar{x}} \right)_{x = \bar{x} = x^*} = 0.$$ (2.29)

This approach is known as the 'direct fitness' approach (Taylor & Frank 1996, Frank 1998), as opposed to 'inclusive fitness'. Clearly, eq. 2.28 can be generalized by letting fitness depend on the behaviour x_i of every individual i in the entire population

$$\frac{dW}{dx} = \frac{\partial W}{\partial x} + \sum_i R_i \frac{\partial W}{\partial x_i},$$ (2.30)

where R_i is the coefficient of relatedness between the focal individual and individual i. In fact, the R_i values measure association between behaviour of different individuals, which may come about by common genealogy but also by other factors such as a shared environment. Hence, the direct fitness approach is more general than the inclusive fitness approach (Frank 1998).

In the next section we give an example of the direct fitness approach, applied to parent–offspring conflict over the sex ratio.

2.5 | Model gallery

In this section we present a series of models with a twofold purpose: first, to illustrate and verify results discussed in section 2.3 and, second, as an overview of the most commonly used ESS techniques. As a prelude to more advanced techniques used later on, we start with some simple models, generating well-known results that apply to organisms with nonoverlapping generations. These models are then extended to include several of the mechanistic aspects discussed before and we use the results to analyse the conflict over sex allocation between parents.

2.5.1 Fisher's principle of equal allocation

Consider a mother with a sex ratio (proportion sons) s among her offspring in a population with sex ratio s^*. We assume nonoverlapping generations, hence the fitness $W(s, s^*)$ of such a mother is a special case of eq. 2.25b, with maternal survival P_f set equal to zero, that is, $W(s, s^*) = m(s)/m(s^*) + f(s)/f(s^*)$. The mother's clutch size is given by $c(s)$, hence the number of sons is given by $m(s) = sc(s)$ and the number of daughters by $f(s) = (1 - s)c(s)$. Thus

$$W(s, s^*) = \frac{c(s)}{c(s^*)}\left[\frac{s}{s^*} + \frac{1-s}{1-s^*}\right]. \tag{2.31}$$

If a mother's clutch size c is independent of the sex ratio s among her offspring, this reduces to

$$W(s, s^*) = \frac{s}{s^*} + \frac{1-s}{1-s^*}, \tag{2.32}$$

which yields an ESS sex ratio $s^* = 1/2$. However, if sons and daughters require different amounts of resource, say E_m units per son and E_f units per

daughter, then for a fixed amount of resource E_T we have $E_T = scE_m + (1 - s)cE_f$, or in other words

$$c(s) = \frac{E_T}{sE_m + (1-s)E_f}. \tag{2.33}$$

Substitution in eq. 2.31 gives the fitness function

$$W(s, s^*) = \frac{s^*E_m + (1-s^*)E_f}{sE_m + (1-s)E_f}\left[\frac{s}{s^*} + \frac{1-s}{1-s^*}\right]. \tag{2.34}$$

The ESS sex ratio s^* is found by solving

$$\left.\frac{\partial W}{\partial s}\right|_{s=s^*} = \frac{(1-s^*)E_f - s^*E_m}{[s^*E_m + (1-s^*)E_f]s^*(1-s^*)} = 0, \tag{2.35}$$

which yields

$$\frac{s^*}{1-s^*} = \frac{E_f}{E_m}. \tag{2.36}$$

This is Fisher's (1930) classic result that the evolutionarily stable ratio of sons to daughters is the inverse of their cost ratio, or, in other words, equal allocation to sons and daughters

$$s^*E_m = (1 - s^*)E_f. \tag{2.37}$$

2.5.2 Fisher's model with costly sex ratio control

Now we extend the previous model by allowing for costs of sex ratio control. Similar models have been studied by Maynard Smith (1980), Eshel and Sansone (1991) and Charnov (1982). Suppose the primary sex ratio is even and parents can adjust the secondary sex ratio by selective abortion. As before, we assume that a son costs E_m units of resource and a daughter E_f units. Without loss of generality we may assume that $E_m > E_f$, so selection favours a sex ratio biased towards daughters when the costs of abortion are small enough. An aborted son costs $0 \le e_m < E_m$ units of resource. Suppose a mother aborts a fraction a of her sons, then for a fixed amount of resources E_T and clutch size c we have $E_T = \frac{1}{2}(1 - a)cE_m + \frac{1}{2}ace_m + \frac{1}{2}cE_f$, or

$$c(a) = \frac{E_T}{\frac{1}{2}[(1-a)E_m + ae_m + E_f]}. \tag{2.38}$$

Note that the secondary (after abortion) sex ratio s is given by $s = (1 - a)/(2 - a)$, or $s/(1 - s) =$

$1 - a$. Substitution of the expression 2.38 for clutch size in eq. 2.31 gives the fitness function

$$W(a, a^*) = \frac{(1 - a^*)E_m + a^*e_m + E_f}{(1 - a)E_m + ae_m + E_f} \left[\frac{1 - a}{1 - a^*} + 1 \right].$$

(2.39)

The ESS sex ratio is found by solving $\partial W/\partial a = 0$, evaluated at $a = a^*$, which gives the evolutionarily stable level of abortion as

$$a^* = \frac{E_m - E_f - 2e_m}{E_m - e_m},$$

(2.40)

or equivalently

$$\frac{s^*}{1 - s^*} = 1 - a^* = \frac{E_f + e_m}{E_m - e_m},$$

(2.41)

which is identical to Fisher's result if and only if $e_m = 0$. Figure 2.1 shows for several values of the relative cost $\rho = E_m/E_f$ of sons how the ESS sex ratio s^* varies with e_m.

The condition that it still pays to abort any sons at all $(a^* > 0)$ is given by

$$e_m < \frac{1}{2}(E_m - E_f).$$

(2.42)

In other words, it doesn't pay to abort if an aborted son costs half the cost difference between the sexes or more. The same result was found by Eshel and Sansone (1991) by means of a population genetical model.

Given E_m, E_f and e_m, we can calculate the proportion p of resources wasted on adjusting the sex ratio as

$$p = \frac{1}{2}a^* e_m = \frac{1}{2}\frac{E_m - E_f - 2e_m}{E_m - e_m} e_m.$$

(2.43)

The value of e_m for which the waste is maximal, $\hat{e}_m$, is found by solving $dp/de_m = 0$, which yields

$$\hat{e}_m = E_m - \sqrt{\frac{E_m(E_m + E_f)}{2}}.$$

(2.44)

If we insert the right-hand side in eq. 2.43, we get an expression for the maximal waste, given the costs E_m and E_f of sons and daughters, respectively. This maximal waste is plotted against E_m for $E_f = 1$ in Figure 2.1, which shows that even for extreme values of $\rho = E_m/E_f$ it never pays to waste a great deal of resources adjusting the sex ratio. For the specific case of $\rho = 2$, $p(\hat{e}_m) = 0.036$.

2.5.3 Parent–offspring conflict

Now we use the model to investigate what the favoured sex ratio is from the offspring's point of view. We use Taylor and Frank's (1996) direct fitness approach to derive Trivers' (1974) classic result. For an individual offspring we can use the same fitness function as eq. 2.34, except that the clutch size is not only a function of its own sex ratio s, but also depends on the sex ratio favoured by the other offspring in the clutch. If we write $\bar{s}$ for the average sex ratio favoured by the offspring in a clutch, and s^* again for the sex ratio in the rest of the population, then we can rewrite eq. 2.34 as

$$W(s, \bar{s}, s^*) = \frac{s^*E_m + (1 - s^*)E_f}{\bar{s}E_m + (1 - \bar{s})E_f} \left[\frac{s}{s^*} + \frac{1 - s}{1 - s^*} \right].$$

(2.45)

The direct fitness equilibrium condition is given by

$$\left(\frac{\partial W}{\partial s} + R \frac{\partial W}{\partial \bar{s}} \right)_{s = \bar{s} = s^*} = 0,$$

(2.46)

where R is the coefficient of relatedness between the offspring. We get

$$E_f - [2E_f - (E_m - E_f)(1 - 2R)]s^* - 2(E_m - E_f)(1 - R)s^{*2} = 0,$$

(2.47)

which is Trivers' (1974) result. For the special case of $R = \frac{1}{2}$ we get

$$\frac{s^*}{1 - s^*} = \sqrt{\frac{E_f}{E_m}},$$

(2.48)

which is less biased than the ESS from the mother's point of view. In general, the larger R, the smaller the disagreement between mother and offspring, and for $R = 1$ the offspring's optimum and maternal optimum coincide (see Figure 2.4). Factors that influence R are the level of inbreeding, the number of fathers per clutch and the clutch size (for a single offspring $R = 1$).

2.5.4 Trivers and Willard's model with two states

Fisher's result is based on the assumption that all mothers are in identical conditions, and they all produce the same sex ratio. Trivers and Willard

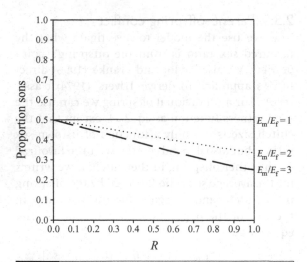

Fig 2.4 ESS sex ratio (proportion sons) under offspring control as a function of the average relatedness R between offspring in a clutch, for several values of the relative cost of a son, E_m/E_f. Note that for $R = 1$ the offspring's ESS coincides with the maternal ESS.

(1973) argued that if mothers in good condition produce fitter offspring than mothers in bad condition, and if sons profit more from good maternal condition than daughters, then mothers in good condition should produce sons and mothers in bad condition should produce daughters. More generally, given the prevailing conditions, a mother should overproduce that sex which yields the highest inclusive fitness from the mother's point of view.

Suppose all mothers have the same amount of resources E_T and sons and daughters are equally costly ($E_m = E_f$), but mothers can be in two possible states (0 or 1), each state occurring with frequency $\frac{1}{2}$. The mother's state affects offspring survival such that the relative survival of sons in state 0 compared to sons in state 1, and that of daughters in state 1 compared to daughters in state 0 is given by $\alpha < 1$. Hence, in state 0 it is better to produce daughters, in state 1 sons. The symmetry of the situation simplifies the analysis because the sex ratio bias in state 0 must equal the bias in state 1. Hence, the average sex ratio s^* will be $\frac{1}{2}$, independently of the sex ratio s_0 produced in state 0 and the sex ratio $s_1 = 1 - s_0$ produced in state 1. For each state we therefore have a fitness function similar to eq. 2.25b,

with

$$m(s^*) = f(s^*) = \frac{1}{2}c\frac{1+\alpha}{2}, \qquad (2.49)$$

where $s^* = 1/2$ is the average sex ratio and c the clutch size. In state 0, the number of sons is given by $m_0(s_0) = \alpha s_0 c$ and the number of daughters by $f_0(s_0) = (1 - s_0)c$; in state 1 $m_1(s_1) = s_1 c$ and $f_1(s_1) = \alpha(1 - s_1)c$. This gives the fitness functions

$$W_0(s_0) = 2\frac{\alpha s_0 + 1 - s_0}{\alpha + 1} \qquad (2.50a)$$

$$W_1(s_1) = 2\frac{s_1 + \alpha(1 - s_1)}{\alpha + 1}. \qquad (2.50b)$$

Hence, the selection differentials are given by

$$\frac{dW_0}{ds_0} = -\frac{dW_1}{ds_1} = \frac{2(\alpha - 1)}{\alpha + 1} < 0, \qquad (2.51)$$

which implies that there is a unique ESS (s_0^*, s_1^*), given by the 'bang-bang' strategy $(0, 1)$.

2.5.5 Trivers and Willard's model with costly sex ratio control

Now we assume that the primary sex ratio is fixed at parity, and the sex ratio can be adjusted secondarily by selective abortion. Let a_0 represent the fraction of sons that are aborted in state 0, and a_1 the fraction of daughters that are aborted in state 1. An aborted offspring costs e units of resource relative to the cost E of an 'accepted' offspring. Given a total amount of resources E_T, the clutch size in state i is given by

$$c_i(a_i) = \frac{E_T}{\frac{1}{2}[2E - a_i(E - e)]}. \qquad (2.52)$$

The fitness functions are then given by

$$W_i(a_i, a_i^*) = \frac{2E - a^*(E - e)}{2E - a_i(E - e)}\frac{(1 - a_i)\alpha + 1}{(1 - a^*)\alpha + 1}. \qquad (2.53)$$

The selection differentials are

$$\left.\frac{\partial W_i}{\partial a_i}\right|_{a_i = a_i^* = a^*} = \frac{-1}{2E - a^*(E - e)}\frac{\alpha(E + e) - E + e}{(1 - a^*)\alpha + 1}. \qquad (2.54)$$

The sign of the right-hand side is independent of the abortion rate a_i. Hence, either all offspring of the undesired sex are aborted, or none. In other words, selection favours the bang-bang sex ratio strategy $(0,1)$ or no sex ratio adjustment at all.

For selection to favour abortion, the right-hand side of eq. 2.54 must be positive, or equivalently

$$e < E \frac{1 - \alpha}{1 + \alpha}. \tag{2.55}$$

This condition is plotted in Figure 2.2.

Two interesting conclusions can be drawn from this analysis. First, costly sex ratio control does not necessarily favour sex ratio trends less extreme than bang-bang strategies. Second, if selection favours conditional abortion, then a fraction $\frac{1}{2}e/E$ of resources is 'wasted' on sex ratio adjustment, which can be considerably more than under unconditional sex ratio adjustment in Fisher's scenario (see Figure 2.2).

2.5.6 Multiple allocation components

In the Appendix we have derived a fitness function for the allocation strategy x^* of a reproducing female, valid for a rather general class of life histories

$$W_f(x, x^*) = P_f(x)v_P(x^*) + \frac{1}{2}\left(\frac{f(x)}{f(x^*)} + \frac{m(x)}{m(x^*)}\right), \tag{2.56}$$

where v_P is the residual reproductive value of a reproducing female, which depends on particular life history details such as the mating system.

We use the fitness function eq. 2.56 to illustrate our approach of section 2.3.2, where we decomposed allocation decisions into multiple components. Accordingly, let s be the primary sex ratio, c clutch size, E_m the investment per son and E_f the investment per daughter. A female's total reproductive effort is then given by $E_T = c[sE_m + (1-s)E_f]$ which determines her survival $P_f = P_f(E_T)$. Survival $\mu = \mu(E_m)$ of a son is assumed to depend only on E_m, not on any other allocation components, and the survival $\phi = \phi(E_f)$ of a daughter only on E_f. The number of surviving daughters and sons produced per mother is then given by

$$m = sc\mu(E_m) \quad \text{and} \quad f = (1-s)c\phi(E_f). \tag{2.57}$$

Inserting s, c, E_f and E_m for x in eq. 2.56 and applying eq. 2.5 yields the four ESS conditions

$$s^* : \frac{1}{2}\left(\frac{1}{s^*} - \frac{1}{1 - s^*}\right) = -c(E_m - E_f)P_f'(E_T^*)v_P \tag{2.58a}$$

$$c^* : \frac{1}{c^*} = -\frac{E_T^*}{c^*}P_f'(E_T^*)v_P \tag{2.58b}$$

$$E_m^* : \frac{1}{2}\frac{\mu'(E_m^*)}{\mu(E_m^*)} = -cs\,P_f'(E_T^*)v_P \tag{2.58c}$$

$$E_f^* : \frac{1}{2}\frac{\phi'(E_f^*)}{\phi(E_f^*)} = -c(1 - s)P_f'(E_T^*)v_P. \tag{2.58d}$$

Note that $-P_f'(E_T^*)v_P$ is positive, assuming that P_f decreases with E_T. Condition 2.58a therefore shows that selection on the sex ratio leads to an overproduction of the cheaper sex. In general, however, evolutionary stability of s, taken in isolation, will not yield equal allocation to sons and daughters

$$s^*E_m = (1 - s^*)E_f. \tag{2.59}$$

In fact, a straightforward calculation yields that eq. 2.58a implies eq. 2.59 only if it is true that $-P_f'(E_T^*)v_P = 1/E_T^*$ or, equivalently, if eq. 2.58b is also satisfied. In other words, Fisher's (1930) equal allocation principle (eq. 2.59) will only hold if sex ratio and clutch size are optimized simultaneously.

Let us now consider several examples. First we have to find an expression for a mother's residual reproductive value v_P, which depends on the mating system. For example, if a breeding female that survives until the next breeding season will breed again, then her residual reproductive value is given by

$$v_P = \frac{1}{1 - P_f}, \tag{2.60}$$

her life expectancy. If a breeding female that survives has the same chance as a surviving daughter to breed in the next season, then their reproductive values are equal, hence $v_P = 1/f$ (see Chapter 3 in Pen 2000). In the numerical examples below we assume that v_P is given by eq. 2.60. In that case, selection on the clutch size gives us condition 2.58b, which for our choice of v_P can be written as

$$\frac{1}{E_T^*} = -\frac{P_f'(E_T^*)}{1 - P_f(E_T^*)} \tag{2.61}$$

Hence, for this choice of v_P, the ESS reproductive effort is completely determined by $P_f(E_T)$, the relationship between reproductive effort E_T and the mother's subsequent survival P_F. Notice that the same total reproductive effort $E_T^* = -(1 - P_f)/P_f'$ can be achieved by various combinations of the allocation components. For clutch size to be evolutionarily stable, only the level of E_T^* is important, not the way it is achieved. In particular, the ESS value of E_T is independent of the sex ratio s. Graphically, condition 2.61 implies that E_T^* can be found by drawing a straight line from $P = 1$ that touches the graph of $P(E_T)$ (see Figure 2.3A). In the case that offspring survival is independent of clutch size, the ESS criteria (eqs. 2.58c and 2.58d) can also be given by a simple graphical interpretation (see Figure 2.3B).

Table 2.1 shows numerical examples of what happens when either all four allocation components are optimized ('no constraints'), or only a subset of allocation components is optimized (due to 'constraints'). If, for instance, the primary sex ratio is fixed at 1/2 (e.g. due to chromosomal sex determination), eq. 2.58a will not hold true and the solution of the remaining equations (eqs. 2.58b to 2.58d) yields a biased sex-allocation ratio, in this specific example male-biased. If clutch size is constrained to fixed value, condition 2.58b will not hold true, and selection on the remaining allocation components results again in biased sex allocation, this time female-biased.

2.5.7 Parent–parent conflict

In the previous section we assumed that all allocation decisions are under the control of one and the same parent. We now extend the model to investigate what happens when sex allocation is under partial control of both parents. We assume there is one father per clutch. The father is in control of the sex ratio s, and the mother controls the clutch size c. Father invests no further in his offspring, and his sex ratio decision does not affect his own subsequent survival. The investments per individual son (E_m) and daughter (E_f) are assumed to be constant now. This does not affect our qualitative conclusions.

As we have seen before, the mother would prefer equal allocation, i.e. eq. 2.36, if she were in control of clutch size and sex ratio. If, on the other hand, the father had full control, he would prefer a 1:1 sex ratio, since his survival is, by assumption, independent of E_T. Hence, in case

of $E_f \neq E_m$ there is a conflict over the sex ratio between the parents. To illustrate possible outcomes of this conflict, we consider two scenarios, according to whether the mother has information about the sex ratio decision of her mate and uses this information to adjust the clutch size.

Analogous to the mother's fitness function (eq. 2.56), the following fitness function holds for a father with allocation strategy y

$$W_m(y, y^*) = P_m(y)v_M(y^*) + \frac{1}{2}\left(\frac{f(y)}{f(y^*)} + \frac{m(y)}{m(y^*)}\right),$$
(2.62)

where v_M is a breeding male's residual reproductive value. Since the male's survival P_m is independent of his sex ratio decision, the first term of eq. 2.62 can be skipped, and the fitness function simplifies to

$$W_m(s, s^*) \propto \frac{sc\mu(E_m)}{s^*c^*\mu(E_m)} + \frac{(1-s)c\phi(E_f)}{(1-s^*)c^*\phi(E_f)}$$
(2.63a)

$$= \frac{c(s)}{c(s^*)}\left(\frac{s}{s^*} + \frac{1-s}{1-s^*}\right).$$
(2.63b)

This yields the ESS condition

$$\left.\frac{\partial W_m}{\partial s}\right|_{s=s^*} = 2\frac{c'(s^*)}{c(s^*)} + \frac{1}{s^*} - \frac{1}{1-s^*} = 0.$$
(2.64)

Now suppose the mother cannot adjust her clutch size to the sex ratio determined by her mate. Then $c'(s^*) = 0$ and it follows from eq. 2.64 that the male's ESS sex ratio is $s^* = \frac{1}{2}$. From section 2.5.6 we know that if the sex ratio is fixed at 1/2, the mother's ESS sex allocation is not equal allocation. Hence, the father 'wins' the conflict.

Now suppose the mother *can* adjust her clutch size according to her mate's strategy. We assume that a breeding female that survives breeds again in the next season. A female's residual reproductive value is then given by eq. 2.60, and selection on clutch size yields condition 2.61 that we derived in the previous section. Since eq. 2.61 says that a female's ESS reproductive effort E_T^* is independent of the sex ratio, a change in the sex ratio by her mate induces a selection pressure on females to change their clutch size in such a way that their reproductive effort stays constant. Since the mother's reproductive effort is given by $E_T = c[sE_m + (1 - s)E_f]$, we can denote the female's clutch size 'response' to the male's sex ratio s as

$$c(s) = \frac{E_T}{s E_m + (1-s)E_f}. \qquad (2.65)$$

Hence

$$\frac{c'(s^*)}{c(s^*)} = \frac{2(E_f - E_m)}{s^* E_m + (1-s^*)E_f}. \qquad (2.66)$$

As a consequence, eq. 2.64 yields $s^* E_m = (1 - s^*)E_f$, equal allocation. Thus, if the mother can adjust her clutch size to the sex ratio decision of her mate, then she 'wins' the conflict.

2.6 | Appendix: Generalized Shaw–Mohler equation

In this appendix we show that the generalized Shaw–Mohler equation (eq. 2.27) holds for a fairly general class of life histories. Suppose individuals of each sex can be in one of two states, a non-breeding or floater state and a breeding or territorial state. The two nonbreeding states, one for females and one for males, can each be regarded as a 'pooling' of many nonbreeding states, which is what makes this description general (Weissing & Pen, in prep.).

The life history parameters that characterize each state are doubly indexed by f or m to denote sex, and by 0 or 1 to denote the breeding state. $P_{ij}(i \in \{f, m\}, j \in \{0, 1\})$ represents the probability of survival from one season to the next. For example, P_{f0} denotes the survival probability of a nonbreeding female, and P_{m1} of a breeding male. The number of surviving daughters produced per breeding female is denoted by f and the number of surviving sons by m. A individual's probability of being in the breeding state in the next season may depend on its current state and is denoted by $\alpha_{ij}(i \in \{f, m\}, j \in \{0, 1\})$. Likewise, the probability that an individual is a nonbreeder is denoted by β_{ij}. For newborns we use the symbols α_i and $\beta_i(i \in \{f, m\})$ to denote the probabilities that they become breeders or nonbreeders in the season after birth.

The numbers of individuals in the four states are stored in the column vector $\mathbf{n} = (n_{f0}, n_{f1}, n_{m0}, n_{m1})$. The population composition $\mathbf{n}'$ in the next season, as a function of the population $\mathbf{n}$ in the current season and the life history parameters defined above, is given by the recurrence relation $\mathbf{n}' = \mathbf{An}$, where $\mathbf{A}$ is the 4×4 matrix

$$\begin{pmatrix} \beta_{f0}P_{f0} & \beta_{f1}P_{f1} + \frac{1}{2}\beta_f f & 0 & \frac{1}{2}\beta_f fQ \\ \alpha_{f0}P_{f0} & \alpha_{f1}P_{f1} + \frac{1}{2}\alpha_f f & 0 & \frac{1}{2}\alpha_f fQ \\ 0 & \frac{1}{2}\beta_m m & \beta_{m0}P_{m0} & \beta_{m1}P_{m1} + \frac{1}{2}\beta_m mQ \\ 0 & \frac{1}{2}\alpha_m m & \alpha_{m0}P_{m0} & \alpha_{m1}P_{m1} + \frac{1}{2}\alpha_m mQ \end{pmatrix}$$

$$(2.67)$$

Reproductive output is multiplied by $\frac{1}{2}$ to account for the genetic share of parents in their offspring (see section 2.4.2). The reproductive output of males is Q times that of females to allow for the possibility that the number of reproducing males, n_{m1}, need not equal the number of reproducing females, n_{f1}. In fact, as we explained before $Q = n_{f1}/n_{m1}$, the ratio of breeding females to breeding males.

The fitness function can be written as

$$W(x, x^*) = P_{f1}v_P + \frac{1}{2}(f v_f + m v_m), \qquad (2.68)$$

where v_P is the (residual) reproductive value of a surviving mother

$$v_P = \alpha_{f1}v_{f1} + \beta_{f1}v_{f0}, \qquad (2.69)$$

v_f the reproductive value of a surviving daughter and v_m the reproductive value of a surviving son

$$v_f = \alpha_f v_{f1} + \beta_f v_{f0} \qquad (2.70a)$$
$$v_m = \alpha_m v_{m1} + \beta_m v_{m0}. \qquad (2.70b)$$

We will now show that

$$\frac{v_f}{v_m} = \frac{m^*}{f^*}, \qquad (2.71)$$

the ratio of sons to daughters at independence. As a consequence, we have

$$W(x, x^*) = P_{f1}v_P + \frac{1}{2}\left(\frac{f}{f^*} + \frac{m}{m^*}\right). \qquad (2.72)$$

The reproductive value of a breeding female in the resident population is given by

$$\lambda^* v_{f1} = P_{f1}^*(\alpha_{f1}v_{f1} + \beta_{f1}v_{f0}) + \frac{1}{2}f^*(\alpha_f v_{f1} + \beta_f v_{f0})$$

$$+ \frac{1}{2}m^*(\alpha_m v_{m1} + \beta_m v_{m0}) \qquad (2.73a)$$

$$= P_{f1}^* v_P + \frac{1}{2}(f^* v_f + m^* v_m), \qquad (2.73b)$$

which implies

$$\lambda^* v_{f1} - P_{f1}^* v_P - f^* v_f = \frac{1}{2}(m^* v_m - f^* v_f). \qquad (2.74)$$

To complete the proof we have to show that the left-hand side is zero or, equivalently, that

$$v_{f1}(\lambda^* - \alpha_{f1} P_{f1}^* - \alpha_f f^*) = v_{f0}(\beta_{f1} P_{f1}^* + \beta_f f^*). \qquad (2.75)$$

The reproductive value of a nonbreeding female in the resident population is given by

$$\lambda^* v_{f0} = P_{f0}(\beta_{f0} v_{f0} + \alpha_{f0} v_{f1}), \qquad (2.76)$$

or equivalently

$$v_{f1} \alpha_{f0} P_{f0} = v_{f0}(\lambda^* - \beta_{f0} P_{f0}). \qquad (2.77)$$

Comparing eq. 2.75 to eq. 2.77, the proof of eq. 2.71 therefore boils down to showing that

$$\frac{\lambda^* - \alpha_{f1} P_{f1}^* - \alpha_f f^*}{\beta_{f1} P_{f1}^* + \beta_f f^*} = \frac{\alpha_{f0} P_{f0}}{\lambda^* - \beta_{f0} P_{f0}}. \qquad (2.78)$$

This equality can be derived as follows. Insertion of n_{f1}/n_{m1} for Q shows that the population dynamics can equivalently be described by the matrix

$$\begin{pmatrix} \beta_{f0} P_{f0} & \beta_{f1} P_{f1} + \beta_f f & 0 & 0 \\ \alpha_{f0} P_{f0} & \alpha_{f1} P_{f1} + \alpha_f f & 0 & 0 \\ 0 & \beta_m m & \beta_{m0} P_{m0} & \beta_{m1} P_{m1} \\ 0 & \alpha_m m & \alpha_{m0} P_{m0} & \alpha_{m1} P_{m1} \end{pmatrix}, \qquad (2.79)$$

which can be written in block matrix notation as

$$\begin{pmatrix} A_{ff} & 0 \\ A_{mf} & A_{mm} \end{pmatrix}. \qquad (2.80)$$

The dominant eigenvalue λ^* is determined by the block matrix A_{ff} which we assume has the largest eigenvalue. This eigenvalue is given by the characteristic equation of A_{ff} (e.g. Lancaster 1969)

$$(\lambda^* - \beta_{f0} P_{f0})(\lambda^* - \alpha_{f1} P_{f1} - \alpha_f f)$$
$$= \alpha_{f0} P_{f0}(\beta_{f1} P_{f1} + \beta_f f), \qquad (2.81)$$

equivalent to eq. 2.78, which is what we wanted to show.

Acknowledgements

We thank Steve Frank, Ian Hardy, Mike Mesterton-Gibbons, Miguel Rodriguez-Girones and Jon Seger for helpful comments on the manuscript.

References

Bull JJ & Charnov EL (1988) How fundamental are Fisherian sex ratios? *Oxford Surveys in Evolutionary Biology*, 5, 96–135.

Caswell H (1989) *Matrix Population Models*. Sunderland, MA: Sinauer.

Charnov EL (1982) *The Theory of Sex Allocation*. Princeton, NJ: Princeton University Press.

Clark AB (1978) Sex ratio and local resource competition in a prosimian primate. *Science*, 201, 163–165.

Conover DO & Voorhees DA (1990) Evolution of a balanced sex ratio by frequency-dependent selection in a fish. *Science*, 250, 1556–1558.

Daan S, Dijkstra C & Weissing FJ (1996) An evolutionary explanation for seasonal sex ratio trends in avian sex ratios. *Behavioural Ecology*, 7, 426–430.

Darwin C (1871) *The Descent of Man and Selection in Relation to Sex*. London: Murray.

Düsing CD (1884) *Die Regulierung des Geschlechtsverhältnisses bei der Vermehrung der Menschen, Tiere und Pflanzen*. Jena: Gustav Fischer Verlag.

Edwards AWF (1998) Natural selection and the sex ratio: Fisher's sources. *American Naturalist*, 151, 564–569.

Emlen ST (1997) When mothers prefer daughters over sons. *Trends in Ecology and Evolution*, 12, 291–292.

Eshel I & Sansone E (1991) Parent-offspring conflict over the sex ratio in a diploid population with different investment in male and in female offspring. *American Naturalist*, 138, 954–972.

Eshel I & Sansone E (1994) Parent-offspring conflict over the sex ratio. II. Offspring response to parental manipulation. *American Naturalist*, 143, 987–1006.

Fisher RA (1930) *The Genetical Theory of Natural Selection*. Oxford: Clarendon.

Frank SA (1987) Individual and population sex allocation patterns. *Theoretical Population Biology*, 31, 47–74.

Frank SA (1998) *Foundations of Social Evolution*. Princeton, NJ: Princeton University Press.

Geritz SAH, Kisdi E, Meszena G & Metz JAJ (1998) Evolutionarily singular strategies and the adaptive growth and branching of the evolutionary tree. *Evolutionary Ecology*, **12**, 35–37.

Hardy ICW, Dijkstra LJ, Gillis JEM & Luft PA (1998) Patterns of sex ratio, virginity and developmental mortality in gregarious parasitoids. *Biological Journal of the Linnean Society*, **64**, 239–270.

Klomp H (1970) The determination of clutch size in birds. *Ardea*, **58**, 1–124.

Komdeur J, Daan S, Tinbergen J & Mateman C (1997) Extreme adaptive modification in sex ratio of the Seychelles warbler's eggs. *Nature*, **385**, 522–525.

Kruuk LEB, Clutton-Brock TH, Albon SD, Pemberton JM & Guinness FE (1999) Population density affects sex ratio variation in red deer. *Nature*, **399**, 459–461.

Lancaster P (1969) *Theory of Matrices*. New York, NY: Academic Press.

Leimar O (1996) Life-history analysis of the Trivers and Willard sex-ratio problem. *Behavioural Ecology*, **7**, 316–325.

Lessard S (1989) Resource allocation in Mendelian populations: further in ESS theory. In: M Feldman (ed) *Mathematical Evolutionary Theory*. Princeton, NJ: Princeton University Press.

Lessells CM (1998) A theoretical framework for sex-biased parental care. *Animal Behaviour*, **56**, 395–407.

Maynard Smith J (1980) A new theory of sexual investment. *Behavioural Ecology and Sociobiology*, **7**, 247–251.

McNamara JM & Houston AI (1996) State-dependent life histories. *Nature*, **380**, 215–221.

Metz JAJ, Nisbet R & Geritz S (1992) How should we define "fitness" for general ecological scenarios? *Trends in Ecology and Evolution*, **7**, 198–202.

Michod RE & Hamilton WD (1980) Coefficients of relatedness in sociobiology. *Nature*, **288**, 694–697.

Myers JH (1978) Sex ratio adjustment under food stress: maximization of quality or numbers of offspring. *American Naturalist*, **112**, 381–388.

Mylius SD & Diekmann O (1995) On evolutionarily stable life histories, optimization and the need to be specific about density dependence. *Oikos*, **74**, 218–224.

Oddie K (1998) Sex discrimination before birth. *Trends in Ecology and Evolution*, **13**, 130–131.

Olivieri I, Couvet D & Slatkin M (1994) Allocation of reproductive effort in perennial plants under pollen limitation. *American Naturalist*, **144**, 373–394.

Parker GA & Maynard Smith J (1990) Optimality models in evolutionary biology. *Nature*, **348**, 27–33.

Pen I (2000) *Sex Allocation in a Life History Context*. Ph.D. thesis, University of Groningen.

Pen I & Weissing FJ (2000) Towards a unified theory of cooperative breeding: the role of ecology and life history re-examined. *Proceedings of the Royal Society of London, Series B*, **267**, 2411–2418.

Pen I, Weissing FJ & Daan S (1999) Seasonal sex ratio trend in the European kestrel: an ESS analysis. *American Naturalist*, **153**, 384–397.

Reiss MJ (1987) Evolutionary conflict over the control of offspring sex ratio. *Journal of Theoretical Biology*, **125**, 25–39.

Seger J & Eckhart VM (1996) Evolution of sexual systems and sex allocation in plants when growth and reproduction overlap. *Proceedings of the Royal Society of London, Series B*, **263**, 833–841.

Shaw RF & Mohler JD (1953) The selective advantage of the sex ratio. *American Naturalist*, **87**, 337–342.

Stubblefield JW & Seger J (1990) Local mate competition with variable fecundity: dependence of offspring sex ratio on information utilization and mode of male production. *Behavioural Ecology*, **1**, 68–80.

Taylor PD (1996) Inclusive fitness arguments in genetic models of behaviour. *Journal of Mathematical Biology*, **34**, 654–674.

Taylor PD & Frank SA (1996) How to make a kin selection model. *Journal of Theoretical Biology*, **180**, 27–37.

Trivers RL (1974) Parent-offspring conflict. *American Zoologist*, **14**, 249–265.

Trivers RL & Willard DE (1973) Natural selection of parental ability to vary the sex ratio of offspring. *Science*, **179**, 90–92.

Werren JH (1980) Sex ratio adaptations to local mate competition in a parasitic wasp. *Science*, **208**, 1157–1159.

Williams GC (1979) The question of adaptive sex ratio in outcrossed vertebrates. *Proceedings of the Royal Society of London, Series B*, **205**, 567–580.

Zhang DY & Wang G (1994) Evolutionarily stable reproductive strategies in sexual organisms: an integrated approach to life-history evolution and sex allocation. *American Naturalist*, **144**, 65–75.

Part 2

Statistical analysis of sex ratio data

Chapter 3

Statistical analysis of sex ratios: an introduction

Kenneth Wilson & Ian C.W. Hardy

3.1 | Summary

In this chapter we discuss how to make best use of sex ratio data. We identify three basic questions that such data can be used to answer: does the sex ratio differ from some theoretically expected *mean* value, does it differ from an expected *distribution* and is variation in sex ratio associated with some measured *explanatory terms*? Our main focus is on the latter question. We discuss analytical methods in order of 'sophistication', starting with *nonparametric* methods (which make few assumptions about underlying statistical distributions), then *classical parametric* methods (which assume that data conform to a normal distribution of deviations from a statistical model) and finally *generalized linear models (GLMs)*. GLMs are semi-parametric methods that encompass models assuming a normal distribution but may also assume other distributions. This is an important advantage as sex ratio data are best expressed as proportions (sex ratio = males/(males + females)) and deviations are expected to conform to a *binomial* distribution. GLMs assuming binomial distributions are often termed *logistic regression models*. Distributions may not conform to the normal or binomial assumptions of classical parametric analyses or logistic GLMs, and we discuss how these problems can be overcome. The statistical approaches we discuss are illustrated with worked examples and case histories from recent sex ratio literature. We also perform simulations to evaluate the relative performances of non-parametric, classical parametric and logistic GLM analyses: GLMs win. A statistical analysis of the sex ratio literature published in 1994–2000 indicates that GLMs are currently being employed in only a small proportion (<30%) of sex ratio analyses and that the proportion does not appear to be increasing. Thus, this chapter serves in part as a manifesto for change, aimed at those who need to be persuaded that the GLM approach is worth learning and who need a short introduction to the subject.

3.2 | Introduction

In this chapter, we present a guide to the statistical analysis of sex ratio data. Our aim is to present a brief introduction to statistical methods that will increase the accuracy and power of sex ratio analyses. As evolutionary ecologists (rather than statisticians), our emphasis here is on the practicalities of analysing sex ratio data and we aim to give an intuitive feel for the different methods, rather than to explore in depth their statistical basis. Readers interested in the formal proofs of the different methods we discuss should consult the following texts and original papers cited therein: Cox and Snell (1989), Hosmer and Lemeshow (1989), McCullagh and Nelder (1989), Collett (1991) and Crawley (1993, 2002). In the remainder of this introductory section we discuss the different ways in which sex ratio data can be expressed (section 3.2.1) and discuss the sorts of questions that empiricists may want to ask about sex ratios (section 3.2.2). We then briefly outline possible analytical approaches used in answering these questions (section 3.3), assuming minimal statistical knowledge (Box 3.1 gives a refresher on statistical terminology). We introduce 'generalized linear models' (GLMs), a family of statistical tools, and we focus on logistic analyses, the sub-class of GLMs appropriate for analysing proportion data (section 3.4). We illustrate the relative merits of the different methods by means of a fictitious example (sections 3.3 & 3.4) and re-analyse some real datasets using this methodology (section 3.5). Although we employ a number of different nonGLM methods to analyse sex ratio data, we do this mainly to illustrate their lack of power and rigour and do not advocate their use except in exceptional circumstances. Finally, we illustrate the relative power of GLMs over alternative methods of analysis using a series of simple simulation studies (section 3.6).

Box 3.1 | A brief introduction to statistical approaches

Statistics is all about differences and associations. Usually, we are asking questions such as 'Is A different from B?' or 'Are changes in C associated with changes in D'. Statistical tests allow us to assess whether differences or associations are **statistically significant** (i.e. whether observed patterns differ from those expected by chance alone). To do this, we generally formulate a **null hypothesis (H_0)**, i.e. we hypothesize that any observed difference or association is due to random effects. **Hypothesis testing** centres around either accepting or rejecting H_0. This decision is usually made by comparing the value of a **test statistic** with some predetermined **critical value** (which can be found in published tables, e.g. Rohlf & Sokal 1995) for a given **significance level**. Traditionally, this significance level is taken to be 0.05 or 5%. This means that one will reject the null hypothesis in favour of the **alternative hypothesis (H_1)** if the probability, P, that observed data could have arisen by chance alone is less than 5%. If it is (i.e. $P < 0.05$), then one may conclude that the difference is 'statistically significant'. Note that the choice of a particular significance level is an ultimately arbitrary convention that dichotomizes a continuum of probabilities. The lower the probability ($0.05 > 0.01 > 0.001$), the more sure one can be that the difference is not just random sampling error with no real underlying difference.

When we are hypothesis testing, we generally test rather general hypotheses (e.g. Does A differ from B or is there an association between C and D?), but at other times we may have *a priori* reasons for testing more specific hypotheses (e.g. Is A larger than B or is C positively correlated with D?). The former type of test is known as a **two-tailed test** and the latter a **one-tailed test**. This is because, in the first instance, we are testing for both positive *and* negative differences and correlations, whereas in the second we are just testing for positive (or negative) differences and associations. As a result, the critical value for rejecting the null hypothesis is increased and the associated P value is reduced (see chapter 7 in Sokal & Rohlf 1985).

If H_0 is rejected, H_1 is supported but not proven. Rejecting a correct H_0 is termed a **type I error** while failing to reject an incorrect H_0 is a **type II error**. The probability of committing a type I error is usually termed α and the probability of committing a type II error is termed β. The **statistical power** of a test (Cohen 1988, Lipsey 1990) is the probability of rejecting a H_0, given that there really is a genuine effect (i.e. given that H_0 is false). In other words, statistical power $= 1 - \beta$. Statistical power generally increases as sample size and effect size increase, and is also dependent on the design of the experiment and the type of test employed (Table 3.1).

Hypothesis testing methods fall into one of two major categories: parametric and nonparametric. **Nonparametric tests** make few assumptions about the underlying statistical distributions and are often used when the errors (residuals) do not conform to the assumptions of a parametric test ('**errors**' are the deviations from the expected values of a statistical model, see below). As a consequence, they are extremely robust to statistical **outliers** (i.e. those data points that are much more extreme than the rest of the measurements in a sample and as a result may cause the sample to seriously violate the underlying assumptions of the statistical model). The main disadvantage of nonparametric tests is that they generally lack the power of equivalent parametric tests (Table 3.1). For example, most nonparametric tests (e.g. Spearman rank correlation, Mann–Whitney U-tests, etc.) arrange data into order according to their value and then use their **rank** positions to test for patterns, trends or associations. As a result, information in the data is lost: for example 10, 11, 1000 and 10, 999, 1000 are ranked identically while 11 and 999 have very different values. Similarly, data may be placed in categories and then the frequencies of these categories analysed. Again, information may go unused. The following books examine nonparametric methodology in detail: Meddis (1984), Neave and Worthington (1988), Siegel and Castellan (1988) and Sokal and Rohlf (1995).

Parametric tests assume that data conform to some underlying **error distribution**. Many methods assume that the errors are normally (**Gaussian**) distributed (these are often referred to as *general* **linear models**, as opposed to *generalized* linear models, see below). When the data do not conform to the normal distribution, **transformations** may be applied to raw data to **normalize** the distribution prior to analysis (e.g. the arcsine-squareroot transformation is often used to normalize proportional data).

Many of the classical statistical tests, such as linear regression and analysis of variance, are simply special cases of the general linear model. For example, when the **explanatory terms** (i.e. those terms that explain variation in the data of interest) include a single factor with two levels or categories with equal variances (e.g. treatments A and B), then the test is referred to as a t-test; when the factor has more than two levels with equal variance (e.g. three or more treatments), it is referred to as an analysis of variance (**ANOVA**); if there is a single explanatory variable or covariate (e.g. distance from point A), it is referred to as **linear regression**; if there is more than one covariate, it is known as **multiple regression**; if there is a single covariate and one or more factors, it is an analysis of covariance or **ANCOVA**, etc. Thus, it is easier to refer to all of these tests as special cases of a general linear model. These models also allow us to determine whether the responses to explanatory terms are additive or interact in some way. If there is a significant **interaction term** (e.g. A*B), then this indicates that the response to covariate A depends on the level of factor B; in this context, A and B are referred to as **main effects**.

Generalized **linear models (GLMs)** are generalizations of the linear models referred to as general. They encompass models with normal errors, but may also assume other error distributions (e.g. **Poisson, binomial, negative binomial,** gamma, etc.). Generalized linear models comprise three components: an **error function**, a **linear predictor** and a **link function**. In sex ratio analyses, we often use GLMs with a *binomial* error function and *logit*-link function. These are often termed **logistic regression models** (see main text).

GLMs, particularly logistic regression models, form the main focus of this chapter. We treat error distributions (i.e. variances) as a consideration during analysis rather than as the focus of the analysis itself; however, sex ratio variances are also of theoretical interest and techniques for their analysis are discussed in Chapter 5.

Some symbols and abbreviations

$-\infty, +\infty$	minus infinity and plus infinity
ln or $\log_e$	natural log (i.e. log to the base e) of x
$\exp(x)$ or e^x	exponent to the power x
χ^2_k	Chi-square (with k degrees of freedom)

3.2.1 Expressing sex ratio data

'Five to one, baby, one in five.
No one here gets out alive'

The Doors

As Jim Morrison's lyric illustrates, *odds* and *proportions* can be used to express the same information (although Morrison stretched poetic licence somewhat since the *odds* 'five to one' are actually equivalent to the *proportion* 'one in six'!). The term 'sex ratio' is commonly used to indicate the numerical relationship between the sexes. However, the quantity of interest is usually expressed as a proportion (conventionally, the number of males divided by the total number of individuals, i.e. males/(males + females)). Here, we conform to this precedent and, unless otherwise stated, we use 'sex ratio' to indicate the proportion of males in a sample, and not a ratio *sensu stricto* (males/females). In Box 3.2 we give an example that shows that analysis of ratios (*sensu stricto*) can lead to errors in interpretation.

3.2.1.1 Proportion data and the binomial distribution

For many organisms, an individual's sex is constrained to be one of two mutually exclusive possibilities: male or female. The data that record this information are said to be *binary*. Other examples include tossing a coin (heads or tails), mortality data (an individual either survives or dies), fertility data (an individual either reproduces or does not) and competition data (an individual either wins or loses). If we examine the sex of one individual and score, for instance, 1 for a male and 0 for a female, the datum is a proportion that has a sample size of one; the sex of the individual is the numerator and the sample size is the denominator, i.e. 1/1 or 0/1 (giving the proportions 1.0 and 0.0).

Often we are interested in determining the average sex ratio or the proportion of males in a group of individuals (e.g. population sex ratio, the sex ratio of the progeny of a given mother, or the sex ratio in a particular brood of offspring). Such data are referred to as *grouped binary data*; the number of males is the numerator, and the total number of individuals sampled is the denominator. For instance, in a brood of six males and seven females the brood sex ratio = 6/(6 + 7) = 0.46.

Grouped binary data are often assumed to conform to the *binomial distribution* (Chapter 5) which describes how frequently different sex ratio values are expected. Ungrouped binary data may conform to the *Bernoulli distribution*, a special case of the binomial distribution for sample sizes of one. Our main focus in this chapter is on grouped binary data, as these are most commonly encountered by empiricists, but where differences in the analysis of grouped and ungrouped binary data are apparent, these are highlighted. We explicitly consider ungrouped binary data in sections 3.4.4.3 and 3.5.2.

Box 3.2 | Analysis of sex ratios *sensu stricto*: a case history

Leonard and Weatherhead (1996) tested the prediction that parents with high dominance ranks will produce more male-biased offspring sex ratios than low-ranking parents using data from domestic chickens, *Gallus gallus domesticus* (a polygynous bird with stable dominance hierarchies in both males and females). Sex ratios were reported as ratios *sensu stricto* (females/males). Classical ANOVA performed on untransformed data (section 3.3) indicated that sex ratios were not affected by maternal or paternal dominance status. However, a significant effect of mating order on sex ratio was found, using paired *t*-tests (which assume normally distributed error variances, section 3.3), for females that mated with a subordinate male first and later with a dominant male, but not for females that mated with a dominant male first. Leonard and Weatherhead (1996) were unable to propose a simple explanation for this result but concluded that chicken sex ratios are not just a function of random assortment of sex chromosomes and that, given the potential economic value of being able to manipulate chicken sex ratios (a few males are needed for breeding stock but the vast majority are superfluous in agricultural egg production), further exploration would be worthwhile.

We questioned the validity of the analysis since classical ANOVA and *t*-tests were performed on untransformed female/male ratios, and there was no mention of whether error variances were normally distributed. Subsequently, Leonard and Weatherhead (1998) re-analysed these data using sex ratio expressed as proportions (males/(males + females)). Errors were normally distributed and ANOVA and *t*-tests were thus employed without transformation. The previously reported effect of mating order on the progeny sex ratio of females first mated to subordinate males was found to be spurious. Other conclusions were unchanged. The biological conclusion is that there is no consistent bias in chicken sex ratios and that poultry farmers are unlikely to be able to increase productivity by manipulating the status of females' mates. The statistical conclusion is that analyses of ratios *sensu stricto* should be avoided: such ratios are asymmetrical and undefined or infinite if only one sex is present in a clutch, hence mean and variance are not finite (see also Chapter 5) and important information on the size of both the numerator (males) and the denominator (males + females) is lost.

3.2.2 Questions in sex ratio data analysis

Before discussing how to analyse sex ratio data, we briefly consider the questions such analyses are likely to be aimed at addressing. First, it may be of interest to compare observed and expected ratios in order to establish whether an organism has control of its progeny sex ratio. Thus, we may want to ask whether the observed sex ratio differs significantly from the even sex ratio (proportion males = 0.5) that is often taken as the 'null' expectation (e.g. under heterogametic sex determination, Chapter 7). Similarly, it may be of interest to compare an observed distribution of group sex ratios (variances) with the binomial (random) expectation, as this can also indicate sex ratio control and the degree of fit to distributions predicted by evolutionary theory. We briefly summarize methods for testing for sex ratio bias in Box 3.3. Box 3.3 also illustrates a method for analysing sex ratio variance, but this issue is dealt with in detail in Chapter 5.

Second, sex ratio data may be used to explore relationships between sex ratio and specific explanatory terms (factors and covariates; see Box 3.1). Sex ratio theory is a rich and important area of evolutionary biology (e.g. Chapters 1,

Box 3.3 | Comparison of observed and expected sex ratios

Before embarking on a large-scale analysis of sex ratio data, it is often informative to begin by asking two simpler questions: first, does the sex ratio differ from the assumed binomial distribution; and second, does the sex ratio differ from some expected value, such as 0.5. Positive results for one or both of these tests could be indicative of nonrandom variation in the sex ratio distribution. Note that an absence of significant deviation does not necessarily mean that there is no nonrandom variation and that significant deviations do not necessarily indicate parental control of sex allocation, as, for example, sex ratios could be biased due to sexually differential developmental mortality. To illustrate these methods, we use the Example 1 data set (Box 3.5).

Deviation from the binomial distribution

If sex ratio data conform to the binomial distribution, then a GLM with binomial errors (and no explanatory terms other than the intercept, i.e. the null model, Table 3.3) should provide a good fit to the data. We can therefore use the goodness-of-fit test for the null model to determine whether the raw sex ratio data deviate from the binomial distribution (section 3.4.3). To do this, we simply compare the null deviance against the χ^2 distribution with df equal to the null degrees of freedom.

Thus, for Example 1, we can ask whether sex ratio distributions for each of the two species (and for both species combined) conform to the binomial distribution

Shirazfish: null deviance = 83.701, null df = 9, $P_{(\chi^2_{9=83.701})} < 0.0001$
Merlotfish: null deviance = 32.475, null df = 9, $P_{(\chi^2_{9=32.475})} = 0.00016$
Both species: null deviance = 117.961, null df = 19, $P_{(\chi^2_{19=117.961})} < 0.0001$

Thus, it appears that both distributions (and the combined distribution) differ significantly from the binomial.

However, when sample sizes are small, this method can severely overestimate the degree of departure from the binomial (Westerdahl et al. 1997, Hartley et al. 1999). Thus, in these circumstances it is wise to test the robustness of the result by performing *randomization tests*. These involve comparing the deviance of the null model with deviances obtained by a series of *randomly generated* datasets in which 'fish' are allocated to 'samples' at random, while maintaining constant sample sizes. In practice, this requires randomizing fish between samples, while maintaining the same distribution of sample sizes and total number of male and female fish. At each iteration, the deviance of the model is noted and the process is repeated 1000 times. The resulting distribution of deviance values then becomes the null distribution of deviance values against which our model is compared. To determine the significance level of departure from the binomial distribution, we simply divide the number of deviance values greater than or equal to our model's null deviance by 1000 (for S-Plus users, a user-defined function for performing these randomizations is available upon request from ken.wilson@stir.ac.uk). However, randomization tests may not perform well when the size of individual samples or the total number of samples is small (Ewen JG, Cassey P & King RAR, unpublished manuscript).

Not surprisingly, given the magnitude of the deviation from the binomial distribution, in this instance, the randomization method confirms the results of our original analysis and all three distributions were found to be significantly different from the binomial ($P < 0.001$). Figure B3.3a illustrates the distribution of the 1000 randomized null deviances for the Merlotfish dataset. As you can clearly see, the observed null deviance (shown by the solid circle) is significantly higher than any of the values obtained via randomization (histogram). Analysis of sex ratio variances is discussed in detail in Chapter 5.

Deviation from sex ratio equality

In Example 1, there were a total of 638 (253 male and 385 female) Shirazfish sampled, and the overall sex ratio is $253/(253 + 385) = 0.397$. While it is clear that this is not an exact match to 0.5, we need to ask whether the difference is statistically significant. Of the possible tests that can be used, we illustrate five and, as these are widely known and described elsewhere, we do this only briefly.

Binomial test

We could calculate the probability of observing a sex ratio as extreme as 0.397 (i.e. 253 or fewer of one sex in a sample of 638 individuals), assuming that the sex ratio is determined by a random (binomial) process with a mean of 0.5. If this probability is less than 0.05, we conclude that the difference between 0.397 and 0.5 is significant; this is a *binomial test* (see e.g. Siegel & Castellan 1988). As sample size increases, the binomial distribution tends towards the normal distribution, and for samples larger than 35 the normal approximation should be used, but 'corrected' for the fact that the normal distribution is continuous while the binomial distribution involves discrete variables (for details see Siegel & Castellan 1988 p38). Using the normal approximation corrected for discontinuity, an observation as extreme as 253 in a sample of 638 individuals gives $z = -19.99$, $P < 0.0001$; Shirazfish sex ratios are significantly female-biased.

Confidence limits

We could look up the confidence limits for binomial proportions, as published in statistical tables (Rohlf & Sokal 1995), which tell us whether our observed sex ratio falls within the 95% (or 99%) confidence bounds of 0.5 for a given sample size (most statistical packages now offer this facility). If it does, then we can be 95% (or 99%) confident that the observed value does not differ from 0.5 purely due to random sampling error. With a sample size of 638, the lower and upper 99% confidence limits for 0.5 are 0.45 and 0.55 respectively. As our observation of 0.39 is outside these bounds, we can be confident that the bias in the Shirazfish population sex ratio is significantly greater than expected by chance alone.

Simulation

Another way to address this same problem is to determine the relative confidence intervals by simulation. In other words, we generate a large number (i.e. > 1000) of simulated datasets comprised of random samples drawn from the binomial distribution with the mean equal to 0.5 and sample size equal to the number of individuals in our dataset (638 in our particular example). If our observed sex ratio lies outside the appropriate confidence intervals for our simulated dataset, then the sex ratio is significantly different from 0.5. This approach is illustrated in Figure B3.3a, in which the histograms represent the simulated dataset and

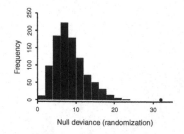

Fig B3.3a Histogram of null deviances obtained from the randomization test. The solid circle represents the observed null deviance.

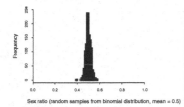

Fig B3.3b Histogram of simulated sex ratios generated by randomly sampling from a binomial distribution with mean equal to 0.5 and sample size equal to 638. The solid circle indicates the observed sex ratio.

the solid circle represents the observed sex ratio. As you can see, none of the 1000 simulated datasets had a sex ratio that was as low as that which we observe, indicating that the probability of observing this sex ratio by chance alone is less than 1 in 1000, i.e. $P < 0.001$.

Chi-square goodness-of-fit test

We could compare the observed numbers of males (253) and females (385) with the number of each sex expected under sex ratio equality ($638/2 = 319$ individuals of each sex) using a chi-square test, which is based on the *deviations* of observed from expected values (for details see e.g. Siegel & Castellan 1988 p45, Sokal & Rohlf 1995 p695). The chi-square, X^2, value computed is also known as *Pearson's statistic* to distinguish it from the chi-square sampling distribution, χ^2, which it approximates. The Shirazfish data generate a Pearson's statistic of 27.31 with $df = 1$, which is greater than the critical value in χ^2 tables for $P = 0.001$, so we conclude that the sex ratio is significantly female-biased. With small samples, and with biased sex ratio expectations, the expected value of one or both sexes may be five or less. In such cases *Fisher's exact test* should be used instead of the chi-square test (e.g. Siegel & Castellan 1988 p103, Crawley 1993 p237).

Likelihood ratio goodness-of-fit test

We could compare the observed and expected numbers of males and females with the numbers expected under sex ratio equality using a *G*-test which is based on the *ratios* of observed and expected values (e.g. Crawley 1993 p234, Sokal & Rohlf 1995 p688, Zar 1999 p505). The Shirazfish data generate $G = 26.87$ with $df = 1$, which is greater than the critical value in χ^2 tables for $P = 0.001$, so again we conclude that the sex ratio is significantly male-biased. Note also that G and X^2 values are generally similar.

Choice of test

Which of these five tests should be preferred is determined by the power function for the class of alternative hypotheses under consideration (E. Meelis pers. comm.). However, the binomial test will usually be the definitive test, the *G*-test is generally preferred over the chi-square goodness-of-fit test (Crawley 1993, Sokal & Rohlf 1995, but see Zar 1999) and the confidence interval and simulation will generally give similar results for large sample sizes.

13, 19 & 20) and there are many predictions that can be tested in this way. Analysis of such relationships forms the main focus of this chapter.

3.3 | Classical analyses of proportion data

A variety of different methods has been used to analyse sex ratio data in recent years (Box 3.4).

In this section we review some of the more traditional methods, highlighting their strengths and weaknesses. We illustrate these points using a fictitious dataset (Box 3.5) on the effects of a pollutant on the sex ratios of two fish species in an Australian creek. The first thing we need to do is to plot the data (Figure 3.1a). The figure appears to indicate that, in both species, populations close to the source of pollution tend to have female-biased sex ratios and that as we get further away from the pollution source the sex ratio

Box 3.4 | Survey of statistical approaches used in recent sex ratio literature

Studies on sex ratio are published in a range of journals within the general field of evolutionary ecology, as well as in taxon-specific journals. To assess which statistical methods are the most commonly used for analyses of sex ratios and other proportional data, we surveyed empirical studies on sex ratio and closely related issues (e.g. sex determination, sex allocation, sex-biased mortality) published in 1994–2000 in four leading evolutionary, ecological and behavioural journals. We found 83 studies, some of which employed more than one approach: see Table B4.3. Part (a) of the table scores the methods used to test for departure from some expected sex ratio value (e.g. 0.5) (see Box 3.3) and part (b) scores methods used to examine trends in sex ratio with explanatory variables, which is the main focus of this chapter.

Table B3.4 | Recently used methods

	Journal (Number of studies)				
Sex ratio analysis method	Animal Behaviour (34)	Behavioral Ecology (16)	Evolution (18)	Oikos (15)	Totals (83)
(a) Deviation from expected sex ratio					
No statistical test			2		2
Binomial test					
Fisher's exact test	2	2	4	3	11
χ^2-test	10	2	4	6	22
G-test	5	1	3	2	11
Other	1		2	3	6
(b) Relationships with explanatory variable(s)					
1. No statistical test	3		1	1	5
2. Nonparametric tests	11	6	3		20
Standard parametric tests:					
3. No transformation	5	3	2	1	11
4. Arcsine squareroot transformation	11	3	7	4	25
5. Other transformation	2	1	1		4
6. Generalized linear modelling (logistic)	4	5	3	4	16

Note that some authors who used standard parametric tests without transformation first tested the appropriateness of the assumption of normal error variances, while others attempted to use GLMs but found a degree of overdispersion to be too large (e.g. heterogeneity factor >4) and opted to use standard parametrics following arcsine squareroot transformation instead (e.g. Flanagan *et al.* 1998). However, some authors employed standard techniques despite using statistical packages (e.g. *SAS*) which are capable of running GLMs; possibly because they were unaware of the advantages of GLM analysis?

We explored our survey data by ranking methods in part (b) in rough order of 'advancement' from 1 (no statistical analysis) to 6 (logistic GLMs) and gave each study a 'sophistication score' equal to the rank of the most advanced method used. We found no evidence that the level of 'sophistication' changed significantly during the surveyed years, or that it is related to the journal in which the study was published (Year, $\chi^2_1 = 0.34$, $P > 0.1$; Journal, $\chi^2_3 = 0.46$, $P > 0.1$; results from log-linear analysis, which is appropriate for count data, Crawley 1993). We also found no relationship with year or journal when we carried out (binary) logistic analyses on: (1) the proportion of parametric tests out of all methods in part (b) (mean = 0.799, Year, $\chi^2_1 = 0.23$, $P > 0.1$; Journal, $\chi^2_3 = 2.77$, $P > 0.1$), (2) the proportion of logistic GLMs out of all methods (mean = 0.277, Year, $\chi^2_1 = 2.37$, $P > 0.1$; Journal, $\chi^2_3 = 4.72$, $P > 0.1$), and (3) the proportion of logistic GLMs out of all parametric methods (mean = 0.346, Year, $\chi^2_1 = 2.53$, $P > 0.1$; Journal, $\chi^2_3 = 3.25$, $P > 0.1$). We conclude that GLMs are underused and that the situation has not recently been improving.

Box 3.5 | Example 1: Pollution and sex ratios in Australian fish

Our first data set is a hypothetical example, in which we examine the effects of a pollutant (alcohol) on sex ratios in two imaginary fish species in Stubbie Creek, Australia: the wide-mouthed Shirazfish and the big-nosed Merlotfish. We imagine that the data were collected by netting fish at 100-m intervals along the creek for a distance of up to 1000 m from the source of the pollutant. The hypothesis we are testing is that the pollutant leads to biased sex ratios in both species.

Table B3.5

Distance from pollution source (m)	Shirazfish		Merlotfish	
	sample size (no. fish)	sex ratio (proportion males)	sample size (no. fish)	sex ratio (proportion males)
100	67	0.30	7	0.14
200	120	0.50	12	0.33
300	21	0.33	30	0.40
400	103	0.13	5	0.20
500	88	0.65	46	0.59
600	34	0.29	67	0.25
700	99	0.31	29	0.31
800	34	0.38	74	0.43
900	67	0.57	35	0.71
1000	5	0.80	134	0.48

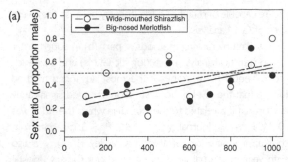

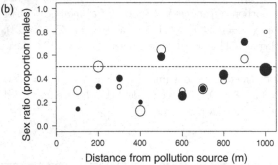

Fig 3.1 Relationship between sex ratio and distance from pollution source for two fish species in Stubbie Creek (Example 1, Box 3.5). The solid line is the least-squares fit to the Merlotfish dataset and the dashed line is the fit to the Shirazfish. (b) Symbol size is proportional to the sample size upon which the sex ratio is based. On both panels the dotted line shows sex ratio equality (0.5).

gets closer to 0.5. We will start with nonparametric analyses (section 3.3.1.2) and then go on to linear models with normal errors (sections 3.3.2 & 3.3.3). In sections 3.4 and 3.5, we analyse these data using generalized linear models.

3.3.1 Nonparametric tests

Nonparametric tests (e.g. Mann–Whitney U-test, Kruskal–Wallis test, Spearman's correlation) are frequently employed in the behavioural sciences because they are simple to implement by hand or by computer and because they make no assumptions about the shape of the underlying error distribution and thus they are extremely robust to outliers (Box 3.1). This does not mean that these tests are 'assumption-free', however, since most nonparametric tests usually assume that the observations are independent and sometimes that the variable under study has underlying continuity or that the distributions have similar shape across groups. Nevertheless, the

assumptions associated with nonparametric tests are fewer and weaker than those associated with equivalent parametric tests (Box 3.1). As a consequence, if all of the assumptions of a parametric test are met, nonparametric tests lack *power* (Box 3.1) and are *wasteful* (Siegel & Castellan 1988).

3.3.1.1 Power-efficiency
The degree of wastefulness of a test can be expressed by its *power-efficiency*, which is concerned with the increase in sample size required to make test B (e.g. a nonparametric test) as powerful as test A (e.g. an equivalent parametric test), when the significance level is held constant and the sample size of test A is held constant. Thus

Power-efficiency of test B (%)
$$= 100 \times N_A/N_B. \qquad \text{(eq. 3.1)}$$

N_A and N_B are the relative sample sizes required to give test B the same power as test A. For example, if test B requires a sample size of $N_B = 25$ to have the same power that test A has when it has a sample size of $N_A = 20$, then test B has a power-efficiency of $100 \times (20/25) = 80\%$ (Siegel & Castellan 1988). In other words, test A would be just as effective with a sample that was 20% smaller than that used in test B. Table 3.1 compares the power-efficiencies of some of the commonly employed nonparametric tests with their most comparable parametric test.

3.3.1.2 Example 1: Fish sex ratios
Now let's return to Example 1 (Box 3.5). We want to investigate whether sex ratio varies consistently with distance from the pollution source in the two fish species. There are a number of ways that we can employ nonparametric tests. If we use Spearman rank-order correlations to assess whether there is an association between sex ratio and distance from the pollution source for each fish species the answer appears to be 'no' (Shirazfish: $r_s = 0.467$, $n = 10$, $P = 0.167$; Merlotfish: $r_s = 0.624$, $n = 10$, $P = 0.063$). Note that if we were testing an *a priori* hypothesis, for example based on data showing that males were more susceptible to the pollutant than females, we could argue that one-tailed probabilities were

Table 3.1 | Power-efficiency of some of the commonly employed nonparametric tests

Nonparametric (NP) test	Parametric (PP) test	Power-efficiency (%) of NP test	Comments
Spearman's rank-order correlation	Pearson's product-moment correlation	91%	Power-efficiency same as for Kendall's rank-order correlation
Wilcoxon signed ranks test	paired t-test	<95.5%	Power-efficiency ~95% even for small sample sizes
Wilcoxon Mann–Whitney U-test	t-test	<95.5%	Power-efficiency ~95% for small sample sizes
Kolmogorov–Smirnov two-sample test	t-test	<95%	Power-efficiency declines slightly with increasing sample size
Kruskal–Wallis one-way ANOVA	One-way ANOVA (F-test)	<95.5%	
Friedman two-way ANOVA	Two-way ANOVA (F-test)	64% ($k = 2$) to 91% ($k \geq 20$)	Power-efficiency dependent on number of matched samples (k)

more appropriate (Box 3.1) and the significance levels would be reduced to $P = 0.083$ and $P = 0.031$, respectively (see also section 3.5.4). Notice that, in both cases, the correlation coefficients are fairly large ($r_s \geq 0.46$), and it seems likely that the lack of significance for these two relationships is due to low statistical power (see Box 3.1 and below). The power-efficiency of the Spearman's rank correlation test is 91% when compared to the most powerful parametric correlation test (Pearson's product-moment correlation; Table 3.1). Re-analysing Example 1 data using Pearson's correlation following arcsine-squareroot transformation to help normalize the error distribution (Section 3.3.3.2, Box 3.6) yields the following correlation coefficients: Shirazfish: $r = 0.498$, $df = 8$, two-tailed $P = 0.143$, Merlotfish: $r = 0.618$, $df = 8$, two-tailed $P = 0.057$. Thus, there does not appear to be a significant relationship between sex ratio and distance from a known pollution source, regardless of whether we use a parametric or nonparametric correlation test. But, was the power of our analysis (Box 3.1) great enough to be able to detect a significant relationship if there was one? Ideally, the power of our test should be greater than 80%. The statistical power of a test is determined by sample size, the amount of variation in the data and the magnitude of the effect one is trying to detect. We can determine the power of our two correlations using the following formula (Cohen 1988, Zar 1999)

$$Z_\beta = (z - z_\alpha)\sqrt{(n - 3)},$$

where z and z_α are the Fisher transformations for r (the correlation coefficient) and r_α (the critical value of r), the Fisher transformation $= 0.5 \ln(1 + r/1 - r)$, $n =$ sample size, and Z_β is the probability of the normal deviate, which can be translated into power $(1-\beta)$ by comparing against the appropriate tabulated value (e.g. Appendix Table B.2 in Zar 1999).

These days, a simper way to determine power is to use a power calculator (such as that which can be found at http://ebook.stat.ucla.edu/calculators/powercalc/). Using this calculator, the power of our two Pearson's correlation tests were determined to be 30.4% and 48.0%, respectively, which are nowhere near the desired 80%. These calculators can also be used to determine the sample sizes required to achieve a given power. In this example, sample sizes of 30 and 19, respectively, are required to achieve 80% power.

Box 3.6 | Effect of arcsine-squareroot transformation

Here we illustrate the effect of arcsine transformation on proportional data. Figure B3.6a shows the relationship between proportions and their transformed values. Arcsine transformation has the effect of stretching out the ends of the distribution, such that the truncation that occurs when the mean of the (binomial) distribution is close to zero or one is reduced. As a consequence, the distribution should become more normalized under transformation.

Figure B3.6b shows the distribution of binomially distributed data, for a range of clutch sizes (CS = 2, 5, 10 and 20) and mean sex ratios (mean SR = 0.5, 0.75 and 0.9). The data show the frequency distribution of 5000 random samples taken from the binomial distribution, using the *rbinom* function in *S-Plus* (open bars) and the effect of arcsine-squareroot transformation on the distribution (closed bars). Note that the effects of arcsine transformation on mean SR = 0.25 and 0.1 are equivalent to those illustrated for mean SR = 0.75 and 0.9, respectively.

Although the nontransformed sex ratios are approximately normal for sex ratios close to 0.5 (especially when clutch sizes are large), the data are severely skewed when sex ratios are heavily biased towards one or other sex (especially when clutch sizes are small). Arcsine transformation tends to make the data more normal (cf. the open and closed bars in the bottom-right figure), though in some cases the effect is to make the data less normal (cf. middle-right distributions). For small clutch sizes and heavily biased sex ratios, arcsine transformation fails to normalize the data.

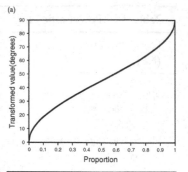

Fig B3.6a Effect of arcsine transformation on proportional data.

Lack of power is not the only problem with the analysis described above. Another is that it fails to take into account the fact that we appear to have the same relationship in both datasets. Ideally, we would want to perform a test in which we utilize information from both species simultaneously and ask whether we get the same relationship in both. Unfortunately, there is no easily accessible nonparametric test that is equivalent to analysis of covariance (but see the Page Test for Ordered Alternatives, Siegel & Castellan 1988). We could perform a Wilcoxon signed ranks test (equivalent to a paired *t*-test) to determine whether the sex ratio variation *within* our two fish species is greater than that *between* them, but this would tell us nothing about their respective sex ratio trends along the creek. An alternative procedure is to perform a Fisher combined probability test (Fisher 1954, section 21.1; Box 18.1 Sokal & Rohlf 1995) that allows us to use the probabilities derived from the correlations we carried out on the two species.

The calculation of the Fisher's combined probability estimate is based on the fact that $-2\ln P$ is distributed as χ^2_2 (see Box 18.1 in Sokal & Rohlf 1995). Thus, by evaluating twice the negative natural logarithm of each of the (k) probabilities we wish to combine, and summing them, we obtain a total ($-2\Sigma \ln P$) that can be compared against the χ^2 distribution with $2k$ (= 4, in this example) degrees of freedom (i.e. χ^2_{2k}). In our example, based on the one-tailed P-values from the Spearman rank correlations

$$
\begin{aligned}
-2\Sigma \ln P &= -2 \times (\ln 0.083 + \ln 0.031) \\
&= -2 \times (-2.4889 - 3.4737) \\
&= -2 \times -5.9626 \\
&= 11.925.
\end{aligned}
\tag{eq. 3.2}
$$

When compared with χ^2_4, this yields a two-tailed probability of $P_{\text{combined}} = 0.0358$. Thus, when we use the information we have on the two fish species, there appears to be a significant trend for sex ratio to increase with distance from the pollution source, but the statistical evidence for such a relationship is far from convincing. As indicated above, this is probably due to a lack of statistical power, because: (1) we are relying

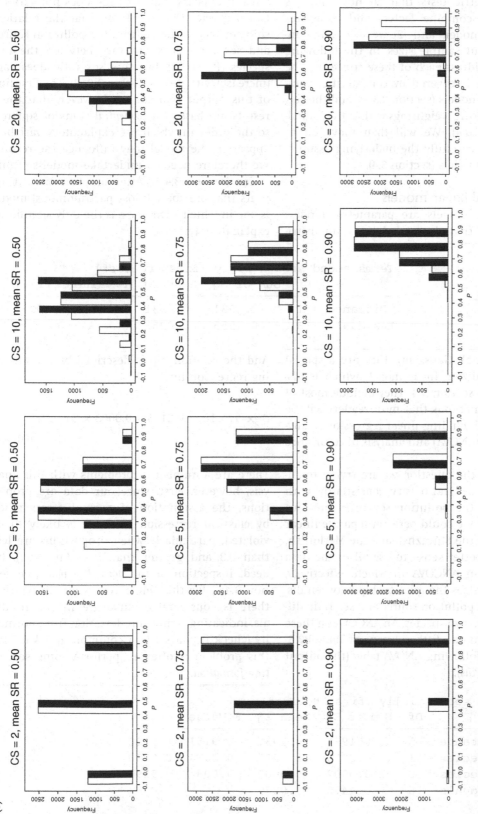

Fig B3.6b Effect of arcsine transformation on randomly generated binomial datasets. Open bars indicate the observed binomial data generated by randomly sampling from a binomial distribution with sample size (i.e. clutch size = CS) equal to 2, 5, 10 or 20, and sample mean (i.e. mean sex ratio = mean SR) equal to 0.5, 0.75 or 0.9. The closed bars are the arcsine-transformed proportions.

(b)

on nonparametric tests that do not allow us to adequately combine factors and covariates in the same model, and (2) we are losing information about sample sizes. In the following sections, we address each of these concerns. The first deficiency is covered by considering (parametric) linear models (section 3.3.2), and the second by considering weightings within these models (section 3.3.3.4). We will then move on to consider more carefully the underlying assumptions of these models (section 3.4).

3.3.2 General linear models

General linear models are parametric models which assume that the underlying error distri-

```
Terms added sequentially (first to last)
                 Df  Sum of Sq  Mean Sq  F Value   Pr(F)

   Distance      1     0.1931   0.1931   7.6337   0.0128 **
   Residuals    18     0.4555   0.0253
```

bution is normal (Gaussian). They are a special type of generalized linear model, which is discussed fully in section 3.4. They include most of the 'classical' methods that most readers will be familiar with, including linear regression, analysis of variance (ANOVA) and analysis of covariance (ANCOVA).

Recall that the question we are trying to address is: does sex ratio vary consistently with distance from the pollution source in our two species of fish? We could perform separate linear regressions for the Shirazfish and the Merlotfish, but it makes better sense to use all of the data and perform an ANCOVA in which, effectively, we are asking: does the relationship between distance from the pollution source and sex ratio differ between our two species. An ANCOVA on these data (first, third and fifth columns of Table B3.5) generates the following ANOVA table (the output comes from *S-Plus*):

Thus, it appears that sex ratio does not vary between species ($P = 0.59$) and that the relationship between distance from the pollution source and sex ratio does not vary between the two species ($P = 0.92$), but that sex ratio does vary (increase) with distance ($P = 0.018$). The first line of this output reminds us, however, that these results are based on *sequential* sums of squares, so the order in which the explanatory variables appear in the model may influence the results. We therefore need to undertake model simplification (section 3.4.5). After simplification, it appears that our 'best' (most parsimonious) model is one in which Distance is the only significant explanatory term:

And the relationship is described by the following regression line

Sex ratio = $0.2173 + 0.0003 \times$ Distance.

(eq. 3.3)

There are a number of problems with this analysis, however. First, since our data are proportions, the assumption of normal errors made by classical regression methods is likely to be violated, particularly when proportions are less than 0.3 and greater than 0.7 (Zar 1999). Indeed, inspection of the *normality plot* (see section 3.4.6) for this linear regression shows that there is considerable curvature in the residuals, indicating significant deviation from normality (check it yourself!). A common 'quick-fix' for this problem is often to perform some sort of *transformation*.

```
Terms added sequentially (first to last)
                  Df  Sum of Sq  Mean Sq  F Value   Pr(F)

       Distance    1    0.1931   0.1931   6.9151   0.0182
        Species    1    0.0083   0.0083   0.2974   0.5930
Distance:Species   1    0.0002   0.0002   0.0081   0.9293
      Residuals   16    0.4470   0.0279
```

3.3.3 General linear models with transformed data

3.3.3.1 Probit transformation

One of the earliest transformations applied to binomial proportion data was the *probit transformation*, which has most commonly been employed in the analysis of dose–response data from bioassays. Probit transformation evolved when such analyses were performed by hand using probit paper. With the advent of desktop computers, this method is now considered rather old-fashioned. For further information see Finney (1971) and Crawley (1993).

3.3.3.2 Arcsine transformation

A much more commonly employed transformation used by sex ratio biologists is known as the *arcsine-squareroot transformation* (also known as the *arcsin transformation* or *angular transformation*). This involves taking the square-root of the proportion, p, and transforming it to its arcsine (i.e. the angle whose sine is $\sqrt{p}$)

$$p' = \arcsin \sqrt{p} \qquad \text{(eq. 3.4)}$$

For proportions between 0 and 1, the transformed values will range between 0 and 90 degrees (some statistical tables and packages present the transformation in terms of radians; a radian is $180°/\pi = 57.2958$ degrees). Note that prior to arcsine transformation, the data must be represented as proportions and not as percentages. Arcsine transforming the fish sex ratio data (Example 1) has little impact on the results of our analysis

i.e. near 0 and 1 (Box 3.6); of course, this can be checked by producing a normality plot (see section 3.4.6). Moreover, arcsine transformation does not get around another major attribute of proportion data, namely that the responses are strictly bounded between 0 and 1 (or 0% and 100%). Thus, the classical linear methods that we used earlier (i.e. linear regression and ANCOVA) could easily predict biologically unrealistic or even impossible results, especially if the variance is high and the data lie close to zero. In Example 1, the linear regression line describing the relationship between sex ratio and distance from the pollutant source indicates that at a distance of 2609 m (untransformed data) or 2956 m (arcsine-transformed data), the creek will comprise only males, and at greater distances the sex ratio will exceed 1! Although extrapolating so far beyond the observed data would be ludicrous, the point remains that classical linear models can predict values that lie outside biologically sensible bounds.

3.3.3.3 Logistic transformation

One way round this problem is to apply the *logistic transformation*, in which our *success probability* p (i.e. proportion of males in our sample) undergoes the following transformation, written as logit (p)

$$\text{logit } (p) = \ln\left(\frac{p}{1-p}\right). \qquad \text{(eq. 3.6)}$$

Thus, for p in the range 0 to 1, logit (p) will range between $-\infty$ and $+\infty$, respectively. If we apply

Terms added sequentially (first to last)					
	Df	Sum of Sq	Mean Sq	F Value	Pr(F)
Distance	1	0.2239	0.2239	7.7734	0.0121 **
Residuals	18	0.5185	0.0288		

And the relationship (in degrees) is described by the following regression line

`Sex ratio = 27.6331 + 0.0211 × Distance.`

$$\text{(eq. 3.5)}$$

Whilst the arcsine transformation often helps to normalize proportion data, it does not work well at the extreme ends of the distribution,

the logit transformation to a simple linear model, we produce the following linear logistic model

$$\text{logit } (p) = \ln\left(\frac{p}{1-p}\right) = a + bx. \qquad \text{(eq. 3.7)}$$

Note that $p/(1-p)$ is the statistical *odds* of success (Jim Morrison's 'five to one'), and so the logistic transformation of p is the *log odds* of success. We can make use of this fact to re-write eq. 3.7 to make p a function of x

$$p = \frac{e^{(a+bx)}}{1 + e^{(a+bx)}}. \qquad \text{(eq. 3.8)}$$

Thus, when $x = -\infty$, $p = 0$ and when $x = +\infty$, $p = 1$, so fulfilling our need for p to be strictly bounded between 0 and 1.

If we apply the logit transformation (eq. 3.7) to the Example 1 sex ratio data and perform an ANCOVA, we arrive at the following result

	Df	Sum of Sq	Mean Sq	F Value	Pr(F)
Distance	1	4.2161	4.2161	7.8672	0.0117 **
Residuals	18	9.6465	0.5359		

And the relationship (*in logits*) is described by the following regression line

```
Sex ratio = −1.3102 + 0.0016
               × Distance.        (eq. 3.9)
```

Back-transforming eq. 3.9 (using eq. 3.8), the predicted sex ratio at the pollution source is $e^{-1.3102}/(1 + e^{-1.3102}) = 0.2124$ and even at 10 000 m away from the pollution source, the predicted sex ratio remains within realistic bounds (e.g. $e^{(-1.3102 + 0.0016 \times 10\,000)}/[1 + e^{(-1.3102 + 0.0016 \times 10\,000)}] = 0.9999996$).

As we shall see in section 3.5, the logit transformation forms the basis for *logistic regression* (i.e. generalized linear modelling with binomial errors and logit link function). We do this within the GLM context (rather than using simple linear regression, as above), because: (1) logistic regression allows for the nonconstant binomial variance (the variance of the binomial distribution equals $np(1 - p)$ and peaks at $p = 0.5$); (2) it deals with the fact that logit(p) values near 0 or 1 are infinite; and (3) it allows for differences between sample sizes by weighting the regression (Crawley 1993, 2002).

3.3.3.4 Weighted linear regression

In all of the models we have considered so far, each data point (i.e. sex ratio) contributes equally. However, it is clear that if we have two sex ratios, and one is based on a sample of five individuals

and the other on 500, we should have much more confidence in the value derived from the larger sample. Thus, in those cases where it is known *a priori* that not all observations contribute equally to the fit of the model, we should *weight* our observations according to the confidence we have in them (usually some function of sample size). This process is called *weighted regression*.

Figure 3.1b shows the data for Example 1 with the size of the symbols reflecting the size of the sample upon which the sex ratio was estimated (i.e. the denominator of the sex ratio; the second and fourth columns in the Table B3.5). It is very clear that there is considerable sample size variation between the data points. A weighted ANCOVA on the logit-transformed data finds that neither Species nor Distance (nor their interaction) is statistically significant (e.g. for Distance, $P > 0.16$). Thus, when we weight sex ratios according to sample size, it appears that there is no consistent relationship between sex ratio and distance from the source of pollution. This is because most of the extreme sex ratios (i.e. those that deviate most from 0.5) are based on small sample sizes.

In section 3.4 we incorporate the ideas of weighted regression and logit transformation into a technique known as generalized linear modelling.

3.4 | Generalized linear models

The *general* linear models we discussed in the previous section are based on the underlying assumption that the distribution of residuals around the fitted model (i.e. the error distribution) is Gaussian (= 'normal'), and that these residuals show no systematic variation with respect to the mean (i.e. that the variance is constant). However, these two assumptions are often violated (as we have seen already for sex ratio data). *Generalized* linear models (GLMs) differ from the *general* linear models encountered previously in allowing one to also specify non-normal error variances, such as Poisson, binomial, negative binomial, gamma and exponential (section 3.4.2.1). We can use GLMs to analyse sex ratio data, but before we can do that we need to understand the rationale and some of the benefits of the GLM approach.

3.4.1 The pros and cons of using GLMs

Generalized linear modelling provides a single theoretical framework for analysing many different types of data. This makes it an extremely powerful and flexible approach. Also, by careful choice of an appropriate *link function* (section 3.4.2.3), the GLM will constrain the predicted values to lie within realistic bounds (as with the logistic transformation, section 3.3.3.3). The main limitation of the GLM approach is that it is restricted to models that are *linear*. This does not mean that GLMs can be used only to describe straight-line relationships, but that it must be possible for the model to be structured in such a way that it describes a linear relationship. For example, the following nonlinear equation

$$y = e^{(a+bx)} \qquad \text{(eq. 3.10)}$$

can be linearized by log-transforming both sides

$$\ln(y) = a + bx. \qquad \text{(eq. 3.11)}$$

Within GLMs, this process is performed by $\log_e$ transforming the dependent variable by specifying a log *link function* (section 3.4.2.3). Some models are intrinsically nonlinear because there is no transformation that can linearize them in all parameters. For example

$$y = a + \frac{b}{c + x}. \qquad \text{(eq. 3.12)}$$

In these circumstances, the GLM is unable to estimate all of the parameters (a, b and c) and we must undertake nonlinear modelling.

3.4.2 Components of a GLM

Generalized linear models have three essential ingredients (Crawley 1993, 2002 provides a fuller explanation).

3.4.2.1 Error structure

The error structure describes the shape of the distribution of residual values around the fitted model. Classical linear models assume a normal (Gaussian) distribution; GLMs allow other error distributions to be defined such as Poisson errors (e.g. for count data), negative binomial errors (e.g. for parasite load data), exponential errors (e.g. survival times) and binomial errors (e.g. sex ratios, mortality and other proportion data).

3.4.2.2 Linear predictor

The linear predictor is a linear equation defining the relationship between the predicted y values and one or more explanatory variables, *on the scale determined by the link transformation* (section 3.4.2.3). The number of terms in the linear predictor is the same as the number of parameters to be estimated from the data. So, for a simple linear regression, there are two terms in the linear predictor (slope and intercept). To determine the fit of a given model, the GLM evaluates the linear predictor for each value of the response variable and compares this with a *transformed* value of y that is determined by the link function. The fitted value is determined by *back-transforming* the predicted values to the original scale (so, for example, with a *log link*, the fitted value is the *antilog* of the linear predictor and with the *reciprocal* link it is the *reciprocal* of the linear predictor). This will become clearer when we go on to examine a specific example (section 3.5.1.1).

3.4.2.3 Link function

Data on proportions, such as sex ratios, are frequently described by the *logistic curve* (Figure 3.2a), because this equation (eq. 3.13) asymptotes at 0 and 1 (or 0% and 100%), and guards against unrealistic values being predicted.

$$p = \frac{e^{(a+bx)}}{1 + e^{(a+bx)}}. \qquad \text{(eq. 3.13)}$$

Clearly, eq. 3.13 describes a nonlinear relationship. However, it can be linearized by applying the *logit transformation* we encountered earlier

$$\ln\left(\frac{p}{1-p}\right) = a + bx. \qquad \text{(eq. 3.14)}$$

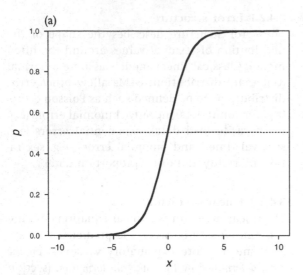

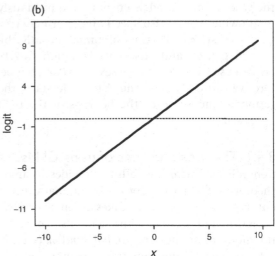

Fig 3.2 The logistic curve (a) can be used to ensure that the fitted values lie between 0 and 1, and the logit transformation (b) can be used to linearize the relationship.

This transformation is known as the *link function*, and it relates the mean value of *y* to its linear predictor (section 3.4.2.2). It is essentially just a transformation that linearizes the model and ensures that the fitted values stay within reasonable bounds (Figure 3.2). As indicated in section 3.4.2.2, the values that emerge from the linear predictor are on the scale of the link function, and predicted values of *y* are generated by back-transforming the linear predictor to the original scale. So, for example, to ensure that predicted count data (e.g. number of beetles per quadrat) never become negative, a *log link* function would be applied. This is because the fitted values would then be antilogs of the linear predictor, and all antilogs are greater than or equal to zero. In the case of proportion data, the *logit link* function is generally applied to ensure that predicted values never exceed one or drop below zero (other link functions include the *identity link* for normal errors, the *reciprocal link* for gamma errors, the *probit link* for bioassays and the *complementary log-log link* for dilution assays). Although the default link function for binary and binomial data is the logit link, the (asymmetrical) complementary log-log link should also be assessed as it will sometimes lead to a lower residual deviance (Crawley 1993, for an empirical example see Petersen & Hardy 1996).

3.4.3 Determining the best-fit model: maximum-likelihood

The 'classical' methods with which most of us are familiar (linear regression, ANOVA, etc.) utilize *least-squares* (LS) methods for determining the best-fit model. In other words, we find the model that *minimizes the sum of squares* of the departures from the observed *y* values from their predicted values. In contrast, generalized linear modelling determines the best-fit model by *maximum-likelihood* (ML) methods. When the GLM has normal errors and an identity link function (as in 'classical' models), ML and LS give identical results (indeed, linear LS methods are a subset of ML, in much the same way as *general* linear models are a subset of generalized linear models). For other kinds of error structure and link functions, LS methods may produce biased parameter estimates, and so ML is generally preferred (e.g. McCullagh & Nelder 1989). Even though ML estimation is relatively straightforward, it is rather laborious, and so most biologists are happy to treat the process as a 'black box'. Those interested in the mechanistic basis to ML estimation in a biological context can find examples in Crawley (1993, 2002) and McCallum (2000).

The basic idea behind any statistical modelling procedure is to determine the parameter values that lead to the best fit of the model to the data. With LS regression, the best-fit model is determined by minimizing the residual sum of squares. With ML, we ask: given our data and our choice of model, *what parameter values*

Table 3.2 A description of deviance terms

Deviance term	Description
Null deviance	Deviance associated with the *null model* ($\equiv$ total sum of squares)
Residual deviance (deviance)	Deviance remaining after some or all terms have been included in the model ($\equiv$ residual sum of squares)
Change in deviance	Deviance associated with inclusion of a particular term in a model ($\equiv$ sum of squares for a particular term)

Table 3.3 Terminology for stages of model simplification

Model	Description
Saturated (full) model	Perfect fit; zero deviance and *df*; one parameter for each observation
Maximal model	Contains all factors, interaction terms and covariates under consideration
Current model	The current model; number of parameters $\leq$ maximal and $\geq$ minimal model
Minimal model	A model with minimal number of terms, in which all parameters are significantly different from zero, and no important terms have been omitted
Null model	Only the grand mean (i.e. one parameter) is fitted; deviance $\equiv$ total sum of squares in 'normal' models

maximize the likelihood of the data being observed (hence the term 'maximum likelihood')? *Likelihood* (or more commonly *log-likelihood*) is used here in a formal sense for assessing the statistical *odds* of producing a particular outcome. The 'best' model is therefore the model that produces the minimal *residual deviance* (Table 3.2), subject to the constraint that all the parameters in the model should be statistically significant (Table 3.3).

Residual deviance is twice the difference between the maximum achievable log-likelihood (i.e. that obtained when the predicted and observed values are identical) and that attained by the model under consideration (McCullagh & Nelder 1989). For most error structures, deviance is distributed asymptotically as *chi-square* (χ^2) and so the *goodness-of-fit* of a model can be determined by calculating the deviance and testing it against the chi-square distribution with the appropriate degrees of freedom (*df*). By convention, if $P > 0.05$, then we usually declare that the model fits the data well (Hardy & Field 1998 give further explanation and examples). A commonly used alternative test statistic is Pearson's X^2, which has the same asymptotic χ^2-distribution as the deviance.

Several analogues of the r^2 measure commonly used in linear models have been proposed (e.g. Hosmer & Lemeshow 1989), but these do not possess the same statistical meaning and are not as widely used: one could, for example, give the percentage of the deviance explained by each term in the model.

3.4.4 Overdispersion and underdispersion

For a well-fitting model, the residual deviance should be approximately equal to the residual degrees of freedom (i.e. the *residual mean deviance* (residual deviance/residual *df*) should be approximately equal to one and certainly less than about 1.5). When this is not the case, either the model does not adequately describe the variation in the data, or the variation in the data is greater than that under binomial sampling. Either way, the most likely result is that the mean deviance will be greater than one. When the model is thought to be correct (i.e. we believe that all important explanatory terms have been included), but the residual mean deviance is greater than one, the data are said to exhibit *extra-binomial variation*, *super-binomial variation* or *overdispersion* (Chapter 5).

3.4.4.1 Causes of overdispersion

There are two main causes of overdispersion in grouped binary data expressed as proportions: either the model is mis-specified in some way, or there is correlation between the responses (i.e. the sexes). Mis-specification could be due to one of the following: (1) a systematic component of the model has been mis-specified (e.g. important variables have not been measured, important interaction terms have been omitted or an explanatory variable needs to be transformed); (2) there are one or more outliers in the observed dataset; (3) an inappropriate link function has been chosen (for proportion data, a complementary log-log link may reduce the degree of overdispersion); or (4) the proportions are based on small numbers of individuals (under these circumstances, the chi-square approximation to the distribution of deviance breaks down and hence a large residual mean deviance may not be problematic).

Once these possible explanations have been eliminated, the most likely explanation for overdispersion in binomial data is *correlation between the binary responses*. In essence, this means that there has been a violation of the assumption that the individual binary observations (i.e. the individual organisms) making up the binomial proportions (i.e. the sex ratios) are *independent* of each other. Since individuals are often grouped together within clutches, broods or families, if sex ratios are biased in any way the individual binary data points (offspring sexes) will be positively correlated, leading to sex ratios that are more variable than they would have been under the assumption that the sexes were distributed binomially. As a consequence, the residual mean deviance will be greater than unity. Overdispersion can be generated not just by inter-clutch variation in sex ratio, but also by any factor that leads to individual binary responses being nonindependent. Overdispersion is also common in mortality data as groups of individuals may tend to survive or die collectively (Chapter 5); Jim Morrison's 'no one here gets out alive'!

3.4.4.2 Correcting overdispersion

Since overdispersion simply means that the variance is greater than that expected under the binomial expectation, the simplest solution is to assume that the variance is not equal to $np(1 - p)$, as assumed by the binomial probability distribution, but is *proportional* to it and equal to $np(1 - p)s$, where s is an unknown scaling factor variously referred to as the *scale parameter*, *dispersion parameter* or *heterogeneity factor*. We can estimate s by dividing the Pearson's X^2 value (or simply the *residual deviance* for the full model) by the residual degrees of freedom. We can then use this estimate of s (usually termed the *empirical scale parameter*) to compare the *scaled deviances* for terms in the model using F-tests, rather than χ^2 tests (in exactly the same way as we would for a conventional linear model). Applying an empirical scale parameter does not affect parameter estimates, but it does inflate their standard errors (which are multiplied by a factor $\sqrt{s}$); thus, type II errors are more likely (and type I errors less likely). This approximation works well, and is the standard method used by most ecologists (indeed, some ecologists would advocate the use of F-tests rather than χ^2 tests for all GLMs with binomial errors whenever there is any overdispersion, especially when sample sizes are small).

Models using empirical scale parameters are prone to inaccuracies when sample sizes (denominators) vary dramatically between proportions. Williams (1982) developed an alternative method that allows for unequal sample sizes by applying an additional weighting function to the data; this method is now known as *Williams' procedure*. A number of statistics packages have the facility to implement Williams' procedure, including *GLIM* and *Genstat*, but not *S-Plus*. F-tests (or t-tests) should be employed to evaluate the significance of variables after using Williams' procedure.

A further method is *quasi-likelihood estimation*. This allows estimation of regression relationships without fully knowing the error distribution of the response variable. Thus, instead of providing an error distribution and link function, one provides a link function and a variance function. For example, perhaps the logit link transformation linearizes the response correctly, but the variance appears to be a linear function of the mean; under these circumstances, both attributes could be incorporated into a quasi-likelihood model. Quasi-likelihood also allows one to estimate the scale parameter in under- and overdispersed regression models. For example, in sex ratio analyses, we can estimate the degree of overdispersion in a logistic regression model by supplying the appropriate link and variance functions for the logistic model and determining the significance of parameter estimates using F-tests.

Yet another method of dealing with overdispersion is to use *generalized linear mixed models* (GLMMs). One of the problems with GLMs is that they allow only one error term; all effects other than the residual error at the lowest level of the data are assumed fixed (McCullagh & Nelder 1989). Therefore, for parameter estimation purposes, each offspring in the sex ratio is treated as an independent data point. However, if variation between 'clusters' (e.g. broods) does not follow the binomial expectation (e.g. due to sex ratio manipulation), then there will be overdispersion and these estimates will be biased. While we might use an empirical scale parameter, Williams' correction or quasi-likelihood to deal with this (see above), an alternative method would be to introduce a second random effect (e.g. the identity of the brood) to deal with the 'between-cluster' variation in sex ratio, in addition to the 'within-cluster' variation, i.e. to use GLMMs (Krackow & Tkadlec 2001). These models will give the same parameter estimates for the fixed effects as the conventional GLMs, but their standard errors will be inflated if the random effect (e.g. nest identity) is influential. Currently, GLMMs are possible in only a few statistical packages (e.g. *Genstat*, but not *S-Plus* or *GLIM*), but their use and availability are likely to grow. For recent examples of the use of GLMMs in sex ratio analyses, see Kruuk *et al.* (1999) and section 3.5.4. Further details of dealing with overdispersion are given by Collett (1991) and Crawley (1993). It is important to emphasize, however, that if overdispersion is very large, then this indicates a badly fitting model and it might be that a different approach would reflect the biology of the system better (e.g. log-linear modelling of the number of males in the clutch).

3.4.4.3 Overdispersion in binary data
Since the deviance for (ungrouped) binary data does not exhibit a χ^2 distribution, its magnitude depends solely on the value of the fitted probabilities. Therefore, large values of residual mean deviance for binary data cannot be taken to indicate overdispersion. Overdispersion may still occur in binary data, but it will not be possible to detect it from the value of the residual mean deviance and it can be modelled only by including a random effect in the model (Collett 1991).

3.4.4.4 Underdispersion
Underdispersion occurs when the variance of a binomial response variable is less than that for the binomial distribution and may be produced when the individual binary observations are *negatively* correlated. Although underdispersion is rare in sex ratio analyses of vertebrates and most invertebrates, it is common among haplodiploid insects and mites (Chapter 5). Despite this, it has yet to receive much attention from statisticians (but see Podlich *et al.* unpublished manuscript). Another reason is that the costs of ignoring underdispersion appear to be relatively small, as it simply leads to conservative tests, i.e. tests in which the chance of a type I error is not increased. On the other hand, ignoring underdispersion reduces the statistical power of the test and hence increases the chances of making type II errors. In the absence of alternative methods, rescaling the data in the same way as for overdispersed data is recommended (Gordon K. Smyth pers. comm.). For example, Hardy and Mayhew (1998) found a significant negative relationship between mean sex ratio and mean clutch size across 26 species of bethylid wasps using classical regression of arcsine-transformed sex ratio data. When we analysed the same data using logistic regression, the relationship appeared to be nonsignificant ($\chi^2{}_1 = 0.86$, $P > 0.1$). However, the model exhibited considerable underdispersion (heterogeneity factor $= 0.178$) and when rescaling was applied, using Pearson's X^2, it transpired that the relationship was indeed significant ($F_{1,24} = 27.0$, $P < 0.001$). Analysis of species-mean data is discussed in Chapter 6.

3.4.5 Model simplification
The aim of statistical modelling is to produce a model that fits the data well while also being as simple as possible: this is known as the

Table 3.4 The sequence of steps in model simplification (after Crawley 1993)

Step	Procedure	Explanation
1	Fit the maximal model	Fit all the factors, interactions and covariates of interest
		Note the residual deviance
		If you are using Poisson or binomial errors, check for overdispersion and rescale if necessary
2	Begin model simplification	Inspect the parameter estimates
		Remove the least significant terms first, starting with the highest order interactions, progressing on to lower order interaction terms and then main effects
		Remember that main effects that figure in significant interactions should not be deleted
3	If the deletion causes an insignificant increase in deviance	Leave that term out of the model
		Inspect the parameter values again
		Remove the least significant term remaining
4	If the deletion causes a significant increase in deviance	Put the term back in the model
		These are the statistically significant terms as assessed by deletion from the maximal model
5	Keep removing terms from the model	Repeat steps 3 or 4 until the model contains nothing but significant terms
		This is the minimal adequate model
		If none of the parameters is significant, then the minimal adequate model is the null model

principle of parsimony or *Occam's razor*; in other words, a model that does not contain any redundant parameters or factor levels. Fitting GLMs is a *journey of exploration!* Often, there is no single best model; several models may adequately fit the data and different modelling procedures may yield very different solutions. But remember that at all times *biology should drive your choice of models*. Indeed, Hosmer and Lemeshow (1989) have argued that 'successful modelling of a complex data set is part science, part statistical methods, and part experience and common sense'.

The first step in the model simplification process is to fit a *maximal model* that contains all of the factors, covariates and interaction terms that might be important in the analysis (Table 3.3). Then, via a series of *step-wise deletion tests* (section 3.4.5.1), any nonsignificant explanatory variables, factors and interaction terms are removed, starting with the highest order terms (e.g. three-way interactions). Once the number of terms in

the model has been reduced such that no more can be removed without reducing the model's explanatory powers (i.e. causing a statistically significant reduction in the amount of variation explained), and none can be replaced that increase the model's explanatory powers, it may be possible to simplify the model still further by grouping together factor levels that do not differ significantly from one another (aggregation) and amalgamate explanatory variables that have similar parameter values (as long as such simplifications make good biological sense). The resultant model is the *minimal model* (Tables 3.3 and 3.4).

Crawley's (1993, 2002) books contain whole chapters on model simplification and it is well worth reading one of these prior to embarking on any GLM exercise. His views on the sequence of steps in the model simplification process are summarized in Table 3.4 but, as Crawley himself is at pains to point out, there are no hard and fast rules.

3.4.5.1 Determining the significance of individual terms in the model

Step-wise deletion tests are χ^2 tests (or *F*-tests) that assess the significance of the increase in deviance that results when a given term is removed from the current model. For example, imagine we have two hierarchical models (i.e. two models for which one of the models contains all of the terms of the other model, plus one or more additional terms) – model 1: $y = a + bx_1 + cx_2 + dx_3$; and model 2: $y = a + bx_1 + cx_2$, which differ in that model 2 does not contain the dx_3 term. To test the significance of the parameter d, we determine the likelihoods for model 1 (l_1) and for model 2 (l_2), and calculate the (change in) *deviance* for the comparison of the two models $[-2 \ln(l_2/l_1)]$, which can then be compared to χ^2, with degrees of freedom equal to the difference in the number of terms in the two models. Here, $P < 0.05$ indicates that the variable was making a significant contribution to the fit of the model and hence should generally be retained. However, for very large sample sizes, or where there are many higher-order interaction terms, statistically significant results may be generated even though the effect sizes are small. In these instances, it may be prudent to increase the critical probability level for retention in the model. For example, a good rule of thumb is that the acceptance probability is set at 5% ($P < 0.05$) for main effects, 1% for two-way interactions, 0.5% for three-way interactions, and so on (MJ Crawley pers. comm.).

A less rigorous method of evaluating the significance of a variable in a statistical model is the *Wald-test*, which tests whether the regression coefficient is significantly different from zero by comparing the estimated coefficient to its standard error. In practice, the Wald-test is usually used as a guide to the sequence in which variables are removed from the model, and the amount of deviance the variable explains is used as the final criterion of its significance.

3.4.6 Model checking

Once the minimal model has been obtained it can be checked using a number of regression diagnostics, discussed in detail by, for example, Hosmer and Lemeshow (1989) and Crawley (1993, 2001). These include assessing the overall fit of the model (section 3.4.3) and producing diagnostic plots. For example, we need to assess whether the standardized residuals exhibit any trends with respect to the explanatory variables or fitted values (Figure 3.3a), and whether the standardized residuals are normally distributed (Figure 3.3b). We use *standardized* residuals when the error distribution is binomial (or Poisson or gamma) because the variance changes with the mean (Crawley 1993). Examples of both a 'residuals plot' and a 'normality plot' are shown in Figure 3.3. A lack of pattern in the residuals plot indicates a well-specified model, while the normality plot should generate a reasonably straight line when the model provides a good fit to the data. However, while these plots are good for detecting extreme observations deviating from a general trend, extreme caution should be exercised in over-interpreting them. This is particularly true for binary data, because all of the points on the residuals plot lie on one of two curves depending on whether the response is 0 or 1. Diagnostic plots are produced as standard in *S-Plus* and some other statistical packages, and Crawley (1993, p. 288) provides a macro for generating them in *GLIM*; as well as an example for binomial data.

3.5 | Logistic analysis of sex ratio data

Having set the GLM scene, we now examine the GLM modelling process as it applies to proportion data in general, and sex ratios in particular. Logistic regression is the term used to describe GLMs in which the error distribution is assumed to be binomial and a logit link function is applied (section 3.4). Many statistical packages now include logistic regression as a special modelling procedure, even if they also have a generic GLM function (e.g. *S-Plus*) or have no other GLM functions (e.g. *Minitab*) (see Apendix 3.1). Logistic regression can be used to model both binary and binomial (grouped binary) data. The statistical methodologies for analysing these two data types are essentially the same. We begin with the

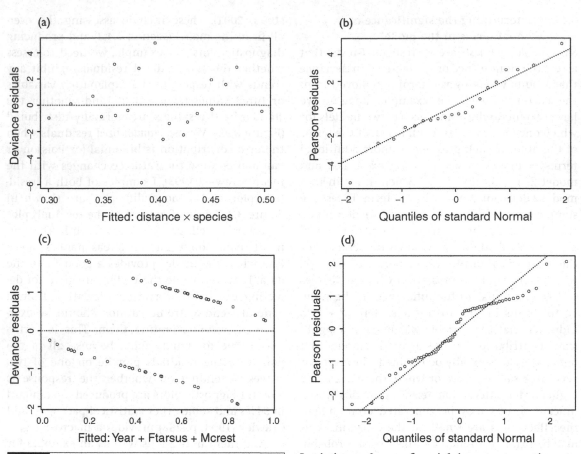

Fig 3.3 Model-checking plots from a logistic regression model. The data are plots derived from a logistic regression model fitted to binomial data from Example 1 (panels a & b) and binary data from Example 2 (panels c & d). In (a) and (c), the deviance residuals are plotted against the fitted values; we refer to these as *residuals plots*. In (b) and (d), the ordered Pearson residuals are plotted against quantiles of the standard Normal distribution; we refer to these as *normality plots*. When the data are binomial (grouped binary), a random scatter of points around zero indicates a well-fitting model, as shown in (a). When the data are (ungrouped) binary, residuals plots are not very useful because all of the points lie on one of two curves depending on whether the response is 0 or 1 (c). For both binomial (b) and binary (d) data, deviation from the line of unity on a normality plot may indicate a poorly fitting model; both of these models appear to fit reasonably well.

analysis of binomial data (section 3.5.1) and follow this up with the analysis of binary data and highlight where the differences lie (section 3.5.2). We give a worked example of analysis of avian sex ratios (section 3.5.3) and discuss a case history of analyses of mammalian sex ratios (section 3.5.4).

Logistic analyses of social insect sex ratios are discussed in Chapter 4.

3.5.1 Analysis of proportions

Because sex ratios are proportions, and involve dividing one integer by another, important information about the size of the sample from which they were calculated is lost. This is one of the main problems with traditional (non-weighted) methods that rely on classical regression or nonparametric statistics (section 3.5.3). When proportions are modelled by logistic regression, this information is regained because information about ratios (e.g. number of males versus number of females or number of successes versus number of failures) *and* sample sizes (i.e. the magnitude of the *binomial denominator*) is included.

In most statistical packages, the data (e.g. sex ratio) are included in the model as two vectors: one describing the ratio, and the other the sample size, or (as in *GLIM*) one describing the numerator (e.g. males) and the other the sample size (i.e. denominator = males + females). In others (e.g. *S-Plus*), the two vectors may be bound together (using the *cbind* command) and represent the raw data that combine to make the ratio (e.g. number of males and number of females). In all cases the method of analysis involves performing a *weighted* regression using the individual sample sizes as weights, and a *logit link* function to linearize the model (section 3.4.2.3).

3.5.1.1 Example 1: Fish sex ratios revisited

We begin by going back to Example 1. You will remember that we wanted to determine whether pollution from a known source resulted in biased sex ratios in two species of fish living in an Australian creek. Having conducted a visual inspection of the data, we can proceed to fitting

logit link is the default option), and the data we want to analyse are in the dataset called Example 1 (data=Example1). The difference lies in how we tell *S-Plus* that we have proportion data with a known denominator. In (1), we give *S-Plus* a term for the proportion of males (SexRatio) and a term for the denominator of the ratio (weights=SampleSize), whereas in (2) we give it the two vectors that together indicate the magnitude of the denominator (cbind (NumMales, NumFemales); cbind simply 'binds' these two vectors together, such that the number of males in the sample is paired to the number of females from the same sample; the denominator = NumMales+NumFemales=SampleSize).

In both cases, we get the following output (the diagnostic plots for this model are shown in Figure 3.3a,b):

Coefficients	Value	Std. Error	t value
(Intercept)	−0.8069	0.1779	−4.5343
Distance	0.0007	0.0002	2.8799
Species	−0.1304	0.1779	−0.7328
Distance:Species	0.0001	0.0002	0.7791

(Dispersion Parameter for Binomial family taken to be 1)
  Null Deviance: 118.0099 on 19 degrees of freedom
Residual Deviance: 107.7797 on 16 degrees of freedom

Terms added sequentially (first to last)

Term	Df	Deviance	Resid.Df	Resid.Dev	Pr(Chi)
NULL			19	118.0099	
Distance	1	9.6185	18	108.3914	0.0019
Species	1	0.0023	17	108.3890	0.9612
Distance:Species	1	0.6092	16	107.7797	0.4350

our maximal model, which in this case includes just three terms, Distance, Species and the Distance:Species interaction. In *S-Plus*, we can specify this model in one of two ways:

The first table in this output tells us, for each of the coefficients, the value of the parameter estimate (Value), its standard error (Std. Error) and a *t* value comparing the estimate against zero

```
(1) model1_glm (SexRatio~Distance*Species, family=binomial,
       weights=SampleSize, data=Example1)
(2) model1_glm (cbind (NumMales,NumFemales) ~Distance*Species,
       family=binomial, data=Example1)
```

In both cases, we tell *S-Plus* that we are creating a generalized linear model (glm) with binomial errors and logit link function (family=binomial;

(t value). The second table tells us the change in the number of degrees of freedom (Df) and change in deviance (Deviance) associated with the sequential inclusion of each of the terms (Term) in the model and the statistical significance of the change in deviance, as determined

tion term. This produces the following output (which has been adapted from *S-Plus* to make it clearer):

```
model2_update(model1,~.-Distance:Species)
anova(model1,model2,test="F")
```

Resid.Df	Resid.Dev	ΔTerms	ΔDf	ΔDeviance	F-value	P(F)
17	108.3672	-Distance:Species	−1	−0.6132	0.0955	0.7612

by chi-square tests (Pr(Chi)). The other values in this table indicate the sequential reduction in the residual degrees of freedom (Resid.Df), and residual deviance (Resid.Dev). Sandwiched between these two tables are three important lines. These tell us that the statistical analysis of this model assumes that the dispersion parameter is taken to be 1; in other words that there is no overdispersion (section 3.4.4). Is this true? We can get a rough idea of this by dividing the residual deviance by the residual degrees of freedom (Resid.Dev/Resid.Df = 107.8/16 = 6.73). Clearly, there is massive overdispersion, whereas our model currently assumes that there is none (i.e. that Resid.Dev/Resid.Df ~ 1). Having checked that we have included all possible terms in our model, and that it has not been mis-specified in any way (e.g. by omitting an important interaction term), that we have not ignored any outliers, and that we have the correct link function (section 3.4.2.3), it seems likely that we have genuine overdispersion. This is perhaps not too much of a surprise given that these data are not real, but we shall not let that worry us at this stage. To proceed as we would do with real data, we need to employ an empirical scale parameter ($s = 6.73$). In *S-Plus*, we do this simply by testing the significance of terms in the model using *F*-tests, rather than *chi-square*

The first command simply removes the interaction term from our maximal model (model 1). The second command asks *S-Plus* to examine the difference between the amount of variation explained by models 1 and 2 (with and without the interaction term) using *F*-tests. In the table, Resid.Df and Resid.Dev are the residual deviance and residual degrees of freedom, respectively, for the model that excludes the terms indicated by ΔTerms (Δ, 'delta', simply means 'change in'). The table tells us that the process of removing the Distance:Species interaction generates a final model that has a deviance of 108.36 and 17 degrees of freedom, and results in a change in deviance of 0.6132 and 1 degree of freedom. But, remember that we are no longer interested in deviances, because our data are overdispersed. We therefore need to concentrate on the *F*-test. This indicates that removing the interaction term from the model does not reduce the amount of variation explained by our model ($F_{1,17} = 0.0955$, $P = 0.7612$). If it did significantly reduce it, then our current model would also be the minimal model (Table 3.3) and the modelling process would be complete for this particular example. However, as it doesn't, we need to go on to test each of the main effects in turn, starting with the term with the lowest *t* value. In *S-Plus*, this is how we would do it:

```
Model3_update (model2,~.-Species)
Model4_update (model2,~.-Distance)
anova (model2,model3,test="F")
anova(model2,model4,test="F")
```

Resid.Df	Resid.Dev	ΔTerms	ΔDf	ΔDeviance	F-value	P(F)	
18	108.3697	-Species	−1	−0.0025	0.0004	0.9839	ns
18	116.1776	-Distance	−1	−7.8104	1.2856	0.2726	ns

tests. We begin the process of step-wise deletion by testing the significance of the interac-

Thus, although Distance appears to have a bigger effect on the fit of the model than

Table 3.5 | Summary of analyses of fish sex ratios (Example 1)

Model	Test	Type of model	Species	P-value (Distance)
1	Spearman's rank order correlation	Nonparametric	Shirazfish	$P > 0.16$ ns
			Merlotfish	$P = 0.063+$
			Combined P	$P = 0.036*$
2	Pearson's product-moment correlation (arcsine-transformed data)	Parametric – normal errors	Shirazfish	$P > 0.14$ ns
			Merlotfish	$P = 0.057+$
			Combined P	$P = 0.029*$
3	General linear model (unweighted, untransformed data)	Parametric – normal errors	Both	$P = 0.013*$
4	General linear model (unweighted, arcsine-transformed data)	Parametric – normal errors	Both	$P = 0.012*$
5	General linear model (unweighted, logit-transformed data)	Parametric – normal errors	Both	$P = 0.012*$
6	General linear model (weighted, arcsine-transformed data)	Parametric – normal errors	Both	$P > 0.16$ ns
7	Generalized linear model (weighted, untransformed data)	Parametric – binomial errors	Both	$P > 0.21$ ns

ns $= P > 0.1$, $+ = 0.05 < P < 0.1$, $* = P < 0.05$.

Species, neither term is statistically significant (Species: $F_{1,18} = 0.0004$, $P = 0.9839$; Species: $F_{1,18} = 1.2856$, $P = 0.2726$). This suggests that there is no consistent effect of pollution on sex ratio in this population. Just to be sure, we should try adding terms back into the model, starting with Distance. When Distance alone is added to the model, no significant variation in sex ratio is explained ($F_{1,18} = 1.675$, $P = 0.21$), even if we employ quasi-likelihood estimation to obtain a better level of compensation for overdispersion ($F_{1,18} = 1.726$, $P = 0.21$).

3.5.1.1.1 SUMMARY OF EXAMPLE 1

In summary, if we compare the performance of the different tests (Table 3.5), we see that the nonparametric tests (Spearman's correlation) gave lower significance values for Distance than the equivalent parametric tests (Pearson's correlation) (cf. models 1 and 2) (section 3.3.1.2). This is almost certainly due to this test's lack

of power. Using a general linear model (in effect, an ANCOVA), we were able to combine a factor and a covariate within a single model, and this improved the significance level associated with Distance, regardless of whether we transformed our sex ratio data or not (cf. model 2 and models 3, 4 and 5) (section 3.3.2). However, when we added a weighting factor to our model, to control for differences in sample size within our dataset, Distance disappeared as a significant term in the model (cf. models 5 and 6) (section 3.3.3). Applying a GLM with binomial errors and logit link function yielded similar results, and confirmed that there was no significant change in sex ratio with distance from the pollution source (section 3.5.1.1). The similarity between the results of models 6 and 7 is probably due to the overriding importance of sample size effects in this analysis (i.e. power, rather than the lack of fit of the data to the normal distribution). Careful examination of Table 3.5 indicates that although the single-species nonparametric tests gave the correct result (i.e. no effect of Distance on sex ratio) it appears to have given it for the wrong reason (i.e. due to lack of power)!

At this point, it is worth emphasizing that although our comparison of the different methods has focused on the statistical significance of the result (i.e. the P value), as biologists we are usually more interested in the biological significance of the result rather than its statistical significance (though journal editors may sometimes disagree!). If sample sizes are large enough, then even a 1% difference between treatment groups will be statistically significant (this is why pollsters question such large numbers of people in the run-up to elections). Thus, it is not sufficient to consider just the statistical significance of any trends in our data, but also their magnitude. Thus, in Example 1, the equation for the (non-significant) logistic regression was

$$\text{Logit(Sex ratio)} = -0.7504 + 0.0006808 \times \text{Distance}.$$

Thus, back on the *original scale*, the sigmoidal relationship between sex ratio and distance from the pollution source is described as follows

$$\text{Sex ratio} = \frac{e^{(-0.7504 + 0.0006808 \times \text{Distance})}}{1 + e^{(-0.7504 + 0.0006808 \times \text{Distance})}}.$$

Thus, the sex ratio was predicted to vary from 0.33 at the source (100 m) to 0.48 at the furthest distance from the source (1000 m). Since this a fairly large increase in the proportion of males (45%) over a relatively short distance, it would be premature to dismiss pollution as a correlate of sex ratio variation at this stage and we might want to gather a new, larger dataset that will increase the power of our analysis.

3.5.2 Analysis of binary data

Often, sex ratio data are best analysed in the form of binary responses. The analysis of binary data using GLMs is exactly the same as for binomial (grouped binary) proportions, except that we do not include any weighting factor because each 'sex ratio' (0 or 1) represents a single individual and we cannot detect or correct for overdispersion (section 3.4.4.3). Effectively, we assume that each data point comes from a binomial trial in which the sample size (n) is equal to 1. In other words, the data are assumed to come from a special, abbreviated form of the binomial distribution, known as the *Bernoulli distribution* (Collett 1991). Whether it is worth analysing data in this format (rather than as sex ratios based on lumping together individuals from similar groupings, e.g. nests or sampling points) is largely dependent on whether each individual in the analysis has unique explanatory variables associated with it (e.g. an individual weight or colour, or individuals are produced one at a time by parents, i.e. brood size = 1, etc.). If it does, then the data are best analysed in binary form; if not then there is little to be gained and the data can be lumped without loss of information.

3.5.2.1 Example 2: Crest size in Crested Auklets

To address this issue, we examine the relationship between chick sex and paternal crest size in the Crested Auklet (Box 3.7 gives background information). Hunter *et al.* (in prep.) collected

---

Box 3.7 | Example 2: Crest size in Crested Auklets

Data on the relationship between crest size and chick sex in the Crested Auklet (*Aethia cristatella*) was compiled by Fiona Hunter and colleagues (Hunter *et al.* in prep.). These small seabirds breed in coastal colonies around the Bering Sea, nest in crevices and produce just one chick each year. The adults are socially monogamous and both sexes prefer mates with a large crest (a sexual ornament sprouting just above the beak). Hunter *et al.* wanted to know whether females were more likely to produce male chicks when they were paired to males with longer crests. Since there is mutual sexual selection in this species, Hunter *et al.* predicted that, provided crest length was heritable, females would produce sons if they were paired to long-crested males, and daughters if they were paired to short-crested males.

Table B3.7a | Sex ratios in Crested Auklets across three years

Year	Number of female chicks	Number of male chicks	Total number of chicks
1993	11	15	26
1994	3	3	6
1995	13	12	25
Total	27	30	57

Data were collected from 57 breeding pairs over three years: 1993, 1994 and 1995 (Table B3.7a). In each year, Hunter et al. recorded the sex and mass of the chick each pair produced, plus the body mass, tarsus (leg) length and crest length for the male and female parents (referred to as the sire and dam, respectively). Table B3.7b shows data for 1993 only.

Table B3.7b | Chick sex and morphometric data in Crested Auklets in 1993

Pair number	Chick sex	Sire Mass (g)	Sire Tarsus (mm)	Sire Crest (mm)	Dam Mass (g)	Dam Tarsus (mm)	Dam Crest (mm)
1	M	293	29.7	43.1	255	26.2	37.7
2	F	278	28.2	40.6	271	26.6	36.2
3	M	258	27.8	42.6	234	27.4	52.0
4	M	289	28.2	41.2	256	27.2	45.5
5	F	276	28.0	38.8	235	27.4	44.4
6	F	256	28.0	36.2	270	28.8	39.2
7	M	306	29.4	39.4	254	28.3	35.0
8	F	248	25.8	31.3	244	27.3	39.5
9	M	254	28.8	38.9	269	28.8	36.2
10	M	264	30.3	49.7	272	29.4	42.0
11	M	267	28.6	36.2	271	29.7	35.2
12	M	309	28.4	40.4	265	28.4	36.4
13	M	308	28.9	44.8	256	27.9	35.4
14	F	271	29.6	34.6	264	28.3	42.6
15	M	248	28.4	38.3	261	29.1	40.2
16	F	282	26.1	36.6	239	27.3	39.2
17	M	271	28.3	36.3	255	29.0	36.5
18	F	241	29.3	35.9	249	28.5	38.0
19	M	262	27.5	33.6	257	29.5	34.1
20	M	244	28.9	40.4	272	28.8	38.3
21	F	274	27.7	39.5	236	27.2	37.0
22	F	258	28.4	41.5	283	29.4	37.4
23	F	275	26.6	41.0	242	26.3	42.4
24	F	277	27.3	30.6	252	28.0	36.0
25	M	281	28.8	38.4	270	28.1	38.0
26	M	271	28.1	43.3	–	28.2	38.8

these data to find out whether females were more likely to produce male chicks when they were paired to males with longer crests. Before addressing this issue (section 3.5.2.1.4) we ask a simpler question, namely: does sex ratio vary between years? There are several ways that we can address this question.

3.5.2.1.1 CONTINGENCY TABLES

Often, the simplest way is to construct a *2 × n contingency table* and calculate *Pearson's chi-square* to test the null hypothesis that individuals are distributed independently with respect to year and sex. A problem with the data in Table B3.7a is that the sample sizes are rather small for 1994. Therefore, in our analyses we shall combine the 1993 and 1994 data (combining 1994 with 1995 gives similar results). This generates a 2×2 contingency table and a chi-square test gives $\chi^2_1 = 0.38$, $P = 0.54$, suggesting that the sex ratio is similar in all years.

3.5.2.1.2 LOG-LINEAR MODELS

A better method for analysing these types of data (often called the *G-test*) extends the contingency table approach and uses *log-linear models*. These are generalized linear models for modelling Poisson-distributed data (as opposed to binomial data). Like the chi-square test, log-linear models yield a χ^2 statistic. Their great advantage is that they can be readily generalized to analyse datasets that are much more complicated than simple 2×2 contingency tables (e.g. Crawley 1993). Moreover, since log-linear models are GLMs, their relation to the other models we have discussed is more easily appreciated.

To analyse the Example 2 data, we need to organize them so that there is a single dependent variable (Count = number of chicks in a given category), and two factors (each with two levels) corresponding to Year and Sex, and then implement a GLM with Poisson errors and log-link function. In *S-Plus*, the resulting model is:

```
model4_glm(Count~Year+Sex, family=poisson, data=Example2a)
```

```
Terms added sequentially (first to last)
        Df   Deviance   Resid.Df   Resid.Dev   Pr(Chi)
NULL                    3          1.4031
Year    1    0.8618     2          0.5413      0.3532
Sex     1    0.1579     1          0.3833      0.6910
```

The significance of the model is tested by comparing its residual deviance (0.3833) with the tabulated χ^2 statistic with 1 degree of freedom (3.841). Since the calculated χ^2 statistic is lower than the critical value in the tables ($\chi^2_1 = 0.3833$, $P > 0.53$), we cannot reject our null hypothesis that the two sexes are distributed randomly across years (i.e. that sex ratio varies between years). Note that if we had a more complicated model, with more factors, we would be better off starting the analysis by constructing a saturated model (Table 3.3), so that we end up with zero deviance and zero degrees of freedom. This would allow us to determine that we had all possible factors in the model before we began the stepwise deletion process (Table 3.4).

3.5.2.1.3 LOGISTIC REGRESSION

A third way of looking at these data is to convert them to *proportions* and analyse them using *logistic regression* (rather than analysing them as counts using log-linear regression). A model in which just the intercept term is fitted yields a residual deviance (0.3833 with 1 *df*) that is equal to that determined by the log-linear model, and again indicates that there is no significant variation in sex ratio between years.

We can perform exactly the same analysis by constructing an unweighted logistic regression model using the *raw* (binary) data (Table B3.7a):

```
Model5_glm(Sex~Year, family=binomial, data=Example2b)
```

In this case, we obtain the following output:

Terms	Df	Deviance	Resid.Df	Resid.Dev	Pr(Chi)
NULL			56	78.8608	
Year	1	0.3833	55	78.4774	0.5358

Again, we find that there is no difference in sex ratio between 1993/94 and 1995 ($\chi^2_1 = 0.3833$, $P = 0.5358$). In fact, log-linear and logistic regression models are exactly equivalent when the response is two level (Aitkin et al. 1989, pp 225–255).

3.5.2.1.4 PATERNAL CREST LENGTH AND CHICK SEX

Here we address the question of whether there is an association between the sex of chick produced by female Crested Auklets and paternal crest length. The usual first step of plotting the data for visual inspection is difficult when the data are binary because the dependent variable has just two states: male and female. In some instances, summarizing the data with respect to sex can be helpful, but in others (particularly more complex models) it is not. In Example 2, the average crest length of males that sired daughters was 38.77 ± 1.18 mm (s.e.), whereas the average for males siring sons was 42.51 ± 0.99 mm ($t = -2.4543$, $df = 55$, $P = 0.0173$). Thus, males siring sons have longer crests than those siring daughters. The focus of the analysis is, however, on the factors that determine offspring sex, and so the dependent variable is chick sex, rather than male crest length.

One way we could address this question would be to perform a simple logistic regression, with chick sex (Sex) as the dependent variable and male crest length (Mcrest) as the only explanatory variable

following linear equation (on the *logit* scale):

$$\text{Logit}(\text{Sex}) = -4.8992 + 0.1231 \times \text{Mcrest}.$$

Thus, back on the *original scale*, the sigmoidal relationship between offspring sex ratio and paternal crest length is described as follows

$$\text{Sex} = p = \frac{e^{(-4.899 + 0.123 \times Mcrest)}}{1 + e^{(-4.899 + 0.123 \times Mcrest)}}.$$

Of course, it is possible that this relationship is spurious, generated by some third factor. For example, perhaps good-quality females produce sons and also find good-quality mates with long crests. Alternatively, perhaps, females produce more sons in 'good years' and males produce longer crests in 'good years', leading to a positive correlation between sex ratio and male crest length across years, which is not present within years. Although we cannot examine all possible confounding variables, we can determine whether any of the other variables we measured are important. Hunter et al. (in prep) measured a variety of morphometric characteristics in addition to male crest size, including tarsus length and body mass, and they did this for both sexes (Table B3.7b). We also know in which year the measurements were made. Therefore, we are in a position to answer our main question while testing for additional factors that might be either accentuating or masking the relationship between paternal crest length and chick sex.

```
Model6_glm(Sex~Mcrest, family=binomial, data=Example2b)
```

Terms	Df	Deviance	Resid.Df	Resid.Dev	Pr(Chi)	
NULL			56	78.8608		
Mcrest	1	6.0653	55	72.7954	0.0137	*

This appears to confirm that the proportion of sons produced increases with increasing paternal crest length; the relationship is described by the

The first problem is which model to begin with. There are seven possible explanatory variables: one factor and six covariates (for a reminder about the difference between a factor and a covariate, refer back to Box 3.1). This means that we have 13 main effects, if we include the six possible quadratic terms. If we also fit all 78 pairwise interactions $(12 + 11 + 10 \ldots + 2 + 1)$, this means that we have 91 parameters to estimate, yet only 56 data points! In these circumstances, it is less obvious what the correct procedure for model simplification is (section 3.4.5). Clearly, a compromise is needed, and this is where the art of statistical modelling comes into its own (and where individual modellers' opinions may differ). The trick is to start at a sensible

quadratic terms. Even with just a single interaction term, our model includes 14 terms! Once we had removed all terms that did not contribute significantly to model fit (stepwise deletion tests), we tried to add more terms, including terms that had previously been rejected and high-order interaction terms. In practice, this involved going through each of the seven main effects, one by one, and testing for inclusion all interaction terms involving that effect. It is important to act systematically. We obtained the following minimal model, based on a series of stepwise-deletion tests

```
Model8_glm(Sex~Year+Ftarsus+Mcrest,family=binomial,data=Example2)
```

Resid.Df	Resid.Dev	ΔTerms	ΔDf	ΔDeviance	Pr(Chi)	
54	69.3214	−Year	−1	−4.8827	0.0271	*
54	69.4760	−Ftarsus	−1	−5.0373	0.0248	*
54	72.1635	−Mcrest	−1	−7.7247	0.0054	**

Coefficients:

Terms	Value	Std.Error	t-value
(Intercept)	−27.0913	10.1588	−2.6667
Year	−0.7745	0.3773	−2.0527
Mcrest	0.1775	0.0744	2.3831
Ftarsus	0.7014	0.3289	2.1325

point and then go back to test those terms that were initially ignored. Crawley (1993) recommends including not more than $n/3$ parameters in an initial model. Thus, in this example, no more than $56/3 = 19$ terms. Starting with the seven main effects plus six quadratic terms leaves room for just six interaction terms. An alternative starting point might be seven main effects plus 12 interaction terms. There are no hard and fast rules, but we started with the following model

```
model7_glm(Sex~Year* (Mcrest+Mmass+Mtarsus+Fmass+Ftarsus+Fcrest),
    family=binomial, data=Example2)
```

This was based on the idea that we could only reasonably allow one interaction term in the initial model and we felt that the most important interaction terms were likely to involve 'year', but we could easily have chosen to start with interactions involving male crest length or the

Thus, a higher proportion of male offspring were produced in 1995 than in 1993/94, and the proportion of male offspring increased with maternal tarsus length (body size) and paternal crest length. The fitted logistic regression lines are shown in Figure 3.4a,b (diagnostic plots are shown in Figure 3.3c,d).

3.5.3 A worked example: Sex ratio manipulation in zebra finches

In this section, we take an example from the literature to illustrate the advantages of logistic regression over traditional methods. Our aim is not to highlight the weaknesses of the published study (which are not atypical, Box 3.4), but rather to highlight the advantages of the GLM approach.

Our example comes from a study by Becky Kilner (1998) on sex ratio manipulation in zebra

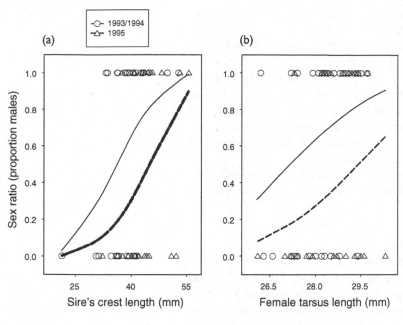

(a)

○- 1993/1994
△- 1995

(b)

Fig 3.4 Relationships between sex ratio and (a) male crest length and (b) female tarsus length in Crested Auklets. Some of the data used to create these plots are given in Table B3.7b. The curves are the fitted partial logistic regression lines.

Our analysis revolves around the data presented in Figure 3.5a. This shows the relationship between the order in which chicks hatch and the proportion of males hatching, for offspring of adult birds fed on either restricted or abundant food. Kilner's analyses (Box 3.8) suggested that sex ratio increased with hatching rank for chicks derived from well-fed pairs, but not for those from pairs given a restricted diet. This conclusion was largely based on a series of nonparametric analyses (outlined in Box 3.8) in which sex ratio was calculated for each of the six hatch ranks separately, resulting in $n = 6$ for each diet. This is despite the fact that the analysis is based on 23 pairs of birds, 42 broods and 162 eggs. This loss of information is particularly important because sample sizes vary considerably across hatch ranks. This point is illustrated in Figure 3.5b, where the size of each

finches. The background to this study and the original analysis are given in Box 3.8. Kilner predicted that sex ratios would be more female-biased when food was abundant and more male-biased when food was restricted. Further, since first-hatched chicks tend to attain heavier fledging weights, she reasoned that within broods females would tend to hatch earlier than males. Here, we address the question: how does diet affect the relationship between hatch order and sex ratio?

Box 3.8 | Example 3: Sex ratio manipulation in zebra finches

This example comes from a study by Kilner (1998) on sex ratio manipulation in the zebra finch (*Taeniopygia guttata*), a small, seed-eating passerine bird that lives throughout the arid and semi-arid zones of Australia and Indonesia. Zebra finches are nomadic and breed opportunistically when there is sufficient food available. There is evidence from other studies that females may manipulate clutch sex ratios in relation to mate quality and food abundance. In wild populations, secondary sex ratios tend to be female-biased when food is abundant, though trends are not consistent between years. In order to test this experimentally, Kilner manipulated the quantity of food available to captive breeding birds and monitored their subsequent primary and secondary sex ratios (here, we restrict our discussion to her analysis of primary sex ratios, i.e. the proportion of males in the brood at hatching). Kilner predicted that sex ratios would be more female-biased when food was abundant and more male-biased when food was restricted. Further, since first-hatched individuals tend to attain heavier fledging weights, she reasoned that within broods females would tend to hatch earlier than males.

Kilner reared 12 pairs of birds under two regimes of food availability. For their first clutch, all birds were reared on a food-restricted regime in which food was rationed via an electronic hopper. Ten of these pairs then produced a second brood of eggs, again on a restricted food regime. Then, for the third brood ($n = 9$ pairs), the hoppers were removed and food was supplied *ad libitum*. A second group of 11 birds was also established at this point, which laid their first batch of eggs under conditions of abundant food. This was to control for any variation in sex ratio due to the number of broods that a pair had previously reared.

Kilner conducted a number of analyses on these data, but here we concentrate on trying to address one question: how does diet affect the relationship between hatch order and sex ratio?

Original analysis

Kilner performed a series of separate tests designed to answer specific aspects of the main question. For example, using Mann–Whitney U-tests she showed that, across all 42 broods, sex ratio was significantly more male-biased when food was restricted than when it was abundant ($P < 0.01$). Using Friedman two-way ANOVAs, she showed that, within the nine pairs of birds for which she had three broods, sex ratios were significantly more male-biased when food was restricted than when it was abundant ($P < 0.05$). Using Wilcoxon signed-ranks tests she showed that, across all 22 'food-restricted' broods, female offspring hatched significantly earlier than male offspring ($P < 0.01$), and a similar relationship was also apparent across the 20 'food-abundant' broods ($P < 0.05$). Using Spearman rank correlations, she showed that within the food-restricted group, the proportion of males hatching increased significantly with increasing hatch order ($P < 0.05$) and a similar, but nonsignificant ($P < 0.1$), trend was apparent in the food-abundant group. Using Wilcoxon signed-ranks tests, she showed that across the six hatch order positions (1–6), the proportion of males hatching was significantly lower when food was abundant ($P < 0.05$). Finally, using the nine pairs for which she had data for three broods, she used Friedman's two-way ANOVA to show that the proportion of males hatching at each rank was significantly lower when food was restricted than when it was abundant ($P < 0.05$). The relationship between diet, hatch order and sex ratio is shown in Figure 3.6a.

While all these tests point to there being a genuine effect of diet and hatch rank on brood sex ratio, this analysis has a number of problems. First, it uses nonparametric tests, which tend to lack power and are susceptible to type II errors (i.e. incorrectly accepting the null hypothesis). Second, while some of the tests underutilize the data, by not weighting sex ratios by clutch or brood size (e.g. the Spearman rank correlations), others appear to be pseudo-replicated (e.g. the Mann–Whitney U-tests, where all of the data are lumped together with respect to diet regime, without taking account the identity of the pair). Third, at least six different tests are used, when one test could do the job; as illustrated in section 3.5.3. A problem associated with this approach is that the probability of generating type I errors increases and so Bonferonni corrections generally need to be applied (Rice 1990).

symbol is proportional to the size of the denominator. By plotting the data in this way, it becomes abundantly clear that the original analysis is likely to be unduly influenced by those data points that are based on small sample sizes (e.g. hatch ranks 5 and 6).

We can re-analyse these data using simple logistic regression in which the dependent variable is the proportion of hatchlings that are male and the explanatory variables are HatchOrder, Diet and their interaction (HatchOrder:Diet). The regression is weighted by clutch size (the denominator of the proportion), so making full use of the available data. Of course, as this is logistic

of hatch order on sex ratio, but that diet and the interaction between hatch order and diet is nonsignificant. However, to test this properly, we need to delete each term from the model in turn and determine whether it results in a significant decrease in the proportion of deviance explained (stepwise deletion tests). Since HatchOrder:Diet is the only interaction term in the model, and is nonsignificant, we can delete it and test the significance of the two main effects (but, clearly, if there was more than a single interaction term, we would perform stepwise deletion tests for each of the interaction terms as well). This produces the following results:

Resid. Df	Resid. Dev	ΔTerms	ΔDf	ΔDeviance	P (χ^2)	
10	8.5487	-Diet	−1	−2.1841	0.1394	ns
10	20.6345	-HatchOrder	−1	−14.2699	0.0002	***

regression, we initially assume a binomial error distribution and a logit-link function.

The results of this first step in the analysis is shown below:

The loss of Diet from the model results in the deviance explained decreasing by a small and nonsignificant amount ($\chi^2_1 = 2.18$, $P > 0.13$). However, the loss of HatchOrder from the model

Coefficients	Value	Std. Error	t value
(Intercept)	−0.7583	0.4835	−1.5682
HatchOrder	0.4357	0.1672	2.6055
Diet	−0.6223	0.7398	−0.8412
HatchOrder:Diet	0.0508	0.2591	0.1961

(Dispersion Parameter for Binomial family taken to be 1)

Null Deviance: 23.61876 on 11 degrees of freedom
Residual Deviance: 6.32614 on 8 degrees of freedom

Terms	Df	Deviance	Resid. Df	Resid. Dev	P(χ^2)
Null model	11	23.6182			
HatchOrder	1	15.0691	10	8.5487	0.0001
Diet	1	2.1841	9	6.3646	0.1394
HatchOrder:Diet	1	0.0385	8	6.3261	0.8444

Checking for overdispersion by calculating the heterogeneity factor (i.e. Resid.Dev/Resid.Df = 6.3261/8 = 0.7907) suggests that there is, in fact, underdispersion, and since it is only slight it is safe to proceed. The output appears to show that there is a significant effect

results in a highly significant decline in the amount of deviance explained ($\chi^2_1 = 14.27$, $P < 0.001$). So, the only significant explanatory variable is HatchOrder, and the analysis of deviance table for this model is shown below (note that when Diet is the only term in the model,

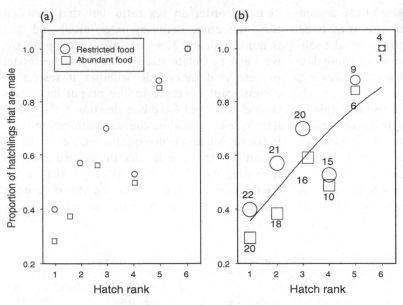

Fig 3.5 Sex ratio at hatching with respect to hatch order, for zebra finch broods reared with abundant (□) and restricted food (○). Data taken from 42 broods, after Kilner (1998, Box 3.8). In (a), the data are shown as they appeared in Kilner (1998); in (b) symbol size is proportional to sample size (which are indicated above or below each symbol) and the line is the fitted logistic regression line to all of the data.

it is marginally nonsignificant: $\chi^2_1 = 2.98$, $P = 0.084$):

Terms	Df	Deviance	Resid. Df	Resid. Dev	P (χ^2)
Null model	11	23.6187			
HatchOrder	1	15.0699	10	8.54877	0.0001 ***

And the summary output for the model (from *S-Plus*) is as follows:

```
Coefficients  Value    Std. Error   t value
(Intercept)   -1.0523  0.3628       -2.9001
HatchOrder     0.4647  0.1268        3.6632
```

(Dispersion Parameter for Binomial family taken to be 1)

```
     Null Deviance:  23.618 on 11 degrees of freedom
Residual Deviance:   8.548 on 10 degrees of freedom
```

This indicates that the intercept for the logistic regression line is significantly different from zero and that there is a significant positive relationship between hatching order and proportion of males in the brood. In other words, by using a weighted analysis of covariance in which we model sex ratio with binomial errors and a logit-link function, it appears that although females tend to hatch before males, this effect is independent of diet, which is nonsignificant.

Remember that the parameter estimates shown here are from the linear predictor (section 3.4.2.2), and so are on a logit scale (logit = $\ln(p/(1-p))$. To back-transform from logits (z) to proportions (p), we apply eq. 3.13 (i.e. $p = e^z/[1 + e^z]$). Thus, the predicted sex ratio for first-hatched eggs is $e^{(-1.052+1\times0.465)}/(1 + e^{(-1.052+1\times0.465)}) = 0.357$, for second-hatched eggs is $e^{(-1.052+2\times0.465)}/(1 + e^{(-1.052+2\times0.465)}) = 0.469$ and for sixth-hatched eggs is $e^{(-1.052+6\times0.465)}/(1 + e^{(-1.052+6\times0.465)}) = 0.850$. The logistic regression line derived from this analysis is shown in Figure 3.5b.

The model output also reminds us that this model assumes that the dispersion (scale) parameter is equal to 1 (i.e. on the logit scale, the variance is independent of the mean). We can re-estimate the scale parameter by dividing the residual deviance by the residual degrees of freedom $8.548/10 = 0.8548$. As this is tolerably close to unity, we need not worry greatly about underdispersion (section 3.4.4.4). One way that we could check the robustness of this result would be by constructing a quasi-likelihood model in which we assumed a logit link, and tested for significance using F-tests instead of chi-square tests (section 3.4.4.2). This produces similar results, with hatching order being the only significant term in the model ($F_{1,10} = 20.791$, $P = 0.0010$). However, diet was close to significance in this model ($F_{1,9} = 4.108$, $P = 0.0733$), suggesting that further experiments or analyses may be justified.

3.5.3.1 Further analyses
While our re-analysis makes better use of the available data and has greater power than those conducted by Kilner, it is not the best analysis possible. This is because our analysis weights all 162 offspring equally, regardless of their parents' identity or which brood they came from. Often, these two factors will lead to overdispersion, but the fact that our model is under- rather than overdispersed suggests that these effects are not biasing our results systematically. However, Kilner's experimental design allows us to test simultaneously for the independent effects of parentage or brood number on sex ratio. For this analysis, we need to employ generalized linear mixed models (GLMMs, section 3.4.4.2) in which these terms are included as random effects (see Krackow & Tkadlec 2001). When we conduct such an analysis (in *Genstat*, using the *irreml* procedure), and determine the significance of terms in the model using F-tests (see Elston 1998), we get results similar to those gained with the GLM: although it is very clear that the Diet:HatchOrder interaction is nonsignificant ($F_{1,14} = 0.07$, $P = 0.80$), and the HatchOrder main effect is significant ($F_{1,14} = 13.04$, $P = 0.0028$), the statistical significance of Diet is once again marginal ($F_{1,14} = 4.07$, $P = 0.063$). After controlling for hatching order, the predicted sex ratios from this model are 0.61 on the restricted diet and 0.45 on the *ad*

libitum diet, suggesting that rationing food leads to a 35% increase in the proportion of males in the brood. Thus, even though the effect of Diet was statistically nonsignificant, the magnitude of the apparent effect suggests that it would be premature to discount the effect of diet on sex ratio in zebra finches.

3.5.3.2 Conclusions
As with all GLMs, logistic regression allows the simultaneous testing of several interacting factors and covariates in a single model. Since the underlying error distribution of sex ratios is known (or presumed) to be binomial, this can be explicitly incorporated into the modelling process, so avoiding *ad hoc* transformations and nonparametric tests which lack power. By weighting sex ratios by their denominators, each individual contributing to the ratio is given equal significance. In contrast to Kilner's (1998) analyses, we found no statistical evidence that diet was a significant determinant of sex ratio in zebra finches (though nonsignificance was marginal). This conclusion was independent of whether hatching order was included or omitted as a covariate in the model. However, in accord with Kilner, we found that females tend to hatch before males. This result was independent of adult feeding regime and, because it utilized information from all of the chicks that hatched successfully, the robustness of our conclusion is illustrated by the high significance of the result ($P < 0.001$).

3.5.4 A case history: Opossum sex ratios
Austad and Sunquist (1986) carried out the first manipulative field test of the Trivers–Willard prediction (Chapter 13) that mothers in relatively good condition will produce more male-biased sex ratios than poor-condition mothers. Females of the common opossum, *Didelphis marsupialis* (a polygynous marsupial producing litters of 2–12 offspring), were given either diet supplements or no supplements (control) and the sexes of subsequent offspring were recorded. Austad and Sunquist analysed these data by comparing the overall sex ratio produced by females in the treatment group with that produced by the control group using a one-tailed binomial test. They found a significant difference ($P = 0.007$)

in sex ratio (but not in litter size) between the two groups.

Subsequently, Wright et al. (1995) correctly pointed out that comparing the overall sex ratios of the two groups was inappropriate since the hypothesis under test predicts an individual level response, not a population (treatment group) level response, to maternal condition. They reanalysed the data (as presented in Sunquist & Eisenberg 1993) using litters as the sampling unit. Litters were classified categorically as 'male-biased' or 'unbiased or female-biased'. Using a χ^2 test they found no significant difference between the proportions of male-biased litters produced by control and supplemented females (18/36 and 20/36 respectively; $\chi^2 = 0.068$, $P > 0.05$).

However, Wright et al.'s categorization of litter sex ratio does not use all of the available information since the actual composition of each litter (the size of the litter and the degree of any sex ratio bias) is overlooked. For example, litters containing six males plus one female are placed in the same 'male-biased' category as litters of four males plus three females, while the degree of male bias is different (see also Box 3.1). Similarly, litters of three males plus one female and litters of six males plus two females have the same sex ratio and are treated equally, despite the fact that larger litters give more trustworthy sex ratio estimates (section 3.3.3.4).

In an attempt to arrive at a more robust conclusion, the opossum data (as obtained from ME Sunquist) were explored using weighted logistic analyses (Hardy 1997). In a first analysis, litters were the sampling unit (D. marsupialis produces two cohorts of litters per season) and litters produced by the same mother were assumed to be statistically independent. No significant influence of cohort was found, so litters produced by the same mother were lumped and a second analysis was performed with mothers as the sampling unit: intuitively, this is more appropriate since the assumption of independent litter sex ratios does not have to be made, and it was mothers, not litters, that received the experimental treatments. Due to overdispersion, Williams' adjustment (appropriate when the binomial denominator varies, section 3.4.4.2) was employed and significance was evaluated with one-tailed t-tests.

Both analyses found that the sex ratio produced by food-supplemented mothers was significantly more male-biased than the sex ratios of control females' offspring (e.g. second analysis, $t = 1.973$, $df = 40$, $P = 0.028$; mean sex ratio of supplemented group = 0.577, control group = 0.488).

One-tailed tests were used because there was an anticipated direction for any difference between treatment groups (i.e. H_0: 'there is no sex ratio difference between the two groups'; H_1: 'the offspring sex ratios of supplemented females are more male-biased than those of control females'). However, not all statisticians agree that one-tailed tests can be used when deviations in the unanticipated direction are possible (Rice & Gaines 1994). Using two-tailed tests would have led to the acceptance of H_0, but would have been suspiciously close to significance at the 5% level (second analysis, $P = 0.0566$). See Hardy (1997) for further discussion, including the use of 'directed tests' (Rice & Gaines 1994) as an alternative intermediate to the extremes of one- and two-tailed testing.

Sven Krackow (pers. comm.) recently reanalysed the opossum data using both GLMs and generalized linear mixed modelling (GLMMs, section 3.4.4.2) which includes a random between-litter effect. For these data, the analysis leads to the same biological conclusion regardless of whether a GLMM or a corrected GLM is employed (Krackow opted for two-tailed testing and concluded lack of significance, $P > 0.061$ for both analyses), while employing uncorrected GLMs led to spurious significance ($P < 0.04$).

Regardless of the degree of statistical significance, the effect of diet supplements on sex ratio is not exceptionally large ('supplemented' litters contained 18% more males), suggesting that more data are probably required before we can reach satisfactorily firm conclusions. Problems in researching mammalian sex ratios are discussed in Chapter 13.

3.6 | Simulation studies

We have argued that GLMs (and their 'offspring') are usually the most appropriate analyses for sex ratio analyses. In this section, we challenge this argument using a series of simulations to ask two questions. First, under what circumstances are errors likely to be generated when analysing sex ratio data? Second, when does the method of analysis really matter?

We use simulated datasets generated using the *rbinom* procedure in *S-Plus* and compare three methods of analysis:

1. A nonparametric method (Wilcoxon Rank Sum Test, equivalent to Mann–Whitney *U*-test).

2. A transformation method (*t*-test on arcsine-squareroot transformed data).

3. A generalized linear model with binomial errors.

In each case, sex ratios are expressed as proportions. (For a comparable analysis for negative binomial data, see Wilson *et al.* 1996 and Wilson & Grenfell 1997.)

3.6.1 Simulation approach

Imagine we want to determine whether females in a population of burying beetles exhibit a clutch sex ratio response to some manipulated variable (e.g. change in day length). We could randomly assign the beetles to one of two treatment groups (increasing day length or decreasing day length) and record the sex ratios of the broods produced (see also section 3.5.4). What is the probability that we will incorrectly *reject* the null hypothesis of no difference between the treatment groups (type I error) or incorrectly *accept* the null hypothesis (type II errors)?

To address this question, we randomly generated two datasets, representing the two treatment groups. In each case, we produced a series of 'virtual' clutches of a given size and sex ratio drawn from the binomial distribution. We then used our three methods to test for a significant difference between the two groups, and repeated this process 1000 times. Thus, the probabilities of type I and type II errors are, respectively, equal to the proportion of simulations in which the analysis indicated a significant difference between the two groups when there wasn't one, or no significant difference when there was one. We performed these simulations for a number of different scenarios. For example, to examine the effect of clutch size on the probability of making errors, we allowed clutch size to vary between 1 and 20 eggs per female, and to determine the effect of sample size we varied the number of

broods included in the analysis between 10 and 50 per treatment group. Sex ratios were allowed to vary between 0 and 1.

3.6.2 Effect of clutch size, sample size and sex ratio

In these simulations, the probability of making a mistake was qualitatively similar for all three methods (K Wilson unpublished analyses). Therefore, in Figure 3.6 we show the results just for the GLM model. This figure comprises 12 graphs, representing the results of the combined effects of clutch size (1, 5, 10 and 20) and sample size ($n = 10$, 20 and 50). Each graph is divided into an 11 × 11 matrix and the axes of the matrix represent the mean sex ratio of each of the two treatment groups (varying between 0.0 and 1.0 in intervals of 0.1). Each cell of the matrix is colour-coded depending on the probability of making an error; the darker the cell, the higher the probability of making a mistake. Thus, white cells represent instances in which there is 0–10% average probability of making a mistake and deep-red cells indicate that the probability of making a mistake is 90–100%. Type I errors are indicated by the colour of cells on the leading diagonal of each matrix (bottom-left to top-right), and type II errors are indicated by the colour of the remaining cells.

Examining just the leading diagonals of each matrix (bottom-left to top-right), it is fairly clear that the probability of making a type I error (i.e. detecting a spurious difference between treatments) remains at less than 10%, regardless of clutch size, sample size or mean sex ratio. The biggest effects are seen in the probability of making a type II error (i.e. failing to detect significant differences between treatments). As expected, the probability of making a type II error is reduced when clutch sizes are large, sample sizes are large and the effect of our manipulation on sex ratio is large. In other words, we make fewer mistakes when the power of our test is high!

3.6.3 Differences between statistical methods

What about quantitative differences between the three methods? Simulations indicated that the nonparametric test and the *t*-test on arcsine-transformed data differ relatively little in their

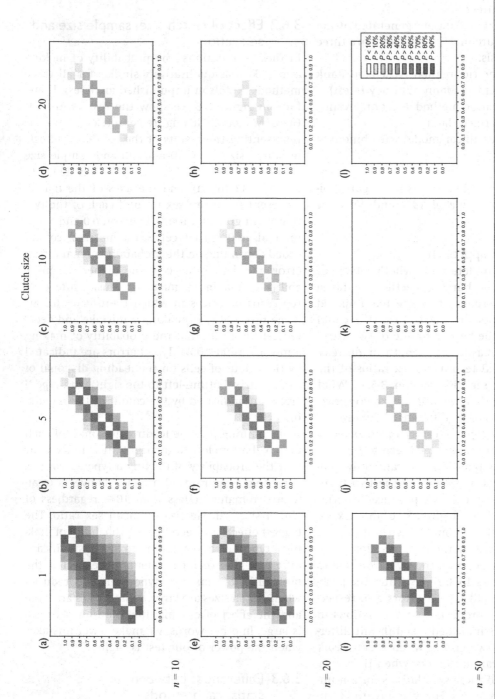

Fig 3.6 Probability of type I and type II errors as a function of difference in sex ratio, clutch size and sample size. Each plot (a – l) represents the results of a series of 1000 simulations examining the ability of a statistical model to discriminate between two datasets comprising 'virtual' clutches of a given size and sex ratio drawn from the binomial distribution. The vertical and horizontal axes on each plot are the mean sex ratios of each of the two clutches being compared (varying between 0 and 1 in steps of 0.1). Clutch size was allowed to equal 1, 5, 10 or 20; sample size equalled 10 clutches, 20 clutches or 50 clutches. The probability of making a mistake was qualitatively similar for all three methods (see text) and so here we show the results just for the GLM model. Each cell of the matrix is colour-coded depending on the probability of making an error; the darker the cell, the higher the probability of making a mistake (see legend attached to plot l). Type I errors are indicated by the intensity of colour in the cells of the leading diagonal of each matrix (bottom-left to top-right), and type II errors are indicated by the intensity of colour in the remaining cells.

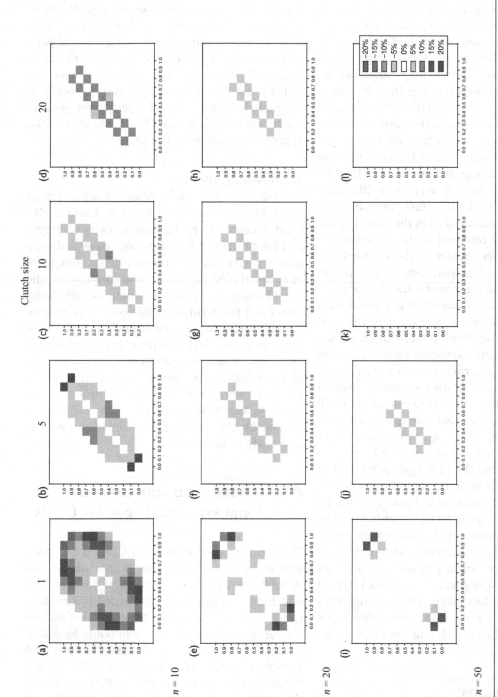

Fig 3.7 The difference between two statistical methods for analysing sex ratio data in their probability of generating type I and type II errors. For details of the simulations, see the legend to Figure 3.6 and the text. In this figure, quantitative differences between the nonparametric and GLM models are shown. Here, the different colours represent differences between the nonparametric method and the GLM in their probability of producing errors (see the legend attached to plot l). When there is little difference between the two methods (i.e. 0–5% difference in the number of errors) the cell is white; when the nonparametric method is better (i.e. produces fewer errors) the cell is coloured pink or red, and when the GLM is better the cell is coloured various shades of blue. In general, there is little difference between the two methods in their probability of producing type I errors (the leading diagonals tend to be white, except when clutch size is one and sample size is very small, i.e. 10), but the GLM method produces significantly fewer type II errors, except when sample sizes are large (outside the leading diagonal, there is much more blue than red).

probability of generating errors (K Wilson un-published analyses), whereas these two methods differ quite markedly from the GLM with binomial errors. This point is illustrated in Figure 3.7. Here, the different colours represent differences between the nonparametric method and the GLM in their probability of producing type I and type II errors. When there is little difference between the two methods (i.e. less than 5% difference in the number of errors) the cell is coloured white; when the nonparametric method is better (i.e. produces fewer errors) the cell is coloured pink or red, and when the GLM is better the cell is coloured various shades of blue. This figure suggests that there is very little difference between the two methods in their probability of generating type I errors (cells in the leading diagonal are generally coloured white). However, the two methods differ greatly in their probability of generating type II errors: genuine differences between treatments are much less likely to be detected using the nonparametric method than when using the GLM (i.e. as expected, the nonparametric method lacks power-efficiency; section 3.3.1.1). It is also apparent that the benefit of using the GLM approach is generally enhanced when clutch and sample sizes are small. Interestingly, it appears that when clutch sizes are small (≤ 5) the difference between the two methods is greatest when both mean sex ratios are close to 0 or 1, whereas when clutch sizes are large (≥ 10) the benefits of using the GLM approach are most evident when both sex ratios are close to 0.5.

Thus, these simulations indicate that sex ratio differences are likely to be difficult to detect in species with small clutch sizes, except when sample sizes are large. Moreover, although type I errors appear to be unlikely in sex ratio analyses regardless of the analytical method used, type II errors are much less likely when using logistic regression than when using alternative methods, especially when clutch and sample sizes are small.

3.7 | Conclusions

The most appropriate approach for analysing sex ratios (and other proportion data) will often be lo-gistic regression (GLM with binomial errors and logit-link function). After all, sex ratios are expressed as proportions and logistic GLMs were developed to analyse proportion data. We hope to have shown that using GLMs is not very much (if at all) more complex than using classical parametric methods (which are currently the most frequently used). The next time you have collected a set of sex ratio data and are ready to begin analysis, ask yourself whether you want to make best use of the data. If you do, your initial approach should be to use GLMs.

Acknowledgements
We thank Tim Benton, Mick Crawley, Simon Gates, Fiona Hunter, Sven Krackow, Kate Lessells, Evert Meelis, Gösta Nachman, Gordon Smyth and Ian Stevenson for help, discussion and comments; any mistakes or misunderstandings are ours not theirs! We thank Darren Shaw for the use of some of his *S-Plus* functions and Becky Kilner and Fiona Hunter (and co-workers) for providing us with some of their raw data. We thank both them and Mel Sunquist, Pat Weatherhead and their co-workers, for tolerating the intrusion of re-analysis. This chapter was written whilst K Wilson was funded by the Natural Environment Research Council, UK. ICW Hardy was funded by the European Commission and the University of Sunderland.

A3.1 | Reference sources and statistical packages for GLMs

The most accessible book on GLMs (i.e. a book written by a biologist rather than a statistician) is probably Crawley's (1993) *GLIM for Ecologists*, which has recently been superseded by Crawley's (2002) *Statistical Computing*. More detailed statistical background can be found in books by Aitkin *et al.* (1989), Cox and Snell (1989), Hosmer and Lemeshow (1989), McCullagh and Nelder (1989), Agresti (1990), Dobson (1990), Collett (1991) and Menard (1995). In addition to these general reference sources, biologist-friendly descriptions of logistic analysis are provided by Shanubhogue and Gore (1987), Trexler and Travis (1993), Sokal and

Rohlf (1995) and Hardy and Field (1998). Some recent examples where GLMs have been used to analyse complex sex ratio datasets are discussed in Hartley *et al.* (1999). Wilson *et al.* (1996) and Wilson and Grenfell (1997) give accounts of GLMs with particular reference to analysing parasite count data.

Logistic analysis is available in at least the following packages: *BIOM-pc, BMDP, EGRET, Genstat, GLIM, GLIMStat, JMP, LOGXACT, MacAnova, SAS, S-Plus, SPSS, SPSSX, STATA, STATISTIX* and *SYSTAT*. Agresti (1990) provides an appendix detailing the options available in various packages and advice on their implementation. The manuals for some of these packages are also excellent reference resources (e.g. SAS Institute Inc. 1995, SPSS 1999). While we de not recommend the *GLIM* manual (Francis *et al.* 1993) for the nonprofessional, the *GLIM* package itself becomes much more user-friendly if you have a copy of Crawley (1993). The following website provides updated information about the most frequently used statistical packages:
http://www.maths.uq.edu.au/~gks/webguide/statcomp.html.

References

Agresti A (1990) *Categorical Data Analysis.* New York: Wiley.

Aitkin M, Anderson D, Francis B & Hinde J (1989) *Statistical Modelling in GLIM.* Oxford: Clarendon Press.

Austad SN & Sunquist ME (1986) Sex-ratio manipulation in the common opossum. *Nature,* **324,** 58–60.

Cohen J (1988) *Statistical Power Analysis for the Behavioural Sciences,* 2nd edn. London: Lawrence Erlbaum Associates.

Collett D (1991) *Modelling Binary Data.* New York: Chapman & Hall.

Cox DR & Snell EJ (1989) *Analysis of Binary Data,* 2nd edn. New York: Chapman & Hall.

Crawley MJ (1993) *GLIM for Ecologists.* Oxford: Blackwell Scientific Publishing.

Crawley MJ (2002) *Statistical Computing.* London: John Wiley.

Dobson AJ (1990) *An Introduction to Generalized Linear Models,* New York: Chapman & Hall.

Elston DA (1998) Estimation of denominator degrees of freedom of *F* distributions for assessing Wald statistics for fixed effect factors in unbalanced mixed models. *Biometrics,* **54,** 1085–1096.

Finney DJ (1971) *Probit Analysis.* Cambridge: Cambridge University Press.

Fisher RA (1954) *Statistical Methods for Research Workers,* 12th edn. Edinburgh: Oliver & Boyd.

Flanagan KE, West SA & Godfray HCJ (1998) Local mate competition, variable fecundity and information use in a parasitoid. *Animal Behaviour,* **56,** 191–198.

Francis B, Green M & Payne C (eds) (1993) *The GLIM System: Release 4 Manual.* Oxford: Oxford University Press.

Hardy ICW (1997) Opossum sex ratios revisited: significant or nonsignificant? *American Naturalist,* **150,** 420–424.

Hardy ICW & Field SA (1998) Logistic analysis of animal contests. *Animal Behaviour,* **56,** 787–792.

Hardy ICW & Mayhew PJ (1998) Sex ratio, sexual dimorphism and mating structure in bethylid wasps. *Behavioral Ecology and Sociobiology,* **42,** 383–395.

Hartley IR, Griffith SC, Wilson K, Shepherd M & Burke T (1999) Nestling sex ratios in the polygynously breeding Corn Bunting *Miliaria calandra. Journal of Avian Biology* **30,** 7–14.

Hosmer DW & Lemeshow S (1989) *Applied Logistic Regression.* New York: Wiley.

Hunter FM, Jones IL, Wilson K, Dawson DA & Burke TA (in prep.) Manipulation of offspring sex in species with mutual sexual selection.

Kilner R (1998) Primary and secondary sex ratio manipulation by zebra finches. *Animal Behaviour,* **56,** 155–164.

Krackow S & Tkadlec E (2001) Analysis of brood sex ratios: implications of offspring clustering. *Behavioral Ecology and Sociobiology,* **50,** 293–301.

Kruuk LEB, Clutton-Brock TH, Albon SD, Pemberton JM & Guinness FE (1999) Population density affects sex ratio variation in red deer. *Nature* **399,** 459–461.

Leonard ML & Weatherhead PJ (1996) Dominance rank and offspring sex ratios in domestic fowl. *Animal Behaviour,* **51,** 725–731.

Leonard ML & Weatherhead PJ (1998) Erratum. *Animal Behaviour,* **55,** 777.

Lipsey MW (1990) *Design Sensitivity: Statistical Power for Experimental Research.* London: Sage Publications.

Mathsoft Inc (1999) *S-Plus Guide to Statistics,* volume 1. Seattle: Mathsoft. Inc.

McCallum H (2000) *Population Parameters: Estimation for Ecological Models.* Oxford: Blackwell Science.

McCullagh P & Nelder JA (1989) *Generalized Linear Models*, 2nd edn. New York: Chapman & Hall.

Meddis R (1984) *Statistics Using Ranks: A Unified Approach*. Oxford: Blackwell.

Menard S (1995) *Applied Logistic Regression Analysis*. Sage University paper series on quantitative applications in the social sciences, 07–106. Thousand Oaks: Sage.

Neave HR & Worthington PL (1988) *Distribution-free Tests*. New York: Routledge.

Petersen G & Hardy ICW (1996) The importance of being larger: parasitoid intruder-owner contests and their implications for clutch size. *Animal Behaviour*, **51**, 1363–1373.

Rice WR (1990) A consensus combined *P*-value test and the family-wide significance of component tests. *Biometrics*, **46**, 303–308.

Rice WR & Gaines SD (1994) 'Heads I win, tails you use': testing directional alternative hypotheses in ecological and evolutionary research. *Trends in Ecology and Evolution*, **9**, 235–237.

Rohlf FJ & Sokal RR (1995) *Statistical Tables*, 3rd edn. New York: WH Freeman & Co.

SAS Institute Inc (1995) *Logistic Regression Examples Using the SAS System*, Version 6, 1st edn. Cary, NC: SAS.

Shanubhogue A & Gore P (1987) Using logistic regression in ecology. *Current Science*, **20**, 933–935.

Siegel S & Castellan NJ (1988) *Nonparametric Statistics for the Behavioural Sciences*, 2nd edn. New York: McGraw-Hill.

Sokal RR & Rohlf FJ (1995) *Biometry*, 3rd edn. New York: WH Freeman & Co.

SPSS (1999) *SPSS Base 9.0 User's Guide*. Chicago: SPSS Inc.

Sunquist ME & Eisenberg JF (1993) Reproductive strategies of female *Didelphus*. *Bulletin of the Florida Museum of Natural History, Biological Sciences*, **36**, 109–140.

Trexler JC & Travis J (1993) Nontraditional regression analyses. *Ecology*, **74**, 1629–1637.

Westerdahl H, Bensch S, Hansson B, Hasselquist D & VonSchantz T (1997) Sex ratio variation among broods of great reed warblers *Acrocephalus arundinaceus*. *Molecular Ecology*, **6**, 543–548.

Williams DA (1982) Extra-binomial variation in logistic linear models. *Applied Statistics*, **31**, 144–148.

Wilson K & Grenfell BT (1997) Generalized linear modelling for parasitologists. *Parasitology Today*, **13**, 33–38.

Wilson K, Grenfell BT & Shaw DJ (1996) Analysis of aggregated parasite distributions: a comparison of methods. *Functional Ecology*, **10**, 592–601.

Wright DD, Ryser JJ & Kiltie RA (1995) First-cohort advantage hypothesis: a new twist on facultative sex ratio adjustment. *American Naturalist*, **145**, 133–145.

Zar JH (1999) *Biostatistical Analysis*, 4th edn. New Jersey: Prentice Hall Inc.

Chapter 4

Analysis of sex ratios in social insects

Jacobus J. Boomsma & Gösta Nachman

4.1 | Summary

Studies of social insects' sex ratios have so far only used statistics analysing (arcsine-transformed) proportions or (log-transformed) ratios. This approach is inaccurate when clutch sizes are highly variable and residuals are overdispersed, as is the case when sex ratios are split. Here we outline a protocol based on logistic regression that simultaneously tests for:

1. Deviations of the total (population-wide) sex ratio from a theoretically predicted equilibrium value (e.g. 1:1 or 3:1).

2. Direct or partial correlations of colony-level sex ratio with discrete or continuous predictor variables or co-variables.

3. The degree of over- or underdispersion of the variance in colony sex ratios compared to a binomial null model, both before analysis and after the inclusion of significant predictor variables.

The tests explicitly take variation in clutch size into account. We use a previously published dataset on sex allocation in the ant *Lasius niger* as a worked example, using *SAS* as the statistical software tool, and evaluate the validity of previous statistical analyses on proportional sex ratio data with the outcome of our current analyses. The present protocol allows evolutionary hypotheses on sex allocation to be tested in the absence of accurate estimates of the female-to-male cost ratio, provided the sample size is large enough.

4.2 | Introduction

Formal analysis of sex ratios in social insects started with the seminal paper by Trivers and Hare (1976). The core of their argument was that sex ratios in social Hymenoptera with worker control were expected to be female-biased because workers are more closely related to their sisters than to their brothers (Hamilton 1964). It is worth stressing that the sex ratio in social Hymenoptera always refers to the new reproductives (gynes and males) produced in a colony, and not any new cohorts of sterile workers (all female), which are produced either in separate clutches or alongside the reproductive brood. The general concept of worker control and worker-queen conflict over sex allocation has received ample support in later studies on social Hymenoptera (see Bourke & Franks 1995 and Crozier & Pamilo 1996 for recent reviews), but the statistical analysis of social insect sex ratios has been problematic from the very start.

In this chapter, we first give a brief historical account of the various problems that have plagued the empirical analysis of social insect sex ratios (section 4.3). We then show that simple logistic regression models that can alleviate similar problems encountered in the analysis of sex ratios in parasitoid wasps, mites and diplodiploid organisms cannot be satisfactorily applied to social insects where both colony sex ratios and clutch size vary considerably, and we formulate explicit criteria for appropriate analysis of sex ratio data in social insects (section 4.4). We then introduce a general method, based on existing procedures in *SAS*, to solve these problems and apply this procedure to an existing dataset of the ant *Lasius niger* (L.) (section 4.5). We provide a worked example of all steps in the analysis and consider various pitfalls, model assumptions and the extent to which previous analyses have led to (in)correct or (in)accurate conclusions (sections 4.6 to 4.9).

4.3 | A brief history of problems in analysis of sex ratios in ants and other social insects

In a first critique, Alexander and Sherman (1977) argued that Trivers and Hare's regression plot (their Figure 4) of the male female numerical sex ratio versus the individual female-male weight ratio was biased, because ratios have the intrinsic property of being highly asymmetrical around unity. A later review by Nonacs (1986) using proportions (males/(females + males)) instead of ratios reproduced the basic Trivers and Hare result, but was criticized because of potential bias in the estimation of the relative costs of producing an average female and male (the cost ratio) and unrecognized effects of colony size (Boomsma 1989). It was further shown that, when analysing the mean sex allocation of populations or species (the total number of females divided by the total number of males from the k colonies in the sample), log-transformed ratios were approximately normally distributed and more straightforward to plot and interpret than the alternative of arcsine transformed proportions.

While the most appropriate method of analysing population-wide sex allocation was being discussed, research emphasis was shifting towards the consideration of colony-level variation in sex ratios. Some studies had shown that sex ratios could differ between conspecific populations and that they were not always normally distributed (e.g. Brian 1979, Boomsma *et al.* 1982), but not until the review by Nonacs (1986) was it recognized that bimodality in colony-level sex ratios was the norm. Three mutually nonexclu-

sive theoretical explanations for these split sex ratios (a term introduced by Grafen 1986) have been proposed, which can be referred to as the resource hypothesis (Nonacs 1986), the constant male hypothesis (Frank 1987) and the variable relatedness asymmetry hypothesis (Boomsma & Grafen 1990, 1991).

The first two of these hypotheses singled out clutch size (often referred to as total colony production) as the decisive variable. It was argued that colonies with small reproductive clutches should specialize in the cheaper sex in terms of body size (usually males, Nonacs 1986) or on the sex whose reproductive success would be most affected by competition with relatives in the large clutches of other colonies (usually males, Frank 1987). The third hypothesis predicted that, after the possible effect of clutch size differences was taken into account, colonies with a relatively low asymmetry in female-to-male relatedness should specialize in males, whereas colonies with a relatively high relatedness asymmetry should specialize in females (Boomsma & Grafen 1990, 1991). Tests of these alternative hypotheses have often involved analysis of the proportional allocation to female or male offspring and its (partial) regression with relatedness asymmetry and total colony production (clutch size) (see Bourke & Franks 1995 and Crozier & Pamilo 1996 for reviews).

Regression analyses with proportions as dependent variables are problematic for a number of reasons. First, clutch size determines the sample size available for the estimation of sex ratio. No researcher would maintain that a clutch of only two males is evidence of a male-biased sex ratio, whereas a clutch of 200 males without any females would justify such a conclusion with very high likelihood. Nonetheless, both kinds of clutches enter an analysis of proportional sex allocation with the same value (1 or 0, depending on whether one uses males or females in the numerator). A partial solution that has been applied to ant sex ratios is to weight the data in proportion to clutch size or total colony production (JJ Boomsma, pers. comm. in Box 5.1 in Bourke & Franks 1995). A more general solution is to use logistic regression (Chapter 3), but this technique has so far not been applied to social insect sex ratios, in contrast to recent studies on other taxa (Chapter 3). The reason may be that logistic regression of overdispersed sex ratios is not straightforward when variation in clutch size (total colony production) is high, as is often the case in social insects. The present chapter explores practical approaches to logistic regression techniques in the analysis of social insect sex ratios. More general theoretical accounts are provided in Chapters 3 and 5.

4.4 | Criteria for appropriate analysis of social insects' sex ratios

There are three different types of question that evolutionary biologists studying sex allocation in social insects would normally like to answer simultaneously. All three are theory driven:

1. Is the average population sex ratio significantly different from a predicted equilibrium value such as 1:1 (50% allocation to females) or 3:1 (75% allocation to females)? The method (outlined in Box 5.1 of Bourke & Franks 1995) of weighting colony sex ratios by their proportional contribution to the mating swarm produces standard errors (s.e.'s) of proportional data when estimating the mating swarm sex ratio. However, using these s.e.'s for parametric tests against single predicted values may be inaccurate when the assumption of normality is violated. As the underlying distribution of the data may not be known *a priori*, a statistical approach that can better handle this problem, while still taking natural clutch size variation into account, is thus to be preferred.

2. Is colony-level sex allocation significantly related to one or several context-specific variables that are predicted to bias the sex ratio? Studies of sex allocation in ants have shown correlations between the colony-level sex ratio (expressed as proportions) and total sexual production or resource-related factors (*e.g.* Boomsma *et al.* 1982, Boomsma 1988, Deslippe & Savolainen 1995, Sundström 1995). The question is whether these correlations hold when effects of varying clutch size and overdispersion have been taken into

account. A proper statistical analysis would answer this question. Chapter 5 provides such general analysis. Here we focus on practical solutions applicable to field studies where we are particularly interested in revealing partial effects when it is likely that the sex ratio is simultaneously affected by several independent variables.

3. Is colony-level variation in sex allocation ratio binomially distributed before and after the inclusion of explanatory variables in the analysis? Or is there underdispersion (i.e. less variation among colonies than expected by chance (a situation expected under strong local mate competition, Hamilton 1967)), or overdispersion (i.e. more variation among colonies than expected by chance)? The latter is expected under various forms of context-dependent sex allocation and produces partially or completely split sex ratios. Overdispersion may be partially explained by independent variables that enter the analysis, but are they sufficient to cause the residuals to be binomially distributed? If they are, this would suggest that there is little variation left to be explained by unmeasured variables.

Overdispersion of sex ratio is a complicated problem, as both known and unknown (or unmeasured) variables may affect the sex allocation ratio. The effects of proximate variables, such as temporary variation in food and weather, are especially difficult to quantify and may prevent the residuals of sex allocation across colonies from being binomially distributed. A proper statistical protocol should therefore quantify the extent to which residuals are overdispersed, both before and after significant covariables have explained their part of the sex ratio variance. Overdispersion before analysis should be explained by measured predictor variables and overdispersion after analysis should be quantified with a measure that indicates how much of the original variation remains to be explained by additional variables.

In section 4.5, we present a statistical protocol that meets these three desirable criteria for the analysis of empirical (usually field) data on sex allocation in social Hymenoptera. It is based on generalized linear models and in particular on multiple logistic regression (e.g.

McCullagh & Nelder 1989, Hosmer & Lemeshow 1989), allows the inclusion of any number of discrete or continuous independent variables and produces statistical tests for the deviation of observed values from predicted values and of the dispersion of residuals. None of the methods that we use is new either with respect to the underlying theory (e.g. Crawley 1993, Hardy & Field 1998) or in the sense that they require special statistical software. Thus, we use a standard statistical procedure available in *SAS* (version 6.12 SAS Institute Inc.), but other statistical software packages may be suitable as well (see Chapter 3). Our protocol aims at identifying those independent variables that contribute most to explain the observed sex ratios. If two or more independent variables are correlated, each of them may have an explanatory power, but in combination they may not be able to improve the predictions significantly. In such cases, it seems most appropriate to focus on the variables that are most likely to have a direct causal influence on the dependent variables, but such choices are based on *a priori* biological knowledge and are not a result of the statistical method *per se*. To identify causal relationships, path analysis would probably be a more sophisticated approach (see Herbers 1990 for pioneering work using path analysis of sex ratios in social insects), but the problem is that path analysis as it has been applied so far (e.g. Herbers 1990, Backus 1995) does not adjust for variation in clutch size in a logistic manner.

4.5 | The statistical protocol

4.5.1 The general model

Since the observed sex ratio, here defined as the number of females divided by the total number of males and females, of a population or any group of individuals reflects the underlying probability that a randomly chosen individual is either a female or male, the variable under study is likely to be binomially distributed with probabilities $P(\text{female}) = \pi$ and $P(\text{male}) = 1 - \pi$. The most obvious statistical method for analysing this kind of data is logistic regression, which assumes that the dependent variable (π) is a probability that can take real values in the range between 0 and 1 (Chapter 3). π may be in-

fluenced by a number of external (independent) variables, and the purpose of a statistical analysis is to identify the most important of these. The Procedure PROC GENMOD in *SAS* (version 6.12 SAS Institute Inc.) provides a powerful tool for such analyses. In this section we describe how we have used this procedure to fit a multiple logistic regression model to empirical data in order to identify possible factors influencing the mean and variance of sex ratio in field studies of sex allocation in social Hymenoptera. It should, however, be emphasized that the statistical tests used in these analyses (χ^2 and F) basically require the dependent variable (in our case the number of males or females sampled per colony) to be continuous and normally distributed. This requirement might be critical if the clutch size (n) is small and/or the proportion (π) of females (or males) in some of the clutches deviates substantially from 0.5. As a rule of thumb, the binomial distribution with parameters (n, π) is approximately normal with parameters ($n\pi, n\pi(1 - \pi)$) if the variance $\sigma^2 = n\pi(1 - \pi) > 9$, but hard criteria to accept or reject the assumption of normally distributed data are not available. As discussed in Chapter 5, the problem of small clutch sizes (n) diminishes with an increase in the number of colonies (N), but even if $N \leq 5$ it may not be of real concern because the statistical tests seemingly become increasingly conservative (Huck *et al.* 1990). In general, however, we recommend that the empirical variances are inspected (using the observed π in the variance formula obtained from each of the N colonies) and that alternative methods (see Chapter 5) are considered if the majority of colonies fail to meet the normal distribution criterion of $s^2 > 9$.

The general model that predicts the sex of an offspring from information about a set of p independent variables, $x_1, x_2, \ldots, x_p$ is

$$\pi = \frac{e^{\beta_0 + \beta_1 x_1 + \beta_2 x_2 + \ldots + \beta_p x_p}}{1 + e^{\beta_0 + \beta_1 x_1 + \beta_2 x_2 + \ldots + \beta_p x_p}} + \varepsilon \quad (4.1)$$

where π is the probability that the offspring is a female. β_j denotes the parameter expressing the relative contribution of the jth independent variable (x_j) to π, while ε is the error term. ε will be binomially distributed with a mean of zero and with variance $\pi(1 - \pi)$ provided that, for a given set of independent variables $x_1, x_2, \ldots, x_p$, each

Table 4.1 | The input file arranged in a format that can be read by the *SAS* program shown in Box 4.1

Colony	Females	Males	Workers	SexProd	Habitat	RA	WorkSize
6	143	655	4711	1571.63	SV	0.736	0.943
14	153	203	20147	1183.73	SV	0.736	1.005
36	603	1282	43437	5147.23	SV	0.736	0.985
40	177	909	9541	2043.57	SV	0.736	0.985
59	93	333	2302	929.13	SV	0.736	1.033
63	360	48	12642	2355.6	SV	0.736	0.950
70	65	93	4995	509.65	SV	0.736	0.993
116	16	13	8537	115.56	SV	0.736	0.978
1	21	136	5038	270.61	KBD	0.603	0.945
4	123	1334	10219	2122.43	KBD	0.603	0.935
5	53	317	5116	656.73	KBD	0.603	0.910
7	129	1358	31402	2184.89	KBD	0.603	0.938
9	64	410	8516	820.24	KBD	0.603	0.903
14	81	594	7014	1113.21	KBD	0.603	0.953
15	63	947	7175	1350.83	KBD	0.603	0.923
21	159	644	4091	1663.19	KBD	0.603	0.883
24	138	2352	37736	3236.58	KBD	0.603	0.890
2	404	299	16159	2888.64	KID	0.696	0.950
5	71	498	6098	953.11	KID	0.696	0.983
6	97	157	5577	778.77	KID	0.696	0.920
12	92	552	10544	1141.72	KID	0.696	0.908
18	29	481	7981	666.89	KID	0.696	0.915
20	9	495	8797	552.69	KID	0.696	0.888
23	86	371	22866	922.26	KID	0.696	0.920

Data are a subset of a larger dataset for the ant species *Lasius niger* (Boomsma *et al.* 1982, Van der Have *et al.* 1988). Each column heading refers to a variable name in the program: 'Colony' = colony number, 'Females' = number of females, 'Males' = number of males, 'Workers' = number of workers, 'SexProd' = sexual production (the sum of the females and males produced, multiplied by their respective costs), 'Habitat' = habitat, 'RA' = relatedness asymmetry, 'WorkSize' = size (mean headwidth) of workers. See text for details.

observation is an independent realization of the dichotomous stochastic variable (female, male) with constant probabilities π (female) and $1 - \pi$ (male). If these conditions do not hold, ε will be either under- or overdispersed. It is important to realize that this condition of independence is not automatically fulfilled in social insects and that its violation may significantly reduce the validity of the analysis, unless an explicit correction for over- or underdispersion is incorporated. The biological mechanisms responsible for over- and underdispersion are briefly discussed in section 4.5.2. The independent variables can be either quantitative (e.g. the number of workers per colony) or qualitative (e.g. habitat type). The purpose of the statistical analysis is to estimate the parameters (β_j) of the model and to decide which factors should be included to provide the best fit of the model to the empirical data.

To illustrate the various steps in this process we use a dataset on the ant *Lasius niger*. The dataset comprises 24 colonies distributed over three habitats on an island in The Netherlands: Strandvlakte (SV), Kobbeduinen (KBD) and Kooiduinen (KID). It is a subset of a larger dataset (Boomsma *et al.* 1982, Van der Have *et al.* 1988), differing in that, for simplicity, we have omitted colonies with missing values for relatedness or worker number. The input file is given in Table 4.1, where rows contain information about the separate colonies and where columns give the respective variables. Note that the name of a variable must not exceed eight letters or digits to be accepted by *SAS*. The procedure (PROC GENMOD) can handle much larger datasets as well as missing values (the missing values have to be replaced by a period (.) in the input file). The dataset contains two dependent variables 'Females' and 'Males'. These can be combined into a single dependent variable, namely the proportion of females π = females/progeny, where progeny = females + males. The logistic regression method takes the number of progeny used to estimate π into consideration, so that large colonies contribute more to the test statistics than small colonies (Chapter 3).

The program used to analyse the data is given in Box 4.1. The central part is the MODEL statement in PROC GENMOD. Here the user can specify the model to be analysed. The independent (or explanatory) variables ('Workers', 'SexProd', 'Habitat', 'WorkSize') are listed at the right side of the = symbol. In principle, the model can contain the main effects of all the qualitative and quantitative variables, the polynomial terms of the quantitative variables (e.g. x^2, x^3, etc.) plus two-way, three-way and higher-order interaction terms of both quantitative and qualitative factors (see e.g. *SAS/STAT User's Guide* volume 2, pp 895–897 for the general notation; SAS 1994). To simplify the analyses and presentation, we have chosen to define the full model as the one that includes only the primary variables and no interactions. A need to include such additional terms may arise when there is specific biological information suggesting that effects of specific variables are nonlinear or that explanatory variables are nonadditive.

Box 4.1 | The SAS program for analysing the sex ratio data given in Table 4.1

Information compiled by *SAS* is shown in bold, while comments are enclosed by /* and */. A file with the program is available upon request at gnachman@zi.ku.dk.

```
/* An example of a SAS program to analyze sex ratio data */
/* The program uses data from Lasius niger as an example */
/* Words written in capital letters are commands to SAS */
/* Words written in lower case letters are specific to the data set and can be
chosen differently by the user */
```

DATA sexratio;

OPTIONS LINESIZE=72;
```
/* specifies length of lines in output */
```

INFILE 'h:\sexratio\data2.cln' FIRSTOBS = 2;
```
/* start reading data from line 2 in the indata file */
```

INPUT Colony $ Females Males Workers SexProd Habitat $ RA WorkSize;
```
/* Specifies order of variables */
/* Variables followed by $ are categorical (qualitative) */
```

Progeny = Males + Females;
```
/* Add total number of progeny produced to data set */
```

```
IF Progeny = 0 THEN DELETE;
/* delete observations where progeny = 0 from dataset */


Variance = Females*(1−Females/Progeny);
/* compute empirical variance */


/* Print empirical variances */


PROC PRINT;
VAR Variance;
RUN;


PROC GENMOD;
TITLE 'Data Set 2: L. niger';
CLASS Colony Habitat; /* Colony and Habitat are qualitative variables */
MODEL Females/Progeny = Habitat Workers SexProd WorkSize
/ DIST=binomial TYPE1 TYPE3 DSCALE OBSTATS;


/* The MODEL statement defines the variables to be included in the model.
If no variables are included at the right side of the equation,
only the intercept will be estimated.
Note that the estimated parameters relate to the logit model */


/* The options after / specify details of the model and the requested output: */


/* DIST=binomial specifies that the residuals are expected to be binomially
distributed */
/* TYPE1 specifies that type 1 Mean Squares are computed */
/* TYPE3 specifies that type 3 Mean Squares are used as test statistics */
/* DSCALE adjusts for overdispersion in the data */
/* OBSTATS tells SAS to print additional information as e.g. the predicted propor-
tions and lower and upper confidence limits. Default is 95% limits, but other limits
can be requested, e. g. 99% limits by adding ALPHA=0.01 */


/* Note that RA is omitted from the model because RA is computed as a habitat
property and therefore is completely correlated with habitat. This would not have
been the case if RA had been computed on a colony basis */


/* linear contrasts test for differences between habitats */
CONTRAST 'KBD vs KID' Habitat 1 −1 0;
/* Difference between habitat KBD and KID */
CONTRAST 'KBD vs SV' Habitat 1 0 −1;
/* Difference between habitat KBD and SV */
CONTRAST 'KID vs SV' Habitat 0 1 −1;
/* Difference between habitat KID and SV */


RUN;
```

Usually the full model includes all independent variables in the input data but in the present example, 'RA' is omitted from the model because this variable does not add any extra information to the dependent variable that is not already contained in the variable 'Habitat'. This is because the allozyme data (Van der Have et al. 1988) only allowed reasonably accurate estimates of the habitat-specific average relatedness asymmetry (RA), implying that 'RA' and 'Habitat' are perfectly correlated so that the first can be used as a substitute for the second or *vice versa*. The only difference is that 'RA', in contrast to the qualitative variable 'Habitat', can be treated as a quantitative variable. If possible, it may often be an advantage to replace a qualitative variable with a quantitative measure, because it increases the generality of the model and reduces the number of parameters in the model, especially if the qualitative variable has many levels. Since 'Habitat' has three levels (SV, KBD, KID), it costs two degrees of freedom (df = levels − 1) to include the variable in the model. These two df can instead be used to include a first- and/or second-order polynomial term of RA (i.e. RA and/or RA^2) in the model (but not a third-order term because this would have required four levels of 'Habitat'). If both the first- and the second-order polynomial term of RA are included in the model, it becomes identical to the 'Habitat' model with respect to how much of the variation in data it can explain, but use of the quantitative variable RA may still be preferred if the purpose is to *predict* the sex ratios in new habitats from which RA has been assessed. Previous publications have analysed both the effect of 'Habitat' (Boomsma et al. 1982) and 'RA' (Van der Have et al. 1988), and here we start with 'Habitat'. The results of the analysis with 'RA' instead of 'Habitat' are given at the end of this section, further illustrating the formal equivalence of the two analyses.

The MODEL statement contains a number of options after the slash sign. DIST = binomial specifies that the residuals are expected to be binomially distributed. TYPE1 and TYPE3 specify that both types of sum of squares (SS) are requested. In contrast to type I SS, type III SS values are independent of the order in which the variables are included in the model (Littell et al. 1991). Type III SS are used as a criterion for deciding whether variables should be included or excluded from the model, because it measures how much a given variable contributes to the explained variation *after* all other variables have already been entered in the model. Neither type II nor type IV SS is available in PROC GENMOD. DSCALE tells *SAS* to compute the 'scale parameter' (also known as the 'heterogeneity factor'). The scaling parameter (s) − which is estimated as the square root of the deviance (D) divided by the degrees of freedom (v), i.e. $s = \sqrt{D/v}$ − is close to unity if data are binomially distributed, larger (>1) if data are overdispersed, and smaller (<1) if data are underdispersed relative to a binomial distribution (see also McCullagh & Nelder 1989). Since overdispersion occurs in most datasets, the DSCALE option is recommended to reduce the risk of committing statistical type I errors, i.e. rejecting a correct null hypothesis. Without scaling, the standard errors of the model's parameters become s times larger and the associated values of χ^2 will be s^2 times smaller than without scaling. It can be shown that if the clutches in all colonies are increased with a factor k, the unscaled χ^2 values will also increase with a factor k, but with scaling they remain the same because the scaling parameter increases with a factor $\sqrt{k}$. In short, scaling ensures that overdispersion will not affect the statistics used for testing the model's parameters. There seem to be no rigorous rules for deciding whether or not to adjust for overdispersion, but if the P values of the relevant tests are much lower without scaling than with scaling, it is prudent to include the scaling option. Also in doubtful cases the more conservative approach is to use DSCALE. Finally, it should be emphasized that the scaling options provided by *SAS* (DSCALE and PSCALE) are less reliable than Williams' adjustment for overdispersion (see Crawley 1993), in particular when brood size among clutches varies strongly and/or sample size is small (Krackow & Tkadlec 2001). The OBSTATS option provides additional output including predicted sex ratios and their confidence limits.

Box 4.2 shows the output produced by the program of Box 4.1. The number of events and number of trials refer to the total number of females and the total number of progeny, respectively.

Hence, the observed sex ratio in the total sample is $3229/17710 = 0.182$. The observed variances for each of the 24 colonies range between 7.172 (colony 8) and 410.1 (colony 3), which means that there are no reasons to reject the assumption of data being normally distributed. Moving to 'LR statistics for type 3 analysis' shows the contribution of

---

Box 4.2 Edited output produced by the SAS program of Box 1

All explanatory variables except relatedness asymmetry (RA) are included in the model (full model). Output produced by the option OBSTATS in PROC GENMOD is not shown here, but see Box 4.3

OBS	VARIANCE
1	117.375
2	87.244
3	410.104
4	148.152
5	72.697
6	42.353
7	38.259
8	7.172
9	18.191
10	112.616
11	45.408
12	117.809
13	55.359
14	71.280
15	59.070
16	127.517
17	130.352
18	171.829
19	62.141
20	59.957
21	78.857
22	27.351
23	8.839
24	69.816

The GENMOD procedure

Model information

Description	Value
Dataset	WORK.SEXRATIO
Distribution	BINOMIAL
Link function	LOGIT
Dependent variable	FEMALES
Dependent variable	PROGENY
Observations used	24
Number of events	3229
Number of trials	17710

Criteria for assessing goodness of fit

Criterion	DF	Value	Value/DF
Deviance	18	1608.7899	89.3772
Scaled Deviance	18	18.0000	1.0000
Pearson Chi-Square	18	1629.9167	90.5509
Scaled Pearson X2	18	18.2354	1.0131
Log Likelihood	.	−86.9073	.

Analysis of parameter estimates

Parameter	DF	Estimate	Std Err	ChiSquare	Pr>Chi
INTERCEPT	1	−2.7005	7.4100	0.1328	0.7155
HABITAT KBD	1	−1.0902	0.6845	2.5370	0.1112
HABITAT KID	1	−0.0562	0.6578	0.0073	0.9319
HABITAT SV	0	0.0000	0.0000	.	.
WORKERS	1	−0.0000	0.0000	1.4482	0.2288
SEXPROD	1	0.0004	0.0003	1.9276	0.1602
WORKSIZE	1	1.3904	7.5608	0.0338	0.8541
SCALE	0	9.4540	0.0000	.	.

Note: The scale parameter was estimated by the square root of DEVIANCE/DOF.

LR statistics for type 1 analysis

Source	Deviance	NDF	DDF	F	Pr>F	ChiSquare	Pr>Chi
INTERCEPT	2894.7773	0	18	.	.	.	.
HABITAT	1799.8812	2	18	6.1251	0.0094	12.2503	0.0022
WORKERS	1799.2139	1	18	0.0075	0.9321	0.0075	0.9311
SEXPROD	1611.8085	1	18	2.0968	0.1648	2.0968	0.1476
WORKSIZE	1608.7899	1	18	0.0338	0.8562	0.0338	0.8542

LR statistics for type 3 analysis

Source	NDF	DDF	F	Pr>F	ChiSquare	Pr>Chi
HABITAT	2	18	2.2288	0.1365	4.4577	0.1077
WORKERS	1	18	1.4988	0.2366	1.4988	0.2209
SEXPROD	1	18	2.0552	0.1688	2.0552	0.1517
WORKSIZE	1	18	0.0338	0.8562	0.0338	0.8542

CONTRAST statement results

Contrast	NDF	DDF	F	Pr>F	ChiSquare	Pr>Chi	Type
KBD vs KID	1	18	3.6092	0.0736	3.6092	0.0575	LR
KBD vs SV	1	18	2.5720	0.1262	2.5720	0.1088	LR
KID vs SV	1	18	0.0073	0.9328	0.0073	0.9319	LR

the four independent variables to the model. A variable is considered significant if the associated P value is less than 0.05. Both F and χ^2 values with their associated P values are provided as test statistics, but the former must be used because it takes into account that the test statistic is derived from empirical variances The factor with the highest P value (here 'WorkSize' with

$P = 0.8562$) is now excluded from the model, and the analysis is repeated with the reduced model. For each step in this backward elimination process, the P values of the remaining parameters tend to decrease. The elimination of variables continues until all remaining variables are significant at the 5% level. In the present example, the elimination procedure stopped when 'Habitat' was the only factor remaining in the model ($F_{2,21} = 6.3873$; $P = 0.0017$) (Box 4.3).

4.5.2 Applying the general model to real data

The parameters associated with a given model are found in Box 4.3 under 'Analysis of parameter estimates'. The full model contains five parameters apart from the intercept. These parameters are associated with the variables followed by 1 df, i.e. 'Intercept' (β_0), 'Workers' (β_1), 'SexProd' (β_2), 'Habitat KBD' (β_3), 'Habitat KID' (β_4), and 'WorkSize' (β_5). The variables x_1 ('Workers'), x_2 ('SexProd') and x_5 ('WorkSize') are quantitative, whereas x_3 ('Habitat KBD') and x_4 ('Habitat KID') are qualitative (dummy) variables, i.e. x_3 is 1 if an observation is from habitat KBD, otherwise 0. Likewise, x_4 is 1 if an observation is from habitat KID, otherwise 0. It means that the predicted values of π in habitat KBD are obtained as $\hat{\pi} = \frac{e^{\beta_0+\beta_1 x_1+\beta_2 x_2+\beta_3+\beta_5 x_5}}{1+e^{\beta_0+\beta_1 x_1+\beta_2 x_2+\beta_3+\beta_5 x_5}}$, in habitat KID as $\hat{\pi} = \frac{e^{\beta_0+\beta_1 x_1+\beta_2 x_2+\beta_4+\beta_5 x_5}}{1+e^{\beta_0+\beta_1 x_1+\beta_2 x_2+\beta_4+\beta_5 x_5}}$ and in habitat SV as $\hat{\pi} = \frac{e^{\beta_0+\beta_1 x_1+\beta_2 x_2+\beta_5 x_5}}{1+e^{\beta_0+\beta_1 x_1+\beta_2 x_2+\beta_5 x_5}}$.

Box 4.3 | Edited output for the reduced model produced by the SAS program of Box 4.1 (after leaving out all explanatory variables except Habitat)

The GENMOD Procedure

Criteria for assessing the goodness of fit

Criterion	DF	Value	Value/DF
Deviance	21	1799.8812	85.7086
Scaled Deviance	21	21.0000	1.0000
Pearson Chi-Square	21	1858.1373	88.4827
Scaled Pearson X2	21	21.6797	1.0324
Log Likelihood	.	−88.8657	.

Analysis of parameter estimates

Parameter	DF	Estimate	Std Err	ChiSquare	Pr>Chi
INTERCEPT	1	−0.7868	0.2783	7.9897	0.0047
HABITAT KBD	1	−1.4892	0.4373	11.5992	0.0007
HABITAT KID	1	−0.4999	0.4651	1.553	0.2824
HABITAT SV	0	0.0000	0.0000	.	.
SCALE	0	9.2579	0.0000	.	.

Note: The scale parameter was estimated by the square root of DEVIANCE/DOF.

LR statistics for type 1 analysis

Source	Deviance	NDF	DDF	F	Pr>F	ChiSquare	Pr>Chi
INTERCEPT	2894.7773	0	21	.	.	.	.
HABITAT	1799.8812	2	21	6.3873	0.0068	12.7746	0.0017

LR statistics for type 3 analysis

Source	NDF	DDF	F	Pr>F	ChiSquare	Pr>Chi
HABITAT	2	21	6.3873	0.0068	12.7746	0.0017

CONTRAST statement results

Contrast	NDF	DDF	F	Pr>F	ChiSquare	Pr>Chi	Type
KBD vs KID	1	21	3.7750	0.0656	3.7750	0.0520	LR
KBD vs SV	1	21	12.3711	0.0020	12.3711	0.0004	LR
KID vs SV	1	21	1.1863	0.2884	1.1863	0.2761	LR

Observation statistics

FEMALES	PROGENY	Pred	Xbeta	Std	HessWgt	Lower	Upper
143	798	0.3129	−0.7868	0.2783	2.0016	0.2088	0.4400
153	356	0.3129	−0.7868	0.2783	0.8929	0.2088	0.4400
603	1885	0.3129	−0.7868	0.2783	4.7281	0.2088	0.4400
177	1086	0.3129	−0.7868	0.2783	2.7240	0.2088	0.4400
93	426	0.3129	−0.7868	0.2783	1.0685	0.2088	0.4400
360	408	0.3129	−0.7868	0.2783	1.0234	0.2088	0.4400
65	158	0.3129	−0.7868	0.2783	0.3963	0.2088	0.4400
16	29	0.3129	−0.7868	0.2783	0.0727	0.2088	0.4400
21	157	0.0931	−2.2760	0.3372	0.1547	0.0504	0.1659
123	1457	0.0931	−2.2760	0.3372	1.4357	0.0504	0.1659
53	370	0.0931	−2.2760	0.3372	0.3646	0.0504	0.1659
129	1487	0.0931	−2.2760	0.3372	1.4653	0.0504	0.1659
64	474	0.0931	−2.2760	0.3372	0.4671	0.0504	0.1659
81	675	0.0931	−2.2760	0.3372	0.6651	0.0504	0.1659
63	1010	0.0931	−2.2760	0.3372	0.9952	0.0504	0.1659
159	803	0.0931	−2.2760	0.3372	0.7913	0.0504	0.1659
138	2490	0.0931	−2.2760	0.3372	2.4536	0.0504	0.1659
404	703	0.2164	−1.2866	0.3726	1.3910	0.1174	0.3644
71	569	0.2164	−1.2866	0.3726	1.1258	0.1174	0.3644
97	254	0.2164	−1.2866	0.3726	0.5026	0.1174	0.3644
92	644	0.2164	−1.2866	0.3726	1.2742	0.1174	0.3644
29	510	0.2164	−1.2866	0.3726	1.0091	0.1174	0.3644
9	504	0.2164	−1.2866	0.3726	0.9972	0.1174	0.3644
86	457	0.2164	−1.2866	0.3726	0.9042	0.1174	0.3644

The estimated value and standard error are provided for each parameter. Confidence limits for each parameter are calculated as

$$P\left[\hat{\beta} - t_\nu \text{s.e.}(\hat{\beta}) < \beta < \hat{\beta} + t_\nu \text{s.e.}(\hat{\beta})\right] = 1 - \alpha \quad (4.2)$$

where $\alpha = 0.05$ corresponds to 95% limits and $\alpha = 0.01$ to 99% limits. The t value for ν de-grees of freedom and for the chosen α is obtained from statistical tables (e.g. Table 12 in Rohlf & Sokal 1981). A parameter is significantly different from 0 if the confidence interval does not contain 0. Another way to test this is to use $t_\nu = \frac{\hat{\beta}}{\text{s.e.}(\hat{\beta})}$ directly or, as done by SAS, use t_ν^2 which is distributed approximately as χ^2 with 1 df. For instance, the intercept (β_0) is estimated to be -2.7005 with s.e. $= 7.4100$, yielding $t_{18} = -2.7005/7.4100 = -0.3644$. χ^2 therefore becomes $t^2 = 0.1328$ $(P = 0.7155)$, implying that β_0 is not significantly different from 0.

A model may contain several parameters that are only just significant, but still the model may be satisfactory to predict the observed pattern. Information about the overall performance of a given model is provided in Box 4.3 under 'Criteria for assessing the goodness of fit'. Deviance (D) expresses the deviation between the observed and predicted values. The lower the deviance, the better the model. Deviance decreases with the number of parameters in the model, so that the best model represents a balance between minimal deviance and maximal number of significant parameters. The simplest model (the null model) is one with no explanatory variables at all, so that the only parameter to be estimated is the overall mean, which in this case is the intercept (β_0). For the present dataset the deviance of the null model is $D_1 = 2894.78$.

The reduced model with 'Habitat' as the only significant explanatory variable (Box 4.3) yields a deviance of $D_2 = 1799.88$. The decline in deviance relative to the null model is $\Delta D = 2894.78 - 1799.88 = 1094.9$. This corresponds to $(\Delta D/D_1)100\% = (1 - D_2/D_1)100\% = 37.8\%$, an estimate which is equivalent to the percentage variance explained in a standard statistical analysis based on normally distributed errors as for example ordinary linear regression or analysis of variance. However, the latter model has two parameters apart from β_0 ($p_2 = 2$; the third habitat provides no independent information) while the former has no extra parameters ($p_1 = 0$). The difference between the null model and the reduced model containing the significant factor(s) is tested by means of an F-test recommended by Manly (1990)

$$F = \frac{(D_1 - D_2)/(p_2 - p_1)}{D_2/(N - p_2 - 1)}, \quad (4.3)$$

where N is the total number of colonies in the dataset (24). F has $p_2 - p_1$ df in the numerator and $N - p_2 - 1$ df in the denominator. In the present example, we get

$$F_{2,21} = \frac{(2894.78 - 1799.88)/(2 - 0)}{1799.88/(24 - 2 - 1)}$$
$$= 547.45/85.71$$
$$= 6.3873,$$

which is the value found under 'LR statistics for type 1 analysis' in Box 4.3. Since $P = 0.0017$, it is justified to include the extra parameters due to 'Habitat' in the model.

If we also include 'Workers', 'SexProd' and 'WorkSize' and compare the result with the model that only includes 'Habitat', deviance for the full model, requiring five parameters, becomes 1608.79. From this we get

$$F_{5,18} = \frac{(1799.88 - 1608.79)/(5 - 2)}{1608.79/(24 - 5 - 1)}$$
$$= 63.7/89.38$$
$$= 0.713.$$

This value cannot be found directly from the *SAS* output, but if the three extra variables ('Workers', 'SexProd' and 'WorkSize') are included in the model after 'Habitat' (note that the order of variables is important) and the χ^2 values for the extra variables found under 'LR statistics for type 1 analysis' are summed and divided by the total number of df for the three variables, it gives $2.1381/3 = 0.713$. The associated $P = 0.502$ is far from being significant. Thus, there is no need to incorporate the extra three variables because they do not improve the model significantly.

After having identified the model with 'Habitat' as being the most appropriate, the next step is to test whether its residuals are binomially distributed. Under the binomial null hypothesis, deviance (D) divided by the appropriate degrees of freedom (ν) should be close to unity. Since the value obtained $(D/\nu = 85.7086)$ is considerably larger than one, it indicates that the data are highly overdispersed, even after explaining 37.8% of the deviance. In the rare cases where we have no *a priori* reasons to assume either over- or underdispersion (e.g. in an ant species which is known to have obligate intranidal mating and local mate competition, but where environmental factors may also influence the colony sex ratio), a two-tailed test for deviation from the null hypothesis (i.e. the residuals are binomially distributed) is most appropriate. The test is based on the fact that D is expected to be distributed as χ^2 with v degrees of freedom. The confidence limits for D/v, assuming that H_0 is true, are found from

$$P\left(\underline{\chi}_v^2/v < \chi_v^2/v < \overline{\chi}_v^2/v\right) = 1 - \alpha, \tag{4.4}$$

where $\underline{\chi}_v^2$ and $\overline{\chi}_v^2$ are the lower and upper limits of χ^2 for v degrees of freedom. These values can be found from a two-tailed χ^2 table (e.g. Table A13 in Campbell 1989). This yields for the 95% and 99% confidence limits, respectively

$$P\left(10.28/21 < \chi^2/v < 35.48/21\right)$$
$$= P\left(0.49 < \chi^2/v < 1.69\right) = 0.95,$$

and

$$P\left(8.03/21 < \chi^2/v < 41.4/21\right)$$
$$= P\left(0.38 < \chi^2/v < 1.97\right) = 0.99.$$

Since the observed value of $D/v = 85.71$ falls far outside these confidence limits, H_0 is rejected both at the 5% and 1% significance level. However, in most cases, including our *Lasius niger* example, the alternative hypothesis is that data exhibit overdispersion, which means that D can be tested one-tailed as the probability of obtaining a χ^2 with v df equal to or larger than the observed value of D. Although the value of D/v demonstrates a very high degree of overdispersion in data, it is considerably less than the value obtained from the null model where $D/v = 125.86$. Thus, the sex ratios are highly heterogeneous to start with and remain so, but to a lesser extent, after explaining 37.8% of the deviance. Biologically, this means that colony-level sex ratios in *L. niger* are split.

When the best model includes a qualitative variable with more than two levels ('Habitat' in the present case) it may be of interest to identify the significance of differences between combinations of these levels. With three levels (SV, KBD, KID), the number of pairwise comparisons is three (KBD versus KID, KBD versus SV and KID

versus SV). With four levels the number will increase to six and with ten levels the number is 45. Tests for pairwise differences are conducted by means of the CONTRAST statement in the procedure PROC GENMOD, but it should be noted that these tests do not control for experiment-wise error, i.e. the risk of committing at least one type I error increases with the number of tests. It is therefore recommended to use a Bonferroni adjustment of the significance level (Sokal & Rohlf 1995). For instance, if $\alpha = 0.05$ is used as the significance level in a single test, and the number of tests (k) is 3, then only values of P less than $\alpha' = \alpha/k = 0.05/3 = 0.0167$ are considered significant. From Box 4.3 it appears that KBD and SV are the only habitats that differ significantly with respect to their sex ratio. It should, however, be noticed that the ordinary Bonferroni adjustment tends to be conservative if the number of simultaneous tests is large; that is, it increases the risk of accepting the null hypothesis even when it is wrong. Therefore, a variety of alternative procedures for multiple hypotheses testing have been suggested (Holm 1979, Simes 1986, Rice 1989). Haccou and Meelis (1992) recommend a sequential Bonferroni procedure for multiple comparisons of k hypotheses based on an ordering of the P values. Applying this procedure to the data in Box 4.3, the ordered P values are: 0.0020, 0.0656 and 0.2884. Let $P_{(i)}$ denote the ith ordered P value associated with the ith null hypothesis (denoted H_{0i}). The overall hypothesis (H_0: all habitats are identical) is rejected if for at least one i ($i = 1, 2, \ldots, k$) we obtain

$$P_{(i)} < \frac{\alpha}{k - i + 1}. \tag{4.5}$$

For $i = 1$, $P_{(1)} = 0.0020 < \frac{0.05}{3} = 0.0167$, implying that H_{01}: KBD = SV has to be rejected. For $i = 2$, $P_{(2)} = 0.0656 > \frac{0.05}{2} = 0.025$, implying that H_{02}: KBD = KID must be accepted. Finally, for $i = 3$, $P_{(3)} = 0.2884 > \frac{0.05}{1} = 0.05$, which of course also leads to acceptance of H_{03}: KID = SV. Hence, in the chosen example the ordinary and the sequential Bonferroni procedures yield the same conclusions but it is obvious that the latter will generally increase the chance of rejecting H_{0i} for $i > 1$, whereas the conclusion will be the same for $i = 1$. If two or more levels (categories) of the same categorical variable are not significantly different, they can be combined (aggregated) into a new category, provided such a new category makes sense. For instance, data from habitats located in different types of forest may be pooled into a single category (called forest), which separates the forest habitats from all the nonforest habitats (Hardy & Field 1998 provide a worked example of data aggregation).

The predicted sex ratios in each of the three habitats can be found by substituting the parameter estimates in the logistic equation. For KBD this yields $e^{-0.7868-1.4892}/(1+e^{-0.7868-1.4892}) = 0.0931$, for KID this gives $e^{-0.7868-0.4999}/(1+e^{-0.7868-0.4999}) = 0.2164$, and for habitat SV one gets $e^{-0.7868}/(1+e^{-0.7868}) = 0.3129$. However, the predicted ratios can more readily be obtained from the 'Observation statistics' (Box 4.3). This also shows that the lower and upper 95% confidence limits for the predicted sex ratio in KBD are 0.050 and 0.1659, respectively. Similarly, the limits for KID are 0.1174 and 0.3644, and for SV 0.2088 and 0.4400. This demonstrates that the numerical sex ratios are significantly male-biased in all habitats, agreeing with earlier conclusions by Boomsma et al. (1982).

It is often relevant to test whether a predicted sex ratio (π_{pred}) differs significantly from one expected from theory (π_{exp}). To perform such a test we need the logit-transformed values of the probabilities, i.e. $z_{exp} = \ln(\frac{\pi_{exp}}{1-\pi_{exp}})$ and $z_{pred} = \ln(\frac{\pi_{pred}}{1-\pi_{pred}}) = \hat{\beta}_0 + \hat{\beta}_1 x_1 + \ldots + \hat{\beta}_p x_p$. Since z_{pred} will follow a normal distribution with standard deviation $s(z_{pred})$ when the number of observations is large enough, the test statistic

$$t_\nu = \frac{z_{pred} - z_{exp}}{s(z_{pred})} \qquad (4.6)$$

will follow a t-distribution with ν degrees of freedom. As an example we compare the predicted sex ratio in habitat SV ($\pi_{pred} = 0.3129$) with an expected value of $\pi_{exp} = 0.135$ (the equivalent of a 1:1 investment ratio; see below). $s(z_{pred})$ is found from 'Observation statistics' (Box 4.3) as 0.2783, so we get

$$t_{21} = \frac{\ln\left(\frac{0.3129}{1-0.3129}\right) - \ln\left(\frac{0.135}{1-0.135}\right)}{0.2783}$$

$$= \frac{-0.787 + 1.857}{0.2783} = 3.845$$

which is highly significant ($P_{two-tailed} = 0.0009$).

4.6 | The sex investment ratio

The study of sex allocation in social insects is complicated by the fact that the typical cost of producing a female (gyne, prospective queen) is often different to that of producing a male and varies considerably across species. As population-wide sex ratio equilibria are expressed in units of investment (Fisher 1958), estimates of the female-to-male cost ratio are needed to test predictions of sex-allocation theory at the population level. The proper estimation of female-to-male cost ratios has been extensively discussed in the literature (Trivers & Hare 1976, Boomsma & Isaaks 1985, Nonacs 1986, Boomsma 1989, Crozier & Pamilo 1992, Boomsma et al. 1995). The proposed 0.7 power conversion of female-to-male dry weight ratios (Boomsma 1989, Boomsma et al. 1995) has proved to be a useful approximation for sexually dimorphic species (e.g. Bourke & Franks 1995). However, this coefficient is only an estimate obtained with error from data across species so that its uncritical use is unwarranted (Crozier & Pamilo 1992). Boomsma et al. (1995) recommend therefore that each published study on sex allocation in a social insect includes an Appendix with the raw numerical sex ratio data per colony, to allow reanalysis if novel procedures become available. The protocol outlined in the present chapter illustrates and reinforces this argument. Our current approach represents an improvement compared to previous analyses, but is unlikely to be the final word.

The statistical protocol presented above may help to solve the cost-ratio problem, because the intercept (β_0) is a numerical sex ratio whose confidence limits are available from the analysis. This observed value can therefore be tested against any expected population- or habitat-specific sex-allocation value, obtained by incorporating any conceivable female-to-male cost ratio. For the *Lasius niger* data analysed above, the means of numerical sex ratio (95% confidence limits in parentheses) were 0.313 (0.209–0.440), 0.093 (0.050–0.166) and 0.216 (0.117–0.364) for SV, KBD and KID, respectively. These mean values are close to those obtained from the larger datasets published previously (Boomsma *et al.* 1982, Boomsma & Isaaks 1985). Using the 0.7 power conversion of the adult dry weight (ADW) sex-allocation ratio (Boomsma *et al.* 1995), the expected numerical sex ratio for complete worker control and 3:1 female bias is 0.319, whereas the corresponding value for queen control and 1:1 sex allocation is 0.135. Clearly, the SV sex ratio is significantly more female-biased than expected under queen control ($t_{21} = 3.845; P = 0.0009$), whereas the KBD sex ratio is significantly less female-biased than expected for full-sib colonies with complete worker control ($t_{21} = 4.429; P = 0.0002$), a conclusion consistent with the previous analyses. It is also clear that the 95% c.l. of the sex ratio in SV and KBD do not overlap and that they are highly significantly different ($P = 0.0023$) (Box 4.3). In studies where it is unclear which female-to-male cost ratio represents the best approximation, tests based on the mean numerical sex ratio (β_0) can thus be performed against expectations based on several alternative cost ratio estimates and the conclusions formulated accordingly.

4.7 | How to explain split sex ratios in *Lasius niger*

The original studies on sex allocation in the monogynous (single queen per colony) ant *Lasius niger* (Boomsma *et al.* 1982, Boomsma & Isaaks 1985, Boomsma 1988, Van der Have *et al.* 1988) demonstrated that sex allocation:

1. Differed among discrete habitats of the same species.

2. Was close to 3:1 in favour of females in a nonresource-limited habitat (SV) where queens were almost invariably single mated and where worker reproduction was absent (except in a few queenless colonies).

3. Was male-biased in a resource-limited habitat (KBD) where double queen mating was common and where worker reproduction occurred.

4. Was intermediate in a third habitat (KID). These results were confirmed by the present analysis.

These results preceded and inspired the formulation of split sex ratio theory (Boomsma & Grafen 1990, 1991). Although these data were interpreted as evidence for conditional local adjustment of sex allocation by workers in correlation with their relatedness asymmetry to the reproductive brood (Van der Have *et al.* 1988, Seger 1991), formal evidence for this correlation was never presented. The procedure of the present analysis allowed us to do this in a simple and straightforward way by replacing the discrete variable 'Habitat' by the continuous variable 'Average Relatedness Asymmetry' (RA) as estimated by Van der Have *et al.* (1988). Entering 'RA' instead of 'Habitat' produced a virtually unchanged result: the positive effect of RA on female bias in the sex ratio was significant ($P = 0.0014$) and explained the same 37.8% of the deviance in the sex ratio, whereas none of the other variables added anything significant. Thus, although evidence for colony-level sex ratio biasing as a function of RA is absent, it is clear that *Lasius niger* has RA-correlated split sex ratios at the habitat level. However, the data do not allow inference of the causes of sex ratio variation at the colony level, something that later studies on other species using more accurate genetic markers have achieved (e.g. Sundström 1995). *Lasius niger* is now known to be a completely panmictic ant throughout much of its European range and to have populations varying in queen mating frequency (Boomsma & Van der Have 1998). Although the habitats SV, KBD and KID were far enough apart to have their own mating swarms, they were not separate breeding populations to the same extent as the populations investigated later by Boomsma and Van der Have (1998) and even those had an F_{ST} value (genotypic variance component at neutral marker loci across populations) very close to zero. Several intriguing questions therefore remain to be answered for this common European ant. First, it remains unclear whether sex ratios are split according to colony-level RA or only at the habitat level. Second, it is still unclear whether differences in habitat quality themselves select for different queen mating frequencies and which of these factors ultimately determines the sex ratio (see also discussion in Van der Have *et al.* 1988).

4.8 | Some earlier conclusions do not hold

The present analysis shows that three additional independent variables did not contribute significantly to explain the observed sex ratios. Total sexual production (SexProd) was shown earlier to be positively correlated with sex allocation after pooling the data from all three habitats (Boomsma 1988), but not in any of the habitats separately. We found that this independent variable has no overall significance when the effect of habitat is controlled for statistically. Worker size (mean headwidth), which was earlier shown to differ among habitats (Boomsma *et al.* 1982) and to be positively correlated with female bias in KBD, was likewise demonstrated to be unimportant as an overall partial effect in the present analysis. Worker number was previously shown to be correlated with both gyne and male production (Boomsma *et al.* 1982), so that an effect on sex allocation seemed unlikely, and this inference was confirmed by the present analysis.

4.9 | Conclusion

The main conclusions from the earlier analyses of sex allocation in *Lasius niger* hold. The habitats differ in both resource availability and RA, and a significant part of the sex ratio variance can be explained by either of these independent variables. However, the present analysis quantifies the overall significance of these main effects much more accurately, shows that the residual sex ratios are still highly split after 37.8% of the deviance has been accounted for, and demonstrates that suggestive additional independent variables (total sexual production, worker number and worker size) do not affect sex allocation independently of habitat or RA.

The statistical analysis based on multiple logistic regression as outlined in this chapter has several advantages above the more traditional analyses that have been applied to sex-allocation data in social insects. It weights the observed sex ratios by their clutch size (i.e. by their contribution to the mating swarm) and allows statistical tests of partial effects due to any number of discrete and/or continuous independent variables, even in cases where the condition of stochastically independent observations is not met, so that sex ratios are over- or underdispersed. In addition, it allows a reliable statistical test of deviations of the observed overall sex ratio or group-wise sex ratio (in the case of a significant discrete independent variable such as 'Habitat') from expected ratios, and finally it allows an analysis of the distribution of residuals. Elsewhere (JJ Boomsma & G Nachman, in prep.) we reanalyse 13 other datasets on sex ratios in ants and bees and show that, although conclusions of previous analyses were usually qualitatively correct, our current approach also produces more convincing statistical significances in these cases and has the potential to discover unexpected partial correlations with variables that were previously considered unimportant.

Acknowledgements

Both authors contributed equally to this chapter. We thank Ian Hardy for inviting us to write it and Andrew Bourke, Sven Krackow, Evert Meelis, Lotta Sundström and an anonymous reviewer for comments. JJ Boomsma was supported by a grant from the Danish Natural Science Research Council.

References

Alexander RD & Sherman PW (1977) Local mate competition and parental investment in social insects. *Science*, **196**, 494–500.

Backus VL (1995) Rules for allocation in a temperate forest ant: demography, natural selection, and queen-worker conflict. *American Naturalist*, **145**, 775–796.

Boomsma JJ (1988) Empirical analysis of sex allocation in ants: from descriptive surveys to population genetics. In: G de Jong (ed), *Population Genetics and Evolution*, pp 42–51. Berlin: Springer-Verlag.

Boomsma JJ (1989) Sex investment ratios in ants: has female bias been systematically overestimated? *American Naturalist*, **133**, 517–532.

Boomsma JJ & Grafen A (1990) Intraspecific variation in ant sex ratios and the Trivers-Hare hypothesis. *Evolution*, **44**, 1026–1034.

Boomsma JJ & Grafen A (1991) Colony-level sex ratio selection in the eusocial Hymenoptera. *Journal of Evolutionary Biology*, **4**, 383–407.

Boomsma JJ & Isaaks JA (1985) Energy investment and respiration in queens and males of *Lasius niger* (Hymenoptera: Formicidae). *Behavioral Ecology and Sociobiology*, **18**, 19–27.

Boomsma JJ & Van der Have TM (1998) Queen mating and paternity variation in the ant *Lasius niger*. *Molecular Ecology*, **7**, 1709–1718.

Boomsma JJ, Van der Lee GA & Van der Have TM (1982) On the production ecology of *Lasius niger* (Hymenoptera: Formicidae) in successive coastal dune valleys. *Journal of Animal Ecology*, **51**, 975–991.

Boomsma JJ, Keller L & Nielsen MG (1995) A comparative analysis of sex ratio investment parameters in ants. *Functional Ecology*, **9**, 743–753.

Bourke AFG & Franks NR (1995) *Social Evolution in Ants*. Princeton: Princeton University Press.

Brian MV (1979) Habitat differences in sexual production by two co-existent ants. *Journal of Animal Ecology*, **48**, 943–953.

Campbell RC (1989) *Statistics for Biologists*, 3rd edn. Cambridge: Cambridge University Press.

Crawley MJ (1993) *GLIM for Ecologists*. Oxford: Blackwell Scientific Publications.

Crozier RH & Pamilo P (1992) Sex allocation in social insects: problems in prediction and estimation. In: DL Wrench & MA Ebbert (eds) *Evolution and Diversity of Sex Ratio in Haplodiploid Insects and Mites*, pp 369–383. New York: Chapman & Hall.

Crozier RH & Pamilo P (1996) *Evolution of Social Insect Colonies*. Oxford: Oxford University Press.

Deslippe RJ & Savolainen R (1995) Sex investment in a social insect: the proximate role of food. *Ecology*, **76**, 375–382.

Fisher RA (1958) *The Genetical Theory of Natural Selection*, 2nd edn. New York: Dover.

Frank SA (1987) Variable sex ratio among colonies of ants. *Behavioural Ecology and Sociobiology*, **20**, 195–201.

Grafen A (1986) Split sex ratios and the evolutionary origins of eusociality. *Journal of Theoretical Biology*, **122**, 95–121.

Haccou P & Meelis E (1992) *Statistical Analysis of Behavioural Data*. Oxford: Oxford University Press.

Hamilton WD (1964) The genetical evolution of social behaviour II. *Journal of Theoretical Biology*, **7**, 17–52.

Hamilton WD (1967) Extraordinary sex ratios. *Science*, **156**, 477–488.

Hardy ICW & Field SA (1998) Logistic analysis of animal contests. *Animal Behaviour*, **56**, 787–792.

Herbers JM (1990) Reproductive investment and allocation ratios for the ant *Leptothorax longispinosus*: sorting out the variation. *American Naturalist*, **136**, 178–208.

Holm S (1979) A simple sequentially rejective multiple test procedure. *Scandinavian Journal of Statistics*, **6**, 65–70.

Hosmer DW & Lemeshow S (1989) *Applied Logistic Regression*. New York: John Wiley & Sons.

Huck UW, Seger J & Lisk RD (1990) Litter size ratios in the golden hamster vary with time of mating and litter size and are not binomially distributed. *Behavioral Ecology and Sociobiology*, **26**, 99–109.

Krackow S & Tkadlec E (2001) Analysis of brood sex ratios: implications of offspring clustering. *Behavioral Ecology and Sociobiology*, **50**, 293–301.

Littell RC, Freund RJ & Spector PC (1991) *SAS System for Linear Models*, 3rd edn. SAS Series in Statistical Applications. Cary, NC: SAS Institute Inc.

Manly BFJ (1990) *Stage-structured Populations. Sampling, Analysis and Simulation*. London: Chapman & Hall.

McCullagh P & Nelder JA (1989) *Generalized Linear Models*, 2nd edn. London: Chapman & Hall.

Nonacs P (1986) Ant reproductive strategies and sex allocation theory. *Quarterly Review of Biology*, **61**, 1–21.

Rice WR (1989) Analyzing tables of statistical tests. *Evolution*, **43**, 223–225.

Rohlf FJ & Sokal RR (1981). *Statistical Tables*, 2nd edn. San Francisco: WH Freeman & Co.

SAS (1994) *SAS/STAT User's Guide*, volumes 1 and 2, 4th printing. Cary, NC: SAS Institute Inc.

Seger J (1991) Cooperation and conflict in social insects. In: JR Krebs & NB Davies (eds) *Behavioural Ecology: An Evolutionary Approach*, 3rd edn., pp 338–373. Oxford: Blackwell.

Simes RJ (1986) An improved Bonferroni procedure for multiple tests of significance. *Biometrika*, **73**, 751–754.

Sokal RR & Rohlf FJ (1995) *Biometry*, 3rd edn. New York: WH Freeman & Co.

Sundström L (1995) Sex allocation and colony maintenance in monogyne and polygyne colonies of *Formica truncorum* (Hymenoptera: Formicidae): the impact of kinship and mating structure. *American Naturalist*, **146**, 182–201.

Trivers RL & Hare H (1976) Haplodiploidy and the evolution of the social insects. *Science*, **191**, 249–263.

Van der Have TM, Boomsma JJ & Menken SBJ (1988) Sex investment ratios and relatedness in the monogynous ant *Lasius niger* (L.). *Evolution*, **42**, 160–172.

Chapter 5

Analysis of sex ratio variances and sequences of sex allocation

Sven Krackow, Evert Meelis & Ian C.W. Hardy

5.1 | Summary

Sex ratio variances can indicate control of sex allocation beyond the production of a certain mean. Furthermore, theory predicts that nonbinomial (nonrandom) offspring-group sex ratio distributions should be selectively favoured in some circumstances. Mechanisms of producing nonbinomial sex ratio variances include the allocation of sex to offspring in fixed or other nonrandom sequences. Variances and sequences are thus linked. In this chapter we review predictions of sex ratio models, consider constraints and expectations arising from particular sex-determination mechanisms and discuss statistical methods for variance and sequence analysis.

5.2 | Introduction

In Chapters 3 and 4 variance was treated as a consideration necessary to ensure appropriate data analysis. However, variance of offspring-group sex ratios are of interest in themselves both because they may be under selection and because they can indicate parental control of sex allocation. We begin by reviewing predictions of sex ratio models (section 5.3) and considering constraints and expectations arising from particular sex-determination mechanisms (section 5.4). We then introduce sex ratio 'precision' and its measurement (section 5.5) and give methods for statistical analysis of sex ratio variance (section 5.6). Sequences of sex allocation, and their relationships with sex ratio variance, are considered in section 5.7.

5.3 | Variance expectations from optimality theory

Optimal sex-allocation theory makes predictions not just about mean sex ratios but also about variances. When mating opportunities extend throughout large populations (panmixis) the sex ratio within a group of offspring is selectively neutral as long as the population sex ratio remains at equilibrium (Kolman 1960). However, in small populations selection favours progeny sex ratios being kept at the optimum value, rather than stochastically fluctuating around an optimal mean (Verner 1965). However, selection for variance reduction may be weak in many organisms, predominantly because random mortality after the production of young can cause sex ratios at reproductive age to vary stochastically, regardless of the variance at birth (Taylor & Sauer 1980). Under some circumstances (e.g. when assumptions of the Trivers–Willard hypothesis apply, Chapter 13), selection may favour increased variance within samples (e.g. Frank 1990). In contrast, when populations are structured into locally mating subgroups ('local mate competition', Hamilton 1967) and premating mortality is low, selection for the precise control of sex allocation and low sex ratio variance can be strong

(Green *et al.* 1982, Nagelkerke & Hardy 1994, Nagelkerke 1996, West & Herre 1998). For instance, when offspring groups are produced by a single mother and mating only occurs between group members, mothers should produce the minimum number of sons necessary to fertilize all daughters: offspring groups containing excess males or only one sex of offspring generate less maternal fitness (Figure 5.1). Figure 5.2

gives some examples of sex ratio distributions in haplodiploid parasitoids. More detailed reviews of the selective pressures acting on sex ratio variance are provided by Frank (1990), Nagelkerke (1996), and Hardy (1992, 1997, Hardy *et al.* 1998).

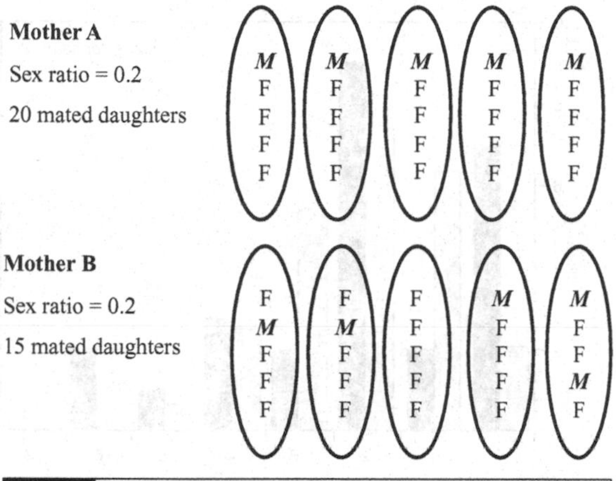

5.4 | Variance expectations from sex determination mechanisms

Optimal sex ratio models assume control of sex allocation. Some sex-determination mechanisms allow a high degree of sex-allocation control while others appear to make control difficult (e.g. Krackow 1995a, 1999). Mechanistic considerations, based on current knowledge of the processes leading to sex determination, can generate expectations for sex ratio variances (null-models) irrespective of adaptive arguments. Genetic or chromosomal sex-determination systems (Chapters 7 & 8) will generate mean sex ratios according to the respective gene or chromosome frequencies and, due to random segregation during meiosis, there will be some stochastic element causing the sampled sex ratios to vary randomly about the expected mean. Given knowledge of the genetic sex-determination system, one can usually deduce the expected distribution of the sexes and, hence, the expected sex ratio variance. Increased or decreased variance is probably due to some degree of control, or the sex specificity of some process, during meiosis, fertilization or at a later stage.

Mother A

Sex ratio = 0.2

20 mated daughters

Mother B

Sex ratio = 0.2

15 mated daughters

Fig 5.1 Illustration of the selective advantage of low sex ratio variance in structured populations. Imagine two mothers A and B; each produces five broods of offspring with an overall sex ratio of 0.2 (five males, *M*, and 20 females, *F*). Mother A allocates sex to offspring precisely (no stochastic element). Males are produced first within each brood and the remaining offspring are females. Sex ratio variance is zero. Mother B allocates sex to individual offspring by a binomial process, with probability, *p*, of producing a male = 0.2 and brood sex ratio variance is nonzero. Assume a male can mate with an unlimited number of females and mating can only occur within broods ('strict local mate competition'). Fitness can be quantified as the number of mated daughters produced. Mother A's daughters all have mates and her fitness is 20. In contrast, five of Mother B's daughters cannot mate as there is no male in their natal brood, and multiple males occur 'unnecessarily' in one brood. Mother B's fitness is 15. We obtained Mother B's brood compositions by generating a random proportion on a pocket calculator for each offspring and defining the offspring as *M* when ≤0.2. Repeating the procedure would probably have generated different brood compositions, but Mother B's fitness can at best equal that of Mother A. Mother B could reduce the occurrence of all-female broods by increasing *p*, but fitness will be maximized by a reduction in variance while maintaining sex ratio at 0.2. See Table 1 in Hardy *et al.* (1998) for further modifications to this basic scenario.

5.4.1 Nonbinomial expectations
Some sex-determination mechanisms, such as environmental or haplodiploid sex determination, may produce no inherent stochastic variability.

In the environmentally sensitive systems of turtles, crocodiles or lizards, sex is determined largely by ambient temperature; hence, variation of the sex ratio may be explained to a large extent in terms of temperature variation (Chapter 7). Thus, sex ratio variation between and within nests will be a consequence of differences in temperature within and between nests; therefore,

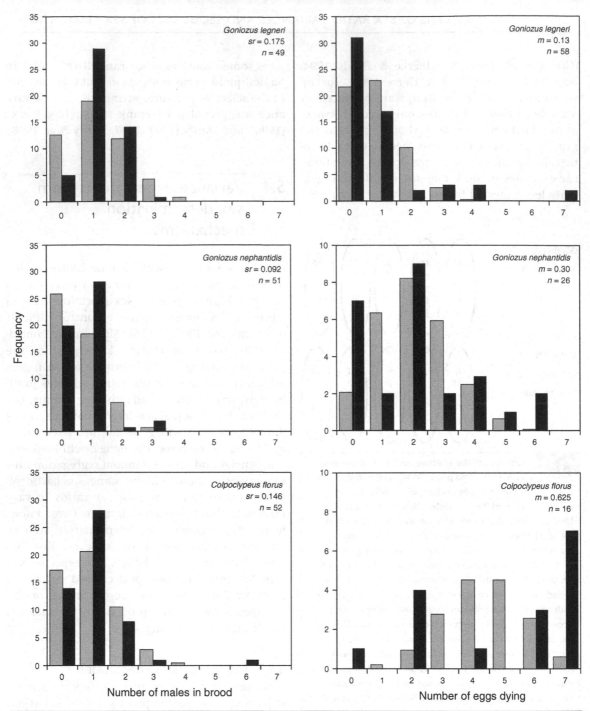

Fig 5.2 Examples of binomial and nonbinomial distributions. Observed distributions of sex ratio and developmental mortality in parasitoid wasps (dark bars) are shown alongside binomial 'expectations' (light bars). The left-hand panels show distributions of brood sex ratio, for 'n' broods containing seven offspring at maturity with mean sex ratio 'sr'. The right-hand panels show mortality distributions for the same species, from 'n' clutches of seven eggs with mean mortality 'm'. Note that sex ratio distributions have more one-male broods than the corresponding binomial distributions and tend to have lower frequencies of other sexual compositions (underdispersion) while the mortality distributions have higher than binomial frequencies of low and high numbers of eggs dying during development (overdispersion). Analyses of the complete datasets (i.e. including other brood and clutch sizes) using Meelis tests (section 5.6.1.2.1) indicate that these differences are significant, except for *C. florus* brood sex ratio which is not significantly different from binomial (in this species sex ratio variance is increased from significantly underdispersed at sex allocation by high developmental mortality). See Hardy and Cook (1995) and Hardy et al. (1998) for further details and examples.

sex ratio distributions could take any form. Evidence for parental control of the sex ratio thus requires a demonstration that parental behaviour manipulates temperature regimes (e.g. by choice of nest site). To elucidate whether there are any additional mechanisms at work, one would have to breed eggs under controlled experimental conditions.

In arrhenotokous haplodiploids (e.g. Hymenoptera), mated mothers store sperm and allocate sex by fertilizing (female) or not fertilizing (male) eggs. Hence, sex ratio variance will reflect this control process (Figure 5.2). In pseudo-arrhenotokous haplodiploids (Chapters 8 & 11), where males are produced by expulsion or inactivation of the paternal chromosome set in a fertilized zygote, sex ratio variance probably reflects variation of physiological factors responsible for chromosome-complement silencing. Maternal 'tagging' of zygotes in haplodiploid organisms has two important implications for analysis of sex ratio variance. First, it allows for precise control of the number of (fe)males at sex determination, so that variation at sampling should reflect variation in maternal optima (potentially modified by developmental mortality). Second, the number of males per clutch is, at least mechanistically, completely independent of the number of females. Hence, sex ratio variance is not a meaningful entity to analyse, as it is a composite of the variance of the number of (fe)males and that of clutch size, and in this case these are determined independently. Rather, one should examine variances of the number of males and females independently, and test against specific expectations that might be quite different for the two sexes. For instance, when we know that a parasitic wasp species has no developmental mortality before mating locally and clutch size never exceeds the number of females that can be fertilized by a single male, and that only one foundress contributes offspring to a clutch, we expect no variation in the number of males (i.e. one male per clutch is the optimal allocation). However, the number of females per clutch may vary (according to host quality, for instance), implying sex ratio variance. However, if there is non-negligible, partial developmental mortality of males, more than one male per clutch should be produced, with the exact value and variance depending on further assumptions of mortality rate, the number of females per clutch and the distribution of mortality (Nagelkerke & Hardy 1994). Depending on the distribution of host qualities and the form of developmental mortality, we might face higher, lower or equal numbers of females per clutch than in the comparable situations without developmental mortality. Again, expected variances of males and sex ratios would differ. Furthermore, if more than one foundress contributes offspring to some clutches, the adaptive number of males per clutch strongly exceeds that of the single foundress case (Hamilton 1967), but the variance of the number of males would depend on whether foundresses can detect other foundresess and the probability of subsequent foundresses adding offspring. At the same time, the optimal number of females per clutch depends on additional life-history parameters, e.g. the potential for intra-specific competition (West et al. 2001). No clear prediction of sex ratio variance can be derived from mere knowledge of the probability of multiple-foundress clutches, whereas variance in male numbers can be predicted.

5.4.2 Binomial expectations

Many genetic sex-determination systems generate the expectation that the probability of becoming male is constant for each zygote due to the random segregation of sex chromosomes during meiosis. For instance, in female heterogametic diploid organisms (such as birds, many snakes, amphibians and lepidopterans) sex is determined at first meiotic division by the segregation of Z and W chromosomes to the first polar body (WZ = female, ZZ = male), a process that appears to be fair and random (e.g. Krackow 1999). Such sex determination can be modelled as a Bernoulli process, i.e. from a statistical perspective it can be considered as a sequence of independent coin tosses, where the number of zygotes in a sample represents the number of tosses and the probability of showing a particular side of the coin is constant (often, but not necessarily, at $p = 0.5$). Given a sample of size n (a clutch) the expected number of males per clutch is $E(m) = np$ and its variance $Var(m) = np(1 - p) = npq$, hereafter termed binomial variance. Hence, in species with meiotic sex determination, clutch sex ratio variance is expected to be binomial around an average of np males. For brevity, we henceforth use the term 'clutch' to refer to all offspring groups, including mammalian litters.

The variance of the number of males per clutch, $Var(m)$, is often used for statistical analyses of binomial sex ratio variance. This is appropriate for a given clutch size where sex ratio variance is proportional to the variance of the number of males, i.e. $Var(m/n) = pq/n = Var(m)/n^2$. Note that results for the number of males (m) also hold for the number of females $(f = n - m)$, as these distributions are symmetrical (in p and q), though the majority of studies employ the proportion of males as the measure of sex ratio. Furthermore, the ratio (*sensu stricto*) of males to females (m/f) is asymmetrical and leads to undefined sex ratios if only males are present in a clutch, and to the value of 0 for any number of females, if there are no males present. This ratio is not a proper statistical measure as its mean and variance are, therefore, not defined.

In male heterogametic animals (such as mammals, nematodes, many amphibians, lizards, insects and spiders), X chromosomes segregate from either no other chromosome (XO systems: XX = female, XO = male; in spiders, up to three X are found) or from Y chromosomes (XY systems: XX = female, XY = male). Males produce equal numbers of X- and Y-chromosome-bearing gametes at meiosis and an ejaculate or sperm package, therefore, contains a large and practically equal number of X- and Y-bearing sperm. Provided sperm are not selected (by mothers or ova) for fertilization on the basis of chromosome complement, and are well mixed, p will be equal (in this case, $p = 0.5$), and independent between (fertilization) trials, generating binomially distributed numbers of males.

In the special case of sperm packages at the fertilization site being small, as may be the case in the *atrium tubae* (at the proximal end of the fallopian tube) of rodents, sex ratio variance has been suggested to be lower than expected from the binomial (Hornig & McClintock 1997). Indeed, if relatively few sperm are present in relation to ova, fertilization of an ovum by an X-sperm increases the probability of further ova being fertilized by Y-sperm, and *vice versa*. If there are equal numbers of X- and Y-sperm in each sperm package of a sample, the distribution of the number of males in the sample of resulting clutches is expected to follow a hypergeometric distribution, which would lead to smaller than binomial variance. However, sperm packages at the fertilization site in rodents for example represent a random sample of sperm from the much larger reservoir within the ejaculate, i.e. the numbers of X- and Y-sperm vary stochastically and differ between fertilizations. Theoretically, if the numbers of X- and Y-sperm in packages are unknown but represent a random sample from a large reservoir, the *unconditional* (i.e. not restricted to specified sperm package sizes) distribution of male zygotes is again expected to be binomial, and not hypergeometric. In other words, in a sample of clutches resulting from fertilizations with low sperm-to-ovum ratio, but with unknown (randomly distributed) effective sperm package composition, the variance of the sex ratio is expected to be binomial.

5.4.3 Causes of deviations from expected binomial variance

Although sex ratio variance roughly follows the binomial expectation in most species with chromosomal sex determination, significantly super-binomial and, more rarely, sub-binomial dispersion has been found (James 1975). Here we discuss some factors that may generate non-binomial variances. Less than binomial variance is termed 'underdispersion' or 'sub-binomial' and greater-than-binomial variance is termed 'overdispersion' or 'super-binomial'. We wish to emphasize that direct experimental evidence for any of the potential mechanisms of variance reduction is scarce, while variance-increasing mechanisms seem to be potentially widespread. Since the effects of mechanisms operating in either direction might cancel, the absence of deviation in any particular sample cannot be taken as hard evidence for the absence of factors affecting sex ratio variance (*sensu* Edwards 1960). Methods for the detection of such deviation are listed in section 5.6. Their usage depends in part on the actual composition of the sample.

5.4.3.1 Overdispersion due to between-clutch variation in p

Many factors that might increase variance after sex allocation are known and may apply to a wide range of organisms (Table 5.1). For instance, polyembryony (including monozygotic twinning) is well known to occur in mammals and some insects (Craig *et al.* 1997). It has been shown by RA Fisher (cited in Edwards 1958) that the proportional increase of sex ratio variance will equal the twinning rate in a population (Table 5.1).

Other factors include segregating sex-linked lethals or sex-ratio distorters and the heterogeneity of physiological causes of sex-specific mortality or fertilization success (Krackow 1995b). Such processes could be viewed as instances of p varying between clutches. The resulting distribution of the number of males is a mixture of binomials and will be overdispersed (super-binomial). It is well known in statistical literature that

$$\text{Var}(m) = \text{E}_p[\text{Var}(m|p)] + \text{Var}_p[E(m|p)]. \quad (5.1)$$

$\text{Var}(m|p)$ denotes the variance of m given the value of p and E_p denotes the expected value over all p values. The second term denotes the variance Var_p over all p values of the expected number of males at each p value, $\text{E}(m|p)$. Clearly, if p did not vary among clutches, $\text{Var}(m|p)$ would equal $\text{Var}(m)$. Hence, $\text{Var}(m)$ must increase

Table 5.1 | Some potential causes constraining sex ratio variance in clutches when unconstrained variance of the number of males is $\text{Var}(m) = npq$ (i.e. the sex-determination mechanism prescribes binomial variance at each clutch size n). Note that there is no empirical evidence for a mechanism causing sub-binomial variation at birth in animals with binomial sex-determination mechanisms, though there are reports of sub-binomial variation of birth sex ratios (see text)

Constraint	Cause	Effect	Var(m)=
p varies *between* clutches with variance $V(p) = pq\phi$	Maternal (family) effects	Increase (Edwards 1960)	$npq + n(n-1)V(p)$
p varies *between* clutches with variance $V(p) = pqt/(n-1)$	Monozygotic twinning rate t	Increase (Edwards 1958)	$npq(1+t)$
p varies *within* clutches with variance $V'(p)$	Time of fertilization or position in sequence	Decrease (Edwards 1960)	$npq - nV'(p)$
correlation (ρ) of sex within clutches[a]	Relative intra-/inter-sexual competition[b]; sex ratio-dependent mortality[c]	Decrease/increase (Brooks *et al.* 1991)	$npq[1 + (n-1)\rho]$

[a] For between-clutch variation, $\rho = \phi$. For within-clutch variation, $\rho = -V'(p)/(n-1)pq$. Hence, for negative correlation, $V(p)$ is undefined, as is $V'(p)$ for positive correlation.

[b] Speculative physiological cause and simulation in Krackow (1997b).

[c] Simulation in Krackow (1992).

when p varies between clutches. A general variance function for binomial distributions with p constant within clutches but varying between clutches is given in Table 5.1.

Note that sex-linked mortality *per se* will not affect the binomiality of a distribution (Fiala 1980) but merely changes the probability of producing a (fe)male. In statistical terms, if sex-linked mortality is constant for experiments under consideration, the distribution of the number of males remains a binomial, although with a different value of p. However, if the expected mortality rate, and hence the value of p, varies between clutches, the resulting distribution is overdispersed.

5.4.3.2 Underdispersion due to within-clutch variation in p

If p takes different values p_i for the ith of n trials, i.e. within clutches of size n, sex ratio variance will be reduced (Table 5.1, Edwards 1960). Although one might conclude from this that within-clutch variation in p leads to underdispersion (James 1975), this does not necessarily hold true for any form of within-clutch variation in p (e.g. positive correlations of the sexes might affect the p_i's as well as increase variance, as outlined below). Underdispersion might be caused by a change in p during the fertilization period in male-heterogametic systems where fertilization spans some time interval (Brooks *et al.* 1991, James 1998; see Gutiérrez-Adán *et al.* 1999 for experimental evidence in sheep), or by p changing during the ovulation sequence in female-heterogametic systems (see section 5.4.2). The latter might originate from sequence-dependent segregation distortion or sex-specific mortality (Krackow 1999).

5.4.3.3 Correlation of the sexes

Edwards (1960) has shown that positive or negative correlations of the sexes within a clutch would lead to increased and decreased sex ratio variance, respectively. This result has been used mainly in toxicological studies but can, of course, be applied to sex ratio data (Table 5.1; Brooks *et al.* 1991). Possible biological mechanisms that would result in such correlations might relate to

inter- and intra-sexual competition. For instance, it has been shown by simulation that intra-sexual competition for resources between blastocysts in a mammalian litter might lead to higher failure at implantation of members of the predominant sex and, thereby, could cause decreased sex ratio variance (Krackow 1997b). Such a process could also be interpreted as representing sex-ratio-dependent mortality (as has been demonstrated to occur in marsh harrier (*Circus aeruginosus*) broods, Dijkstra *et al.* 1998). If sex-ratio-specific mortality is either dome- or U-shaped with respect to the original sex ratio, it leads to over- and underdispersed sex ratios, respectively (Krackow 1992).

5.5 | Precise sex allocation and sex ratio precision

If the fundamental sex-determination mechanism leads to an expectation of binomial variation (section 5.4.2), the ratio of observed variance to expected binomial variance (i.e. the ANOVA χ^2 introduced in section 5.6.1.1 divided by its degrees of freedom; sometimes termed the variance ratio, R) is expected to be unity (Nagelkerke & Sabelis 1998). $R < 1$ indicates underdispersion (for sex ratio data this is termed 'precision') and $R > 1$ indicates overdispersion. Provided effects unrelated to the parental strategy can be excluded, $R \neq 1$ might indicate a degree of sex ratio control (Box 5.1). Furthermore, a correlation of R with another variable may suggest the mechanism by which control is achieved. Although this approach has not been used to date in 'binomial' species, a study using a related rationale (James' method; section 5.6.1.2.2) found that variance is increased in laboratory mice by multiple mating of a sire, suggesting that the effect of relative time of fertilization on p might be a cause of sub-binomial variance in mammals (relative time of insemination was assumed to vary more strongly in multiple than singly mating sires; Krackow 1997a). The effect of relative time of insemination was later confirmed more directly (Krackow & Burgoyne 1998).

Box 5.1 | Quantifying sex ratio variance deviation

Three methods of quantifying the deviation of sex ratio variance from binomial expectation have been used in sex ratio studies, the rationale being to compare between groups in order to draw inferences about factors that promote or restrict deviations. Since variances do not reflect some measure of a quantity but reflect some parameter of probability distributions, we point out some caveats.

1. The ratio R gives the observed variance of the number of males divided by the variance expected from a binomial distribution, taking clutch size for the number of trials (e.g. Nagelkerke & Sabelis 1998). If this ratio is smaller than 1, sex ratio is said to be precise. In fact, R multiplied by the degrees of freedom is the analysis of variance χ^2 value given in section 5.6.1.1 (sometimes called Pearson's goodness-of-fit χ^2: McCullagh & Nelder 1989) and can be used for statistical inference with large clutch sizes (in which case the number of males approaches the normal distribution). In cases where the normal approximation does not hold (i.e. small clutches), the distribution of R depends on the sample size and clutch size distribution. Hence, a larger departure from $R = 1$ in one sample need not imply a lower likelihood of the variance deviation than in a sample where R departs less. For non-normal data, statistical inference must, therefore, be based on tests especially designed for small samples and small clutch sizes. Hence, comparing R between groups or correlating the values with some variable may not always be meaningful. (See Hardy *et al.* 1998 for an attempt to explore the relationship between R and mean developmental mortality across bethylid wasp species.)

2. Some authors are concerned with separating clutch size effects on the mean sex ratio (which would obviously increase sex ratio variance when the overall sex ratio is used for estimating p) from other effects on the sex ratio variance. While estimating p at each clutch size separately would achieve this in appropriately large samples (section 5.6.1.2.2), cases with only one or a few clutches per clutch size, but large clutches, led Green *et al.* (1982) to propose a measure derived from regression analysis. Taking the residual squared error from analysis of variance (the observed residual variance multiplied by its degrees of freedom) and dividing it by the binomial expectation (sum of npq for all clutches where p is estimated from the regression) is supposed to yield an analysis of variance χ^2 value (similar to R multiplied by its degrees of freedom). However, this approach is not fully valid as the expected variance (npq) takes the expected p value at each clutch size from the slope of the regression, and this regression estimate is also used to calculate the residual variance. Hence, the measures are inter-dependent and the ratio is not amenable to straightforward statistical inference. In other words, any variance deviation might be caused by either a variance-affecting factor or by a misfit of the regression, when, for example, the true relationship is not linear, or the data are non-normal. The interpretation of group differences, therefore, is not straightforward regarding inferences on sex ratio precision. (See West & Herre 1998 for an attempt to explore the relationship between this measure of variance (termed 'Green variance' by West & Herre) and the proportion of single foundress broods across fig wasp species.)

3. A related approach takes the residual deviance from a logistic generalized linear model (GLM, Chapter 3) maximum-likelihood approach in order to quantify deviations from binomial variation, after 'fitting' parameters such as clutch size (as in Chapter 4). This approach takes into account the binomial expectation of the error variance in parameter estimation (in contrast to Green's method). The residual deviance in such a model is defined as 'twice the difference between the maximum achievable log-likelihood and that attained under the fitted model' and is asymptotically χ^2 distributed for normal data (McCullagh & Nelder 1989). The deviance divided by the degrees of freedom is expected to be 1 if the error is binomial, hence this factor (sometimes termed scale, s^2, (Chapter 3) or heterogeneity factor, HF) does indicate super-binomial variation if HF > 1 and sub-binomial (precise sex ratios) if HF < 1, similar to R. However, if the number of (fe)males is non-normal, its distribution again depends on details of the sample, hence straightforward statistical inference is not possible. (West & Herre 1998 used HF as well as the 'Green variance' to explore fig wasp sex ratio variances across species.)

In conclusion, for normally distributed data (large clutch size), R times ν equals the ANOVA χ^2 given in section 5.6.1.1 (which is equal to the Pearson's goodness-of-fit χ^2 from a GLM 'null model'), HF times ν (generalized deviance) is asymptotically equal to the generalized Pearson's χ^2, and Green's statistic is similar to the generalized Pearson's χ^2, but biased. If data are non-normal (clutch size not large), all statistics are biased depending on details of the sample. If datasets are, additionally, not very large, sizes of test statistics might even distort differences, i.e. larger values might not correspond to stronger deviations from expected variances.

Hence, in many cases the analysis of relationships between sex ratio variance and other (explanatory) variables can only be tentative and conclusions should be drawn with caution. Moreover, if the numbers of males and females are independently determined, sex ratio variances are not a meaningful entity for analysis (section 5.4.1). Rather, variances in numbers of males and females should be analysed independently, and expectations may even differ for each distribution (section 5.4.1). In some cases, the assumption of a Poisson distribution might be a reasonable, general null expectation (Krackow & Tkadlec 2001). Deviances from GLM with log-link functions may then replace those from logit-linked GLM and techniques applied as indicated above, subject to similar caveats.

On the other hand, for large samples, it is known that the difference of the deviances of two hierarchical models is distributed as χ^2. Hence, specified alternative hypotheses on the variance structure could be tested with GLM based on deviances, as indicated in 5.6.2.1. The hypotheses tested with this approach are, of course, not limited to the binomial distribution.

In species with sex-determination mechanisms that do not lead to an expectation of binomial variance (e.g. haplodiploid insects, section 5.4.1), calculation of the variance ratio, R, is, strictly, of limited value. This is because binomial variance cannot be taken as the expected value, hence a different value of R in two samples might reflect differences in expected variation as well as in stochastic variation. Imagine a haplodiploid species producing a maximum of two males at the beginning of an egg laying sequence (with mothers fertilizing all subsequent eggs). Due to unwitnessed mortality and/or underlying differences in the adaptive number of males, there may sometimes be 1 or 0 males in a clutch. This would lead to a variance of m, which would not be related in any way to the number of females produced later on. But R would then decrease with clutch size, erroneously leading to the conclusion that sex-allocation precision increases with clutch size (indeed, theory predicts that the selective advantage of sex ratio precision under local mate competition (LMC) is greater at smaller clutch sizes, Green *et al.* 1982).

Of course, sex ratio variance in haplodiploids may be influenced by stochastic components of fertilization control (imperfectly precise sex allocation) and/or developmental mortality (e.g. Hardy *et al.* 1998, Ratnieks & Keller 1998, Figure 5.2). In the first case, however, the expected distribution need not be a function of the overall clutch size, but may depend on the mode of determining the number of males to be produced. For example, if mothers withhold stored sperm after laying the penultimate egg in a clutch, but there is a low probability x of some carry-over effect of preceding sperm supply, the expected variance of the number of males is $x(1 - x)$, regardless of total clutch size, rather than npq.

Developmental mortality, on the other hand, can be of several types (Nagelkerke & Hardy 1994, Hardy & Cook 1995, Hardy *et al.* 1998). All-or-nothing mortality of clutches will obviously not affect any measure of sex ratio variance, as the sex ratios of clutches without mortality are unchanged and the sexual compositions of clutches with complete mortality are irrelevant. If there is a clutch-specific mortality risk (e.g. due to variation in the quality of developmental resources), stochastic variation will be added and the number of (fe)males will approach a hypergeometric distribution at each clutch size, provided the initial sex composition is fixed (Green *et al.* 1982, Nagelkerke & Hardy 1994). If the risk of mortality of individuals is independent of others in a clutch, binomial variation of the number of (fe)males will result, with the initial number of (fe)males, not the resulting clutch size, representing the number of trials (n).

In conclusion, R cannot tell us anything quantitative about the precision of sex allocation in species with potentially perfect control of the sex ratio, apart from the fact that variance is not binomial (but this value and related measures of variance have nevertheless been used to explore relationships between variance and other variables in such species, Box 5.1). Indeed, if the binomial has been ruled out as a potential null hypothesis, the variance of the number of each sex would have to be looked at independently. To investigate further the stochastic component in that variation due to the underlying sex-determination process and the pattern of developmental mortality, careful experiments are needed, e.g. by assessing variance in clutch composition in parasitoid wasps before any developmental mortality has occurred (Hardy *et al.* 1998).

5.6 | Analyses of sex ratio variance

In this section we give an overview of methods for testing whether data conform to a binomial distribution of the number of males per clutch and comment on the validity of each method. We also discuss tests to differentiate between specified alternative distributions. Mathematical notation is defined in Box 5.2. We do not, however, give a general overview of methods for testing deviations from nonbinomial expectations (section 5.4.1) as these will fully depend on the details of the mechanisms involved. We instead mention some important cases when they are of interest.

5.6.1 Detecting deviations from binomial expectation

For testing against general departures from the expectation of binomiality, both when p is known and unknown, standard goodness-of-fit tests can be applied. Examples are the χ^2 test for goodness of fit or the likelihood ratio test known as the *G*-test (these tests can be found in virtually every statistical textbook, e.g. Sokal & Rohlf 1995). Such testing requires reasonable sample sizes at each clutch size/sex ratio combination to accord to the test's requirements (e.g. Harmsen & Cook 1983). For testing the more specific assumption that the probability of becoming male is constant, i.e. the numbers of males in clutches follow a binomial distribution with common parameter p (known or unknown), a variance or dispersion test can be

Box 5.2 | Notation

α level of significance

χ^2 statistic with a χ^2 distribution

E expectation

f number of females in a clutch

ϕ probability distribution

G log-likelihood ratio χ^2 test statistic

i index (e.g. of a clutch in a sample)

k order of a Markov chain

$l(..)$ $l(\theta; m_1, \ldots, m_N) = \prod_{i=1}^{N} \phi(\theta; m_i)$, the likelihood function

$L(..)$ $\log l(\theta; m_1, \ldots, m_N)$, the log-likelihood function

m number of males in a clutch

M $E \sum_{i=1}^{N} m_i^2$

n $m + f$, i.e. clutch size

n_{tot} $\sum_{i=1}^{N} n_i$

N number of clutches in a sample

ν number of degrees of freedom

p probability of a male in a clutch

q $1 - p$

R $\dfrac{\text{sample variance}}{npq} = \dfrac{\text{observed variance}}{\text{binomial variance}}$

s $\sum_{i=1}^{N} m_i$

θ parameter (a vector)

U standardized test statistic from the Meelis test

V $\text{Var} \sum_{i=1}^{N} m_i^2$

Var variance

z statistic with standard normal distribution

viations (i.e. the observed variance times ν) and the expected variance is approximately χ^2 distributed with ν degrees of freedom ($\nu = N$)

$$\chi^2_{(\nu)} = \frac{\sum_{i=1}^{N} (m_i - np)^2}{npq}, \qquad (5.2)$$

where n equals the clutch size and m_i denotes the observed number of males in the ith sample, $i = 1, 2, \ldots N$. The parameter p is replaced by the maximum likelihood estimator

$$\hat{p} = \frac{\sum_{i=1}^{N} m_i}{Nn} \qquad (5.3)$$

if p is unknown. In that case the number of degrees of freedom is reduced by one ($\nu = N - 1$).

With variable clutch size n_i and when p is known the test statistic is

$$\chi^2_{(\nu)} = \sum_{i=1}^{N} \frac{n_i^2 (p_i - p)^2}{n_i pq} = \sum_{i=1}^{N} \frac{n_i (p_i - p)^2}{pq}, \qquad (5.4)$$

where $p_i = m_i / n_i$ and $\nu = N$.

If p is unknown, then

$$\hat{p} = \frac{\sum_{i=1}^{N} m_i}{\sum_{i=1}^{N} n_i} \qquad (5.5)$$

is substituted for p, where the number of degrees of freedom is reduced by one, as before. These tests are locally most powerful for testing homogeneity, i.e. for testing for a common probability p.

If there are effects of independent variables on sex ratios within a dataset, the above tests would, of course, indicate overdispersion (section 5.4.3.1). Several methods have been proposed to evaluate residual variance deviations after correcting for the effect of covariates on the sex ratio (Box 5.1, points 2 & 3). Tests based on deviances from generalized linear models (Chapters 3 & 4) give an impression of how sex ratio variance relates to a binomial distribution after correction for clutch size and other covariate effects, but, strictly, are valid only if the number of (fe)males can be assumed to be normally distributed, i.e. for large clutch sizes (Box 5.1).

applied. Tests of this type are called homogeneity tests.

5.6.1.1 Large clutches

Statistical theory shows that the number of males (m) in a sample can be assumed to follow a normal distribution in large clutches of size n (rule of thumb: $np(1 - p) > 9$; Johnson *et al.* 1992, p 114). To test for deviation of the observed variance from binomial expectation for the case of a known probability (p), we can apply another result from statistical theory which states that the ratio of the sum of the observed squared de-

5.6.1.2 Large samples of small clutches

When mean clutch size is small, the binomial distribution will depart from normality and the homogeneity test statistics of the previous section cannot be approximated sufficiently accurately by a χ^2 distribution. Despite this, Huck et al. (1990) used the χ^2 test outlined above, arguing that the test statistic will asymptotically approach a distribution close to the χ^2 distribution when the number of different clutch sizes in a sample is larger than five (Huck et al. 1990) and that asymptotic χ^2 estimates will lead to 'conservative' levels of significance (i.e. underestimate the type I error). As this method has not been approved in the statistical literature, we do not recommend unguided usage, i.e. asymptotic properties need to be verified with the given sample parameters.

5.6.1.2.1 THE MEELIS TEST FOR EQUAL CLUTCH SIZES

More appropriately, in cases of a large number of small clutches of equal size, we can apply the approximation developed by Meelis (see Nagelkerke & Sabelis 1998 for a derivation of the formulas). The expected values of the mean (M) and variance (V) of the sum of the squared number of males per clutch, for a given clutch size n, are defined by

$$M = \frac{s[s(n-1) + n(N-1)]}{Nn - 1} \qquad (5.6)$$

and

$$\begin{aligned} V = {} & \frac{s^{(4)}(n-1)[Nn(n-1) - 4n - 6)]}{(Nn - 1)^{(3)}} \\ & + \frac{4s^{(3)}(n-1)^{(2)}}{(Nn - 1)^{(2)}} + \frac{2s^{(2)}(n-1)}{Nn - 1} \\ & - \frac{s^2(s - 1)^2(n - 1)^2}{(Nn - 1)^2} \end{aligned} \qquad (5.7)$$

where s is the sum of males over the N different clutches. The notation $s^{(i)}$ denotes $s(s - 1)\ldots(s - i + 1)$. The test statistic U is then defined by

$$U = \frac{\sum_{i=1}^{N} m_i^2 - M}{\sqrt{V}}. \qquad (5.8)$$

Under the hypothesis of homogeneity, and if N is large enough ($N > 10$), U approximately follows a standard normal distribution: large negative values indicate underdispersion and large positive values overdispersion, and significance can be looked up in tables of the standard normal deviate z. Examples of applications of this method to sex ratio variance and variance in developmental mortality can be found in Hardy and Cook (1995), Hardy et al. (1998), Nagelkerke and Sabelis (1998) and Rabinovich et al. (2000). However, these studies expand the statistics to cases of varying clutch size with unapproved methods.

5.6.1.2.2 JAMES' METHOD FOR UNEQUAL CLUTCH SIZES

A method slightly less powerful for a single clutch size than the previous method, but appropriate for analysing samples of small clutches of unequal sizes, is given by James (1975). He suggests usage of the 'simplified maximum likelihood estimate' outlined by Robertson (1951). For a given clutch size n he defines

$$K_i = \tfrac{1}{2} \left[\frac{f_i(f_i - 1)}{\hat{q}^2} + \frac{m_i(m_i - 1)}{\hat{p}^2} - \frac{2f_i m_i}{\hat{p}\hat{q}} \right] \qquad (5.9)$$

Note that a typo replacing the minus in front of the third term by a plus sign has been perpetuated in the literature in the formulation of K_i (Krackow 1992, 1997a; cf. Soede et al. 2000).

It can be shown that

$$\frac{\sum_{i=1}^{N} K_i}{\sum_{i=1}^{N} I_i} \qquad (5.10)$$

is approximately normally distributed with mean zero and variance $1 / \sum_{i=1}^{N} I_i$, where

$$\sum_{i=1}^{N} I_i = \frac{Nn(n - 1)}{2\hat{p}^2\hat{q}^2} \qquad (5.11)$$

for N clutches of size n.

These results per clutch sizes can easily be combined by calculating $\sum \sum_{i=1}^{N} K_i / \sum \sum_{i=1}^{N} I_i$ (where the second summation is over clutch sizes), which is normally distributed with mean zero and variance $1 / \sum \sum_{i=1}^{N} I_i$. Hence, the first quantity divided by the square root of the second follows a standard normal distribution. This result can then be used to assess the significance level of any departure of Var($\hat{p}$) from zero (Table 5.1): significant positive test statistics (z) indicate overdispersion of the sex ratio and negative values indicate underdispersion. By taking the estimated sex ratio $\hat{p}$ at each clutch size, clutch size effects on the mean sex ratio will not affect the measure of sex ratio dispersion.

This method is specific in that it tests whether p fluctuates randomly between clutches. James (1975) further suggests that negative values of $\sum_{i=1}^{N} K_i$ might indicate that p varies within clutches, but this cannot be straightforwardly concluded from sex ratio underdispersion.

5.6.1.3 Small numbers of small clutches

If the sexual compositions of only a limited number of small clutches are known and the normal or χ^2 approximations are not sufficiently accurate, an exact test based on $\sum_{i=1}^{N} m_i^2$ must be applied. The critical values must be determined for each set of sample sizes $n_1, n_2, \ldots, n_N$, hence it is not feasible to tabulate all these values. Deriving exact critical values certainly requires statistical expertise; a related example of how to achieve this can be found in Haccou and Meelis (1994; p. 234).

A less powerful technique is given by Green et al. (1982). This method is based on the occurrence of a particular number of males in clutches of size n and is therefore not especially powerful for detecting under- or overdispersion, i.e. only very extreme cases would lead to significant departures. Examples of usage can be found in Green et al. (1982), Morgan and Cook (1994) and Rabinovich et al. (2000).

5.6.2 Modelling specified alternatives

Apart from testing for deviations from binomial expectation, specific alternative hypotheses can also be evaluated. This approach is meaningful if there is reason to believe that a certain set of factors must, at least predominantly, be responsible for deviations from the binomiality or the homogeneity hypothesis.

5.6.2.1 Specifying alternative hypotheses

The likelihood ratio approach is generally applicable to comparisons of several different under-

lying mechanisms affecting sex ratio variance (e.g. Brooks et al. 1991). Under the assumptions of the alternative hypothesis, the probability distribution of the number of males in a sample is denoted by ϕ, parameterized by θ (possibly a vector). It is assumed that the binomial distribution is a special case of ϕ, i.e. for a certain value (or combination of values) of θ, say θ_0, ϕ is a binomial distribution. For N clutches containing m males, the likelihood function for given clutch size n is defined by

$$l(\theta; m_1, m_2, \ldots, m_N) = \prod_{i=1}^{N} \phi(\theta; m_i). \quad (5.12)$$

The maximum of the logarithm of the likelihood function

$$
\begin{aligned}
L(\theta; m_1, m_2, \ldots, m_N) \\
= \log l(\theta; m_1, m_2, \ldots, m_N) \quad (5.13) \\
= \sum_{i=1}^{N} \log \phi(\theta; m_i)
\end{aligned}
$$

gives the maximum likelihood estimators of parameters for each of the two models. The maximum log-likelihood under the null hypothesis of binomiality, $L(\hat{\theta}_0; m_1, m_2, \ldots, m_N)$, and the maximum under the alternative hypothesis, $L(\theta^*; m_1, m_2, \ldots, m_N)$, can then be compared (note that the estimates of θ are in general unequal). The likelihood ratio test statistic is defined by

$$
\begin{aligned}
G = 2[L(\theta^*; m_1, m_2, \ldots, m_N) \\
- L(\hat{\theta}; m_1, m_2, \ldots, m_N)]. \quad (5.14)
\end{aligned}
$$

For large values of N, the statistic G is approximately χ^2 distributed, with the number of degrees of freedom equal to the difference in the number of parameters of the two models. This only holds when one class of models is a submodel of the other (otherwise techniques such as Akaike's rule (Akaike 1974) would need to be applied; an outline of these is beyond the scope of this chapter).

Maximum likelihood estimates can be calculated using software packages such as *SAS*, *GLIM*, *GENSTAT* or *S-Plus*. However, extensive help from statisticians with formulation of hypotheses, derivation of likelihood functions and implementation into software is needed by most experimental biologists.

5.6.2.2 Simulating datasets to examine variances

When analytical approaches become too complicated and intractable, estimating the significance of sex ratio variance deviations by Monte-Carlo simulations may give straightforward provisional insights. A detailed introduction to such randomization techniques can be found in Manly (1991). In principle, to simulate the binomial expectation for a sample, one can take the clutch size (n) distribution from the sample and estimate p from the marginal totals (i.e. sum of males and females at each clutch size). By repeatedly generating random numbers of (fe)males from a binomial distribution with parameters n and p for as many clutches as found in the sample, one derives a simulated distribution of sex ratios. If this is repeated many times (more than 5000 is often recommended), the proportion of simulated distributions that exhibit variance greater than the observed variance will roughly represent the significance of the deviation of observed from expected variance of the sample.

As an example, Hartley et al. (1999) took the clutch size distribution and total number of males and females from the sample, and re-allocated individuals 1000 times, at random. The deviance from a logistic model (Box 5.1) with only the mean sex ratio as parameter ('null model') was calculated in each case, and significance defined as the proportion of more extreme deviances than in the real sample. Hence, this is the same approach used by Huck et al. (1990, section 5.6.1.2), but with the distribution of the test statistic simulated rather than assumed to be χ^2 distributed. By taking the overall sums of males and females to estimate p, those tests included any effect of clutch size for overdispersion. In a similar simulation model, Lambin (1994) sets the p value a priori, i.e. assumes it to equal 0.5, as expected from chromosomal sex determination. However, one should keep in mind that the overall bias of the sex ratio would affect the expected variance and, hence, differences of

mean sex ratios would affect variance deviations in such a simulation.

Furthermore, nonbinomial distributions could be devised for generating the random datasets, and specific parameters may be added, e.g. sex-ratio-dependent mortality or nonlinear effects of clutch size and other covariates. One could then compare variance deviations observed and those generated by the simulations to tentatively conclude if the importance of the presumed relationships for the variance deviations hold. Such an approach was used by Avilés et al. (1999, 2000) to explore social spider sex ratios. They tested not only for departure from binomial expectation, but also from the outcome of exact sex ratio determination with the observed level of mortality before sexing (Avilés et al. 1999), and from the outcome of a random mechanism restrained to yielding at least one male per egg sac (Avilés et al. 2000).

5.7 | Allocation sequences

As well as evaluating the variance of the number of males in samples of clutches, the within-clutch production sequence of the sexes can be examined. Similar to analyses of sex ratio variance, sequence analyses assess whether observations deviate from the null expectation that the sex of offspring is not sequentially controlled. In addition, when sequential control of sex is established, techniques using additional information inherent in sex-allocation sequences may generate further insights into the actual mechanism of sequential control.

5.7.1 Control of sex-allocation sequence

When zygotes are produced sequentially, the probability of becoming male, p, might depend on the position of a zygote within a production sequence. Sequence effects can be viewed as variations of p within clutches, which may lead to underdispersion (Table 5.1, section 5.4.3.2). However, in some animals subgroups of offspring cannot readily be distinguished as 'clutches'. For instance, in pseudo-arrhenotokous mites, females lay eggs more or less continuously (Nagelkerke &

Sabelis 1998, Chapter 11). One way of testing for binomial variance would be to artificially break down the sequence into subsamples; for example, by taking all eggs laid per female during a specified observation period. As long as p does not influence n (the 'clutch' size), valid conclusions can still be drawn. Nagelkerke and Sabelis (1998) examined the eggs laid per one or more days by individuals within groups in two species of pseudo-arrhenotokous mites, and tested for deviation from the binomial using the normal distribution approach for large 'clutches' and U-statistics (section 5.6.1.2.1) for small 'clutches'. As the variance was highly significantly subbinomial, they concluded that mothers must have an effective means of controlling sex-allocation sequence.

5.7.1.1 Runs of allocated sex

If sequences are long, and the sex of each individual in the sequence is known, one can compare an observed sequence to the sequence expected from a binomial distribution using a runs test (Siegel & Castellan 1988). However, sometimes the structure of data does not allow straightforward testing. For instance, Heinsohn *et al.* (1997) report sequences of clutch composition in *Eclectus roratus* parrots. These parrots lay two-egg clutches, though frequently only one young is raised and, eventually, sexed. Hence, the sampled clutches consisted of mostly one male or one female, but two male, two female, and mixed-sex clutches were also found. Simulating one-sex and mixed-sex clutch sequences by taking the number of each category and the length of the sequences from the actual dataset, the authors estimated the probability of sequences containing the observed long runs of same-sex clutches to be far below 0.01. Furthermore, assuming sex-biased infanticide to be the cause of sex ratio distortion (see Heinsohn *et al.* 1997 for details) showed that sequences would still not adhere to a random model, i.e. there was insufficient infanticide to account for the long same-sex runs observed. On the premise that the cases reported were a random sample of *E. roratus* progenies, the authors safely conclude that these parrots must exert considerable control of the sex of their young prior to egg laying (this does not preclude the possibility of additional subsequent control).

5.7.2 Exploring the mechanism of sequential allocation

A sample of long sequences may also contain information that could suggest the underlying mechanism of sex ratio determination. In this case, the sequence may be seen as a discrete Markov chain, i.e. the probability of producing a male (or a female) depends on the sex of the preceding individuals. The sequence is a discrete Markov chain of order one (Kemeny & Snell 1976) if the probability of an individual at a specific position in the sequence being a male (or female) depends on the sex of the preceding individual. The sequence is a discrete Markov chain of order two if that probability depends on the sex of the two preceding individuals, and the sequence is of order k if the probability depends on the k preceding individuals, etc. In the case of complete independence, k equals zero. With a χ^2 test or a likelihood ratio test one can test, for example, whether the probability of a male depends solely on the sex of the preceding individual or on the sex of the two preceding individuals (Kemeny & Snell 1976, Haccou & Meelis 1994).

Nagelkerke and Sabelis (1998) applied this approach when analysing egg-laying sequences of pseudo-arrhenotokous mites (which have control over the sex ratio and allocation sequence, section 5.7.1, Chapter 11). Sex is determined by chromosome complement-silencing after laying fertilized eggs and mothers might somehow tag specific eggs by a physiological signal that would induce chromosome silencing (by expulsion or some other inactivation process). In order to control sex ratio over the egg-laying sequence, a mother might keep track of the tagging process and change p appropriately. Nagelkerke and Sabelis (1998) modelled the form of dependency of a variance function on clutch size. Increasing clutch size was achieved by either breaking down the known sequence into sub-sequences of one egg, two eggs, etc. or by dividing the sequences into 0.5-, 1-, 2-day, etc. intervals. The variance function was chosen to be the variance ratio R (section 5.5; see Box 5.1 for some caveats). The authors conclude that the form of the plot of R versus 'clutch size' is best approximated assuming an underlying Markov chain with memory window size including all previous eggs laid. One might, therefore, concede that a simple, discrete Markovian process is not an appropriate model to describe the structure of the data.

5.7.2.1 Exploring the form of sequential allocation

Once control of sex allocation has been demonstrated (e.g. by analysis of sex ratio variance) the form of the sequence in a sample of clutches can be of interest (e.g. Chow & Mackauer 1996). For instance, in haplodiploids with a high degree of sex-allocation control, there must be a mechanism by which mothers produce a certain number of sons per clutch, and vary this number in relation to clutch size, host quality, number of other mothers adding offspring to the clutch, etc. A common strategy is to produce a male egg first or early in the clutch (i.e. with higher probability at the beginning of the sequence) and then further males at intervals during the sequence, though other types of sequence (such as 'male egg last') have been identified (Hardy 1992).

The forms of such effects have, as yet, been explored largely visually. Up to now, researchers have characterized sequence effects in one of three ways: percentage of males for all eggs at a given absolute position in an egg-laying sequence, i.e. first eggs, second eggs, etc. (e.g. Dijkstra *et al.* 1990, Kilner 1998); sex ratio versus relative position in sequence, i.e. first eggs, last ones and those in between (Lessells *et al.* 1996); cumulative sex ratio over sequence, i.e. first eggs, second plus first eggs, and so on (Chow & Mackauer 1996). Wajnberg (1993) extracted specified features of the ordering of the sexes within clutches by deriving proxy variables. For instance, the mean position of males in the laying sequence was represented by the sum of rank positions of males, ranks being the number of an egg within a sequence. The exact probabilities of each value per clutch were then used in standard statistical tests to explore the influence of independent factors, such as host size, on the form of sequential production of the two sexes (Colozza & Wajnberg 1998).

5.7.2.2 A caution for sequential effect identification

Effects of the fertilization/laying sequence and clutch size *per se* will hardly be separable in many observational/survey studies. This is because larger clutches have more sequence positions, hence the effects of sequence and size will be correlated (Figure 5.3). This can be illustrated by imagining a population producing clutches of up to three eggs. When exclusively eggs laid third are distorted, only three-egg clutches can exhibit sex ratio bias. Alternatively, if the sex ratio of three-egg clutches were biased by an overall (rather than sequence-specific) change in p, then in the population as a whole the sex ratio of first-laid eggs would be least affected (because many one- and two-egg clutches would contribute unbiased eggs to the subset of data on first eggs). Data on second eggs would derive from two- and three-egg clutches, hence being more strongly affected, and data on third eggs would reflect the three-egg whole-clutch sex ratio. In consequence, sequence effects could only be separated from overall clutch size effects if one had complete knowledge of egg sequence per clutch and a sufficiently large sample to analyse sequences by clutch size (Cooke & Harmsen 1983).

Gathering complete knowledge of the sequence and original clutch size is often hampered by unwitnessed mortality occurring between clutch production and sex ratio sampling. In this case a prediction of sex ratio variance effects could be used to differentiate between the two mechanisms. If parents manipulate sex ratios according to overall clutch size, then p differs between clutches. Futhermore, any random mortality implies clutches with divergent p at each clutch size. Hence, even methods correcting for clutch size effects (sections 5.6.1.1 & 5.6.1.2.2) will exhibit increased variance. On the other hand, contingent sequence effects will reduce sex ratio variance. If variance is reduced, sequence effects are thus more likely than overall clutch-size-dependent ones. However, the variance-increasing shuffling effect of unwitnessed mortality will also be present if sex ratio is distorted according to ova sequence. Hence, if sex ratio variance does not deviate from binomial expectation, or is slightly overdispersed,

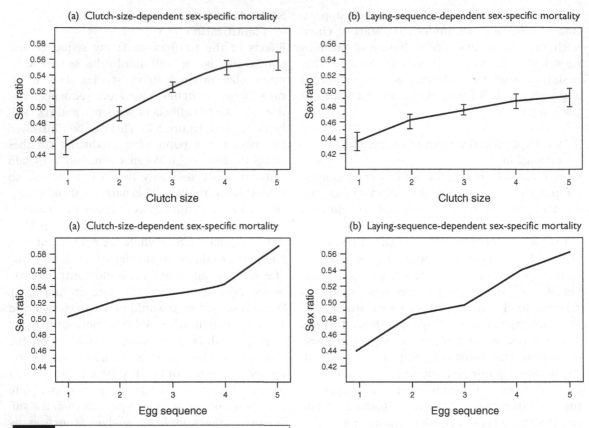

Fig 5.3 Intricacies of sequence effect analysis. In two simulations, 10 000 clutches were produced with clutch size drawn from a Poisson distribution with mean three, discarding clutch size zero and limiting larger ones to clutch size five. Sex was then sequentially assigned, at random, before mortality took place. Mortality was imposed on the eggs, with sex-specificity depending on clutch size or on laying sequence. Both kinds of mortality led to an apparent dependency of the sex ratio on clutch size as well as sequence position of viable eggs. (a) Clutch-size-specific sex-differential mortality was simulated by taking the survival rate s of male eggs as $s = 0.8$, and female ones at each clutch size n as $s = 0.8 + (3 - n)/10$. This leads to sex ratio increasing with clutch size (upper left panel) as well as laying sequence (lower left panel). (b) Sequence-specific sex-differential mortality was generated by setting the survival rate s of male eggs to $s = 0.8$, and that of female eggs at each sequence position seq to $s = 0.8 + (3 - seq)/10$. This causes the sex ratio, again, to correlate with clutch size (upper right panel) as well laying sequence (lower right panel).

sequence effects cannot be ruled out. Detailed (experimental) studies are more likely to distinguish the causes of bias than more sophisticated statistical techniques.

In birds, egg sequence has long been suggested to have some bearing on p (see citations in Krackow 1995b, 1999), but not without some scepticism (Fiala 1981, Cooke & Harmsen 1983). In the light of the above argument, extreme scrutiny is necessary when drawing inferences of sequence effects versus overall clutch size effects on offspring sex ratio from observational data. Taking relative sequence positions (Lessells *et al.* 1996, see above) would certainly remove the possibility that clutch size causes artifactual sequence effects. However, parameter estimates would be strongly biased, for example underestimated when sex ratios depend monotonically on sequence and 'last eggs' are composed of eggs from positions exhibiting varying skews. Furthermore, if clutch size had any bearing on the sex ratio, the direction of the slope with sequence would not be affected, but parameter estimates and variances would intricately depend on the clutch size effect and clutch size distribution. A recently proposed method to test for sequence effects while correcting for possible between-clutch overdispersion (which would result from clutch size effects) is to use clutch identity as a random effect in a two-level generalized linear mixed model on absolute sequence position (Krackow & Tkadlec 2001).

5.8 | Concluding remarks

Optimality theory identifies sex ratio variance as being under selection in certain circumstances; generally, low variance is adaptive. In contrast, mechanistic considerations lead to expected variances ranging from random (binomial) variation about a mean in species with genetic or chromosomal sex determination to no inherent stochastic variation in species with potentially perfect control of sex determination (e.g. arrhenotokous insects). While it may appear that measuring sex ratio variance can test functional explanations and identify sex-determination mechanisms, differences in sex ratio variance between samples or deviations of observed values from an expectation may have many explanations: adaptive, mechanistic and also stochastic. Moreover, the absence of significant deviations does not imply that there are no mechanisms of sex ratio variance manipulation at work, as different mechanisms can cancel each other's effects. When inferences from variance analyses are drawn with care, the utility of this approach is corroborated by fundamental insights that can be derived (e.g. Nagelkerke & Sabelis 1998).

We have given approved analytical methods for detecting sex ratio variance deviations from (mainly binomial) expectation, though these do not cover all potential empirical questions concerning sex ratio variance. Hence, some questions have been tackled by simulation methods. We would like to point out that while simulation methods are generally more easily applicable than analytical approaches, the differences between the two approaches should be kept in mind. With analytical methods one can prove a statement for a specified model. But, of course, if the model assumptions are not fully justified (e.g. when binomial variation cannot clearly be considered as the null expectation) any conclusion from such analysis is artifactual and may be wrong. Although simulation methods do not rely on potentially erroneous model assumptions, they are not performed for the whole parameter space (e.g. they consider a particular clutch size distribution where others would have been possible). The validity of any conclusion then depends on whether the simulations have been performed for a representative set of parameter values, which can only be investigated by robustness analyses. Provided such scrutiny is applied, simulation methods should, and will, become more widely applied when testing for specified effects on sex ratio variation in complex situations.

Acknowledgements

We thank Kees Nagelkerke and Richard Green for helpful comments.

References

Akaike H (1974) A new look at the statistical model identification. *IEEE Transactions on Automatic Control*, AC-19, 716–723.

Avilés L, Varas C & Dyreson E (1999) Does the African social spider *Stegodynus dumicola* control the sex of individual offspring? *Behavioral Ecology and Sociobiology*, **46**, 237–243.

Avilés L, McCormack J, Cutter A & Bukowski T (2000) Precise, highly female-biased sex ratios in a social spider. *Proceedings of the Royal Society, London, B*, **267**, 1445–1449.

Brooks RJ, James WH & Gray E (1991) Modelling sub-binomial variation in the frequency of sex combinations in litters of pigs. *Biometrics*, **47**, 403–417.

Chow A & Mackauer M (1996) Sequential allocation of offspring sexes in the hyperparasitoid wasp, *Dendrocerus carpenteri. Animal Behaviour*, **51**, 859–870.

Colozza S & Wajnberg E (1998) Effects of host egg mass size on sex ratio and oviposition sequence of *Trissolcus basalis* (Hymenoptera: Scelionidae). *Environmental Entomology*, **27**, 329–336.

Cooke F & Harmsen R (1983) Does sex ratio vary with egg sequence in lesser snow geese? *The Auk*, **100**, 1–8.

Craig SF, Slobodkin LB, Wray GA & Biermann CH (1997) The 'paradox' of polyembryony: a review of the cases and a hypothesis for its evolution. *Evolutionary Ecology*, **11**, 127–143.

Dijkstra C, Daan S & Buker JB (1990) Adaptive seasonal variation in the sex ratio of kestrel broods. *Functional Ecology*, **4**, 143–147.

Dijkstra C, Daan S & Pen I (1998) Fledging sex ratios in relation to brood size in size-dimorphic altricial birds. *Behavioral Ecology*, **9**, 287–296.

Edwards AWF (1958) An analysis of Geissler's data on the human sex ratio. *Annals of Human Genetics*, **23**, 6–15.

Edwards AWF (1960) The meaning of binomial distribution. *Nature*, **186**, 1074.

Fiala KL (1980) On estimating the primary sex ratio from incomplete data. *American Naturalist*, **115**, 442–444.

Fiala KL (1981) Sex ratio constancy in the red-winged blackbird. *Evolution*, **35**, 898–910.

Frank SA (1990) Sex allocation theory for birds and mammals. *Annual Review of Ecology and Systematics*, **21**, 13–55.

Green RF, Gordh G & Hawkins BA (1982) Precise sex ratios in highly inbred parasitic wasps. *American Naturalist*, **120**, 653–665.

Gutiérrez-Adán A, Pérez-Garnelo J, Granados JJ, Garde M, Pérez-Guzmán B, Pintado & De La Fuente J (1999) Relationship between sex ratio and time of insemination according to both time of ovulation and maturational state of oocyte. *Zygote*, **7**, 37–43.

Haccou P & Meelis E (1994) *Statistical Analysis of Behavioural Data. An Approach Based on Time-Structured Models.* Oxford: Oxford University Press.

Hamilton WD (1967) Extraordinary sex ratios. *Science*, **156**, 477–488.

Hardy ICW (1992) Nonbinomial sex allocation and brood sex-ratio variances in the parasitoid Hymenoptera. *Oikos*, **65**, 143–158.

Hardy ICW (1997) Possible factors influencing vertebrate sex ratios: an introductory overview. *Applied Animal Behaviour Science*, **51**, 217–241.

Hardy ICW & Cook JM (1995) Brood sex ratio variance, developmental mortality and virginity in a gregarious parasitoid wasp. *Oecologia*, **103**, 162–169.

Hardy ICW, Dijkstra LJ, Gillis JEM & Luft PA (1998) Patterns of sex ratio, virginity and developmental mortality in gregarious parasitoids. *Biological Journal of the Linnean Society*, **64**, 239–270.

Harmsen R & Cooke F (1983) Binomial sex ratio in the lesser snow goose: a theoretical enigma. *American Naturalist*, **121**, 1–8.

Hartley IR, Griffith SC, Wilson K, Shepherd M & Burke T (1999) Nestling sex ratios in the polygynously breeding Corn Bunting *Miliaria calandra. Journal of Avian Biology*, **30**, 7–14.

Heinsohn R, Legge S & Barry S (1997) Extreme bias in sex allocation in *Eclectus* parrots. *Proceedings of the Royal Society, London, Series B*, **264**, 1325–1329.

Hornig LE & McClintock MK (1997) Sex ratios are multiply determined: a reply to James. *Animal Behaviour*, **54**, 467–469.

Huck UW, Seger J & Lisk RD (1990) Litter sex ratios in the golden hamster vary with time of mating and litter size and are not binomially distributed. *Behavioral Ecology and Sociobiology*, **26**, 99–109.

James WH (1975) The distribution of the combinations of the sexes in mammalian litters. *Genetical Research, Cambridge*, **26**, 45–53.

James WH (1998) Hypotheses on mammalian sex ratio variation at birth. *Journal of Theoretical Biology*, **192**, 113–116.

Johnson NL, Kotz S & Kemp AW (1992) *Univariate Discrete Distributions.* New York: John Wiley.

Kemeny JG & Snell JL (1976) *Finite Markov Chains.* New York: Springer-Verlag.

Kilner R (1998) Primary and secondary sex ratio manipulation by zebra finches. *Animal Behaviour*, **56**, 155–164.

Kolman WA (1960) The mechanism of natural selection for sex ratio. *American Naturalist*, **94**, 373–377.

Krackow S (1992) Sex ratio manipulation in wild house mice: the effect of fetal resorption in relation to the mode of reproduction. *Biology of Reproduction*, **47**, 541–548.

Krackow S (1995a) The developmental asynchrony hypothesis for sex ratio manipulation. *Journal of Theoretical Biology*, **176**, 273–280.

Krackow S (1995b) Potential mechanisms for sex ratio adjustment in mammals and birds. *Biological Reviews, Cambridge*, **72**, 225–241.

Krackow S (1997a) Effect of mating dynamics and crowding on sex ratio variance in mice. *Journal of Reproduction and Fertility*, **110**, 87–90.

Krackow S (1997b) Further evaluation of the developmental asynchrony hypothesis of sex ratio variation. *Applied Animal Behaviour Science*, **51**, 243–250.

Krackow S (1999) Avian sex ratio distortions: the myth of maternal control. In: NJ Adams & RH Slotow (eds)

Proceedings of the 22nd International Ornithological Congress, pp 425–433. Johannesburg: BirdLife South Africa.

Krackow S & Burgoyne PS (1998) Timing of mating, developmental asynchrony and the sex ratio in mice. *Physiology and Behavior*, **63**, 81–84.

Krackow S & Tkadlec E (2001) Analysis of brood sex ratios: implications of offspring clustering. *Behavioral Ecology and Sociobiology*, **50**, 293–301.

Lambin X (1994) Sex-ratio variation in relation to female philopatry in townsend voles. *Journal of Animal Ecology*, **63**, 945–953.

Lessells CM, Mateman AC & Visser J (1996) Great tit hatchling sex ratios. *Journal of Avian Biology*, **27**, 135–142.

Manly BFJ (1991) *Randomization and Monte-Carlo Methods in Biology.* London: Chapman & Hall.

McCullagh P & Nelder JA (1989) *Generalized Linear Models.* London: Chapman & Hall.

Morgan DJW & Cook JM (1994) Extremely precise sex ratios in small clutches of a bethylid wasp. *Oikos*, **71**, 423–430.

Nagelkerke CJ (1996) Discrete clutch sizes, local mate competition, and the evolution of precise sex allocation. *Theoretical Population Biology*, **49**, 314–343.

Nagelkerke CJ & Hardy ICW (1994) The influence of developmental mortality on optimal sex allocation under local mate competition. *Behavioral Ecology*, **5**, 401–411.

Nagelkerke CJ & Sabelis MW (1998) Precise control of sex allocation in pseudo-arrhenotokous phytoseiid mites. *Journal of Evolutionary Biology*, **11**, 649–684.

Rabinovich JE, Torres Jordá MT & Bernstein C (2000)

Local mate competition and precise sex ratios in *Telenomus fariai* (Hymenoptera: Scelionidae), a parasitoid of triatomine eggs. *Behavioral Ecology and Sociobiology*, **48**, 308–315.

Ratnieks FLW & Keller L (1998) Queen control of egg fertilisation in the honey bee. *Behavioral Ecology and Sociobiology*, **44**, 57–61.

Robertson A (1951) The analysis of heterogeneity in the binomial distribution. *Annals of Eugenics*, **16**, 1–15.

Siegel S & Castellan NJ (1988) *Nonparametric Statistics for the Behavioral Sciences.* New York: McGraw-Hill.

Soede NM, Nissen AK & Kemp B (2000) Timing of insemination relative to ovulation in pigs: effects on sex ratio of offspring. *Theriogenology*, **53**, 1003–1011.

Sokal RR & Rohlf FJ (1995) *Biometry: The Principles and Practice of Statistics in Biological Research.* New York: Freeman.

Taylor PD & Sauer A (1980) The selective advantage of sex-ratio homeostasis. *American Naturalist*, **116**, 305–310.

Verner J (1965) Selection for sex ratio. *American Naturalist*, **99**, 419–421.

Wajnberg E (1993) Genetic variation in sex allocation in a parasitic wasp: variation in sex pattern within sequence of oviposition. *Entomologia Experimentalis et Applicata*, **69**, 221–229.

West SA & Herre EA (1998) Stabilizing selection and variance in fig wasp sex ratios. *Evolution*, **52**, 475–485.

West SA, Murray MG, Machado CA, Griffin AS & Herre EA (2001) Testing Hamilton's rule with competition between relatives. *Nature*, **409**, 510–513.

Chapter 6

Comparative analysis of sex ratios

Peter J. Mayhew & Ido Pen

6.1 | Summary

Comparative studies use the characteristics of different taxa as a source of data, and such studies have made important contributions towards the understanding of sex ratios. Comparative data require special methods for statistical analysis because not all the variance in taxon characteristics is evolutionarily independent. Solving this problem requires explicit phylogenetic and evolutionary assumptions that create challenges at each stage of a comparative study. Here we review the essentials of a comparative approach: asking questions, collecting data, choosing methods, analysing data and drawing conclusions. We include a worked example of a recent sex ratio study on New World nonpollinating fig wasps (West & Herre 1998a), which we analyse using phylogenetically independent contrasts and by simulation methods. Finally we review the relevant software for comparative analysis of sex ratios and how to obtain it.

6.2 | Introduction

I find that in Great Britain there are 32 indigenous trees[:] of these 19 or more than half... have their sexes separated, – an enormous proportion compared with the remainder of the British flora: nor is this wholly owing to a chance coincidence in some one family having many trees and having a tendency to separated sexes: for the 32 trees belong to nine families and the trees with separate sexes to five families.

Charles Darwin, from Stauffer (ed) (1975).

In recent years an abundance of reviews has dealt with the theoretical problems of conducting a comparative study (Ridley 1983, Brooks & McLennan 1991, Harvey & Pagel 1991, Martins 1996a, Pagel 1999). Many biologists today are, as Darwin was, aware of the problem of non-independence of species characters and how a knowledge of phylogeny and evolutionary processes can in principle help to overcome it, as well as make full use of the data (Harvey 1996). For researchers, this awareness does not in itself solve the problem for two reasons. First, the detail of how best to use phylogeny is still a matter of hot debate (see Harvey & Nee 1997, Price 1997a), and this makes comparative methods a potential minefield for the uninitiated who need to know succinctly the different analysis options and their implications. Second, the major problem for the empiricist remains how to gain experience with comparative methods. This chapter differs from all the aforementioned reviews, in that we try to provide practical as well as theoretical information to help readers conduct a good comparative analysis in general, and a good comparative analysis of sex ratios in particular.

The following sections work the reader through the stages that will most commonly characterize a comparative study of sex ratios. At each step we discuss whether and how comparative techniques can be useful, and the problems and challenges associated with them. We provide information on the most relevant software (Appendix 6.1), but encourage readers to follow the intermediate sections too. Possession of software can be very useful, but to perform good analyses you need to apply and interpret the software appropriately. If we have a single message it is that a good comparative study addresses that challenge, whereas a poor comparative study ignores it. We hope to impress upon the reader that care in choosing and using statistics applies here as for any other empirical study.

6.3 | Before starting

6.3.1 Why do a comparative analysis?

For present purposes, comparative studies are those which use the variation between different taxa as a source of data to help answer a biological question. Comparative studies are one of a number of empirical approaches to answering a question, and we distinguish studies in which the data consist of the properties of individuals within a species (observational, experimental and 'two-species comparative studies', see Garland & Adolph 1994, Price 1997b), and those in which data consist of the results of a number of separate studies addressing an identical question in one or more species (meta-analyses, Arnqvist & Wooster 1995).

Many readers will already be set on a comparative approach. Hypotheses may grow from informal comparative observations which then beg more formal investigation: 'where the experimental biologist predicts the outcome of experiments, the evolutionary biologist retrodicts the experiment already performed by nature; he teases science out of history' (Wilson 1994, p 167). In other cases, scientists may begin with a question and then ask if a comparative approach can help answer it. We see four reasons for choosing the comparative approach.

1. In addressing questions across different taxa, comparative analyses are likely to produce results of interest to more scientists compared to studies of a single species.

2. They do not suffer from the problem of extrapolating results from a single species to other species, which so often characterizes stud-

ies of individual 'model organisms': comparative studies address across-taxon variation, with real data.

3. Different taxa often show character variation that is difficult to obtain by experimental or other means: natural selection has performed manipulations and worked over timespans that are not possible in experiments.

4. As a result, cross-species variation is often much larger than within-species variation.

In summary, variation across taxa is often large, easily obtained, widely interesting and widely applicable: four good reasons for choosing a comparative approach.

Although a comparative approach can be very illuminating (Harvey 1996), it rarely tells a complete story. Comparative studies can be excellent ways to demonstrate correlation, but are much less adept at showing causation (but see Richman & Price 1992). For the latter, experiments may provide an answer. We also stress that the answer to any question may depend on the taxonomic level within which the question is phrased (e.g. Mayhew & Hardy 1998), so may depend on whether a comparative or other approach is taken. Thus, comparative and other kinds of analyses can be, but are not always, complementary.

What specific comparative questions do sex ratio researchers ask? By questioning the contributors to this book, we compiled 20 published comparative sex ratio studies (Table 6.1). Comparative studies have been used to investigate the causes and effects of sex ratio evolution in a wide range of taxa; from nematodes and Protozoa to birds and primates. Studies of Hymenoptera are particularly common. The Hymenoptera are species rich, often practical to study, and display particularly diverse sex ratios; they are ideal material for comparative study. Studies of plants are notably absent from our list, though not from lack of searching. Only a small proportion of plant species are dioecious and can therefore display sex ratios (Chapter 17), but perhaps this review will stimulate some comparative work on these. If we had expanded our search to include sex allocation in general, including monoecious as well as dioecious taxa, then plants would be better represented, not least by inclusion of

Table 6.1 | Some comparative studies of the sex ratio

Reference	Null hypotheses tested	Taxa studied (number of independent contrasts)	Methods used	Null hypotheses rejected/not rejected?	Comments and conclusions
Waage (1982)	Sex ratio is unrelated to number of hosts in a patch	31 species of scelionid wasp	Cross-species comparisons	Rejected: proportion of males increases with patch size	No phylogeny estimate available, analysis preceded modern comparative techniques. Evidence is consistent with effects of local mate competition.
Slagsvold et al. (1986)	Sex ratio of juveniles is unrelated to sexual size dimorphism in adults	23 species of bird	Cross-species comparisons	Rejected: sex ratio is biased towards the smaller sex	Authors also compared within and between the two major taxa: passerines and raptors, repeating their overall findings. Evidence mostly consistent with Fisher's argument for equal investment in both sexes.
Herre (1987)	Sex ratio is unrelated to inbreeding levels or frequency with which a given foundress level is encountered	13 species of fig wasp	Cross-species comparisons	Both hypotheses rejected: sex ratio is more male biased under lower inbreeding levels, and species that rarely encounter single foundress broods produce more males than expected	No phylogeny estimate available. Evidence consistent with the notion that inbreeding promotes female bias in haplodiploid species, and that adaptation is poorer in circumstances which are less commonly encountered, and so are weak selective forces.

Source	Prediction	Taxa	Method	Result	Comments
Griffiths & Godfray (1988)	Sex ratio is unrelated to clutch size	23 species of bethylid wasps	Cross-species comparisons	Rejected: proportion of males decreases with increasing clutch size	Phylogenetic evidence limited, and no within-generic resolution. The relationship is not significant when genus is controlled for, suggesting that a phylogenetic analysis would be worthwhile (see Hardy & Mayhew 1998a). Evidence is consistent with effects of local mate competition.
Johnson (1988)	Sex ratio unaffected by degree of local resource competition (LRC)	15 primate genera	Cross-genera comparisons	Rejected: more male-biased sex ratios found when LRC higher	Used analysis of higher nodes (see Harvey & Pagel 1991) after identifying appropriate taxonomic levels for comparison. Evidence consistent with effects of LRC.
Gowaty (1993)	Sex ratio is unrelated to the sex which disperses most	40 bird species, including 6 Anseriformes and 12 Passeriformes	Comparison of species and family mean sex ratios in Passeriformes and Anseriformes	Rejected: passerines (where females disperse) have female-biased sex ratios and anserines (where males disperse) have male-biased sex ratios	Evidence consistent with effects of LRC. Follow-up analysis and criticism by Weatherhead and Montgomerie (1995).
Read et al. (1995)	Sex ratios are unrelated to prevalence of infection	Parasitic Protozoa (taxa mostly unidentified) from 13 bird species	Simple graphical data plot, no statistics used	Rejected: sex ratios more female biased when prevalence low	Taxonomic difficulties precluded incorporation of phylogenetic aspects. Evidence consistent with effects of local mate competition.

Table 6.1 (cont.)

Reference	Null hypotheses tested	Taxa studied (number of independent contrasts)	Methods used	Null hypotheses rejected/not rejected?	Comments and conclusions
Shutler et al. (1995)	Sex ratios are unrelated to variables that could influence the probability of selfing	*Haemoproteus* blood parasites from 11 passerine bird populations	Cross-population comparisons	Not rejected	Phylogenetic issues not discussed, but presumably phylogenetic information within the genus is poor.
Weatherhead & Montgomerie (1995)	Sex ratio is unrelated to the sex which disperses most	16 bird species	Maddison's (1990) concentrated changes test	Not rejected	Analysed data in Gowaty (1993) using different methods with lower power, and concluding that the null hypothesis is not convincingly supported or rejected. Evidence suggests that local resource competition is not the primary influence on sex ratios.
Mitani et al. (1996)	Sexual dimorphism unrelated to operational sex ratio	18 primate species (16 independent contrasts)	Independent contrasts	Rejected: relative male size increases with operational sex ratio	Evidence supports the notion that primate testes size evolves in response to the degree of competition for mates among males.
Herre et al. (1997)	Sex ratio unaffected by degree of outbreeding, or in single-foundress broods by brood size	22 fig wasp species	Cross-species comparisons	Both rejected: proportion of males decreases with brood size and increases with degree of outbreeding	Authors state that their analysis is preliminary (presumably due to incomplete phylogenetic information), and that available information does not suggest phylogenetic constraint in characters of interest. Evidence consistent with the effects of inbreeding and local mate competition.

Reference	Hypothesis	Data	Method	Result	Conclusion
Poulin (1997a)	Sex ratios are unrelated to prevalence and intensity of infection	193 populations/species of nematode (maximum 51 contrasts) and acanthocephalan (18 contrasts)	Cross-population/species comparisons, independent contrasts	Rejected for intensity of infection in laboratory populations of nematodes, where increasing intensity decreases male bias. Otherwise not rejected	Weak evidence connecting the characteristics of parasite populations with their sex ratios.
Poulin (1997b)	Sexual dimorphism is unrelated to sex ratio	46 natural populations from 41 species (29 independent contrasts) and 30 experimental populations from 21 species (21 independent contrasts) of parasitic nematode	Cross-population comparisons, independent contrasts	Rejected: as proportion of males increases, relative male size increases	Evidence consistent with competition for mates among males as an agent of sexual size dimorphism.
West et al. (1997)	Sex ratios are unrelated to virginity levels	33 species of nonpollinating fig wasp and 22 species of pollinating fig wasp (11 and 10 independent contrasts respectively)	Cross-species comparisons, independent contrasts	Not rejected by both analyses	Sex ratio is used as an indicator of the degree of local mate competition. These results suggest that species with sons that compete locally for mates are not less likely to be unmated.
Dijkstra et al. (1998)	Fledgling sex ratio is unrelated to brood size	Seven bird species	Cross-species analysis	Rejected – the smaller sex is under-represented at fledging when brood size is small or large.	Data are consistent with sex-ratio-dependent mortality as an agent of fledging sex ratio bias in sexually dimorphic birds.

Table 6.1 (cont.)

Reference	Null hypotheses tested	Taxa studied (number of independent contrasts)	Methods used	Null hypotheses rejected/not rejected?	Comments and conclusions
Hardy & Mayhew (1998a)	Sex ratio is unrelated to clutch size and sexual dimorphism	26 species of bethylid wasp (maximum 12 independent contrasts)	Cross-species comparisons (independent contrasts)	Clutch size hypothesis rejected by both cross-species regression and independent contrasts: proportion of males decreases with clutch size. Sexual dimorphism hypothesis rejected only by cross-species regression: proportion of males increases with relative male size	Comparative analysis leads to considerable loss of power due to poorly resolved phylogeny. Evidence consistent with the notion that local mate competition is an agent of sex ratio bias and that as the degree of nonlocal mating increases, sex ratios become less biased, provided that sexual size dimorphism is a measure of the degree of nonlocal mating.
Hardy et al. (1998)	Sex ratio variance is unrelated to developmental mortality	16 estimates in 11 species of parasitoid Hymenoptera (9 independent contrasts)	Cross-population/ species comparisons (independent contrasts)	Rejected by cross-species analyses: variance higher when mortality higher. Not rejected by independent contrasts	Phylogenetic analysis leads to considerable loss of power. Evidence consistent with the notion that locally mating species lay initially precise sex ratios that are then exposed to developmental mortality that increases sex ratio variance in the emerging adults.

Study	Prediction	Taxa	Method	Result	Conclusion
West & Herre (1998a)	Sex ratio is unrelated to proportion of fruit infested and presence/absence of male wings	17 species of New World nonpollinating fig wasp	Cross-species comparisons and independent contrasts	Both hypotheses rejected for cross-species regression: proportion of males increases with proportion of fruit infested, and species with winged males have less biased sex ratios. Contrast analysis only performed for proportion of fruit infested, where the null hypothsis is also rejected	Only one independent contrast possible between a winged and nonwinged clade, but the value of that contrast is as predicted by theory. Evidence is consistent with local mate competition as an agent of sex ratio bias, and with the notion that sex ratios become less biased as the degree of nonlocal mating increases.
West & Herre (1998b)	Sex ratio variance is unrelated to proportion of broods produced by a single foundress	16 Panamanian pollinating fig wasps (15 independent contrasts)	Cross-species comparisons, independent contrasts	Rejected: variance lower as proportion of single-foundress broods higher	Supports the notion that the phenotypic variance of a trait under stabilizing selection is correlated with the trait's impact on fitness.
Fellowes et al. (1999)	Sex ratio is unrelated to male wingedness, proportion of fruit infested	44 species of Old World nonpollinating fig wasp (5 contrasts)	Cross-species analysis, independent contrasts	Rejected for male wingedness by both analyses: possession of winged males correlates with higher proportion of males. Rejected for proportion of fruit infested by cross-species analyses (contrast analysis not performed); proportion of males increases with proportion of fruit infested	Consistent with local mate competition as an agent of sex ratio bias, and with the notion that sex ratios become less biased as the degree of nonlocal mating increases.

Darwin's observations quoted at the start of the chapter.

Within the taxa represented, comparative studies have provided some of the most important data contributing to sex ratio research: questions about the effects of mating stucture are very common, and the comparative data are consistent with the notion that local mate competition influences the evolution of female-biased sex ratios (e.g. Waage 1982, Griffiths & Godfray 1988, Read *et al.* 1995, Poulin 1997a, Hardy & Mayhew 1998a, West & Herre 1998a, Fellowes *et al.* 1999, although see Shutler *et al.* 1995). A number of studies address inbreeding as an agent of sex ratio bias (Herre 1987, Read *et al.* 1995, Shutler *et al.* 1995, Herre *et al.* 1997, Poulin 1997a, Hardy & Mayhew 1998a, West & Herre 1998a, Fellowes *et al.* 1999). Three studies address the influence of local resource competition (Johnson 1988, Gowaty 1993, Weatherhead & Montgomerie 1995). Two of these (Gowaty 1993, Weatherhead & Montgomerie 1995) study the same set of data on brood sex ratios in birds but reach different conclusions. Two studies address sex ratio variances rather than just average sex ratios (Hardy *et al.* 1998, West & Herre 1998b). Comparative data relating to the effects of intensity of selection (e.g. Herre 1987, West & Herre 1998b) are especially valuable because information on this subject is hard-won. Two studies (Slagsvold *et al.* 1986, Dijkstra *et al.* 1998) test the causes of sex ratio bias in sexually dimorphic birds. Finally, sex ratio has sometimes been included in comparative studies as an explanatory variable, for example as a proxy for the intensity of sexual selection (Mitani *et al.* 1996, Poulin 1997b), or the degree of local mate competition (West *et al.* 1997). Although comparative studies have made important contributions to sex ratio research, there is also scope to broaden the range of issues and taxa which might be addressed with such techniques; we hope this chapter will stimulate that development.

6.3.2 Comparative analysis and phylogeny

Comparative analyses, like other statistical analyses, face two important challenges: to not reject a null hypothesis if it is true and to reject it if it is false. Failure to meet these challenges is called type I and type II error, respectively. If a test avoids type I error it has validity, and if it avoids type II error it has power. The validity of statistical tests lends us confidence in the inferences we make from data. Power is obviously important if we are to conclude all we might from our observations, but if we never conclude anything we are also not drawing incorrect conclusions. If a valid test can also be powerful, so much the better. The power and validity of a comparative test can be demonstrated by using the test to analyse dummy datasets in which the true evolutionary relationships between characters are controlled, and observing how often they reject the null hypothesis under consideration (e.g. Grafen 1989, Purvis *et al.* 1994, Grafen & Ridley 1996).

The major obstacle to finding valid comparative tests arises from the nonindependence of related taxa. Closely related species are likely to be similar because they inherit characters from a common ancestor. This means that not all the variance in species characters is the result of independent evolutionary events: species can only have been independent of other species in the dataset after the last moment they shared an ancestor with one of them (Felsenstein 1985). If related taxa share variables causing evolution to proceed in the same way on related parts of the evolutionary tree (so called 'third variables', Ridley 1989), taxa can only be treated as evolutionarily independent if we have controlled for them statistically (Price 1997b). If we treat species as fully independent, for example by taking raw species characters as data for analysis, we may overestimate the amount of evolutionarily independent variation, and will artificially inflate the significance of any test used, a problem known as phylogenetic overcounting.

Statistical tests currently attempt to deal with phylogenetic overcounting in three ways: the first and most common approach is to try to partition out the variation in the raw species dataset that is evolutionarily independent, and use that for analysis instead (Figure 6.1). The second possible approach is to continue using the raw species data but adjust the significance value of the test statistic to account for the nonindependence of species (Figure 6.1). A third possible approach is to ask how well alternative evolutionary

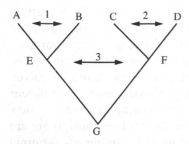

Fig 6.1 A schematic representation of some comparative methods. Using just raw species data (A, B, C, D) gives four data points but may be invalid due to the nonindependence of species. Comparing extant sister taxa allows us to partition out variation in A and B that is independent of E (comparison 1) and C and D that is independent of F (comparison 2), but only leaves the two data points instead of four. Comparing ancestors also allows us to partition out the variation in E and F that is independent of G (comparison 3) if we are prepared to assume an evolutionary model to estimate E and F. That same evolutionary model also allows us to simulate a distribution of null species data and hence use the raw species data to compare with that distribution (four points). Alternatively, we can try to fit an evolutionary model of correlated character evolution onto the data and the phylogeny and compare the likelihood of this model with that of an alternative model assuming independence of characters.

models fit the data across a phylogeny, and to compare the likelihood of those models, instead of analysing the data themselves (Figure 6.1). All three such tests require assumptions about the phylogeny and evolutionary processes that link the taxa in our dataset together: 'Tell us how to reconstruct the past, and we shall perform the comparative analysis with precision' (Harvey & Pagel 1991).

Although validity has been the major motivation for incorporating phylogenetic knowledge into comparative tests, there are other benefits and costs involved. Another benefit is the satisfaction of making full use of the data (Harvey 1996). Taking a phylogenetic perspective can allow one to ask whether evolutionary correlations apply equally well across different parts of the tree and this can allow one to estimate how the extant character states in a dataset have arisen (Mayhew & Hardy 1998). Furthermore, statistics may be used not just to estimate significance but also coefficients of association and regression slopes. Again, if the assumed model of evolution is correct, phylogenetic techniques allow more

accurate estimates of such parameters than cross-species comparisons (Harvey & Pagel 1991). In particular, if correlated evolution varies slightly between different taxa, slopes calculated from cross-species analyses might be biased according to the relative representation of different taxa in the dataset.

We identify three costs of phylogenetic awareness: time to do the test, loss of power (higher type II error) if the phylogeny is not fully resolved, and lower validity (higher type I error) if incorrect evolutionary assumptions are made.

The time taken to perform a phylogenetically based test is often very small, because appropriate software is now readily available. Time may be required to familiarize oneself with each application, and to assemble phylogenetic knowledge about the group in question, but this is a worthwhile investment.

Although loss of power in a 'phylogenetically aware' test might be great, it need not always be so. If phylogeny and evolutionary processes are well known, more independent variation can be partitioned out, and very little power is lost. Greater power can be gained by comparing ancestors as well as extant taxa, or by simulation methods (Figure 6.1), and if the phylogeny is poorly resolved one can now relatively easily test the effect of different possible phylogenies (e.g. Martins 1996b). However, increased power often demands more evolutionary assumptions. If these assumptions are inappropriate, there is a higher risk of type I error (Price 1997b).

A consensus on how to tackle the problem of incorrect assumptions has yet to emerge. One school of thought recommends abandoning phylogeny altogether (e.g. Ricklefs 1998), and especially if the characters are thought to be phylogenetically labile (e.g. Westoby et al. 1995). One superficially compelling reason is that in studies where both cross-species and phylogenetic methods have been used, the results usually agree, suggesting that phylogenetic methods are not necessary (Ricklefs & Starck 1996). A more moderate approach is to test for phylogenetic independence (Abouheif 1999) or lability (Björklund 1997) prior to conducting a test, and to use phylogenetic methods only if the characters are not labile or not phylogenetically independent.

One point of concern about trait lability is that even highly labile traits can be affected by taxa-specific traits and hence be subject to phylogenetic overcounting (see Ridley 1989). Probably the most sensible option is simply to use phylogenetic methods carefully. If power is not a problem, one can use methods, such as pairwise comparisons, which make fewer evolutionary assumptions than others. Tests are also becoming available which incorporate the uncertainty surrounding our evolutionary assumptions into the analysis (e.g. Martins 1996b, Garland & Ives 2000). With other methods it is often possible to infer from the data an evolutionary signature which can be used to choose appropriate assumptions (see Pagel 1999). This may include retesting for phylogenetic independence once the comparative method has been applied (Freckleton 2000).

We argue that careful use of phylogenetic methods allows one to reap the benefits of phylogenetic awareness, including reduced type I error, with little extra type II error. But even if you have decided on a phylogenetically based test, exploring cross-species data is informative. The former tells you about evolutionary correlations, but the latter tells you about trends in extant characters that are the product of those evolutionary correlations. Most biologists are interested in pattern as well as process, and many of the studies in Table 6.1. have done this. In short, although phylogenetic awareness can add much at little cost, cross-species comparisons also have their place if used carefully (Price 1997b).

6.4 | Collecting comparative data

6.4.1 Taxon characteristics

Data from comparative analyses generally derive from two sources: first-hand collection by original researchers using standardized methods (e.g. West & Herre 1998a), or second-hand collection by 'data-miners' from several studies previously published by different authors (e.g. Griffiths & Godfray 1988). Because data collection on several species tends to be time consuming and costly, whilst second-hand data may often be plentiful and easy to collect, the latter are more common. This leads to several challenges: first, comparative analyses are often forced by available literature-based data to use species as terminal taxa, but to what extent is sex ratio, or another variable, a species character? Second, error is introduced into the estimates of terminal taxon characteristics by different methods used by different researchers, and by typographical errors. How should this error be dealt with? Third, there are several possible ways of summing up terminal taxon properties with a single number. Which should be analysed? Fourth, more than one independent estimate of terminal taxon characters may exist. How should these be treated? Finally, how far should we extend our data collection, and which taxa should we be targeting?

Some characteristics of taxa have a high spatial or temporal variance, such that the average is very poorly estimated by studies conducted at a single location at a single time. For such characteristics, comparative literature-based studies may be useless, and indeed in direct contradiction of statistical assumptions (independent contrast analyses, for example, assume that within-species variation is negligible and that characteristics are measured without error). The only way directly around this problem is to sample more intensively. Repeatability analyses, which describe the variability of interest from large datasets, allow one to test the relevance of an estimated average (e.g. Arneberg et al. 1997). However, in most studies this is impractical and the most we can hope to do is use our background knowledge of characters to try to identify the scale of the problem. With sex ratios conclusions will vary, for different taxa have different sex determination mechanisms, which may be more or less constraining (Chapters 7 & 8). We know that sex ratios in many species are sensitive to local conditions. In such species, populations may be the appropriate taxon for comparison.

One possible solution to the problem of measurement error is to set criteria for including a study in the analysis to try to reduce the noise in the data. Purvis and Harvey's (1995) comparative test of Charnov's mammal life-history model illustrates how this might be done. They assembled a dataset of 80 mammalian life tables after rejecting studies conforming to any one of seven criteria, including those based on captive populations and those based on small samples (<30 individuals). All data were examined to check for exceptional values which might represent conversion or typographical errors, or data derived by chance from the tails of the real biological distributions.

Rejecting large quantities of data is a luxury affordable only with intensively studied taxa. More often studies will be forced by paucity of data to include most published information. Examining trends at different taxonomic levels can then suggest whether there was a large amount of measurement error in the species-level data. When there is, relationships may hold better at deeper levels of the phylogeny, because, at least with some methods, measurement error can be eliminated in the estimation of ancestral states.

The problem of summarizing terminal taxon properties with a single statistic is common to most comparative studies. Usually some kind of average is used for continuous variables; commonly the arithmetic mean, the median, or the mid-point of the range. What can be used depends on what is reported in the literature, but if raw data are available it may be wise to examine them graphically to generate the best possible estimate. If the data are right skewed, as is common for ecological or life-history characters, the median or mean of the log-transformed data may be more appropriate than the arithmetic mean. If possible, it is worth trying to standardize the measure used within each study to reduce the error variance in the data.

How to treat independently collected estimates of variables depends on the phylogenetic independence of each study. If studies of distinct populations allow all the relevant variables to be estimated within each population, these can be included as separate taxa within the analysis. Møller and Birkhead (1992) tried this for populations of barn swallows, which nest solitarily or communally (see Harvey & Nee 1997 for a critique). More commonly it will not be clear if estimates are drawn from distinct populations, and not all relevant variables may be collected from a single study. Such cases may be treated as independent estimates of the same population and an average value calculated from both as above.

In principle, within-species variation can be incorporated into the statistical model by specifying a standard error of each species estimate (Martins & Hansen 1997), though this is not possible with many common methods. One way to accommodate it into independent contrasts is to extend the branch lengths of taxa for which estimates are uncertain so that these are weighted less in the calculation of contrasts.

Taxon characteristics may require transformation prior to analysis. If sex ratio is the response variable and the data are assumed to be fully continuous, arcsine square-root transformation is appropriate. Whilst logistic regression is generally preferred to data transformation in other, non-comparative, sex ratio analyses (Chapter 3), this is presently difficult to achieve in many comparative tests because analyses are not carried out on the raw data but on some other calculated values (such as contrasts), which have their own analytical demands (Garland et al. 1992). Logistic analyses are however theoretically possible in this context (Martins & Hansen 1997). Simulation methods are an exception which do use the raw data, and do allow logistic regression to be used (sections 6.5.5 & 6.6.2). Other continuous data are often right skewed and may need to be log-transformed. A given change in a variable usually carries higher biological significance if it occurs when the absolute values are small than if they are large. Log-transformation has the effect of transforming absolute into relative differences, and so tell us more meaningful biology. Some tests, such as independent contrasts, assume a Brownian (random walk) model of evolution, and logged data are more likely to conform to this model (Garland et al. 1992, Freckleton 2000).

How far should one go with data collection? One consideration is that more data can add power. The extent to which new data add power depends both on the intended method and the phylogeny used. Limitations may be reached due to paucity of relevant studies. If there are ample data in the literature, one might profit from both collecting more estimates of species already covered, to reduce the error variance in the dataset prior to analysis, and including new taxa that bring to the dataset novel information about correlated evolution. A datum on a different genus may be worth many data on genera already covered if it allows more evolutionarily independent variation to be partitioned out.

Which taxa may reasonably be incorporated into a comparative analysis? Taxa must be sufficiently closely related, or have sufficiently similar biologies, for a single question to be relevant to them all, but sufficiently distantly related for evolution to have produced variation of interest (Pagel 1994a). Sometimes this may include distant relatives, and sometimes populations of the same species. One situation to be wary of is when many of the included taxa are very close relatives, such as species in the same genus, but a few are more distantly related, such as species from different families (e.g. Hardy *et al.* 1998). In such cases it is worth examining whether comparisons among or between different families give very different results. If the analysis makes assumptions about the branch lengths of the phylogeny, checking the validity of and modifying the assumed branch lengths can be particularly illuminating (Garland *et al.* 1992).

A final worthwhile step is to make the full dataset available via a published table, appendix or journal website. Comparative datasets can take much effort to compile. Future researchers will be grateful not to have to start from scratch.

6.4.2 Phylogenies

In most cases, comparative analyses will require a two-pronged approach to data collection: collection of taxon character estimates and collection of phylogenetic information. Well-defined species-level trees with branch length estimates exist for some groups such as primates (Purvis 1995) and carnivores (Bininda-Emonds *et al.* 1999). More often branch length information is lacking, different species appear on different phylogeny estimates or cladograms, and some on none. Then, estimates of phylogeny must be assembled according to the species in each study. Even where no cladograms are available, taxonomy may still allow some best guess at phylogenetic relationships (e.g. Mayhew & Blackburn 1999), although one should be aware that taxonomies are sometimes not derived from cladistic principles and hence may not even attempt to estimate phylogeny. One can rarely afford

the luxury of ignoring taxonomy. In exceptional cases, such as when species in the same genus are studied, there may be both no cladistic or taxonomic estimate available. In that case one might consider investigating the effect of different possible phylogenies.

Where different cladograms or taxonomies are available, one has several options (Bininda-Emonds *et al.* 1999). First, one might consider computing a consensus tree, which simply describes on which relationships the available phylogeny estimates agree. Most phylogeny reconstruction software will do this (Appendix 6.1). An alternative is to combine all the raw morphological or molecular data into one large matrix and reanalyse it. Reanalysis and consensus estimates are feasible only if previous analyses consider the same set of taxa. If there is some overlap but also much discrepancy in the taxa considered, and many phylogeny estimates exist to cover the taxa of interest, an alternative option is to compute a 'supertree' (Sanderson *et al.* 1998). An example of the latter is Purvis' composite estimate of primate phylogeny (Purvis 1995), which presents an estimate for all 203 extant primate species, resolved into 160 nodes of which 90 are dated, and derived from 112 source publications. This composite tree was calculated by recoding the nodes on the published trees as binary characters and then combining the datasets for different trees. The combined data were then analysed by standard parsimony algorithms. The benefits and drawbacks of these alternatives are still debated, but there are cases, as above, where one option is clearly preferable (Sanderson *et al.* 1998, Bininda-Emonds *et al.* 1999).

Such studies, though laudible, are time consuming and require familiarity with techniques for phylogeny reconstruction, so short cuts may be preferred. Luckily, in many cases trees can be combined by eye because there is little discrepancy between them, or little overlap in the taxa studied. A large drawback is that such estimates are not explicit about their assumptions, so different researchers may come to different solutions. Nonetheless, if this is the preferred option, we suggest at least consulting an expert taxonomist on the relevant group before attempting analysis. It is extremely easy to make big mistakes when combining trees by eye if one is not very familiar with the groups concerned.

An alternative, if a few relatively complete but conflicting estimates are available, is to do a separate analysis on each estimate (e.g. Seehausen *et al.* 1999). Discrepancies in the analysis outcome then may allow identification of the most important parts of the tree for further taxonomic work, and the robustness of the outcome under present phylogenetic uncertainties.

Branch length information is not always required in comparative tests (e.g. Maddison 1990), but if it exists methods are available that use it (e.g. Purvis & Rambaut 1995). Normally though, branch lengths are unknown and, if required, must be assumed. Adjusting branch lengths can often be a useful way of incorporating appropriate statistical assumptions when it is difficult to differentiate between evolutionary assumptions (section 6.6.1).

Whatever the final approach, the phylogeny used should be made available so that researchers can repeat the analysis if necessary, and the methods used to construct it made explicit, so that readers can judge its worth. No single estimate of phylogeny will ever be the last word, so facing up to its faults displays good scientific sense.

6.5 | Choosing comparative methods

The analysis method chosen will depend on the way the data are coded, the computer system available, the extent of phylogenetic knowledge, and the evolutionary and statistical assumptions which one is prepared to make. This section is a quick guide to those requirements for the most common methods.

6.5.1 Cross-species comparisons

Because they may give biased or invalid estimates of the degree of correlated evolution, cross-species comparisons must be used with extreme care, or as some would advocate 'kept at the other end of a barge pole' (Ridley & Grafen 1996). However, cross-species comparisons are still the most common comparative method used (15 of the 20 sex ratio studies in Table 6.1.). They do not require phylogenetic information, but that is exactly why they are often invalid. They can provide a useful way of illustrating trends in extant character states, but drawing conclusions about correlated evolution is risky. We recommend that comparative researchers at the very least explore cross-species trends graphically or with contingency tables, but also that they go on to further analyses if at all possible. Both categorical and continuous variables and any combination of the two can be analysed easily. If sex ratio is the response variable, logistic regression is recommended; transformation (arcsine square-root) of sex ratio prior to using other parametric tests is an alternative (Chapter 3). Most statistics software packages allow this.

6.5.2 Independent contrasts

The most popular phylogenetically based method, used in 9 of the 20 studies in Table 6.1, is that of independent contrasts. There are several possible implementations, based on different assumptions. The method is due to Felsenstein (1985), who noted that differences between two daughter lineages across a node (contrasts) represent variation acquired since those taxa last shared an ancestor, and so are phylogenetically independent in the absence of confounding third variables, which they can but may not always control for (Price 1997b, Freckleton 2000). Contrasts can be calculated between ancestral taxa as well as extant taxa if a given model of evolution is assumed, which allows those ancestral states to be calculated. Most commonly, the assumed evolutionary model is Brownian, which means that evolution proceeds in a random walk fashion, and therefore that ancestral states are a weighted average of their descendent states. The suitability of this model should be questioned, and appropriate prior transformation can result in data that do evolve as assumed and are phylogenetically independent (Freckleton 2000, section 6.6.1). Contrasts may be analysed parametrically by linear regression through the origin or nonparametrically by a sign test (Garland *et al.* 1992). Regressions on contrasts must be forced through the origin because the average change in the dependent variable, in response to zero change in the independent variable, must itself equal zero (Garland *et al.* 1992). The original implementation is in Felsenstein's *PHYLIP* software package (for MS-DOS or Apple Macintosh), but the phylogeny used must be fully resolved.

Subsequent implementations of Felsenstein's approach allow analysis even if parts of the phylogeny are uncertain. One way is to contrast across soft polytomies (nodes representing uncertain relationships of more than two sister taxa). The most popular package to do this is *CAIC* (Purvis & Rambaut 1995) for Apple Macintosh, which outputs the contrasts (as well as some limited data analysis and model checking in the newest version) which should then be analysed using a standard statistics package. It does allow contrast analysis of categorical characters, using Burt's (1989) method, if they are ordered explanatory variables, but otherwise variables must be coded as continuous. This means that bounded response variables, such as sex ratios, should be transformed (usually arcsine square-root) prior to analysis. In addition, it makes sense to transform other continuous variables prior to analysis to conform to the Brownian motion model of evolution, which is often achieved by a logarithmic or other power transformation (Garland *et al.* 1992, Freckleton 2000).

Phylogenetic regression (Grafen 1989) is a macro run in the *GLIM* statistical package (and hence is MS-DOS, UNIX and Apple Macintosh compatible), and there are different versions corresponding to the different versions of *GLIM*. The output is perhaps less useful than *CAIC* – an *F* statistic, *P* value and slope of the relationship. Contrasts are not output, and neither are reliable values of *R*-squared. Phylogenetic regression has been used to analyse categorical as well as continuous characters, ordered and unordered, both as dependent and independent variables, and can simultaneously control for several independent variables, and so is highly flexible (see Mayhew 1998, Mayhew & Blackburn 1999). Its major drawbacks are the output content and format, and the fact that it is user-unfriendly; for example, the documentation contains the helpful statement, 'If you do not know how to get into *GLIM*, ask somebody'!

If the phylogeny is completely unknown, Martins' (1996b) implementation of Felsenstein's method in *COMPARE* (implemented over the World-Wide Web) will allow contrasts to be calculated over a range of phylogenetic hypotheses to see if significant relationships are possible or likely. A mix of Felsenstein's and Martin's approaches, based on Losos (1994), has been developed by Blomberg (*Fels-Rand*, for Apple Macintosh, Appendix 6.1). The method randomly resolves polytomies in the tree many times and produces a distribution of statistics used to determine significance, but known nodes are allowed to remain fixed unlike in Martin's method.

A method developed by Garland *et al.* (1999) and Garland and Ives (2000) (executed in the *PDAP* software for PC) incorporates error in ancestral state reconstruction into the comparative analyses.

Independent contrast techniques are explicit about branch lengths in the phylogeny. If they are known they can be incorporated, but if they are unknown they must be assumed. The two commonest assumptions are equal branch lengths (Pagel 1992), equivalent to a punctuated model of evolution, and Grafen's (1989) assumption that the age of a lineage is proportional to the number of taxa it contains. Phylogenetic regression uses the latter assumption as default, *CAIC* uses the former, although any set of lengths may be specified. If branches are unknown, branch lengths may be adjusted to provide the best fit to statistical or evolutionary assumptions (Grafen 1989, Purvis & Rambaut 1995, section 6.6.1).

6.5.3 Other methods for continuous data

A popular method which, to our knowledge, has yet to be applied to sex ratio studies is Phylogenetic Autocorrelation (Cheverud *et al.* 1985, Gittleman & Kot 1990). It can be implemented using the *COMPARE* software over the World-Wide Web, and a version for Apple Macintosh is also available. Like independent contrasts, Phylogenetic Autocorrelation is designed for use with continuous data, and works by partitioning the data for each species into phylogenetic and specific components, then using the specific components to test for evolutionary correlations (Gittleman & Luh 1992). Because of this explicit partitioning of variance, Phylogenetic Autocorrelation can provide interesting comparative data on evolutionary lability (see Gittleman *et al.* 1996). It is not always as valid as independent contrasts, although it can be more powerful (Purvis *et al.* 1994).

If internal phylogeny is poorly known, it is often still possible to carry out phylogenetically aware tests by comparing pairs of sister taxa, so-called pairwise comparisons (Møller & Birkhead 1992). No special software is required and simple nonparametric tests are often used.

6.5.4 Methods for categorical data

Although most comparative analyses of sex ratios will use some continuous measure of sex ratios as the response variable, it is possible that sex ratio could be coded categorically, such as male-biased versus female-biased (e.g. Weatherhead & Montgomerie 1995), or that sex ratio is used as one of the explanatory variables in an analysis of other categorical variables such as habitat type. For mixtures of categorical and continuous variables, some of the above methods, such as phylogenetic regression, are still useful, but additional tests are available which especially consider the correlated evolution of one categorical, usually binary, character on another (Ridley 1983, Burt 1989, Maddison 1990, Harvey & Pagel 1991, Møller & Birkhead 1992, Sillén-Tullberg 1993, Pagel 1994b, Grafen & Ridley 1996). Tests for correlated evolution of categorical characters are more problematic than those for continuous characters because change in categorical characters will tend to be rarer than change in continuous characters, leaving large areas of the tree in which no variation occurs. Different tests tend to treat these areas of the tree in different ways, and there is much controversy about how best to do it (Harvey & Pagel 1991, Ridley & Grafen 1996).

The statistical techniques used in categorical methods are variable: Ridley (1983), Burt (1989) and Møller and Birkhead (1992) attempt to identify phylogenetically independent events or comparisons. Maddison (1990) uses a randomization technique to estimate the probability of change in one character occurring in areas of the tree where another character is constant, and Sillén-Tullberg (1993) uses a similar technique in the

Contingent States Test. Harvey and Pagel's (1991) method involves the calculation of a set of standardized scores which describe the degree of change in each character along each branch of the phylogeny relative to that expected by chance. It is these standardized scores which become the data for analysis. Pagel (1994b) uses maximum likelihood techniques to fit alternative evolutionary models to the data and the phylogeny, and then compares the likelihood of the alternative models. Software is available to perform some of these (Appendix 6.1), but others, such as Ridley's test and Møller and Birkhead's test, can easily be done by hand. The *MacClade* software package (Maddison & Maddison 1992) maps the evolution of categorical characters across a tree using parsimony and is useful for identifying appropriate evolutionary events in Ridley's test, and as a precursor to Maddison's test and in the Contingent States Test.

6.5.5 Simulation methods

Instead of using analytical methods to partition the variance in the data into dependent and independent components, it is also possible to use phylogenetic information in computer (Monte-Carlo) simulations to adjust the significance levels of cross-species analyses (Garland *et al.* 1993). The simulation approach proceeds in two steps. First, a large number (typically 1000) of dummy datasets is generated by simulating the evolution of characters across a tree using a null model of uncorrelated evolution. Second, a cross-species analysis is performed on each of the dummy datasets, yielding distributions of the relevant statistics. By comparing the statistics (e.g. *F* values) obtained in the cross-species analysis of the real data with the distributions obtained from simulations, one can judge the likelihood of the null hypothesis being true.

The first step, simulation of character evolution, requires assumptions about a model of character evolution and an estimate of the ancestral character value at the root of the phylogenetic tree. Whereas analytical methods such as independent contrasts are nearly always based on the Brownian model of evolution, simulation methods allow much greater flexibility in the choice of an evolutionary model. For (arcsine square-root transformed) sex ratio data, the assumption of Brownian motion may sometimes be a poor one, especially for taxa with chromosomal sex determination (e.g. mammals, birds), where there is likely to be some evolutionary 'pull' towards a sex ratio of 0.5. In such a case, the Ornstein–Uhlenbeck process (Felsenstein 1988) might be a more appropriate choice. In an Ornstein–Uhlenbeck (or rubber-band) process, the expected change of a character towards a fixed point (e.g. 0.5) is at a rate proportional to the character's distance from this point.

Ancestral states in simulations can either be guessed (e.g. 0.5 for bird or mammal sex ratios), or estimated (e.g. with independent contrasts if the assumed evolutionary model is Brownian). By using a range of different ancestral states, the robustness of this assumption can be checked. Another advantage of the simulation approach is that unresolved nodes (at least three taxa originating from the same node) do not lead to a loss of degrees of freedom, as is the case in most analytical methods. Unresolved nodes are simply kept as the best available information and it is assumed that at least three taxa have originated simultaneously from the same ancestor.

The second step in the simulation approach, a cross-species analysis of the dummy datasets (e.g. by means of logistic regression), offers the possibility of accounting for the difference in sample size between studies. In this way, measurement error due to small sample sizes can be controlled for. This is particularly useful for sex ratio studies, because average sex ratios based on only a few individuals are more likely to be biased and may hide real biological trends in the data. In the simulated datasets, each species can be assigned the same sample size as in the real dataset to observe if a biased sex ratio is to be expected by chance or not. We give an example of how this can be done in section 6.6.2.

Several of the computer packages listed below offer the possibility of generating simulated datasets based on a number of different evolutionary models (step 1). The output of these programs can be analysed (step 2) using standard statistical software. One of us (IP) has produced a user-unfriendly program which performs both steps (simulation and logistic regression); it is available upon request, as is assistance in operating the software. Furthermore, a recent free software package (*PDAP*) developed by Garland and coworkers is now available for PC (see Appendix 6.1).

6.6 | Analysing comparative data: a worked example

Once you have your question, dataset, phylogeny and method, it is time to analyse. In this section we present a worked example to illustrate how this might be done. We have chosen the study of sex ratios in New World nonpollinating fig wasps by West and Herre (1998a) because the data are amenable to different forms of analysis, and illustrate well some of the challenges and problems involved (see also Hardy & Mayhew 1998b).

The data consist of sex ratios of 17 species of nonpollinating fig wasp collected from figs in Panama, together with the proportion of host fruit collected which contained each wasp species, and whether the wasp species has males that are winged or unwinged. The data are used to test two related hypotheses. First, that foundress number is positively correlated with sex ratio (proportion of males). This is a prediction of Hamilton's theory of local mate competition (Hamilton 1967). Second, that as the proportion of nonlocal mating increases, sex ratio increases. This prediction is made by several models that extend Hamilton's theory (see Hardy 1994). Both foundress number and the degree of nonlocal mating are measured by proxy variables because of the difficulty of making accurate field observations: nonpollinating fig wasps, unlike pollinating wasps, oviposit from the outside of the fruit, so foundress numbers cannot be observed directly without watching the fruit continuously or using molecular techniques. Instead, West and Herre (1998a) used the proportion of fruit parasitized as a proxy for foundress number, because both are likely to be related to wasp density. They used a model to show that this relationship holds under different assumptions about wasp distributions amongst fruit. The degree of nonlocal mating is also difficult to observe, but

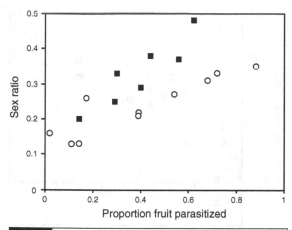

Fig 6.2 Fig wasp sex ratios. The raw species data on proportion fruit parasitized (i.e. contrasts in the proportion of fruit parasitized) and sex ratio plotted for fig wasps with wingless males (closed squares) and winged males (open circles). Reproduced from West and Herre (1998a) with permission from Birkhäuser Verlag.

because some species have wingless males which never exit the natal fig, and others are winged and can potentially fly to mate with females from other figs, male wingedness may be used as a proxy for the degree of nonlocal mating.

West and Herre began by using the raw species data in a logistic regression (Chapter 3) to explore whether the proportion of fruit parasitized and male wingedness were correlated with sex ratio among extant taxa. Both relationships were as predicted from theory: species which parasitized a higher proportion of available fruit also had a greater proportion of males, and although the intercept did not significantly depend on male wingedness, the relationship was steeper for those species that had winged males, which therefore had a greater proportion of males than unwinged species (Figure 6.2). Proportion fruit parasitized (i.e. contrasts in the proportion of fruit parasitized) and male wingedness together accounted for 94.5% of the sex ratio deviance.

6.6.1 An analysis using independent contrasts

Is the cross-species relationship due to correlated evolution between the proportion of fruit parasitized and sex ratio, and between male wingedness and sex ratio? Or does it represent a his-

torical legacy of a few evolutionary incidents? One way to find out is to attempt to partition out the character variance that is evolutionarily independent. West and Herre used Purvis and Rambaut's (1995) implementation of Felsenstein's independent contrasts (CAIC) to achieve this. They based their analysis on a preliminary estimate of phylogeny constructed using molecular data (for methods see Machado et al. 1996), which was not fully resolved and led to just nine contrasts from the original 17 species. Subsequent work (CA Machado pers. comm.) has resolved some parts of the tree, allowing us to do a more complete analysis here. The topology is illustrated in Figure 6.3. Like West and Herre, we use two different branch length assumptions: uniform branch lengths (Pagel 1992) and assuming that taxon age is proportional to the number of species it contains (Grafen 1989). The average sex ratio for each species was calculated by summing the total number of males collected and dividing by the total number of individuals of each species. Sex ratio was arcsine square-root transformed prior to calculating the contrasts (CAIC requires the dependent variable to be fully continuous), but proportion fruit was left untransformed. Male wingedness was coded as a

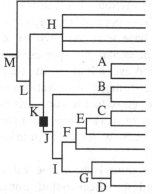

Fig 6.3 The evolutionary relationships of the nonpollinating fig wasps. An estimate for the species in West and Herre's (1998a) study from Machado et al. (1996, CA Machado pers. comm.). Provisional species names (in inverted commas) are assigned according to the *Ficus* species infested. Nodes are assigned letters referring to the contrasts in Figure 6.4. The black square represents the likely moment when male wings were lost since only species in the *Idarnes flavicollis* and *Idarnes carme* species groups have wingless males.

factor with possession of wings coded as 1, and winglessness coded as 0. The above and following procedures are all detailed in the *CAIC* documentation, which is a useful guide to independent contrast analysis.

To begin, contrasts on arcsine square-root sex ratio were calculated against proportion fruit parasitized using the CRUNCH algorithm in *CAIC*, the standard algorithm for calculating contrasts between continuous variables. From 17 species in the original dataset, 13 independent contrasts could be calculated. We lost two potential contrasts among the *Aepocerus* genus which is unresolved, and one potential contrast because *Idarnes* 'columbrinae' has not been subjected to phylogenetic investigation and must be excluded. Examination of the contrasts showed that differences between closely related species and those at the tips of the tree account for the majority of the evolutionarily independent variation; contrasts made deep in the phylogeny all clustered around the origin whereas those at the tips had greater magnitude (Figure 6.4.a,b). This suggests little phylogenetic constraint in both variables, which is consistent with the flexibility and control over sex ratios normally attributed to Hymenoptera. The contrasts made under Grafen's branch length assumptions were slightly different; more variance was attributed to the tips of the tree than to the base because the branches originating from the base were assumed to be longer and therefore the variance in species characters must have evolved over a longer time than those at the tips. The most negative sex ratio contrast in both analyses refers to the comparison between the *Idarnes incerta* species group and the other *Idarnes* species groups, which differ in male wingedness.

Where possible we should check the validity of evolutionary and statistical assumptions, and *CAIC* allows us to do both by examining the output directly. Felsenstein's assumption of random walk evolution of continuous characters will be contradicted if the absolute value of contrasts depends on the value of estimated ancestral states at nodes. If the value of contrasts depends on nodal values, the contrasts depend on ancestry, defeating the aim of the analysis. However, neither of these re-

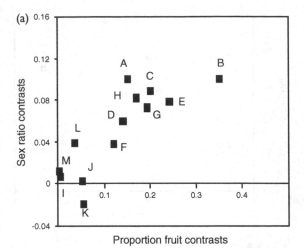

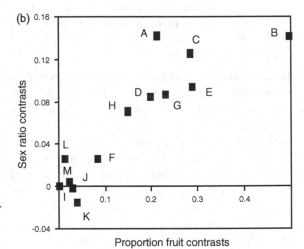

Fig 6.4 Independent contrast plots. Independent contrasts of arcsine square-root sex ratio on proportion fruit infested calculated in *CAIC* using (a) uniform branch lengths and (b) Grafen branch lengths. Letters refer to the contrasts across nodes in Figure 6.3.

lationships was significant (arcsine square-root sex ratio – uniform branches: $F_{1,11} = 0.003$, $P = 0.961$; Grafen: $F_{1,11} = 0.278$, $P = 0.608$. proportion fruit – uniform branches: $F_{1,11} = 0.06238$, $P = 0.8074$; Grafen: $F_{1,11} = 0.046$, $P = 0.834$), showing that the Brownian motion model had been correctly applied and that our contrasts were phylogenetically independent.

A second assumption, if we want to apply a parametric analysis such as regression to our data, is that the variance in the contrasts must be homogeneous. This can be tested by regressing the absolute values of the standardized contrasts against the square-root of the variance in the raw contrasts (Garland *et al.* 1992), which *CAIC* provides as output for the dependent variable. If a significant relationship is observed, the variance will be related to the mean, and regression is inappropriate. Simple regression was inappropriate under both branch length assumptions (uniform branches: $F_{1,11} = 25.86$, $P = 0.004$, Grafen: $F_{1,11} = 28.06$, $P = 0.003$). One option was to apply a nonparametric test such as a sign test. Another was to try transforming the data differently before analysis, or to try a different branch length assumption. Because we had already transformed the data and explored two branch length assumptions, a third alternative was now preferable: a weighted regression through the origin to take account of the relative uncertainty over the different data points. Since we wanted the magnitude of the contrasts to be independent of the square-root of their variances, we used (1/square-root contrast variance) as the weight variable.

The weighted regressions were significant for both branch length assumptions. For uniform branch lengths the estimated slope was 0.364, $F_{1,12} = 85.77$, $P < 0.0001$. For Grafen branch lengths the estimated slope was 0.359, $F_{1,12} = 75.53$, $P < 0.0001$. Thus, positive differences in proportion fruit parasitized were significantly associated with positive differences in sex ratio.

The only evolutionarily independent comparison which could be made between taxa that differed in male wingedness was between the *Idarnes incerta* species group and the other *Idarnes* species groups. Calculated using *CAIC*'s BRUNCH algorithm (implementing Burt's (1989) method of contrast analysis for categorical variables), this contrast had the value 0.019 for uniform branch lengths and 0.015 for Grafen branch lengths. We note two points from this: first, that the direction of the contrast is as predicted from theory: the presence compared to the absence of wings in males is associated with an increase in sex ratio. Second, we cannot reject the null hypothesis that wingedness and sex ratios are evolutionarily correlated solely using contrasts between sister taxa which differ in state because this gives only one data point.

It was possible to calculate contrasts in wingedness whilst controlling for proportion fruit parasitized by using residual values of arcsine square-root sex ratio for each species. Because the contrast regression slopes are less biased than the species regression slopes, residuals are calculated as the species arcsine square-root sex ratio value minus the product of the contrast slope and the species value of proportion fruit parasitized. Contrasts on these residuals can then be calculated against contrasts in wingedness using the BRUNCH algorithm of *CAIC*. Such contrasts confirm the direction of the trend: under equal branch lengths the contrast on wingedness has the value 0.078; under Grafen branch lengths, 0.046. These are both quite substantial and positive differences in arcsine square-root sex ratio (compared with values in Figure 6.4), showing that after accounting for proportion fruit parasitized, the difference in wingedness between the two clades is associated with sex ratio differences in the direction predicted by theory, with the fully winged clade having less biased sex ratios.

6.6.2 An analysis by comparison of real with simulated null data

Table 6.2 contains statistics and test results based on logistic regression analysis applied to the raw data of West and Herre. In addition to the standard significance levels not taking into account the underlying phylogeny, the table contains P values based on F distributions generated by simulations. These tests were implemented by our own programs, but the more user-friendly *PDAP* package is also now available (see Appendix 6.1). Results are shown for the assumption of uniform branch lengths and for branch lengths computed according to Grafen's (1989) method. Simulations were carried out using Brownian motion of arcsine square-root-transformed sex ratios. For the ancestral sex ratio (proportion males) at the root of the tree, an estimate of 0.31 was obtained using Felsenstein's (1985) approach. The variance of the Brownian motion process was scaled by trial and error in such a way that the variance in simulated sex ratios at the tips of the tree matched the observed variance in the real data. For each of the 1000 simulated sex ratio datasets,

Table 6.2	Logistic regression analysis of fig wasp sex ratio data from West and Herre (1998a)					
Variable	Deviance	Df	F	P	Pa	Pb
Constant	216.1	16				
Proportion parasitized	102.7	15	16.57	0.001	0.009	0.01
Winged	17.81	14	66.72	<0.0001	0.001	0.002

P values are based on regular F distributions without taking phylogeny into account and on F distributions obtained from simulated sex ratio evolution across the phylogenetic tree (Figure 6.3).
[a] Uniform branch lengths.
[b] Branch lengths according to Grafen's (1989) method.

a random number of males was drawn for each species from a binomial distribution $B(n, p)$ where p is the expected proportion of males from the simulation and n the sample size of the real study. Logistic regression was carried out and F statistics were stored. The resulting F distribution was used to compute a P value for the F statistic obtained from the analysis of the real data. Although the P values based on simulations are about tenfold higher than those based on standard F distributions, they still suggest that sex ratios have not evolved independently of foundress number and the degree of local mating.

In the above example, both logistic regression on the raw species values, *CAIC*, and simulations based on Brownian motion suggest that increasing foundress number leads to less female-biased sex ratios. However, we have to be more careful over our conclusions about the degree of local mating. The simulation analysis found a significant relationship between sex ratio and male wingedness, but the contrast analysis did not. The difference is probably due to the way that these methods treat areas of the tree where male wingedness is constant. The contrast analysis ignores these areas, and this leaves only one phylogenetically independent event. The simulation recognizes this event but it also recognizes and takes into account areas where wingedness is constant. We conclude that the raw species values are significantly different from those expected if sex ratio were independent of male wingedness, but, because only a single change in male wingedness is probably involved, any character which differs between the two clades involved could also be responsible. Clearly phylogenetic

awareness is especially valuable in this study. It is worth noting that other comparative studies of the effect of partial local mating have had to confront similarly difficult issues of methodology (Hardy & Mayhew 1998a, Fellowes *et al.* 1999; see Hardy & Mayhew 1999). Combining cross-species analyses with phylogenetic analyses has usually been enlightening.

6.7 | Drawing conclusions from comparative data

Care is needed in drawing conclusions from comparative tests because different tests, using different assumptions, can give different results (section 6.6.). We have to know what assumptions we have used and their implications. Sometimes, and perhaps often, this will lead to no simple conclusion: fair enough. It is better to be informed about our lack of knowledge and withhold judgement than to ignore it and draw erroneous conclusions. Doing so increases our confidence in the inferences we do make from data, and that is what comparative methods are trying to achieve.

A6.1 | A comparative biologist's toolbox

This section provides websites relevant to obtaining the most useful software for comparative analyses. The addresses are valid at the time of writing, but websites and people do come and go! Most include the relevant documentation.

A6.1.1 Software for analysing comparative data

A website which provides useful information on comparative software is: http://evolution. genetics.washington.edu/phylip/software.html.

The following paragraphs detail some of the most useful implementations.

CAIC. An Apple Macintosh implementation of independent contrasts; very practical and user-friendly. A PC version may be on the way. Can be downloaded free of charge from: http://www.bio.ic.ac.uk/evolve/software/caic/index.html.

COMPARE. A range of applications for analysing comparative data and phylogenetic trees executable over the Internet. Includes two independent contrast methods (the tree must be completely resolved for Felsenstein's method, and for Martin's method the tree must be completely unknown) and phylogenetic autocorrelation. May be executed from: http://darkwing.uoregon.edu/~compare4/.

CoStar. Applies the Contingent States Test (Sillén-Tullberg 1993) for discrete characters. Works in PC environment but requires MacClade files as input; MacClade runs on Apple Macintosh computers. Available free from: http://www.zoologi.su.se/personal/patrik/costa.html.

DISCRETE. A maximum likelihood method for the analysis of discrete data for PC, Silicon Graphics and Sun Ultra. Distributed free by the author: Dr Mark Pagel, School of Animal and Microbial Sciences, University of Reading, RG6 6AJ, UK, m.pagel@reading.ac.uk. A maximum likelihood method for continuous data is also available for Apple Macintosh users, but accompanying documentation is not ready at the time of writing.

Fels-Rand. An Apple Macintosh application of independent contrasts allowing bootstrap-type analyses over areas of phylogenetic uncertainty whilst retaining known phylogenetic information. Developed by Simon Blomberg (University of Queensland, Australia). Available free of charge from: http://dingo.cc.uq.edu.au/~ansblomb/.

MacClade. A user-friendly package for exploring phylogenetic trees for Apple Macintosh. Main strength is parsimony reconstruction of the evolution of categorical variables, which is useful for identifying independent events in Ridley's (1983) method. Also performs Maddison's (1990) concentrated changes test for comparative analysis of two-state variables. Published by Sinauer: http://phylogeny.arizona.edu/macclade/macclade.html and http://www.sinauer.com/Titles/frmaddison.htm.

PDAP. A collection of comparative software for PC, distributed free from: http://www.wisc.edu/zoology/faculty/fac/Garland/PDAP.html.

It runs simulation analyses as described above, and can perform independent contrast analyses incorporating information about the error in ancestral state reconstruction.

Phylogenetic autocorrelation. An Apple Macintosh application of Gittleman and Kot's method (also available in COMPARE, above), available by anonymous file transfer protocol (FTP) from: ftp.math.utk.edu. under the directory pub/luh.

Phylogenetic independence. A windows-executable package available free at http://life.bio.sunysb.edu/ee/ehab, which conducts tests for phylogenetic independence of both discrete and continuous cross-species data. If the data are not phylogenetically independent, the author recommends pursuing analysis with a phylogenetic comparative method, but if they are phylogenetically independent, cross-species analysis may be sufficient.

Phylogenetic regression. A macro written for the *GLIM* statistics package, distributed free by the author: Dr Alan Grafen, St John's College, Oxford, UK (alan.grafen@st-johns.oxford.ac.uk). For information on *GLIM* see: http://www.nag.co.uk/stats/GDGE_soft.asp. Crawley (1993) provides a useful introductory *GLIM* text.

A6.1.2 Obtaining phylogenetic information

A useful starter website is the tree of life website: http://phylogeny.arizona.edu/tree/. This links to many other sites on phylogenies and taxonomy.

A6.1.3 Other software for exploring phylogenies

Extensive lists of phylogenetics software are available at:

http://evolution.genetics.washington.edu/phylip/software.html.

zttp://www.cladistics.org/education.html
http://evolve.zps.ox.ac.uk/

Three widely used general purpose packages for reconstructing phylogenies are:

PHYLIP. For MS-DOS and Apple Macintosh. May be downloaded free of charge from: http://evolution.genetics.washington.edu/phylip.html.

PAUP. For Apple Macintosh and UNIX. Published by Sinauer: http://www.sinauer.com/Titles/frswofford.htm.

MEGA. A PC package for use with molecular data. Available free from: http://www.MEGAsoftware.net.

Acknowledgments

We are grateful to Andrea Gillmeister, Ian Hardy, Allen Herre, Scott Johnson, Carlos Machado, Andrew Read, Stuart West and two anonymous referees.

References

Abouheif E (1999) A method for testing the assumption of phylogenetic independence in comparative data. *Evolutionary Ecology Research*, **1**, 895–909.

Arneberg P, Skorping A & Read AF (1997) Is population density a species character? Comparative analyses of the nematode parasites of mammals. *Oikos*, **80**, 289–300.

Arnqvist G & Wooster D (1995) Metaanalysis – synthesizing research findings in ecology and evolution. *Trends in Ecology and Evolution*, **10**, 236–240.

Bininda-Emonds ORP, Gittleman JL & Purvis A (1999) Building large trees by combining phylogenetic information: a complete phylogeny of the extant Carnivora (Mammalia). *Biological Reviews*, **74**, 143–175.

Björklund M (1997) Are 'comparative methods' always necessary? *Oikos*, **80**, 607–612.

Brooks DR & McLennan DA (1991) *Phylogeny, ecology and behavior.* Chicago: University of Chicago Press.

Burt A (1989) Comparative methods using phylogenetically independent contrasts. *Oxford Surveys in Evolutionary Biology*, **6**, 33–53.

Cheverud JM, Dow MM & Leutenegger W (1985) The quantitative assessment of phylogenetic constraints in comparative analyses: sexual dimorphism in body weight among primates. *Evolution*, **39**, 1335–1351.

Crawley MJ (1993) *GLIM for Ecologists.* Oxford: Blackwell Scientific.

Dijkstra C, Daan S & Pen I (1998) Fledgling sex ratios in relation to brood size in size-dimorphic altricial birds. *Behavioral Ecology*, **9**, 287–296.

Fellowes MDE, Compton SG & Cook JM (1999) Sex allocation and local mate competition in the Old World non-pollinating fig wasps. *Behavioral Ecology and Sociobiology*, **46**, 95–102.

Felsenstein J (1985) Phylogenies and the comparative method. *American Naturalist*, **125**, 1–15.

Felsenstein J (1988) Phylogenies and quantitative characters. *Annual Review of Ecology and Systematics*, **19**, 445–471.

Freckleton RP (2000) Phylogenetic tests of ecological and evolutionary hypotheses: checking for phylogenetic independence. *Functional Ecology*, **14**, 129–134.

Garland T & Adolph SC (1994) Why not to do two species comparative tests – limitations on infering adaptations. *Physiological Zoology*, **67**, 797–828.

Garland T & Ives AR (2000) Using the past to predict the present: confidence intervals for regression equations in phylogenetic comparative methods. *American Naturalist*, **155**, 346–364.

Garland T, Harvey PH & Ives AR (1992) Procedures for the analysis of comparative data using phylogenetically independent contrasts. *Systematic Biology*, **41**, 18–32.

Garland T, Dickerman AW, Janis CM & Jones JA (1993) Phylogenetic analysis of covariance by computer simulation. *Systematic Biology*, **42**, 265–292.

Garland T, Midford PE & Ives AR (1999) An introduction to phylogenetically based statistical methods, with a new method for confidence intervals on ancestral states. *American Zoologist*, **39**, 374–388.

Gittleman JL & Kot M (1990) Adaptation: statistics and a null model for estimating phylogenetic effects. *Systematic Zoology*, **39**, 227–241.

Gittleman JL & Luh H-K (1992) On comparing comparative methods. *Annual Review of Ecology and Systematics*, **23**, 383–404.

Gittleman JL, Anderson CG, Kot M & Luh H-K (1996) Comparative tests of evolutionary lability and rates using molecular phylogenies. In: PH Harvey, AJ Leigh Brown, J Maynard Smith & S Nee (eds) *New*

Uses for New Phylogenies, pp 289–307. Oxford: Oxford University Press.

Gowaty PA (1993) Differential dispersal, local resource competition, and sex ratio variation in birds. *American Naturalist*, **141**, 263–280.

Grafen A (1989) The phylogenetic regression. *Philosophical Transactions of the Royal Society of London, Series B*, **326**, 119–157.

Grafen A & Ridley M (1996) Statistical tests for discrete cross-species data. *Journal of Theoretical Biology*, **193**, 255–267.

Griffiths NT & Godfray HCJ (1988) Local mate competition, sex ratio and clutch size in bethylid wasps. *Behavioural Ecology and Sociobiology*, **22**, 211–217.

Hamilton WD (1967) Extraordinary sex ratios. *Science*, **156**, 477–488.

Hardy ICW (1994) Sex ratio and mating structure in the parasitoid Hymenoptera. *Oikos*, **69**, 3–20.

Hardy ICW & Mayhew PJ (1998a) Sex ratio, sexual dimorphism and mating structure in bethylid wasps. *Behavioral Ecology and Sociobiology*, **42**, 383–395.

Hardy ICW & Mayhew PJ (1998b) Partial local mating and the sex ratio: indirect comparative evidence. *Trends in Ecology and Evolution*, **13**, 431–432.

Hardy ICW & Mayhew PJ (1999) Reply to M.D. Drapeau. *Trends in Ecology and Evolution*, **14**, 235.

Hardy ICW, Dijkstra LJ, Gillis JEM & Luft PA (1998) Patterns of sex ratio, virginity and developmental mortality in gregarious parasitoids. *Biological Journal of the Linnean Society*, **64**, 239–270.

Harvey PH (1996) Phylogenies for ecologists. *Journal of Animal Ecology*, **65**, 255–263.

Harvey PH & Nee S (1997) The phylogenetic foundations of behavioural ecology. In: JR Krebs & NB Davies (eds) *Behavioural Ecology, An Evolutionary Approach*, pp 334–349. Oxford: Blackwell Scientific.

Harvey PH & Pagel MD (1991) *The Comparative Method in Evolutionary Biology*. Oxford: Oxford University Press.

Herre EA (1987) Optimality, plasticity, and selective regime in fig wasp sex ratios. *Nature*, **329**, 627–629.

Herre EA, West SA, Cook JM, Compton SG & Kjellberg F (1997) Fig-associated wasps: pollinators, sex-ratio adjustment and male polymorphism, population structure and its consequences. In: J Choe & B Crespi (eds) *Social Competition and Cooperation in Insects and Arachnids*. volume 1. *The Evolution of Mating Systems*, pp 226–239. Cambridge: Cambridge University Press.

Johnson CN (1988) Dispersal and the sex ratio at birth in primates. *Nature*, **332**, 726–728.

Losos JB (1994) An approach to the analysis of comparative data when a phylogeny is unavailable or incomplete. *Systematic Biology*, **43**, 117–123.

Machado CA, Herre EA & Bermingham E (1996) Molecular phylogenies of fig pollinating and non-pollinating wasps and the implications for the origin and evolution of the fig-fig wasp mutualism. *Journal of Biogeography*, **23**, 531–542.

Maddison WP (1990) A method for testing the correlated evolution of two binary characters: are gains or losses concentrated on certain branches of the tree? *Evolution*, **44**, 539–557.

Maddison WP & Maddison DR (1992) *MacClade. Analysis of Phylogeny and Character Evolution*. Sunderland, Massachusetts: Sinauer.

Martins EP (ed) (1996a) *Phylogenies and the Comparative Method in Animal Behavior*. Oxford: Oxford University Press.

Martins EP (1996b) Conducting phylogenetic comparative studies when the phylogeny is not known. *Evolution*, **50**, 12–22.

Martins EP & Hansen TF (1997) Phylogenies and the comparative method: a general approach to incorporating phylogenetic information into the analysis of interspecific data. *American Naturalist*, **149**, 646–667.

Mayhew PJ (1998) The evolution of gregariousness in parasitoid wasps. *Proceedings of the Royal Society of London, B*, **265**, 383–389.

Mayhew PJ & Blackburn TM (1999) Does development mode organize life history traits in the parasitoid Hymenoptera? *Journal of Animal Ecology*, **68**, 906–916.

Mayhew PJ & Hardy ICW (1998) Nonsiblicidal behavior and the evolution of clutch size in bethylid wasps. *American Naturalist*, **151**, 409–424.

Mitani JC, Gros-Louis J & Richards AF (1996) Sexual dimorphism, the operational sex ratio, and the intensity of male competition in polygynous primates. *American Naturalist*, **147**, 966–980.

Møller AP & Birkhead TR (1992) A pairwise comparative method as illustrated by copulation frequency in birds. *American Naturalist*, **139**, 644–656.

Pagel MD (1992) A method for the analysis of comparative data. *Journal of Theoretical Biology*, **136**, 361–364.

Pagel M (1994a) The adaptationist wager. In: P Eggleton & RI Vane-Wright (eds) *Phylogenetics and Ecology*, pp 29–51. London: Academic Press.

Pagel M (1994b) Detecting correlated evolution on phylogenies: a general method for the comparative analysis of discrete characters. *Proceedings of the Royal Society of London, B*, **255**, 37–45.

Pagel M (1999) Inferring the historical patterns of biological evolution. *Nature*, **401**, 877–884.

Poulin R (1997a) Covariation of sexual size dimorphism and adult sex ratio in parasitic nematodes. *Biological Journal of the Linnean Society*, **62**, 567–580.

Poulin R (1997b) Population abundance and sex ratio in dioecious helminth parasites. *Oecologia*, **111**, 375–380.

Price T (1997a) Book review: phylogenies and the comparative method in animal behavior (EP Martins (ed)). *Animal Behaviour* **54**: 235–238.

Price T (1997b) Correlated evolution and independent contrasts. *Philosophical Transactions of the Royal Society of London, B*, **352**, 519–529.

Purvis A (1995) A composite estimate of primate phylogeny. *Philosophical Transactions of the Royal Society of London, B*, **348**, 405–421.

Purvis A & Harvey PH (1995) Mammal life-history evolution: a comparative test of Charnov's model. *Journal of Zoology*, **237**, 259–283.

Purvis A & Rambaut A (1995) Comparative analysis by independent contrasts (CAIC); an Apple Macintosh application for analyzing comparative data. *Computer Applications in the Biosciences*, **11**, 247–251.

Purvis A, Gittleman JL & Luh H-K (1994) Truth or consequences: effects of phylogenetic accuracy on two comparative methods. *Journal of Theoretical Biology*, **167**, 293–300.

Read AF, Anwar M, Shutler D & Nee S (1995) Sex allocation and population structure in malaria and related parasitic protozoa. *Proceedings of the Royal Society of London, B*, **260**, 359–363.

Richman AD & Price T (1992) Evolution of ecological differences in the old world leaf warblers. *Nature*, **355**, 817–821.

Ricklefs RE (1998) Evolutionary theories of aging: confirmation of a fundamental prediction, with implications for the genetic basis and evolution of lifespan. *American Naturalist*, **152**, 24–44.

Ricklefs RE & Starck JM (1996) Applications of phylogenetically independent contrasts: a mixed progress report. *Oikos*, **77**, 167–172.

Ridley M (1983) *The Explanation of Organic Diversity: the Comparative Method and Adaptations for Mating*. Oxford: Oxford University Press.

Ridley M (1989) Why not to use species in comparative tests. *Journal of Theoretical Biology*, **136**, 361–364.

Ridley M & Grafen A (1996) How to study discrete comparative methods. In: EP Martins (ed) *Phylogenies and the Comparative Method in Animal Behavior*, pp 70–103. Oxford: Oxford University Press.

Sanderson MJ, Purvis A & Henze C (1998) Phylogenetic supertrees: assembling the trees of life. *Trends in Ecology and Evolution*, **13**, 105–109.

Seehausen O, Mayhew PJ & van Alphen JJM (1999) Evolution of colour patterns in East African cichlid fish. *Journal of Evolutionary Biology*, **12**, 514–534.

Shutler D, Bennett GF & Mullie A (1995) Sex proportions of *Haemoproteus* blood parasites and local mate competition. *Proceedings of the National Academy of Science, USA*, **92**, 6748–6752.

Sillén-Tullberg B (1993) The effect of biased inclusion of taxa on the correlation between discrete characters in phylogenetic trees. *Evolution*, **47**, 1182–1191.

Slagsvold T, Røskaft E & Engen S (1986) Sex ratio, differential cost of rearing young, and differential mortality between the sexes during the period of parental care: Fisher's theory applied to birds. *Ornis Scandinavica*, **17**, 117–125.

Stauffer RC (ed) (1975) *Charles Darwin's Natural Selection*. Cambridge: Cambridge University Press.

Waage JK (1982) Sib-mating and sex ratio strategies in scelionid wasps. *Ecological Entomology*, **7**, 103–112.

Weatherhead PJ & Montgomerie R (1995) Local resource competition and sex ratio variation in birds. *Journal of Avian Biology*, **26**, 168–171.

West SA & Herre EA (1998a) Partial local mate competition and the sex ratio: a study on non-pollinating fig wasps. *Journal of Evolutionary Biology*, **11**, 531–548.

West SA & Herre EA (1998b) Stabilizing selection and variance in fig wasp sex ratios. *Evolution*, **52**, 475–485.

West SA, Herre EA, Compton SG, Godfray HCJ & Cook JM (1997) A comparative study of virginity in fig wasps. *Animal Behaviour*, **54**, 437–450.

Westoby M, Leishman MR & Lord JM (1995) On misinterpreting the 'phylogenetic correction'. *Journal of Ecology*, **83**, 531–534.

Wilson EO (1994) *Naturalist*. Washington: Island Press.

Part 3

Genetics of sex ratio and sex determination

Chapter 7

Sex-determining mechanisms in vertebrates

Sarah B.M. Kraak & Ido Pen

7.1 | Summary

Vertebrates have various sex-determining mechanisms. These have been broadly classified as either genotypic sex determination (GSD) or environmental sex determination (ESD). This terminology, however, may obscure the facts that mixtures between genotypic and environmental sex determination exist, and that genotypic and environmental sex determination may themselves be the extremes of a continuum. Sex ratio evolution plays an important role in the evolution of sex-determining mechanisms.

7.2 | Introduction

This chapter starts with the proximate aspects of sex-determining mechanisms (section 7.3). We introduce the traditional classification of sex-determining mechanisms that exist in vertebrates (section 7.3.1) and the distribution of mechanisms among extant vertebrate taxa (section 7.3.2). At phylogenetically shallow levels, different mechanisms are present. We describe how the existence of either male or female heterogamety, or ESD is usually established for individual species or taxa (section 7.3.3). Cases of mixed sex determination, i.e. combinations of GSD and ESD, are also observed (section 7.3.4) and we caution that this phenomenon has implications for sex identification by molecular markers (section 7.3.5). We stress that phenotypic sex

generally has environmental and genetic components and discuss a model that attempts to unify sex determination by stating that sex determination in all vertebrates is mediated by differential growth of the embryo (section 7.3.6).

In the second part of the chapter we discuss the evolution of sex-determining mechanisms. Evolution from one system to another can be quite rapid (section 7.4). We stress that sex ratio selection plays an important role in the evolution of sex-determining mechanisms (section 7.4.1). This usually leads to sex-determining mechanisms that produce an unbiased sex ratio, but under some conditions mechanisms that bias the sex ratio are favoured. We conclude the chapter with an illustration of how one can investigate verbal models of the evolution of sex determination by means of mathematical models. We present a simulation model with which we analyse a hypothesis for the evolution from ESD to GSD, attempting to account for male heterogamety in some taxa and female heterogamety in others (section 7.4.2).

7.3 | Proximate aspects of sex determination

7.3.1 Traditional classification of sex-determining mechanisms

Sex determination is traditionally classified as either 'genotypic' or 'environmental'. The term genotypic sex determination (GSD) signifies that the sex of a zygote is determined entirely by its genotype; the sex of an individual is fixed at fertilization. The most common type of GSD involves sex chromosomes. If the male is the sex with two different sex chromosomes, this is termed male heterogamety, and the sex chromosomes are referred to as X and Y (females are XX, males are XY). Likewise, if the female is the sex with two different sex chromosomes, this is termed female heterogamety and the sex chromosomes are Z and W (females are ZW, males are ZZ). In polygenic sex determination, which is less common, sex is determined by a number of genes, each with minor effect, distributed throughout the chromosome complement.

The term environmental sex determination (ESD) signifies that the sex of an individual is determined irreversibly by the environment experienced during early development. Where the decisive environmental factor is temperature, we refer to this as 'temperature-dependent sex determination' (TSD). Sex determination may also be influenced by pH (in fish: Römer & Beisenhertz 1996), and by social conditions or relative juvenile size (in fish: Francis & Barlow 1993, Holmgren & Mosegaard 1996). Although we restrict this chapter to the discussion of primary sex determination, we mention here that in many fish species sex change is part of their natural life history, and is often induced by environmental stimuli (Francis 1992).

7.3.2 Distribution of sex-determining mechanisms among vertebrate taxa

Figure 7.1 shows the phylogeny of extant vertebrate taxa along with the reported sex-determining mechanisms. Sex determination by sex chromosomes is universal in birds (female heterogamety) and mammals (male heterogamety) and is present in both forms (male and female heterogamety) among reptiles, amphibians and fish. ESD is common among reptiles, and also exists in amphibians and fish. These data should not be treated as final, since the interpretation of sex-specific markers is not entirely clear (sections 7.3.3.1 & 7.3.4), and some of the studies reporting ESD were conducted at temperatures outside the range normally experienced by the species under study (Hayes 1998). For some of these cases it remains to be shown to what extent sex is

environmentally determined in the wild (sections 7.3.3.2 & 7.3.4). Polygenic sex determination (not in figure) in some fish species has been reported, e.g. *Xiphophorus helleri* (Price 1984) and in *Menidia menidia* (Lagomarsino & Conover 1993).

At phylogenetically shallow levels, different mechanisms may be present. For example, male and female heterogamety occur in the amphibian sister families Hylidae and Bufonidae respectively. Moreover, ESD and both male and female heterogamety exist within the reptilian family Gekkonidae. None of these mechanisms appears to have evolved only once. Instead of the conservatism of such a basic function as sex determination, as might have been intuitively expected, sex-determining mechanisms seem to be evolutionarily flexible (Chapter 8). Even mammals, in which the X and Y sex chromosomes are generally supposed to be conserved, variation in sex-determining mechanisms occurs (reviewed in Fredga 1994, Jiménez *et al.* 1996, McVean & Hurst 1996, Mittwoch 1996a).

7.3.3 Evidence for various sex-determining mechanisms

7.3.3.1 Genotypic sex determination: male or female heterogamety

Evidence for male or female heterogamety in a species traditionally comes from investigating the karyotype. Sometimes the different sex chromosomes can be recognized by their size. If they seem similar, cytological techniques, such as C-banding (e.g. Schmid *et al.* 1988, 1992, 1993), are used. However, in many species sex chromosomes appear to be morphologically indistinguishable from autosomes. In these cases, breeding experiments may indirectly establish heterogamety, e.g. if sex-linked marker genes exist, as in the guppy (Winge 1932). An alternative approach is the analysis of offspring sex ratios of artificially induced gynogenetic females, of artificially induced triploids, or of crosses between two individuals of the same genetic sex of which one is artificially 'sex-reversed' by hormone treatment (Beatty 1964, Richards & Nace 1978, Price 1984). In the 1980s, H-Y antigen, a minor histocompatibility antigen specific for the heterogametic sex, was proposed as a tool to identify the heterogametic sex (Engel & Schmid 1981, Engel *et al.*

(a)

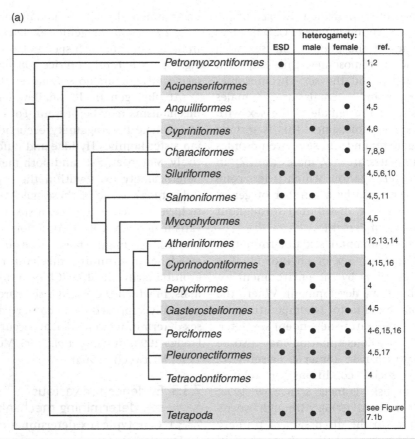

	ESD	heterogamety: male	heterogamety: female	ref.
Petromyozontiformes	●			1,2
Acipenseriformes			●	3
Anguilliformes			●	4,5
Cypriniformes		●	●	4,6
Characiformes			●	7,8,9
Siluriformes	●	●	●	4,5,6,10
Salmoniformes	●	●		4,5,11
Mycophyformes		●	●	4,5
Atheriniformes	●			12,13,14
Cyprinodontiformes	●	●	●	4,15,16
Beryciformes		●		4
Gasterosteiformes		●	●	4,5
Perciformes	●	●	●	4-6,15,16
Pleuronectiformes	●	●	●	4,5,17
Tetraodontiformes		●		4
Tetrapoda	●	●	●	see Figure 7.1b

Fig 7.1 Phylogenetic distribution of reported sex-determining mechanisms within extant vertebrate taxa. (a) Fishes, at the level of orders. (b) Tetrapods, at the level of families. Taxa homogeneous for sex-determining mechanisms (e.g. birds and mammals) are collapsed. Mechanisms are classified into environmental sex determination (ESD) or genotypic sex determination (GSD) with male or female heterogamety. The evidence of heterogamety comes from breeding experiments, karyotypes, cytogenetics, H-Y antigen, or sex-specific DNA (e.g. Y-associated *Sry* or *Zfy*, W-associated Bkm). Evidence of ESD comes from laboratory studies that manipulated rearing conditions. These data should not be treated as final, since the interpretation of sex-specific markers is not entirely clear, and some of the studies reporting ESD have been conducted at temperatures outside the range normally experienced by the species under study (Hayes 1998). Only a minority of existing species has been investigated for their sex-determining mechanisms (Janzen & Paukstis 1991a); in particular relatively few fish species have been investigated. The relative rarity of ESD among, for example, amphibian families may therefore reflect the lack of studies on ESD in those taxa. Fish phylogeny is based on Nelson (1994), Lundberg (1996), Janvier (1996a,b). Tetrapod phylogeny is based on Ford & Cannatella (1993), Gaffney & Meylan (1988), Gauthier *et al.* (1988), Hillis & Green (1990), Janzen & Paukstis (1991b), Larson & Dimmick (1993), Laurin (1996), Laurin & Reisz (1995), Laurin *et al.* (1996), and Rieppel (1988). The classification of sex-determining mechanisms is based on (numbers in last column): (1) Beamish (1993), (2) Docker & Beamish (1994), (3) VanEenennaam *et al.* (1999), (4) Chourrout (1986), (5) Sola *et al.* (1981), (6) Moreira-Filho *et al.* (1993), (7) Bertollo & Cavallaro (1992), (8) Molino *et al.* (1998), (9) Maistro *et al.* (1998), (10) Patiño *et al.* (1996), (11) Craig *et al.* (1996), (12) Conover & Heins (1987), (13) Strüssman *et al.* (1996a), (14) Strüssman *et al.* (1996b), (15) Francis (1992), (16) Römer & Beisenhertz (1996), (17) Goto *et al.* (2000), (18) Schmid *et al.* (1993), (19) Hillis & Green (1990), (20) Schmid & Haaf (1989), (21) Witschi (1929), Pieau (1975), Richards & Nace (1978), (22) Schmid *et al.* (1988), (23) Schmid *et al.* (1992), (24) Mahony (1991), (25) Duellman & Trueb (1986), (26) Dorazi *et al.* (1995), (27) Janzen & Paukstis (1991a), (28) Engel *et al.* (1981), (29) Janzen & Paukstis (1991b), (30) Nakamura *et al.* (1987), (31) Demas *et al.* (1990), (32) Wellins (1987), (33) Olmo (1986), (34) Viets *et al.* (1994), (35) Caputo *et al.* (1994), (36) Gorman (1973), (37) Olmo *et al.* (1990), (38) Volobouev *et al.* (1990), (39) Moritz (1990), (40) King & Rofe (1976), (41) Ganesh *et al.* (1997), (42) Cole (1971), (43) Cree *et al.* (1995).

(b)

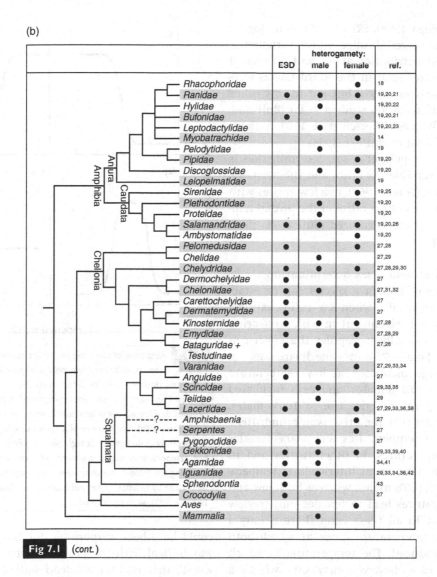

Fig 7.1 (cont.)

1981). This gave conflicting results in at least one case: in the turtle *Siebenrockiella crassicollis* the female is H-Y positive (Engel *et al.* 1981) while a cytogenetic study identifies the male as the heterogametic sex (Carr & Bickham 1981). During the 1990s researchers tried to establish heterogamety by searching for sex-specific DNA. This has involved screening for the sex-specific presence of Bkm-related satellite DNA, characterized by repetitive GATA sequences (e.g. Demas *et al.* 1990, Nanda *et al.* 1992). Bkm (banded krait minor) was originally isolated from the W chromosome of the snake *Bungarus fasciatus* (banded krait), and high concentrations of Bkm-related sequences appear to be linked to the W or Y chromosome in

many species (Jones & Singh 1981, 1985, Singh & Jones 1982). Species have also been screened for the sex-specificity of genes related to the human Y-linked genes *SRY* and *ZFY* (e.g. Ganesh *et al.* 1997). ZFY (zinc finger Y, Page *et al.* 1987) and *SRY* (sex-determining region Y, Sinclair *et al.* 1990; *Sry* in mouse, Gubbay *et al.* 1990) have both been proposed as candidates for the male-determining gene TDF (testis-determining factor). ZFY eventually fell out of favour (Palmer *et al.* 1989), but an important role for *SRY*/*Sry* in mammalian sex determination has been well established (Koopman *et al.* 1991, Goodfellow & Lovell-Badge 1993). However, there are some exceptions: the gene appears to be absent in the mole voles *Ellobius lutescens* and

E. tancrei (Fredga 1994). *SRY* and *ZFY* homologues have been conserved throughout the vertebrates. A male-biased distribution of *SRY*- and *ZFY*-related genes has been found in the lizard *Calotes versicolor* (Ganesh *et al.* 1997), but not in any other nonmammalian species studied so far (Bull *et al.* 1988, Griffiths 1991, Tiersch *et al.* 1991, Valleley *et al.* 1992, Coriat *et al.* 1993, 1994).

Evidence for polygenic sex determination is provided by variable sex ratios and the heritability of this trait (Scudo 1967). In a few turtles with ESD heritabilities of sex ratios at the pivotal temperature have been measured (Bull *et al.* 1982, Janzen 1992).

7.3.3.2 Environmental sex determination

The existence of ESD in a species can be established experimentally when sex ratios vary according to the environment in which offspring are reared. Especially in reptiles the effect can be extreme (Figure 7.2): in some lizards and in alligators, eggs incubated at low temperature give rise to 100% females, and eggs incubated at high temperatures give rise to 100% males. In many turtles it is the other way round: 100% males at low temperatures and 100% females at high temperatures. In other turtles and in crocodiles, incubation at intermediate temperatures leads to 100% males, whereas both low and high temperatures lead to females only (review in Bull 1983). In all these cases there is only a very narrow temperature range at which both sexes are produced. The temperature at which this is the case, however, may vary within a species and is heritable (Bull *et al.* 1982, Janzen 1992). In fishes the sex ratios usually vary less extremely with temperature, but nevertheless temperature-dependent sex determination has been established in various species (e.g. Conover & Heins 1987, review in Francis 1992, Römer & Beisenhertz 1996, Goto *et al.* 2000).

A difficulty that arises with the interpretation of experiments that test for environmental sex determination is the possibility of differential mortality. If a biased sex ratio is found as a result of an experimental manipulation of environmental conditions at rearing, e.g. incubation temperature, and one wants to conclude that sex is determined environmentally, it needs to be assessed to what extent biased mortality

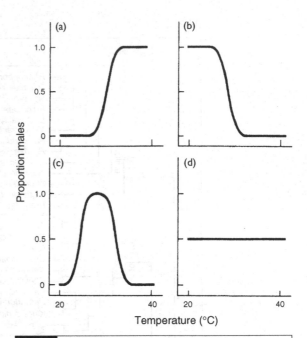

Fig 7.2 Response of sex ratio to incubation temperature in reptiles. These graphs represent only the approximate form of the response and are not drawn according to any single species. There are four patterns recognized at present. (a) Females develop at low temperature, males at high temperature. (b) The reverse of a; males develop at low temperatures and females at high ones. (c) Females at low and high temperatures, males at intermediate ones. (d) The hatchling sex ratio of some species is not significantly influenced by incubation temperature. From Kraak and de Looze (1993), with permission.

could have been responsible for the result. One way to deal with the problem is to 'assume the worst', that is, that all dead individuals are of the sex least favourable for the hypothesis. These 'data' can then be included when testing statistically for sex ratio bias. However, this procedure is often too conservative. Researchers should take care that the conditions of the experiment (other than the experimental manipulations) are as conducive to the survival of the tested individuals as possible, and identify the sex of the individuals before much mortality has taken place.

Although the possibility of ESD has been established for many species by manipulation of rearing conditions in the laboratory, very few studies have investigated to what extent ESD operates in the wild. The European pond turtle *Emys orbicularis* has been shown to exhibit ESD in the laboratory (Zaborski *et al.* 1982). However, a study of wild populations of this turtle revealed that the sex of only 17% of wild individuals was determined by the temperature (Girondot *et al.* 1994) (see also section 7.3.4).

7.3.4 Mixed sex determination

A combination of ESD and GSD, sometimes with major genetic factors, can be present within the individual. Examples are the fish species *Menidia menidia* (Conover & Heins 1987) and *Limanda yokohamae* (Goto *et al.* 2000), and the turtle *Emys orbicularis* (Zaborski *et al.* 1988, Girondot *et al.* 1994). Some other reptiles with ESD also show signs of heterogamety (e.g. Engel *et al.* 1981, Nakamura *et al.* 1987, Wellins 1987, Demas *et al.* 1990, Ewert *et al.* 1990). Recently, a form of temperature-dependent sex determination has been reported in poultry (Ferguson 1994a,b), while all birds are known to have ZZ/ZW sex chromosomes.

In a study of the pond turtle *Emys orbicularis*, all individuals from eggs incubated at 25–26°C became males, and all individuals from eggs incubated at 30–30.5°C became females (Zaborski *et al.* 1982). The gonadal cells of all males typed H-Y negative, as did blood cells of half of them; blood cells of the other half typed H-Y positive. In all females, the gonadal cells typed H-Y positive, but blood cells typed H-Y positive in only half of them, and negative in the other half. Zaborski *et al.* (1982) consider the animals with H-Y-negative blood cells as genotypic males, and the animals with H-Y-positive blood cells as genotypic females. Thus, the H-Y-blood-cell-negative phenotypic females at the high temperature are 'sex-reversed' genotypic males, and the H-Y-blood-cell-positive phenotypic males at the low temperature are 'sex-reversed' genotypic females. Apparently this turtle has a form of GSD, probably with female heterogamety, since the female is the H-Y-positive sex. The phenotypic sex, however, does not correspond with the genotypic sex in half of the individuals reared at the two extreme temperatures. This implies that the genetic status can be totally overruled by the influence of temperature, and the H-Y type of the gonadal cells can be completely reversed in accordance with the developing sex of the gonad. Girondot *et al.* (1994) found that, in a natural population of *Emys orbicularis*, both 'sex-reversed' individuals

and individuals whose phenotypic and genotypic sex match occur among males as well as females. There are indications that the fish *Scardinius erythrophthalmus* has a similar sex-determining system; Koehler *et al.* (1995) found all males and half of the females to be homogametic, whereas the other half of the females were heterogametic. Moreover, a similar situation might exist in other turtle species in which heterogamety has been inferred from H-Y typing (Engel *et al.* 1981, Nakamura *et al.* 1987, Wellins 1987) while TSD has also been inferred, from laboratory experiments with eggs incubated at different temperatures (Bull *et al.* 1982, Yntema 1976, 1979).

In poultry, a comparable situation may exist. In the experiments described by Ferguson (1994a,b), poultry eggs were treated with abnormally high or low temperatures during incubation. He found that approximately 10% of the hatched birds had a sexual phenotype (confirmed by macroscopic and histological examination) that was different from their sexual genotype (confirmed by a W-specific molecular marker). Apparently, the influence of temperature can overrule the influence of the sex-determining genes in at least some individuals. It is not known whether any 'sex-reversed' poultry would naturally occur or, if so, at what frequencies.

In the examples described above, sex determination seems to be governed by sex chromosomes (i.e. a major genetic factor) as well as by an influence of temperature. A study of the silverside *Menidia menidia* (Lagomarsino & Conover 1993) suggests that, in this fish, sex determination is controlled by an interaction between major genetic factors, polygenic factors and temperature, and that the relative importance of each component differs with latitude. This study examined family sex ratios at two different temperatures for two different populations. In the high-latitude population, the sex ratios tended to fall into distinct classes, as expected from Mendelian segregation of a major sex factor(s). In this population temperature had no influence on sex ratios. In the southern population, temperature had a highly significant influence on sex ratios, and sex ratios did not conform to Mendelian ratios. High-latitude populations appear to have evolved a major sex-determining factor(s) that overrides the effect of temperature, and this factor(s) is lacking in low-latitude populations.

7.3.5 Consequences for measuring sex ratios

The finding of mixtures between ESD and GSD has implications for the practice of identifying the sex of individuals, of species that supposedly exhibit GSD, using molecular markers. For example, when behavioural ecologists are confronted with biased sex ratios, they want to know whether the bias is caused by differential mortality of the sexes or whether the primary sex ratio is biased. It is desirable to know the sex of individuals at as young an age as possible, long before the sex can be identified by external morphology, and without having to sacrifice the individuals. Various molecular methods have recently been developed to establish primary sex ratios in behavioural ecological studies of birds (Griffiths *et al.* 1996, Ellegren & Sheldon 1997). A molecular marker is judged to be sex specific (or even W- or Y-linked) if it is consistently present in one sex and absent in the other in a large enough sample of known males and females. If, however, naturally 'sex-reversed' individuals occur under the influence of certain environmental conditions, this method does of course not apply. Markers should therefore be tested under a wide range of environmental conditions. In Ferguson's (1994a,b) experiments, 10% of poultry were 'sex-reversed' when exposed to pulses of lower temperature during incubation. In some bird studies on sex ratios in nature (Daan *et al.* 1996), the deviation from a 1:1 sex ratio was of the same order of magnitude. It is possible that, especially in the case of adaptively biased sex ratios, temperature-induced 'sex reversal' may be the very mechanism that parent birds use to control offspring sex ratios. Incubating parents may expose their eggs to pulses of different temperatures. If this is the case, primary sex ratios cannot be established with molecular markers. It is, nevertheless, reassuring that one bird study in fact demonstrated extremely biased primary sex ratios with molecular techniques (e.g. Komdeur *et al.* 1997), implying biased ratios of genetic sex, and not 'sex reversal'. Although no evidence exists to support the notion that TSD is operating in birds in the field, until we know more about temperature-induced 'sex reversal' in birds, caution is recommended.

Another area where researchers have tried to establish a nonfatal method of identifying sex at an early age is in endangered species of sea turtles. Sex cannot be identified by external morphology before four years of age. Wellins (1987) found that the blood cells of males typed H-Y positive, consistent with male heterogamety. These turtles, however, are known to have temperature-dependent sex determination, implying that a situation similar to that of the European pond turtle (section 7.3.4) may exist. Therefore, one cannot be sure whether the H-Y status of blood cells of an individual always corresponds to its phenotypic sex. The frequency of natural 'sex reversals' should first be determined. Another study (Demas *et al.* 1990) found male-specific Bkm-related DNA in sea turtles. The sample was small, however, and the natural frequency of 'sex-reversed' individuals has not been investigated. Moreover, the relationship between Bkm-related DNA and phenotypic sex is not clear. Demas *et al.* (1990) mention the possibility that the DNA is altered in accordance with the developing sex, as induced by the temperature. More information is needed before it can be decided whether sex can be reliably identified by molecular markers in these species.

In fish aquaculture, it is also desirable to identify the sex of individuals at an early age. While molecular techniques have become much more commonplace (e.g. Coughlan *et al.* 1999), we again stress that hopes of relying on molecular markers may be too high. Fish are notorious for having labile sex determination (Francis 1992): ESD, socially induced sex determination, and sex change induced by various stimuli have been documented. Reports of sex-specific molecular markers in salmon (Devlin *et al.* 1994), for example, are alternated with reports of ESD in salmon of the same genus (Craig *et al.* 1996). Thorough study is needed of the relationship between phenotypic sex and genetic constitution. Laboratory studies, with controlled environmental conditions, are powerful. Knowledge of the situation in the field, however, is indispensable.

7.3.6 A universal mechanism: a model

Here we describe a model, put forward by Kraak and de Looze (1993), in which we view ESD and GSD as two extremes of a continuum. Both environment and genes determine phenotypic sex, but the extent of their contribution varies. When genes dominate, sex is said to be genetically determined, and when environmental influences dominate, sex determination is called environmental. Several authors have suggested that growth rate may be a universal organizing principle of sex determination (Mittwoch 1971, 1996a, Kraak & de Looze 1993) by acting as the main trigger of sexual differentiation in a critical period during early development. This idea may contribute to viewing ESD and GSD as part of the same mechanism. Mittwoch was the first to propose that the sex chromosomes give rise to quantitative phenotypic differences in growth rate that result in two qualitatively different classes of individuals, i.e. females and males (e.g. Mittwoch et al. 1969, Mittwoch 1989, 1996a). More specifically, she suggested that the mammalian Y chromosome carries growth-enhancing allele(s), and that for testis development to occur the embryonal gonad will need to reach a threshold size by a critical time in development, failing which the gonad will become an ovary (Figure 7.3a) (Mittwoch et al. 1969, 1996a). Figure 7.3b illustrates the opposite threshold mechanism, which might operate in birds: fast growing gonads become ovaries, and the W chromosome may carry the growth promotors

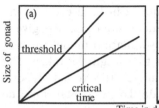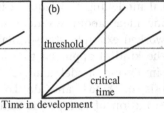

Fig 7.3 Threshold model for growth-dependent sex determination. If a threshold gonadal size or growth rate is reached by a critical time in development, the genes that are responsible for sex differentiation into (a) males or (b) females are activated and expressed. If the threshold is not reached in time, the other sex develops. From Kraak and de Looze (1993), with permission.

(Mittwoch 1971, 1986). Kraak and de Looze (1993) suggested the unification of sex-determining mechanisms for all vertebrates, by proposing that also in other vertebrates one or the other of these threshold mechanisms is operating. The proximate cue for differentiation of the gonads into either testes or ovaries, in all vertebrates, is thought to be the size or stage reached at a critical time in development. Growth rate, in turn, is a quantitative phenotypic trait caused by environmental influences (e.g. temperature) and genetic factors, with ESD at one end of the continuum and GSD at the other. In species with ESD the relationship between sex and growth is thought to be adaptive: the sex that benefits more from fast growth should arise under fast growth conditions (Charnov & Bull 1977, Head et al. 1987, Ewert et al. 1994, Shine 1999, but see Janzen & Paukstis 1991b). Sex determination can then be viewed as a condition- or state-dependent strategy (sensu McNamara & Houston 1996).

Any environmental influence on growth at the proper time is, in this view, sex-determining. Any gene that has an effect on growth in this period, be it minor or major, is a (minor or major) sex-determining gene. It could be the case that *Zfy* or *Sry* acts as a growth factor (but see Burgoyne et al. 1995). Sex determination is in principle polygenic. Heterogamety may be caused by the linkage of several growth-promoting alleles on one of the chromosomes in a pair (Kraak & de Looze 1993). Or, one or a few genes may have strongly sequestered the process of gonadal differentiation, as in mammals, by influencing growth rate at the right time and the right place. Even in the latter case, effects of minor growth genes and/or environmental effects may contribute to a resulting growth rate that induces 'sex reversal'. The term 'sex reversal' is not strictly appropriate. The term is used to indicate that the phenotypic sex of an individual is not in accordance with its genotype. But here, 'genotype' refers only to sex chromosome constitution; if it referred to all sex-chromosomal and autosomal genes, the individual's phenotypic sex could, in fact, be in accordance with the genotype. This idea is supported by the finding that autosomal deletions resulting in slow growth can give rise to XY females in mice (Cattanach et al. 1995). Furthermore, in true hermaphroditism in humans (i.e. the presence of ovarian and testicular tissue in the same individual), the ovarian tissue occurs more often on the left side while testicular tissue is more often present on the right side, and in normal mammalian embryos right gonads grow faster than left gonads (Mittwoch 1996b,c). Whatever the reason ('environmental'?) for this asymmetry in growth rate, this may mean that in rare individuals the size of the left gonad has remained just below the critical threshold, while the right has just exceeded the threshold, at the critical time.

Evidence supporting this model of sex determination, e.g. that early mammalian embryonal growth is related to the presence of Y-linked genes, has been extensively reviewed by Hurst (1994), Mittwoch (1996a) and Erickson (1997); see also Roldan and Gomiendo (1999). Others present evidence and propose models for reptiles with TSD, in which the effect of temperature on asynchronous (heterochronic) development plays a role in sex determination, and speculate on the universal validity of such models for vertebrate sex determination (Haig 1991, Smith & Joss 1994, and see Johnston et al. 1995). Several studies on ESD in fish implicate a relationship between growth and phenotypic sex (Blázquez et al. 1999, Goto et al. 2000). Models that do not focus on the influence of temperature on growth have been proposed by Deeming and Ferguson (1988, 1991). In their view, the dose of a particular molecule determines sex. In GSD the dose is genetically specified. In ESD, the efficiency of gene transcription or translation, or the stability of the mRNA or gene product, or the activity of the gene product is determined by the environmental conditions. Some evidence against the importance of growth in sex determination is the fact that numerous studies of reptile eggs show that water availability during development significantly influences embryonic growth rate (reviewed by Packard 1991), yet no effect of water availability on sex determination has been demonstrated in these species (Packard et al. 1989).

The model of sex determination outlined above can easily account for all combinations of ESD and GSD. According to this view, growth genes anywhere in the genome influence sex determination, and thus tend to be sex specific; but not consistently so, due to the additive genetic and environmental effects on growth. It can explain that even in birds, where sex determination is strongly canalized by factor(s) on the sex chromosomes, temperature can sometimes override the genetic status. It can potentially explain the as yet unexplained 'sex-reversed' humans, e.g. females with an intact SRY gene but a deletion at the short arm of chromosome 9 (Bennett et al. 1993, Raymond et al. 1999), or other XY females and XX males that cannot be accounted for by SRY mutations (Kusz et al. 1999). This is because any factor that sufficiently disturbs the normal growth of embryonal tissues may influence whether the gonad reaches the threshold or not, and hence gonadal differentiation.

7.4 | Evolution of sex-determining mechanisms

Sex-determining mechanisms can evolve quite rapidly, even though some of the genes involved in the process are quite conserved (reviewed by Marín & Baker 1998, Chapter 8). Several models have been proposed to account for the evolution from one system of sex determination to another (reviewed by Bull 1983, Werren & Beukeboom 1998). In section 7.4.1 we stress the importance of sex ratio selection, which plays a decisive role in all such models. In section 7.4.2 we address the question of how mathematical modelling techniques can be used to examine ideas about sex determination. We give a worked example of a simulation model to show the kind of approach that might be taken. The simulation model analyses the verbal hypothesis of Kraak and de Looze (1993) for the evolution of sex-specific heterogamety in vertebrates. We will not deal with evolutionary processes that take place after the establishment of heterogamety, such as the degeneration of Y chromosomes and the evolution of dosage compensation, because these have been treated elsewhere (Charlesworth 1996, Charlesworth & Charlesworth 2000).

7.4.1 The importance of sex ratio selection

Because sex-determining mechanisms control the inheritance of sex, they also determine the primary sex ratio among offspring. A sex ratio of 0.5 is usually advantageous (Chapters 1 & 2), hence systems of sex determination tend to be most stable when they lead to an even sex ratio (Bull 1983, Karlin & Lessard 1986). Nur (1974) provided a simple one-locus-two-allele model to illustrate this. Consider a locus, with alleles A and a, that affects sex determination, but not fertility or survival. Allele A has frequency x in females and y in males, and a proportion M of the offspring become male. Thus, the frequency of A is given by $p = (1 - M)x + My$. Because females and males contribute equally to the next generation (every offspring has one mother and one father), the frequency of A in the next generation will be $p' = (x + y)/2$, hence the change in frequency from one generation to the next is given by

$$\Delta p = p' - p = (y - x)(1/2 - M). \tag{7.1}$$

The frequencies of genes that are involved in sex-determining systems often differ between males and females ($x \neq y$), hence eq. 7.1 tells us that in equilibrium ($\Delta p = 0$), the sex ratio is even ($M = 1/2$). The beauty of this argument is that it holds regardless of how x and y affect the sex ratio M.

Sometimes sex ratios other than 0.5 are selected for (Chapters 1 & 2), and then sex-determination mutations that bias the sex ratio may have an advantage. For example, in several species of lemming a mutant X chromosome, designated X*, causes X*Y individuals to develop as females instead of males (Fredga et al. 1976), thus causing a female-biased sex ratio. It has been argued that the X* chromosome has a selective advantage because the high rate of inbreeding in lemmings favours a female-biased sex ratio (Maynard Smith & Stenseth 1978).

There may also be conflicts of interest over the sex ratio (Chapters 2 & 8) between parent and offspring, between parents or between nuclear and cytoplasmic genes, and this may be an important driving force of evolutionary changes in sex determination (Werren & Beukeboom 1998). For example, cytoplasmic elements are nearly always transmitted via eggs (not via sperm) and therefore favour strongly female-biased sex ratios, unlike autosomal nuclear genes that usually favour a balanced sex ratio (Chapter 9). However, we do not know of any vertebrate examples.

Sex ratio selection is also thought to explain the evolution of ESD (Bull 1983). All else being equal, selection favours a low sex ratio variance among offspring rather than a sex ratio that fluctuates with environmental conditions (Charnov 1982), as would be the case with ESD. However, if fitness varies with environmental conditions in a sex-specific way, then selection favours overproduction of the sex that benefits most given the prevailing condition (Trivers & Willard 1973, Charnov & Bull 1977). ESD is a mechanism that achieves such condition-dependent sex ratios (see Chapter 8 for invertebrate examples). ESD, in turn, influences the population sex ratio. Models have shown that when sex depends on environment rather than genotype, the sex expressed under relatively unfavourable conditions will be more abundant (Charnov 1982, Bull 1983, Frank & Swingland 1988).

7.4.2 Simulations of a scenario for the evolution of ESD to GSD

Kraak and de Looze (1993) have suggested that whichever of the two threshold mechanisms of Figure 7.3 is present in taxa with sex chromosomes is historically determined. In a verbal model they proposed a transition from adaptive ESD to GSD with sex chromosomes, in which genes take over the role of the environment in bringing about differential growth. They assumed vertebrate sex determination to be growth dependent, as argued above (section 7.3.6), and ESD to be ancestral (as supported by Bull 1980, Janzen & Paukstis 1991a, Cree et al. 1995). According to their evolutionary scenario the sex that grows fastest and has a size advantage under ESD will be the sex with heterogametic sex chromosomes. Verbal arguments, however, are not very transparent with respect to their dependence on implicit assumptions. Often a more formal, mathematical, treatment is required in order to see on what assumptions the predicted outcome depends. We previously analysed part of the verbal argument using a two-locus simulation study

Fig 7.4 Crossing-over between the two homologous chromosomes that carry the eight growth loci with alleles 'o' and '*'. The chromosomes break at a location randomly chosen in the sequence of loci and the chromosome parts are swapped.

(Kraak *et al.* 2000). Here we present a multi-locus simulation study that investigates the argument (Kraak & de Looze 1993) that selection would favour the strong linkage of growth-accelerating alleles on one chromosome that would subsequently become the Y or W chromosome. This situation, in which growth genes are sex determining and at the same time have a differential effect on female and male fitness, is a special case of the situation in which selection favours sexually antagonistic genes becoming linked to a sex-determining locus (Rice 1987).

We consider a diploid, randomly mating population of constant size ($N_f + N_m = 500$) but varying sex ratio. The simulations start with a sex ratio of 0.5. Generations are discrete and nonoverlapping. The sex of each individual is determined by the value of a phenotypic trait P relative to a threshold value T. We arbitrarily label the sex developed for $P > T$ 'male' and that for $P \leq T$ 'female' (as in Figure 7.3a). An individual's trait value results from the additive interaction of genetic and environmental factors: $P = G + E$. E corresponds to random individual variations in environmental conditions and is drawn at random from a normal distribution with mean zero and variance VE. We considered $VE = 0.25$, $VE = 0.05$ and $VE = 0.01$. G reflects the additive genetic effects and is the average of the 16 allelic values at eight loci that are located on one pair of homologous chromosomes. An individual's threshold value T is the average of the two allelic values at an unlinked threshold locus, and an individual's recombination rate R between the growth loci similarly results from the allelic values at an unlinked recombination locus.

At the threshold and recombination loci a broad spectrum of 250 alleles is feasible, the allelic values ranging from -1 to $+1$ and 0 to 0.5 respectively. At the growth loci only two alleles are feasible, the allelic values being either 0 and 0.5 (case A), or -0.25 and $+0.25$ (case B). At the start of each simulation, all individuals are homozygous for T alleles with value zero and homozygous for R alleles with value 0.5. In case A all individuals at the start are homozygous at each G locus for alleles with value 0. This situation corresponds to ESD since an individual's sex is purely determined by its environment. Here only growth-accelerating mutations are possible (and back mutations to the growth-neutral allele). In case B the individuals at the start have a random sequence of allelic values at their G loci. This situation corresponds to polygenic sex determination; genetic effects can be growth enhancing or growth inhibiting. Genetic variation is generated by mutation. At the T locus, a given allele T_i changes with probability μ_T into a new allele T_i', where the new value is chosen at random from the interval $[T_i - \Delta T, T_i + \Delta T]$. The same holds true at the R locus with mutation rate μ_R and maximal mutation step size ΔR. At the G loci a given allele changes with probability μ_G into the alternative allele. We keep $\Delta = 0.1$ and $\mu_T = \mu_R = \mu_G = 0.01$. The recombination rate R of an individual determines the probability that crossing-over takes place between the two homologous chromosomes that carry the G loci, at one location randomly chosen from the seven locations between the eight loci in the sequence. By such a crossing-over event parts of the two allele sequences are swapped (Figure 7.4).

In addition to its role in sex determination, the phenotypic value P also has a direct effect on viability: the probability W of survival to reproduction of an individual is linearly related to P, $W(P) = 0.5 + \alpha P$. It is a crucial assumption of our model that α, the slope of the linear relation of W to P, is sex specific: males are more positively affected by a high value of P than females. In our simulations $\alpha_m = 1$, but various values of α_f were considered: $\alpha_f = -1$, $\alpha_f = -0.5$, $\alpha_f = 0$, and $\alpha_f = +0.5$ (Figure 7.5). For each parameter combination ten simulations were carried out, running through 50 000 generations.

The prediction of Kraak and de Looze (1993) is confirmed only when size benefits differ maximally between males and females ($\alpha_f = -1$, Figure 7.5a), and only if we start with pure ESD, i.e. all individuals being homozygous at each G locus for neutral alleles, and the alternative alleles are growth accelerating (case A). Figure 7.6 depicts the results of a typical case of this kind ($VE = 0.01$). The sex ratio remains close to 0.5 throughout the 50 000 generations (Figure 7.6a). The mean recombination rate R (of both males and females) drops at about generation 10 000 and then fluctuates around a low value (Figure 7.6b). At the same time when the recombination rate drops, mean male size P goes up (Figure 7.6c) as well as mean male heterozygosity (= fraction of G loci at which an individual is heterozygous, Figure 7.6e). Mean threshold T remains rather constant in males and females. At generation 25 000 most females have zero or one growth-accelerating allele on both chromosomes (Figure 7.6h), whereas most males have seven or eight growth-accelerating alleles tightly linked on one homologue and zero or one on the other (Figure 7.6g). We interpret this as females having two X chromosomes, and males having one Y and one X chromosome, which recombine at low rates. The Y chromosome carries growth-enhancing alleles; hence, the fast-growing sex that benefits most from large size became the heterogametic sex. This result was replicated ten times for $VE = 0.05$, and ten times for $VE = 0.01$ (at which parameter value the low recombination rate remained more stable). With $VE = 0.25$ there is almost no selection for growth-accelerating alleles, and sex determination remains almost purely environmental. When starting conditions are polygenic (case B), the outcomes are slightly different. Both males and females stay heterozygous at G loci, but R goes down while males accumulate many growth-accelerating alleles linked together on one homologue and females accumulate many growth-inhibiting alleles linked together on one homologue, the other homologue being variable in both males and females.

(a)
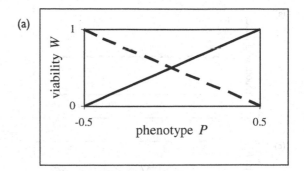

(b)
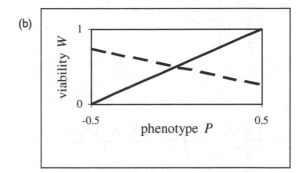

(c)
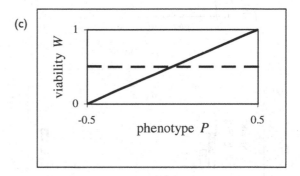

(d)
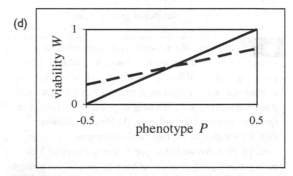

Fig 7.5 Growth-dependent survival until reproduction. (a) $\alpha_m = 1, \alpha_f = -1$; (b) $\alpha_m = 1, \alpha_f = -0.5$; (c) $\alpha_m = 1, \alpha_f = 0$; (d) $\alpha_m = 1, \alpha_f = +0.5$. Male values are shown by solid lines and female values by broken lines.

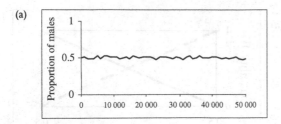

(a)

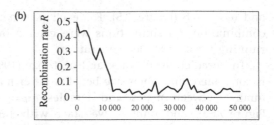

(b)

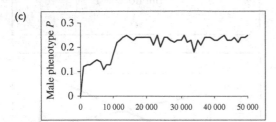

(c)

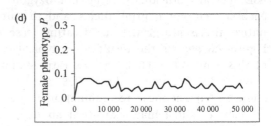

(d)

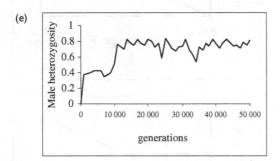

(e)

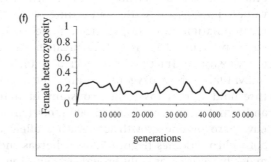

(f)

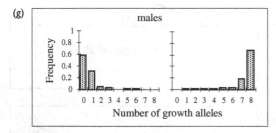

(g)

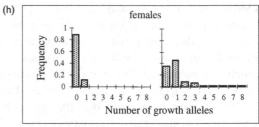

(h)

Fig 7.6 Outcome of a simulation run. The simulation was carried out for 50 000 generations with $N = 500$, $\Delta = 0.1$, $\mu_T = \mu_R = \mu_G = 0.01$, $VE = 0.01$, $\alpha_m = 1$, $\alpha_f = -1$, starting with each G locus being homozygous for allelic value 0 and the alternative allele having value 0.5. (a) The sex ratio (proportion males) through time. (b) Mean recombination rate R through time. (c) Mean male phenotype P through time. (d) Mean female phenotype P through time. (e) Mean male heterozygosity (fraction of heterozygous G loci) through time. (f) Mean female heterozygosity through time. (g) Frequency distribution of growth-accelerating alleles on the two homologues in males at generation 25 000. (h) Frequency distribution of growth-accelerating alleles on the two homologues in females at generation 25 000.

With a slightly smaller difference in size benefits ($\alpha_f = -0.5$, Figure 7.5b) both sexes become/stay heterozygous at the G loci, and with even smaller fitness differentials ($\alpha_f = 0$ and $\alpha_f = +0.5$, Figure 7.5c and 7.5d) growth-accelerating alleles tend to approach fixation in both sexes, while R fluctuates randomly. In case $\alpha_f = 0$ or $\alpha_f = +0.5$ males become heterozygous at the T locus and get a lower mean T than do females. These patterns are similar for case A and case B.

In conclusion, if the difference in fitness effects of size between the sexes is large enough in a species with ESD, selection may favour the linkage of growth-accelerating genes on one homologue of a pair of autosomes in the fast-growing sex. This pair of autosomes will then effectively become a pair of sex chromosomes, with the fast-growing sex being the heterogametic sex. However, this occurs only under certain restrictive conditions, and it is not clear how often these conditions are met in nature. For example, a starting situation without genetic variation for growth seems unlikely. Therefore, we cannot yet conclude that the proposed scenario provides a sufficient explanation for the presence of male versus female heterogamety.

7.5 | Conclusions

We started this chapter with the traditional classification of sex-determining mechanisms as either GSD, with female or male heterogamety (or polygenic sex determination), or ESD. However, both the establishment of GSD and that of ESD appear sometimes to be ambiguous. ESD has often been established in the laboratory, but sometimes with experimental conditions that are outside the range naturally experienced by the species. Few studies exist to show that ESD operates in the wild. One study on a turtle that exhibits ESD in the laboratory suggests that it does not occur at high frequencies naturally. More studies should investigate ESD in the wild. GSD, and in particular the heterogametic sex, is often established with molecular techniques. However, the relationship between the phenotypic sex of an individual on the one hand, and the genetic constitution or presence of a molecular marker in an individual on the other hand is not clear. We recommend that when molecular markers are to be used for sex ratio studies, they should be tested with sufficiently large samples and under a wide range of environmental conditions. The discrepancies between phenotypic sex and genotypic sex should be the subject of study as they might shed light on the nature of sex determination. The apparent existence of mixtures of ESD and heterogamety challenges the traditional classification. We discuss a model of sex determination that attempts to account for these cases. Other models may be

plausible too. Ultimately, an evolutionary model should explain not only the existing modes of sex determination, but also their phylogenetic distribution.

Acknowledgements

We thank James Cook, Ian Hardy, Ian Swingland, Franjo Weissing and, in particular, an anonymous referee for their critical comments. We also want to thank Ellen de Looze, mother both of some of the ideas in this chapter and of the first author.

References

Beamish FW (1993) Environmental sex determination in southern brook lamprey, *Ichthyomyzon gagei*. *Canadian Journal of Fisheries and Aquatic Sciences*, **50**, 1299–1307.

Beatty RA (1964) Chromosome deviations and sex in vertebrates. In: CN Armstrong & AJ Marshall (eds) *Intersexuality in Vertebrates Including Man*, pp 17–143. London: Academic Press.

Bennett CP, Docherty Z, Robb SA, Ramani P, Hawkins JR & Grant D (1993) Deletion 9p and sex reversal. *Journal of Medical Genetics*, **30**, 518–520.

Bertollo LAC & Cavallaro ZI (1992) A highly differentiated ZZ/ZW sex-chromosome system in a Characidae fish, *Triportheus guenteri*. *Cytogenetics and Cell Genetics*, **60**, 60–63.

Blázquez M, Carrillo M, Zanuy S & Piferrer F (1999) Sex ratios in offspring of sex-reversed sea bass and the relationship between growth and phenotypic sex differentiation. *Journal of Fish Biology*, **55**, 916–930.

Bull JJ (1980) Sex determination in reptiles. *The Quarterly Review of Biology*, **55**, 3–21.

Bull JJ (1983) *Evolution of Sex Determining Mechanisms*. Menlo Park: Benjamin/Cummings.

Bull JJ, Vogt RC & Bulmer MG (1982) Heritability of sex ratio in turtles with environmental sex determination. *Evolution*, **36**, 333–341.

Bull JJ, Hillis DMH & O'Steen S (1988) Mammalian ZFY sequences exist in reptiles regardless of sex-determining mechanisms. *Science*, **242**, 567–569.

Burgoyne PS, Thornhill AR, Kalmus Boudrean S, Darling SM, Bishop CE & Evans EP (1995) The genetic basis of XX-XY differences present before gonadal sex differentiation in the mouse. *Philosophical Transactions of the Royal Society of London, series B*, **350**, 253–261.

Caputo V, Odierna G & Aprea G (1994) A chromosomal study of *Eumeces* and *Scincus*, primitive members of the Scincidae (Reptilia, Squamata). *Bollettino di Zoologia*, **61**, 155–162.

Carr JL & Bickham JW (1981) Sex chromosomes of the Asian black pond turtle *Siebenrockiella crassicollis* (Testudines: Emydidae). *Cytogenetics and Cell Genetics*, **31**, 178–183.

Cattanach BM, Rasberry C & Beechey CV (1995) XY sex reversal associated with autosomal deletions. *Mouse Genome*, **93**, 426.

Charlesworth B (1996) The evolution of chromosomal sex determination and dosage compensation. *Current Biology*, **6**, 140–162.

Charlesworth B & Charlesworth D (2000) The degeneration of Y chromosomes. *Philosophical Transactions of the Royal Society of London, series B*, **355**, 1563–1572.

Charnov EL (1982) *The Theory of Sex Allocation*. Princeton, NJ: Princeton University Press.

Charnov EL & Bull JJ (1977) When is sex environmentally determined? *Nature*, **266**, 828–830.

Chourrout D (1986) Revue sur le déterminisme génétique du sexe des poissons téléostéens. *Bulletin de la Société Zoologique de France*, **113**, 123–144.

Ciofi C & Swingland IR (1997) Environmental sex determination in reptiles. *Applied Animal Behaviour Science*, **51**, 251–265.

Cole CJ (1971) Karyotypes of the five monotypic species groups of lizards in the genus *Sceloporus*. *American Museum Novitates*, **2450**, 1–17.

Conover DO & Heins SW (1987) Adaptive variation in environmental and genetic sex determination in a fish. *Nature*, **326**, 496–498.

Coriat A-M, Muller U, Harry JL, Uwanogho D & Sharpe PT (1993) PCR amplification of *SRY*-related gene sequences reveals evolutionary conservation of the *SRY*-related motif. *PCR Methods and Applications*, **2**, 218–222.

Coriat A-M, Valleley E, Ferguson MWJ & Sharpe PT (1994) Chromosomal and temperature-dependent sex determination: the search for a conserved mechanism. *The Journal of Experimental Zoology*, **270**, 112–116.

Coughlan T, Schartl M, Hornung U, Hope I & Stewart A (1999) PCR-based sex test for *Xiphophorus maculatus*. *Journal of Fish Biology*, **54**, 218–222.

Craig JK, Foote CJ & Wood CC (1996) Evidence for temperature-dependent sex determination in sockeye salmon (*Oncorhynchus nerka*). *Canadian Journal of Fisheries and Aquatic Sciences*, **53**, 141–147.

Cree A, Thompson MB & Daugherty CH (1995) Tuatara sex determination *Nature*, **375**, 543.

Daan S, Dijkstra C & Weissing FJ (1996) An evolutionary explanation for seasonal sex ratio trends in avian sex ratios. *Behavioral Ecology*, **7**, 426–430.

Deeming DC & Ferguson MWJ (1988) Environmental regulation of sex determination in reptiles. *Philosophical Transactions of the Royal Society of London, series B*, **322**, 19–39.

Deeming DC & Ferguson MWJ (1991) Physiological effects of incubation temperature on embryonic development in reptiles and birds. In: DC Deeming & MWJ Ferguson (eds) *Egg Incubation: Its Effects on Embryonic Development in Birds and Reptiles*, pp 147–171. Cambridge: Cambridge University Press.

Demas S, Duronslet M, Wachtel S, Caillouet C & Nakamura D (1990) Sex-specific DNA in reptiles with temperature sex determination. *The Journal of Experimental Zoology*, **253**, 319–324.

Devlin RH, McNeil BK, Solar II & Donaldson EM (1994) A rapid PCR-based test for Y-chromosomal DNA allows simple production of all-female strains of chinook salmon. *Aquaculture*, **128**, 211–220.

Docker MF & Beamish FWH (1994) Age, growth, and sex ratio among populations of least brook lamprey, *Lampetra aepyptera*, larvae – an argument for environmental sex determination. *Environmental Biology of Fishes*, **41**, 191–205.

Dorazi R, Chesnel A & Dournon C (1995) Opposite sex determination of gonads in two *Pleurodeles* species may be due to a temperature-dependent inactivation of sex chromosomes. *Journal of Heredity*, **86**, 28–31.

Duellman WE & Trueb L (1986) *Biology of Amphibians*. New York: McGraw-Hill.

Ellegren H & Sheldon BC (1997) New tools for sex identification and the study of sex allocation in birds. *Trends in Ecology and Evolution*, **12**, 255–259.

Engel W & Schmid M (1981) H-Y antigen as a tool for the determination of the heterogametic sex in Amphibia. *Cytogenetics and Cell Genetics*, **30**, 130–136.

Engel W, Klemme B & Schmid M (1981) H-Y antigen and sex determination in turtles. *Differentiation*, **20**, 152–156.

Erickson RP (1997) Does sex determination start at conception? *BioEssays*, **19**, 1027–1032.

Ewert MA, Etchberger CR & Nelson CE (1990) An apparent co-occurrence of genetic and environmental sex determination in a turtle. *American Zoologist*, **30**, 56A.

Ewert MA, Jackson DR & Nelson CE (1994) Patterns of temperature-dependent sex determination in turtles. *The Journal of Experimental Zoology*, **270**, 3–15.

Ferguson MWJ (1994a) Temperature dependent sex determination and growth in reptiles and manipulation of poultry sex by incubation temperature. In: *Proceedings of the 9th European Poultry Conference in Glasgow*, pp 380–382.

Ferguson MWJ (1994b) Method of hatching avian eggs. Patent WO 94/13132.

Ford LS & Cannatella DC (1993) The major clades of frogs. *Herpetological Monographs* **7**, 94–117.

Francis RC (1992) Sexual lability in teleosts: developmental factors. *The Quarterly Review of Biology*, **67**, 1–18.

Francis RC & Barlow GW (1993) Social control of primary sex differentiation in the Midas cichlid. *Proceedings of the National Academy of Sciences, USA* **90**, 10673–10675.

Frank SA & Swingland IR (1988) Sex ratio under conditional sex expression. *Journal of Theoretical Biology*, **135**, 415–418.

Fredga K (1994) Bizarre mammalian sex-determining mechanisms. In: RV Short & E Balaban (eds) *The Difference Between the Sexes*, pp 419–431. Cambridge: Cambridge University Press.

Fredga K, Gropp A, Winking H & Frank F (1976) Fertile XX- and XY- females in the wood lemming *Myopus schisticolor*. *Nature* **261**, 225–227.

Gaffney ES & Meylan PA (1988) A phylogeny of turtles. In: MJ Benton (ed) *The Phylogeny and Classification of the Tetrapods*, volume 1, *Amphibians, Reptiles, Birds*, pp 157–219. Oxford: Clarendon Press.

Ganesh S, Mohanty J & Raman R (1997) Male-biased distribution of the human Y chromosomal genes SRY and ZFY in the lizard *Calotes versicolor*, which lacks sex chromosomes and temperature-dependent sex determination. *Chromosome Research*, **5**, 413–419.

Gauthier J, Estes R & de Queiroz K (1988) A phylogenetic analysis of Lepidosauromorpha. In: R Estes & G Pregill (eds) *Phylogenetic Relationships of the Lizard Families*, pp 15–98. Stanford: Stanford University Press.

Girondot M, Zaborski P, Servan J & Pieau C (1994) Genetic contribution to sex determination in turtles with environmental sex determination. *Genetical Research, Cambridge*, **63**, 117–127.

Goodfellow PN & Lovell-Badge R (1993) The SRY and sex determination in mammals. *Annual Review of Genetics*, **27**, 71–92.

Gorman GC (1973) The chromosomes of the reptilia, a cytotaxonomic interpretation. In: AB Chiarelli & E Capanna (eds) *Cytotaxonomy and Vertebrate Evolution*, pp 349–424. London: Academic Press.

Goto R, Kayaba T, Adachi S & Yamauchi K (2000) Effects of temperature on sex determination in marble sole *Limanda yokohamae*. *Fisheries Science*, **66**, 400–402.

Griffiths R (1991) The isolation of conserved DNA sequences related to the human sex-determining region of Y gene from the lesser black-backed gull (*Larus fuscus*). *Proceedings of the Royal Society of London, series B*, **244**, 123–128.

Griffiths R, Daan S & Dijkstra C (1996) Sex identification in birds using two CHD genes. *Proceedings of the Royal Society of London, series B*, **263**, 1251–1256.

Gubbay J, Collignon J, Koopman P, Capel B, Economou A, Münsterberg A, Vivian N, Goodfellow P & Lovell-Badge R (1990) A gene mapping to the sex-determining region of the mouse Y chromosome is a member of a novel family of embryonically expressed genes. *Nature*, **346**, 245–250.

Haig D (1991) Developmental asynchrony and environmental sex determination in alligators. *Journal of Theoretical Biology*, **150**, 373–383.

Hayes TB (1998) Sex determination and primary sex differentiation in amphibians: genetic and developmental mechanisms. *Journal of Experimental Zoology*, **281**, 373–399.

Head G, May RM & Pendleton L (1987) Environmental determination of sex in the reptiles. *Nature*, **329**, 198–199.

Hillis DM & Green DM (1990) Evolutionary changes of heterogametic sex in the phylogenetic history of amphibians. *Journal of Evolutionary Biology*, **3**, 49–64.

Holmgren K & Mosegaard H (1996) Implications of individual growth status on the future sex of the European eel. *Journal of Fish Biology*, **49**, 910–925.

Hurst LD (1994) Embryonic growth and the evolution of the mammalian Y chromosome. I. The Y as an attractor for selfish growth factors. *Heredity*, **73**, 223–232.

Janvier P (1996a) Jawed vertebrates. In: D Maddison & W Maddison (eds) *The Tree of Life*, http://phylogeny. arizona.edu/tree/eukaryotes/animals/chordata/ gnathostomata.html.

Janvier P (1996b) Vertebrata. In: D Maddison & W Maddison (eds) *The Tree of Life*, http://phylogeny.

arizona.edu/tree/eukaryotes/animals/chordata/
vertebrata.html.

Janzen FJ (1992) Heritable variation for sex ratio
under environmental sex determination in the
common snapping turtle (*Chelydra serpentina*).
Genetics, **131**, 155–161.

Janzen FJ & Paukstis GL (1991a) Environmental sex
determination in reptiles: ecology, evolution, and
experimental design. *The Quarterly Review of Biology*,
66, 149–179.

Janzen FJ & Paukstis GL (1991b) A preliminary test of
the adaptive significance of environmental sex
determination in reptiles. *Evolution*, **45**, 435–440.

Jiménez R, Sanchez A, Burgos M & Díaz de la
Guardia R (1996) Puzzling out the genetics of
mammalian sex determination. *Trends in Genetics*,
12, 164–166.

Johnston CM, Barnett M & Sharpe PT (1995) The
molecular biology of temperature-dependent sex
determination. *Philosophical Transactions of the Royal
Society of London, series B*, **350**, 297–303.

Jones KW & Singh L (1981) Conserved repeated DNA
sequences in vertebrate sex chromosomes. *Human
Genetics*, **58**, 46–53.

Jones KW & Singh L (1985) Snakes and the evolution
of sex chromosomes. *Trends in Genetics*, **1**, 55–61.

Karlin S & Lessard S (1986) *Theoretical Studies on Sex
Ratio Evolution*. Princeton, NJ: Princeton University
Press.

King M & Rofe R (1976) Karyotypic variation in the
Australian gekko *Phyllodactylus marmoratus* (Gray)
(Gekkonidae: Reptilia). *Chromosoma*, **54**, 75–87.

Koehler MR, Neuhaus D, Engel W, Schartl M & Schmid
M (1995) Evidence for an unusual ZW/ZW'/ZZ
sex-chromosome system in *Scardinius erythrophtalmus*
(Pisces, Cyprinidae), as detected by cytogenetic and
H-Y-antigen analyses. *Cytogenetics and Cell Genetics*, **71**,
356–362.

Komdeur J, Daan S, Tinbergen J & Mateman C (1997)
Extreme adaptive modification in sex ratio of the
Seychelles warbler's eggs. *Nature*, **385**, 522–525.

Koopman P, Gubbay J, Vivian N, Goodfellow P &
Lovell-Badge R (1991) Male development of
chromosomally female mice transgenic for *Sry*.
Nature, **351**, 117–121.

Kraak SBM & de Looze EMA (1993) A new hypothesis
on the evolution of sex determination in
vertebrates: big females ZW, big males XY.
Netherlands Journal of Zoology, **43**, 260–273.

Kraak SBM, Pen I & Weissing FJ (2000) Joint evolution
of environmental and genetic sex determination. In:
I Pen *Sex Allocation in a Life History Context*, pp 149–

160. Ph.D. thesis University of Groningen
(http://www.biol.rug.nl/theobio/main/
research/ido/bsex_rat.htm).

Kusz K, Kotecki M, Wojda A, Jaruzelska J,
Szarras-Czapnik M, Ruszczynska-Wolska A,
Latos-Bielenska A & Warenik-Szymankiewicz A
(1999) Incomplete masculinisation of XX subjects
carrying the *SRY* gene on an inactive X
chromosome. *Journal of Medical Genetics*, **36**,
452–456.

Lagomarsino IV & Conover DO (1993) Variation in
environmental and genotypic sex-determining
mechanisms across a latitudinal gradient in the
fish, *Menidia menidia*. *Evolution*, **47**, 487–494.

Larson A & Dimmick WW (1993) Phylogenetic
relationships of the salamander families: an
analysis of congruence among morphological and
molecular characters. *Herpetological Monographs*, **7**,
77–93.

Laurin M (1996) Terrestrial vertebrates. In: D
Maddison & W Maddison (eds) *The Tree of Life*,
http://phylogeny.arizona.edu/tree/eukaryotes/
animals/chordata/terrestrial_vertebrates.html.

Laurin M & Reisz RR (1995) A reevaluation of early
amniote phylogeny. *Zoological Journal of the Linnean
Society*, **113**, 165–223.

Laurin M, Gauthier JA & Hedges SB (1996) Amniota.
In: D Maddison & W Maddison (eds) *The Tree of Life*,
http://phylogeny.arizona.edu/tree/eukaryotes/
animals/chordata/amniota.html.

Lundberg JG (1996) Actinopterygii. In: D Maddison &
W Maddison (eds) *The Tree of Life*, http://phylogeny.
arizona.edu/tree/eukaryotes/animals/chordata/
actinopterygii/actinopterygii.html.

Mahony MJ (1991) Heteromorphic sex-chromosomes in
the australian frog *Crinia bilingua* (Anura,
Myobatrachidae). *Genome*, **98**, 334–337.

Maistro EL, Mata EP, Oliveira C & Foresti F (1998)
Unusual occurrence of a ZZ/ZW sex-chromosome
system and supernumerary chromosomes in
Characidium cf. fasciatum (Pisces, Characiformes,
Characidiinae). *Genetica*, **104**, 1–7.

Marín I & Baker BS (1998) The evolutionary dynamics
of sex determination. *Science*, **281**, 1990–1994.

Maynard Smith J & Stenseth NC (1978) On the
evolutionary stability of the female biased sex ratio
in the wood lemming (*Myopus schisticolor*): the effect
of inbreeding. *Heredity*, **41**, 205–214.

McNamara JM & Houston AT (1996) State-dependent
life histories. *Nature*, **380**, 215–221.

McVean G & Hurst LD (1996) Genetic conflicts and the
paradox of sex determination: three paths to the

evolution of female intersexuality in a mammal. *Journal of Theoretical Biology*, **179**, 199–211.

Mittwoch U (1971) Sex determination in birds and mammals. *Nature*, **231**, 432–434.

Mittwoch U (1986) Males, females and hermaphrodites. *Annals of Human Genetics*, **50**, 103–121.

Mittwoch U (1989) Sex differentiation in mammals and tempo of growth: probabilities vs. switches. *Journal of Theoretical Biology*, **137**, 445–455.

Mittwoch U (1996a) Genetics of sex determination: an overview. *Advances in Genome Biology*, **4**, 1–28.

Mittwoch U (1996b) Unilateral phenotypic manifestations of bilateral structures: which phenotype matches the genotype? *Frontiers in Endocrinology*, **16**, 121–129.

Mittwoch U (1996c) Sex-determining mechanisms in animals. *Trends in Ecology and Evolution*, **11**, 63–67.

Mittwoch U, Delhanty JDA & Beck F (1969) Growth of differentiating testes and ovaries. *Nature*, **224**, 323–325.

Molino WF, Schmid M & Galetti PM (1998) Heterochromatin and sex chromosomes in the Neotropical fish genus *Leporinus* (Characiformes, Anastomidae). *Cytobios*, **94**, 141–149.

Moreira-Filho O, Bertollo LAC & Galetti PM Jr (1993) Distribution of sex chromosome mechanisms in neotropical fish and a description of a ZZ/ZW system in *Parodon hilarii* (Parodontidae). *Caryologia*, **46**, 115–125.

Moritz C (1990) Patterns and processes of sex chromosome evolution in gekkonid lizards (Sauria: Reptilia). In: E Olmo (ed) *Cytogenetics of Amphibians and Reptiles*, pp 205–220. Basel: Birkhäuser Verlag.

Nakamura D, Wachtel SS, Lance V & Beçak W (1987) On the evolution of sex determination. *Proceedings of the Royal Society of London, series B*, **232**, 159–180.

Nanda I, Schartl M, Feichtinger W, Epplen JT & Schmid M (1992) Early stages of sex chromosome differentiation in fish as analysed by simple repetitive DNA sequences. *Chromosoma*, **101**, 301–310.

Nelson JS (1994) *Fishes of the World*, 3rd edn. New York: John Wiley & Sons.

Nur U (1974) The expected changes in the frequency of alleles affecting the sex ratio. *Theoretical Population Biology*, **5**, 143–147.

Olmo E (1986) *Animal Cytogenetics*, volume 4, *Chordata 3, A, Reptilia*. Berlin: Gebrüder Borntraeger.

Olmo E, Odierna G, Capriglione T & Cardone A (1990) DNA and chromosome evolution in lacertid lizards. In: E Olmo (ed) *Cytogenetics of Amphibians and Reptiles*, pp 181–204. Basel: Birkhäuser Verlag.

Packard GC (1991) Egg incubation: its effects on embryonic development in birds and reptiles. In: DC Deeming & MWJ Ferguson (eds) *Egg Incubation: Its Effects on Embryonic Development in Birds and Reptiles*, pp 213–228. Cambridge: Cambridge University Press.

Packard GC, Packard MJ & Birchard GF (1989) Sexual-differentiation and hatching success by painted turtles incubating in different thermal and hydric environments. *Herpetologica*, **45**, 385–392.

Page DC, Mosher R, Simpson EM, Fisher EMC, Mardon G, Pollack J, McGillivray B, de la Chapelle A & Brown LG (1987) The sex determining region of the human Y chromosome encodes a finger protein. *Cell*, **51**, 1091–1104.

Palmer MS, Sinclair AH, Berta P, Ellis NA, Goodfellow PN, Abbas NE & Fellous M (1989) Genetic evidence that ZFY is not the testis-determining factor. *Nature*, **342**, 937–939.

Patiño R, Davis KB, Schoore, JE, Uguz C, Strüssman CA, Parker NC, Simco BA & Goudi CA (1996) Sex differentiation of channel catfish gonads: normal development and effects of temperature. *Journal of Experimental Zoology*, **276**, 209–218.

Pieau C (1975) Effets des variations thermique sur la différentiation du sexe chez les vertébrés. *Bulletin du Société Zoologique Français*, **100**, 67–76.

Price DJ (1984) Genetics of sex determination in fishes – a brief review. In: GW Potts & RJ Wootton (eds) *Fish Reproduction*, pp 77–89. London: Academic Press.

Raymond SC, Parker ED, Kettlewell JR, Brown LG, Page DC, Kusz K, Jaruzelska J, Reinberg Y, Flejter WL, Bardwell VJ, Hirsch B & Zarkower D (1999) A region of human chromosomes 9p required for testis development contains two genes related to known sexual regulators. *Human Molecular Genetics*, **8**, 989–996.

Rice WR (1987) The accumulation of sexually antagonistic genes as a selective agent promoting the evolution of reduced recombination between primitive sex chromosomes. *Evolution*, **41**, 911–914.

Richards CM & Nace GW (1978) Gynogenetic and hormonal sex reversal used in tests of the XX-XY hypothesis of sex determination in *Rana pipiens*. *Growth*, **42**, 319–332.

Rieppel O (1988) The classification of the Squamata. In: MJ Benton (ed) *The Phylogeny and Classification of the Tetrapods*, volume 1, *Amphibians, Reptiles, Birds*, pp 261–293. Oxford: Clarendon Press.

Roldan ERS & Gomiendo M (1999) The Y chromosome as a battle ground for sexual selection. *Trends in Ecology and Evolution*, **14**, 58–62.

Römer U & Beisenhertz W (1996) Environmental determination of sex in Apistogramma (Cichlidae) and two other freshwater fishes (Teleostei). *Journal of Fish Biology*, **48**, 714–725.

Schmid M & Haaf T (1989) Origin and evolution of sex chromosomes in Amphibia: the cytogenetic data. In: SS Wachtel (ed) *Evolutionary Mechanisms in Sex Determination*, pp 37–56. Boca Raton, FL: CRC Press, Inc.

Schmid M, Steinlein C, Feichtinger W, Almeida CG de & Duellman WE (1988) Chromosome banding in Amphibia, XIII. sex chromosomes, heterochromatin and meiosis in marsupial frogs (Anura, Hylidae). *Chromosoma*, **97**, 33–42.

Schmid M, Steinlein C & Feichtinger W (1992) Chromosome-banding in Amphibia. 17. 1st demonstration of multiple sex-chromosomes in amphibians – *Eleutherodactylus maussi* (Anura, Leptodactylidae). *Chromosoma*, **101**, 284–292.

Schmid M, Ohta S, Steinlein C & Guttenbach M (1993) Chromosome-banding in Amphibia. 19. Primitive ZW/ZZ sex chromosomes in *Buergeria buergeri* (Anura, Rhacophoridae). *Cytogenetics and Cell Genetics*, **62**, 238–246.

Scudo FM (1967) Criteria for the analysis of multifactorial sex determination. *Monitore Zoologia Italia (N. S.)*, **1**, 1–21.

Shine R (1999) Why is sex determined by nest temperature in many reptiles? *Trends in Ecology and Evolution*, **14**, 186–189.

Sinclair AH, Berta P, Palmer MS, Hawkins JR, Griffiths BL, Smith MJ, Foster JW, Frischauf AM, Lovell-Badge R & Goodfellow PN (1990) A gene from the human sex-determining region encodes a protein with homology to a conserved DNA-binding motif. *Nature*, **346**, 240–244.

Singh L & Jones KW (1982) Sex reversal in the mouse (*Mus musculus*) is caused by a recurrent nonreciprocal crossover involving the X and an aberrant Y chromosome. *Cell*, **28**, 205–216.

Smith CA & Joss JMP (1994) Sertoli cell differentiation and gonadogenesis in *Alligator mississippiensis*. *The Journal of Experimental Biology*, **270**, 57–70.

Sola L, Cataudella S & Capanna E (1981) New developments in vertebrate cytotaxonomy III. Karyology of bony fishes: a review. *Genetica*, **54**, 285–328.

Strüssman CA, Cota JCC, Phonlor G, Higuchi H & Takashima F (1996a) Temperature effects on sex-differentiation of 2 South-American atherinids, *Odontesthes argentinensis* and *Patagonia hatcheri*. *Environmental Biology of Fishes*, **47**, 143–154.

Strüssman CA, Moriyama S, Hanke EF, Cota JCC & Takashima F (1996b) Evidence of thermolabile sex determination in pejerrey. *Journal of Fish Biology*, **48**, 643–651.

Tiersch TR, Mitchell MJ & Wachtel SS (1991) Studies on the phylogenetic conservation of the SRY gene. *Human Genetics*, **87**, 571–573.

Trivers RL & Willard DE (1973) Natural selection of parental ability to vary the sex ratio of offspring. *Science*, **179**, 90–92.

Valleley EMA, Muller U, Ferguson MWJ & Sharpe PT (1992) Cloning and expression analysis of two ZFY-related zinc finger genes from *Alligator mississippiensis*, a species with temperature-dependent sex determination. *Gene*, **119**, 221–228.

VanEenennaam AL, VanEenennaam JP, Medrano JF & Doroshov SI (1999) Evidence of female heterogametic sex determination in white sturgeon. *Journal of Heredity*, **90**, 231–233.

Viets BE, Ewert MA, Talent LG & Nelson CE (1994) Sex-determining mechanisms in squamate reptiles. *The Journal of Experimental Zoology*, **270**, 45–56.

Volobouev V, Pasteur G, Bons J, Guillaume CP & Dutrillaux B (1990) Sex-chromosome evolution in reptiles – divergence between 2 lizards long regarded as sister species, *Lacerta vivipara* and *Lacerta andreansky*. *Genetica*, **83**, 85–91.

Wellins DJ (1987) Use of an H-Y antigen assay for sex determination in sea turtles. *Copeia*, **1987**, 46–52.

Werren JH & Beukeboom LW (1998) Sex determination, sex ratios and genetic conflict. *Annual Review of Ecology and Systematics*, **29**, 233–261.

Winge Ø (1932) The nature of sex chromosomes. *Proceedings of the 6th International Congress of Genetics*, **1**, 343–355.

Witschi E (1929) Studies on sex differentiation and sex determination in amphibians. III. Rudimentary hermaphroditism and Y chromosome in *Rana temporaria*. *The Journal of Experimental Zoology*, **54**, 157–223.

Yntema CL (1976) Effects of incubation temperatures on sexual differentiation in the turtle, *Chelydra serpentina*. *Journal of Morphology*, **150**, 453–462.

Yntema CL (1979) Temperature levels and periods of sex determination during incubation of eggs of *Chelydra serpentina*. *Journal of Morphology*, **159**, 17–28.

Zaborski P, Dorizzi M & Pieau C (1982) H-Y antigen expression in temperature sex reversed turtles *Emys orbicularis*. *Differentiation*, **22**, 73–78.

Zaborski P, Dorizzi M & Pieau C (1988) Temperature-dependent gonadal differentiation in the turtle *Emys orbicularis*: concordance between sexual phenotype and serological H-Y antigen expression at threshold temperature. *Differentiation*, **38**, 17–20.

Chapter 8

Sex determination in invertebrates

James M. Cook

8.1 | Summary

Invertebrates display a great variety of different sex-determining mechanisms. While sex determination may be quite conserved in some
taxa, in others it differs between closely related species or even between different populations of the same species. Sex determination
and sex allocation tend to evolve interactively
through the common medium of sex ratio selection. However, sex ratio conflicts can occur due
to differential selection upon genes acting in different individuals (e.g. parent versus offspring)
or with different transmission patterns (e.g.
nuclear versus cytoplasmic). Sex ratio selection,
especially when there is conflict, is probably of
key importance in generating the turnover of
sex-determination systems to produce the great
diversity observed. Most sex-determining mechanisms can be classified as primarily genetic
(GSD) or environmental (ESD) but in some cases
both influences are important. In addition, cytoplasmic sex factors may be present (Chapter 9).
Selection for adaptive sex ratio patterns may
sometimes achieve similar endpoints using different raw material. For example, patterns of association between offspring gender and patch
quality may be achieved through zygotic ESD
or via genes influencing maternal behaviour in
certain GSD cases. At a different level, common
GSD systems, such as male heterogamety, are
known to be underpinned by different molecular
mechanisms.

8.2 | Introduction

Sex determination refers to the genetic and environmental basis of an individual's gender and
is consequently of great relevance to considerations of the sex ratio. Since the division of reproduction into separate male and female sexes
occurred well before the origin of animals, and is
such a fundamental aspect of their biology, one
might expect the genetic basis of sex to be similar
in most invertebrates. However, there is instead
a great variety of invertebrate sex-determining
mechanisms (reviewed by Bull 1983). Furthermore, a great deal of the variation actually
occurs between closely related species (rather
than higher taxa), and even between different populations of the same species (Bull 1983,
Rigaud 1997). This low-level variation offers good
opportunities to identify the factors responsible
for variation in the way that sex is inherited and
how sex determination and sex ratios coevolve.

Recent work, especially in molecular genetics,
is helping us to gain an overview of the evolution of sex determination and some interesting
patterns have emerged. First, there are several
cases where sex determination varies between
different populations of the same species, e.g. the
woodlouse *Armadillidium vulgare* (Rigaud 1997),
the housefly *Musca domestica* (Milani *et al.* 1967,
Kerr 1970, Bull 1983) and the shrimp *Gammarus
duebeni* (Bulnheim 1969, Watt & Adams 1994,
Dunn & Hatcher 1997). Second, the key regulatory gene *sex-lethal* has conserved function in

different *Drosophila* species (Bopp *et al.* 1996) but has diverged in function in some other Diptera (the housefly *Musca domestica*, Meise *et al.* 1998, and the medfly *Ceratitis capitata*, Saccone *et al.* 1998). Third, some sex-determining genes may be essentially conserved in function across very distantly related animals; recently Raymond *et al.* (1998) inferred homology of the genes *doublesex* in *Drosophila* and *mab-3* in the nematode worm *Caenorhabditis elegans* (and slightly weaker links with the human gene *DMT1*). Functionally, both *doublesex* and *mab-3* are involved in sex-specific neuroblast differentiation and yolk protein gene transcription (Raymond *et al.* 1998). There seems to be an emerging pattern that

primary sex-determining signals (Box 8.1) may evolve rapidly, while genes further downstream (e.g. *doublesex* and *mab-3*), which control phenomena such as germ-line and somatic sex differentiation, are more likely to be conserved in function (Wilkins 1995).

Considerable debate surrounds the question of what drives these changes in sex determination and there is growing evidence of a key role for sex ratio selection in general (Bull 1983, Chapter 7), and sex ratio conflicts in particular (Werren & Beukeboom 1998, Chapter 9). In recent years, detailed studies of a range of invertebrates (notably parasitic wasps, woodlice, ladybird beetles and freshwater shrimps) have

---

Box 8.1 | Glossary of sex determination terms

- **Sex factors.** These are genetic entities that influence the inheritance of sex (Bull 1983), e.g. X and Y sex chromosomes in many species, and cytoplasmic sex factors (Chapter 9).
- **Sex tendency.** Most sex factors have a sex tendency. For example, the human Y chromosome has a male tendency while the X has a female tendency. The complementary sex determination (CSD) sex alleles of Hymenoptera are an exception, with no particular sex tendency.
- **Primary sex-determining signal.** This 'tells' an individual zygote what sex to become. Examples include the ratio of X chromosomes to autosomes in *Drosophila*, and the presence/absence of a Y chromosome in mammals.
- **Zygotic sex-determining genes** (Werren 1987, Werren & Beukeboom 1998) are expressed in the zygote and include the genes involved in the important X:A ratio in *Drosophila*, as well as the male-determining *SRY* gene in mammals.
- **Parental sex ratio genes** (Werren 1987, Werren & Beukeboom 1998) are expressed in the parent and influence the sex ratio of offspring. Examples include driving chromosomes in several species, selection of oviposition patch in species with environmental sex determination (ESD) and egg fertilization rate in haplodiploid species (see Chapter 9).
- **Parental effect sex-determining genes** (Werren & Beukeboom 1998) depend on the parental genotype but influence the sex of individual offspring. Most examples are of maternal effects, such as the *daughterless* gene in *Drosophila* and a wide variety of cytoplasmic sex factors that distort the sex ratio in order to enhance their own transmission (see Chapter 9).
- **Haplodiploidy.** Several taxa have haploid males but diploid females. This can be achieved in different ways: under **arrhenotoky** females develop from fertilized eggs while males arise from unfertilized eggs; under **paternal genome loss** both sexes arise from fertilized eggs but the paternal set of chromosomes is degenerate in males.

Table 8.1 | Major categories of invertebrate sex-determining mechanisms

System	Females	Males	Examples
Genetic sex determination:			
Chromosomal			
• Male heterogamety	XX	XY	Very widespread, e.g. *Drosophila*, *C. elegans*
• Female heterogamety	ZW	ZZ	Common, e.g. Lepidoptera, Schistosomatid worms
Haplodiploidy			
• Arrhenotoky	diploid (fertilized eggs)	haploid (unfertilized eggs)	Hymenoptera, Thysanoptera, some nematodes
• Paternal genome loss (PGL)*	diploid	haploid after PGL	*Sciara* gnats, coccid bugs, some mites
Environmental sex determination:			
• temperature	high	low	Some mosquitoes, e.g. *Aedes stimulans*
• nutrition level	high	low	Mermithid (parasitic) nematodes
• photoperiod	short	long	*Gammarus duebeni* (crustacean)
• mate availability	default	if female present	*Bonellia viridis* (echiurid worm)

*Some authors use the terms pseudo-arrhenotoky (Chapter 11) or parahaploidy to refer to PGL or similar systems.

revealed that sex determination is often mediated not only by nuclear autosomal genes but also by the genes of cytoplasmic parasites, such as *Wolbachia* bacteria and microsporidia (Chapter 9). While different genomes may influence sex determination of the same individual, they do not necessarily favour the same sex ratio. An important generalization is that most intracellular parasites are transmitted only through the egg cytoplasm and, as a consequence, are selected to bias the sex ratio towards females (Chapter 9). On the other hand, selection on autosomal genes (transmitted through both sexes) favours an unbiased sex ratio, leading to genetic conflict over the sex ratio. This emphasizes the fundamental point that sex determination and sex ratios interact and coevolve (Bull 1983, Werren & Beukeboom 1998), such that analyses must often consider both simultaneously.

The primary interest of most readers of this book probably lies in understanding sex ratios, so my discussion of sex determination is framed with this in mind. Bull (1983) provided a detailed review of the diversity and evolution of sex-determining mechanisms, while White (1973)

surveyed many aspects of cytology and sex determination. Recent brief overviews of sex determination are available for animals (Mittwoch 1996) and plants (Juarez & Banks 1998), and the role of genetic conflict in the evolution of sex determination was reviewed by Werren and Beukeboom (1998).

8.3 | Diversity and its genetic basis

The diversity of sex-determining mechanisms is manifest at two different levels. First, we can recognize a wide range of different ways in which sex is inherited (Table 8.1) and these classic genetic models of sex determination are the main focus of this chapter. Second, we are also becoming more aware of molecular diversity underlying apparently similar mechanisms. For example, consider the familiar case of male heterogamety, in which females have two X chromosomes while males are either XY (humans, *Drosophila*) or XO (*C. elegans*). The similarity of inheritance of sex masks several underlying differences (Hodgkin 1990, Cline & Meyer 1996, Mittwoch 1996). First, sex depends on X to autosome (A) balance in the worm and fruitfly but on the presence of a dominant Y factor in humans. Second, the X:A principle may apply to worm and fly but the actual ratios and outcomes of disruptions to normal genotypes are different (Hodgkin 1990, Cline & Meyer 1996). Third, the molecular pathways in *Drosophila* and *C. elegans* may have some common components (Raymond *et al.* 1998) but are more notable for their differences (Hodgkin 1990, Cline & Meyer 1996). Although the classic models of sex determination are most relevant to sex ratio studies, it is important to realize that a common system like male heterogamety may be an emergent property of a range of different underlying molecular mechanisms.

Classic sex-determining mechanisms can be categorized in different ways (see Bull 1983) and Table 8.1 gives an overview of common systems known to occur in invertebrates. Although we know nothing about the mode of sex determination in most invertebrate species, it seems likely that most cases will fall into one of the diverse categories already described. In general a system involves either primarily genetic sex determination (GSD) or environmental sex determination (ESD). However, there is usually at least a minor role for environmental effects in GSD and vice versa (Bull 1983). In addition, some systems (polyfactorial/polygenic) include substantial inputs from both genetics and environment and one can think of particular cases as lying along a continuum between pure ESD and pure GSD (Chapter 7). GSD systems often involve one sex being homomorphic and the other sex heteromorphic for a key sex factor (chromosome or allele) but can also involve the sexes differing in overall ploidy (male haploid systems). ESD systems tend to involve a threshold value of a key environmental variable (e.g. nutrition) at which the expected gender of the individual switches from male to female (or vice versa). Cytoplasmic sex factors may also be involved and are addressed in detail in Chapter 9.

The great diversity of sex-determining systems begs the questions of how and why evolutionary transitions occur. Proposed causes of turnover include genetic conflict arising from opposing (sex ratio) selection pressures on different genes (Cosmides & Tooby 1981, Werren & Beukeboom 1998). These can arise due to the different transmission patterns of genes (nuclear versus cytoplasmic, autosomes versus sex chromosomes) or expression in different individuals (parent versus offspring, male versus female). Hitchhiking of sex-determination genes could also be important (Bull 1983, Wilkins 1995); this occurs when sex-determination genes are not selected directly themselves but are tightly linked to other genes under strong selection. For example, sex determination in the housefly *Musca domestica* has changed because sex-determination genes linked to a DDT-resistance allele were brought under strong selection by DDT insecticide use (Milani *et al.* 1967, Kerr 1970). Population structure and ecology can also influence sex ratio selection, as in the case of local mate competition (Hamilton 1967), and may also select for sex determination changes (Hamilton 1967, Bull 1983, section 8.5.2). Finally, an Addition-Attrition model has been proposed for sex chromosome systems (mammals in particular), in which translocations onto the Y chromosome (which has limited recombination and repair) lead to the degeneration (and consequent turnover) of sex-determining genes (Graves 1995). Transitions between GSD and ESD may not pose any special problems, especially if key regulatory genes (e.g. *sex-lethal* in *Drosophila*) are temperature sensitive (see Johnston *et al.* 1995). Bull (1983, his Chapter 10) noted that most GSD systems contain some degree of response to the environment, and several genes (e.g. *fem-1,2,3*) involved in sex determination in *C. elegans* have temperature-sensitive mutations (Hodgkin 1988), but this area clearly deserves more empirical work.

Werren (1987) and Werren and Beukeboom (1998) drew attention to three different kinds of genes directly involved in sex determination: (1) zygotic sex-determining genes, (2) parental sex ratio genes and (3) parental effect sex-determining genes (Box 8.1). In addition to direct effects one must also consider modifier and suppressor genes, which are expected to evolve in conflict situations. For example, suppressors of meiotic drive are known in *D. simulans* (Lyttle 1977, Atlan *et al.* 1997, Capillon & Atlan 1999) and other *Drosophila* supecies (Jaenike 1996, Carvalho *et al.*

Table 8.2	Diverse modes of sex determination in the insect order Diptera			
System	Species	Females	Males	Reference
GSD:				
Dominant male factor	*Megaselia scalaris*	m/m	M/m	Mainx 1964
Dominant female factor	*Musca domestica*	F/f	f/f	Milani *et al.* 1967
X:A ratio	*Drosophila melanogaster*	XX/AA	X/AA	Hodgkin 1990
Dominant Y chromosome	*Musca domestica*	X/X	X/Y	Milani *et al.* 1967
Maternal genotype	*Chrysomya rufifacies*	F*/f	f/f	Ullerich 1984
ESD:				
Temperature	*Aedes stimulans**	high	low	Horsfall & Anderson 1963
Maternal nutritional status	*Heteropeza pygmaea*	good	poor	Went & Camenzind 1980

*This is only partial ESD since presumptive males (R/r) are feminized by high temperatures but presumptive females (r/r) develop as females regardless of temperature (Horsfall & Anderson 1963, Nöthiger & Steinmann-Zwicky 1985).

1998), while autosomal resistance genes against the *Wolbachia* feminizing action have been reported in the woodlouse *Armadillidium vulgare* (Rigaud & Juchault 1992).

The general interplay of sex ratios and sex determination is illustrated by cases where male and female fitness payoffs are different. In many species the relationship between individual fitness and resource or patch quality differs between the sexes such that selection favours the production of one sex in 'good patches' and the other in 'bad patches' (Charnov & Bull 1977, Charnov 1982). Adaptive evolution of an appropriate sex-allocation/sex-determination combination has arisen repeatedly but has been achieved in different ways in different species. In the echiurid marine worm *Bonellia viridis* juveniles experience a planktonic dispersal phase and then settle on the substrate. Sex determination occurs at the settling point and follows a simple rule (Leutert 1975): individuals become males if they settle upon a conspecific female but otherwise become female. An analogous situation occurs in many species of solitary parasitoid wasps, where females generally develop in large host insects and males in small host insects (Charnov *et al.* 1981, Charnov 1982, Godfray 1994). In both cases there is evidence that selection has shaped the 'appropriate' sex expression in the two types of patch (with or without female, small or large host), according to differential fitness consequences of patch type for males and females. However, the

genetics of sex determination are quite different. In the worm, sex determination is by ESD and depends upon sex-determining genes acting in the zygote. In the wasps, sex is determined by GSD (haplodiploid arrhenotoky) and the patch-sex matching is a consequence of parental sex ratio genes acting in the mother, who controls offspring sex by fertilizing (or not) the eggs.

8.4 | Determination of sex determination

Before embarking upon experimental studies of sex ratio in a given species, it is important to know something about the mode of sex determination. In some species sex determination (above the molecular level) is sufficiently well understood to make further investigations unnecessary as precursors to certain sex ratio studies. In far more species, however, there have been no direct studies of sex determination. Unfortunately, we can only extrapolate from information about closely related species with appropriate caution. While it would be reasonable to assume that an unstudied mammal has male heterogamety, there is insufficient knowledge of variation in sex determination in most higher taxa of invertebrates, and we know that closely related invertebrate species (and even populations) can vary dramatically in their sex-determination systems (Table 8.2). The determinants of sex can be divided into GSD, ESD and cytoplasmic sex factors. Since the latter are reviewed in Chapter 9, I concentrate here on experimental studies of GSD and ESD.

8.4.1 Genetics

How does one begin to investigate sex determination in an unstudied species? If little or nothing is known, cytological study of males and females is a valuable first step since sex chromosome differences may suggest one of the common male or female heterogamety systems. However, sex chromosomes are poorly differentiated, or even indistinguishable in some species (e.g. *A. vulgare*, Rigaud 1997). Cytogenetic data can also provide valuable evidence of other GSD systems. For example, demonstration of haploid males and diploid females suggests haplodiploid arrhenotoky or paternal genome loss (Box 8.1, Table 8.1). In addition, cytogenetic study offers the possibility of observing B chromosomes or cytoplasmic parasites (e.g. bacteria or microsporidia) that might influence sex determination.

There are several practical issues to bear in mind when considering cytological studies: (1) freshly killed individuals are usually required, (2) many invertebrates have tiny chromosomes, (3) it is important to identify appropriate life stages (often immature, e.g. larvae) in order to find actively dividing cells and (4) to identify appropriate tissues that have active cell division but not somatic polyploidy.

Examples of practical considerations include Quicke's (1997) recommendation to use prepupal nerve cord or cephalic ganglia from larval parasitoid wasps, while Gokhman and Quicke (1995) found that ovarian tissue from freshly eclosed adult females was also suitable for cytological study. Beukeboom and Pijnacker (2001) dissected insect hosts to collect recently laid parasitoid wasp eggs. These were then fixed and stained over a time series of 30-min intervals as it was not known *a priori* when active cell division would be evident. Some valuable techniques may be quite taxon-specific; for example, Dunn and Hatcher (1997) used a syringe of brackish water to flush *Gammarus duebeni* shrimp embryos from the mother's marsupium (brood pouch) and

then freeze-fractured them, prior to fixation and staining.

In most cases, it is necessary to dissect or manipulate tissues before fixation and it will usually be best to do this under an appropriate aqueous solution, such as Ringer's. Solutions that include colchicine are often used to stimulate cell division (e.g. Gokhman & Quicke 1995). Some tissues or organs may also need flushing or cleaning to remove extraneous material before fixation. Fixation procedures vary somewhat between taxa and studies but often involve the use of a form of Carnoy's solution (e.g. 3:1 ethanol:acetic acid was used by Beukeboom & Pijnacker 2001). In terms of staining, the DNA-specific 4′, 6′- DAPI (diamidinophenolidone) dye, which fluoresces under ultraviolet light, is both easy to use and gives good resolution. DAPI is now used widely on all sorts of invertebrates but other stains, such as Feulgen's (visible under bright field illumination), can also be used. As starting points, fixation and staining procedures for nematode worms are given by Sulston and Hodgkin (1988: pp 598–601) and for parasitoid wasps by Gokhman and Quicke (1995: Appendix 1).

Practical limitations are reduced in taxa that have conveniently evolved giant polytene chromosomes and in some Diptera it is possible to stain the polytene chromosomes to observe chromosome inversions. *Chironomus* midges do not have well-differentiated sex chromosomes but their sex factors have often been identified via chromosome inversions, which have facilitated the study of their variable sex-determining systems (Thompson & Bowen 1972, Bull 1983 p 32).

Few taxa are known to have polytene chromosomes but all are amenable to population genetic studies and clues to sex determination can come from a range of different genetic markers, e.g. sex-linked genes suggest the possibility of sex chromosomes. Indeed, sex linkage in *Drosophila* was actually one of the key early observations (Bridges 1922) supporting the existence and nature of chromosomes themselves! A recent example is provided by a study of the opossum shrimp *Mysis relicta* (Vainola 1998). Crustaceans show a great variety of sex-determining mechanisms (ESD, GSD, cytoplasmic factors) but nothing was known about the order Mysida. However, Vainola (1998) found strong sex-linkage at the *Mpi* (mannose-6-phosphate isomerase) alloenzyme locus, with all females homozygous and nearly all males heterozygous, suggesting possible male heterogamety. In another case, Wilson *et al.* (1997) studied *Sitobion* aphids using X-linked microsatellite markers. It was already known that male aphids (XO) are produced by X chromosome loss but the use of polymorphic microsatellite markers confirmed that the identity of the X chromosome lost is effectively random.

While detailed molecular investigations will only rarely be justified for sex ratio studies, classic genetics can be used to investigate a number of possible effects. For example, isofemale lines were used to demonstrate between-line genetic variation in offspring sex ratio in the parasitoid wasp *Nasonia vitripennis* (Orzack & Parker 1986, 1990). The general principles of classic genetic crossing experiments are illustrated in Griffiths *et al.* (2000), and Legrand *et al.* (1987) describe a wide range of such applications in investigations of crustacean sex determination.

Detailed investigation of genes and molecules involved in sex determination generally depends upon detecting or generating appropriate mutants, which often acquire self-explanatory names such as *sisterless, doublesex* and *sex-lethal* in *Drosophila*. However, genetic markers unrelated to sex determination *per se* are often valuable for investigating inheritance patterns. For example, eye colour mutants have been used to discriminate between haploid and diploid males in parasitoid wasps and show that diploid males are biparental (Whiting 1943, Cook *et al.* 1994, Butcher *et al.* 2000a).

Genetic markers are available for few invertebrate species but it is possible to generate them deliberately using mutagens. For example, Butcher *et al.* (2000a) fed freshly eclosed males of the parasitoid wasp *Diadegma chrysostictos* with a solution containing ethyl/methylsulphonate (0.38% (w/v) in 0.4% (w/v) glucose, 0.4% (w/v) fructose). These males were then mated to virgin females and used to establish isofemale lines. Screening of four generations revealed unusual phenotypes, and standard backcross analyses revealed two single locus recessive markers – *rosy* (red eye colour) and *curl* (curled forewing edge) –

which were then used to detect biparental, diploid males (Butcher *et al.* 2000a). This approach was highly successful but the generation of mutants is probably only a reasonable option in species that have short generations and can be easily mass-cultured.

Tissue inoculation and gland manipulation have been valuable tools in the investigation of crustacean sex determination (Legrand *et al.* 1987). For example, removing the androgen gland from young males of the decapod *Macrobrachium* leads to a female sexual phenotype, while androgenic hormone (released by an implant) is able to induce androgen gland development in a female. These and other experiments (reviewed by Legrand *et al.* 1987) suggest that individuals of the main species investigated have most or all of the genes necessary for male or female sex differentiation and that the presence and action of the androgen gland is the key to sexual development.

8.4.2 Environment

ESD occurs when the sex of an individual cannot be predicted from its genotype due to overriding effects of the environment. It is less common than GSD and many of the well-studied animal cases involve reptiles (Chapter 7); however, there are also plenty of invertebrate cases. Environmental variables commonly involved include temperature, photoperiod, salinity, pH, mate availability (see Table 8.1) and nutrition level (Chapter 9 in Bull 1983). There appears to be a consistent (though rarely well-tested) link (Chapter 10 in Bull 1983) between the existence of ESD and sex differential consequences of the environment (patch quality), as predicted by the model of Charnov and Bull (1977). In testing environmental effects, it is clearly important to bear in mind the natural range of temperature, for example, experienced under field conditions (Chapter 7).

Many issues concerning ESD investigations are well-illustrated by studies of the shrimp *Gammarus duebeni*. Initial evidence for ESD may often come from observation or geographical surveys and the sex-allocation patterns of *G. duebeni* reportedly show geographical variation (influenced by photoperiod). The sex-determining system in this species has both genetic and

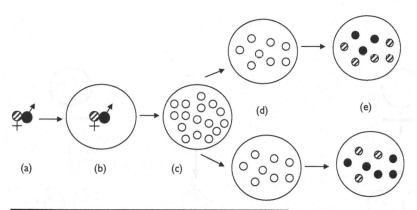

Fig 8.1 Split-brood design for investigating ESD in *Gammarus duebeni*. Male *G. duebeni* guard the females with which they have mated, and mated pairs (a) can be collected from natural populations and established in large Petri dishes in controlled-environment chambers (b). After the brood is produced (c), it can then be split and the sexually undifferentiated young allocated to different environmental treatments (d) (short and long day photoperiods). When the offspring are old enough to be sexed (female oostegites visible/male penial papillae visible), the sex ratio can be measured (e). The ESD effect is estimated from the difference between halves of the same brood, while the strength of the GSD is indicated by sex ratio variation between broods (see section 8.4.2 and Watt & Adams 1994 for further details).

environmental components; the latter involves a tendency to overproduce males in long day situations. This response may be adaptive because, in more seasonal sites, offspring produced in long day situations experience a greater period for growth and this has greater fitness benefits for males than for females (McCabe & Dunn 1997). Watt and Adams (1994) argued that if the degree of ESD is adaptive, it should be higher in more seasonal populations. They therefore contrasted ESD levels in *G. duebeni* populations in northern (more seasonal) and southern (less seasonal) England. An appropriate experimental design involved splitting single broods (Bulnheim 1967, 1969, Watt & Adams 1994) of presexual offspring into two treatments (long and short day) (Figure 8.1). This allowed measurement of the treatment effect (degree of ESD), which could then be compared between samples from different sites. Importantly, by using paired samples from the same family the genetic component of sex determination can also be controlled for statistically. Watt and Adams (1994) found that family also accounted for a signifi-

cant proportion of sex ratio variation, indicating a nontrivial GSD component.

The problem of controlling for different influences on sex determination is crucial but can be difficult. The *G. duebeni* story is further complicated by the demonstration that the same populations studied with respect to ESD also vary in the prevalence and intensity of infection of a feminizing microsporidial infection (Dunn & Hatcher 1997)! Another example is provided by Rigaud *et al.* (1991, 1997), who tested for an effect of temperature on sex determination in woodlice. While effects were found, these were best explained by the destruction of feminizing *Wolbachia* bacteria. Indeed, high temperatures have also been used to recover males from *Wolbachia*-induced asexual lineages of parasitic wasps (Stouthamer *et al.* 1990) and great care must be taken to investigate the effect of temperature on sex determination.

Having introduced approaches to the study of GSD and ESD, I now focus on two case studies while emphasizing some general issues: (1) the inherent link between sex determination and sex ratios, (2) approaches to the study of sex determination and linked topics and (3) the common need for detailed species-level biology to understand and predict outcomes.

8.5 | Complementary sex determination in Hymenoptera

Insects in the order Hymenoptera (ants, bees, wasps and sawflies) hold a special place in sex

(a)

(b)

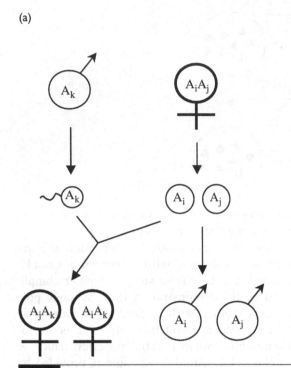
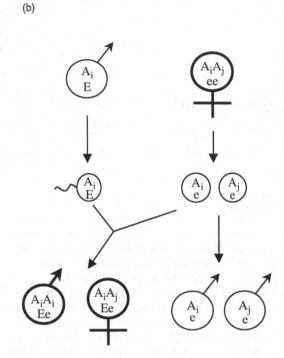

Fig 8.2 Matched and unmatched matings under single locus CSD. In an unmatched mating (a), the parents have different sex alleles so all diploid offspring are heterozygous females (A_iA_k or A_jA_k). In contrast, in a matched mating (b), the parents share one sex allele (A_i), leading to homozygous diploid sons (A_iA_i). Use of an eye colour marker with recessive (e) and dominant (E) alleles allows distinction of haploid and diploid sons according to eye colour. Bold symbols denote diploid individuals.

ratio studies, providing numerous diverse examples of adaptive sex allocation (Charnov 1982, Godfray 1994). One reason for this is that their genetic system (arrhenotokous haplodiploidy) offers females proximate control of the sex of each offspring via fertilization of the egg. Under arrhenotoky unfertilized eggs become *haploid* males while fertilized eggs become *diploid* females. However, all is not quite so simple and Whiting (1939, 1943) demonstrated complementary sex determination (CSD) in the parasitoid wasp *Bracon hebetor*. Under CSD sex is determined by multiple alleles at a single locus. Sex locus heterozygotes (A_iA_j) are females while sex locus homozygotes (A_iA_i) and hemizygotes (A_i or A_j) are diploid and haploid males respectively. The diploid males produce diploid sperm and are consequently generally sterile (Cook 1993a, Cook &

Crozier 1995). A multilocus form of CSD, in which diploids are female if heterozygous at one of several sex loci, has also been proposed (Crozier 1971, 1977). However, as it has not been demonstrated to date, I consider only the single locus case.

The fact that CSD has been demonstrated in some sawflies, bees, ants and parasitoid wasps suggests that it may be ancestral and widespread in the Hymenoptera. However, it is clear that some Hymenoptera do not have CSD (e.g. Cook 1993b, Beukeboom *et al.* 2000) and, despite much debate on possible mechanisms (Crozier 1971, 1977, Luck *et al.* 1992, Stouthamer *et al.* 1992, Cook 1993a,b, Beukeboom 1995, Dobson & Tanouye 1998), there has been little empirical progress with nonCSD species.

8.5.1 Detecting CSD and diploid males

Under CSD matings can be matched or unmatched (Figure 8.2). Diploid males are produced only in matched matings, where they comprise 50% of the diploid offspring. Recessive eye colour markers provide a convenient way to test whether diploid males occur when predicted and that they are indeed biparental (e.g. Whiting

1939, 1945, Cook *et al.* 1994, Butcher *et al.* 2000a) and a body colour marker has also been used for this purpose (Periquet *et al.* 1993). Direct cytogenetic determination of male diploidy is also possible, though rarely easy.

Other phenotypic characters may be used to recognize diploid males (but not to demonstrate that they are biparental). For example, in the sawfly *Athalia rosae* haploid and diploid males have nonoverlapping size distributions (Naito & Suzuki 1991). Unfortunately, this does not seem to apply to most other Hymenoptera (Cook & Crozier 1995, Butcher *et al.* 2000a,b). Recently, Butcher *et al.* (2000a) developed a flow cytometry method that gives nonoverlapping data for haploid and diploid males of several species of parasitoids in the superfamilies Ichneumonoidea and Chalcidoidea, as well as three sawfly species (Butcher *et al.* 2000b). The technique requires preparing single-cell brain neuronal suspensions and then using propidium iodide fluorescence. In addition to its probable general applicability, Butcher *et al.* (2000b) stated that, '. . . this method requires no significant expertise and allows complete processing of up to 300–500 samples a day relatively inexpensively . . .'.

In addition to targeted studies, initial evidence for diploid males in a given hymenopteran species may come serendipitously from population genetic studies using genetic markers, e.g. allozymes in the 'sweat' bee *Lasioglossum zephyrum* (Kukuk & May 1990). Allozyme markers can then be used to study diploid males (Periquet *et al.* 1993, Butcher *et al.* 2000a) as described above for eye colour markers. One problem with allozymes is their generally low level of variability; however, the increasing use of highly polymorphic microsatellite markers is likely to provide many further examples of hymenopteran diploid males.

Inbreeding experiments provide useful tests of CSD and rely on the fact that inbreeding increases the proportion of matched matings (and consequent diploid male production). Experiments generally require the presence of both matched and unmatched matings so that a sex ratio contrast can be made (Figure 8.2). If individuals are sampled from the wild and several lines set up this should present no problem (e.g. Beukeboom *et al.* 2000). However, many parasitoid

wasps are kept in culture and some populations may contain only two sex alleles (Stouthamer *et al.* 1992, Cook 1993c), precluding any unmatched matings. Once CSD is demonstrated, reciprocal crosses of multiple isofemale lines can be used to estimate the number of sex alleles within or between populations (e.g. Whiting 1943, 1945, Naito & Suzuki 1991, Heimpel *et al.* 1998, Butcher *et al.* 2000a,b). Alternatively, the number of sex alleles can be estimated from the fraction of matings that are matched.

In many Hymenoptera, an unmated female will still lay unfertilized eggs (Godfray 1994, Quicke 1997), which become normal haploid males. A one-way test to rule out CSD can be applied to species amenable to virgin mother-son matings (Cook 1993b). After a virgin female has laid eggs, she can be kept at 10–12°C with a supply of 50% honey solution to prolong her life, while her (larval) sons are raised at 20–30°C to speed their development to sexual maturity (e.g. Cook 1993b, Beukeboom *et al.* 2000, Butcher *et al.* 2000a). Each female can then be mated to a son to generate matched matings and the offspring sex ratio and developmental mortality must then be measured (Cook 1993b).

Under CSD, all mother-son crosses are matched matings and 50% of the fertilized eggs give rise to diploid males (Figure 8.2). The predicted offspring sex ratio depends on the ratio of haploid to diploid eggs (i.e. the egg fertilization rate). Since diploid males arise from fertilized eggs, detection is easier when there is a high proportion of eggs fertilized (female-biased sex ratio). Fortunately, several gregarious parasitoid wasp species produce female-biased sex ratios due to local mate competition (e.g. Cook 1993b). In addition, many other solitary parasitoid wasps tend to lay fertilized eggs in large hosts (Charnov 1982, Godfray 1994), allowing the experimental manipulation of conditions to make detection easier. Following a mother-son cross a strongly female-biased offspring sex ratio at emergence is only consistent with CSD if diploid males have high juvenile mortality. Consequently, it is possible to rule out CSD by demonstrating CSD-inconsistent values of (1) low brood mortality and (2) female-biased brood sex ratios (for calculations see Cook 1993b, Beukeboom *et al.* 2000).

Controlling for mortality is a problem in many sex-ratio and sex-determination studies and is probably worst when studying endoparasites. However, the problem can often be minimized by ensuring a high host:parasite ratio (e.g. Beukeboom *et al.* 2000, Butcher *et al.* 2000a).

8.5.2 How CSD interacts with the sex ratio

Here I illustrate interactions between sex determination and sex ratios by contrasting hymenopteran species with and without CSD. Sex determination is not well understood in the nonCSD species but they do not inherently produce sterile diploid males. The issues discussed here make it clear why it is important to know whether a haplodiploid species has CSD when studying its sex allocation and reproductive behaviour (see also Cook & Crozier 1995, Godfray & Cook 1997, Ode *et al.* 1997).

CSD has an intrinsic genetic load due to sterile diploid males and selection favours an infinite number of sex alleles because any rare allele occurs less commonly in the homozygous (sterile) situation. The actual number in a population is limited (to say k) by drift and population size. At equilibrium in a random mating population, sex allele frequencies are equal at $1/k$ and a proportion $1/k$ of diploid individuals (fertilized eggs) are sterile males (instead of normal females). Diploid male production has several consequences for the sex ratio. First, it effectively converts some females into nonfunctional males and so increases the population sex ratio, regardless of mating structure. Second, there is a strong effect in inbreeding species because the probability of a matched sibling mating is 0.5 (independent of k in the population at large). Consequently, it is not surprising that characteristically inbreeding species, such as many members of the large superfamily Chalcidoidea, do not have CSD (Luck *et al.* 1992, Stouthamer *et al.* 1992, Cook 1993a, Beukeboom 1995, Cook & Crozier 1995).

The exact influence of CSD on the sex ratio depends on details of diploid male biology, which differs between species. For example, because larval mortality of diploid males is unusually high in *Bracon hebetor*, the effect on the sex ratio is less marked (but mortality is correspondingly higher) (Godfray & Cook 1997). A more extreme case is the honey bee (*Apis mellifera*), whose workers cannibalize diploid male larvae in presumptive (female) worker cells (Woyke 1963). In several other species, diploid males survive and mate normally. Their diploid sperm lead to only occasional triploid or aneuploid offspring in most species (but see Naito & Suzuki 1991) but diploid male matings more often result in 'pseudovirgin' females that produce only unfertilized eggs (normal haploid sons) (Godfray & Cook 1997). Thus diploid male matings can increase the proportion of haploid males in the population! Selection then acts upon mated females to alter their egg fertilization rate in response to the proportion of pseudovirgin females (Godfray 1990, Godfray & Cook 1997), but the pressure is weak in large outbred populations due to the large number of sex alleles and corresponding rarity of diploid males. On the other hand, both diploid males and their matings may have substantial effects on small or inbred laboratory cultures of Hymenoptera with CSD, and could cause male-biased sex ratios and even stochastic extinctions in extreme cases (Stouthamer *et al.* 1992, Cook 1993c).

8.6 | Evolution of sex-determining mechanisms in *Armadillidium vulgare*

The woodlouse *Armadillidium vulgare* provides an intriguing example of the evolution of sex determination through the interaction of nuclear and cytoplasmic sex factors (for a detailed review see Rigaud 1997). The basic system is female heterogamety (ZW females, ZZ males) but some mothers produce a marked excess of daughters ('thelygeny'). Experimental studies by Legrand and Juchault (1970) showed that this thelygenic phenotype could be induced in normal ZW females by inoculation of tissue from thelygenic females. The explanation turned out to be an intracellular bacterium found only in females (Martin *et al.* 1973), and later identified (Rousset *et al.* 1992) as *Wolbachia* (see Chapter 9). Further studies (reviewed by Rigaud 1997) suggested

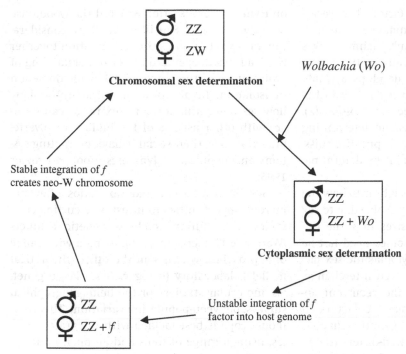

Fig 8.3 A putative scheme for the evolution of sex determination in *Armadillidium vulgare* (after Rigaud 1997). Sex determination in *A. vulgare* depends on both chromosomal and cytoplasmic factors and different populations can have different modes of sex determination (represented by details in boxes). Transitions between states may be driven by genetic conflicts between the different sex factors and might result in cyclic changes (see section 8.6 and Rigaud 1997 for further details).

that *Wolbachia* converts ZZ males into functional females. In theory, feminization of genetic males by a cytoplasmic sex factor has dramatic population effects by causing selective elimination of the W chromosome. Field data from *A. vulgare* support this prediction since *Wolbachia*-infected phenotypic females are always feminized ZZ individuals and several field populations contain no ZW females (Juchault *et al.* 1992, 1993).

Selection on *Wolbachia* favours a female-biased host sex ratio due to their cytoplasmic inheritance. This creates conflict because selection on host autosomal genes favours a balanced sex ratio. If parasite vertical transmission is imperfect, there is resultant selection on uninfected females to produce a male-biased sex ratio (Werren 1987, Hatcher & Dunn 1995). There is an enduring debate (Chapter 7) over the extent to which heterogamety systems (especially in mammals) constrain parental sex ratio manipulation and this case may be an example. A parental response to selection to overproduce offspring of one sex is surely achieved more easily with the maternal control (via fertilization) offered by haplodiploid arrhenotoky (section 8.5.2 for an analogous haplodiploid case) (Godfray 1990, Godfray & Cook 1997). However,

an alternative host 'response' is the evolution of autosomal genes that resist the effects of the parasite sex ratio distortion and Rigaud and Juchault (1992) describe a possible example in *A. vulgare*.

Another twist to the *A. vulgare* sex-determination story followed the discovery of female-biased strains that did not harbour *Wolbachia* bacteria (Juchault & Legrand 1976); another, as yet uncharacterized, non-Mendelian feminizing factor, labelled *f*, is responsible. Some key differences from the feminizing *Wolbachia* suggest that it may be a mobile nuclear element, such as a transposon (Legrand & Juchault 1984), and the series of observations outlined below suggest that sex determination in *A. vulgare* might evolve in a sequential, or even cyclic, fashion (Figure 8.3).

The *f* factor appeared spontaneously in a lineage derived from a single ZW female inoculated with *Wolbachia* (Legrand & Juchault 1984). Importantly, it appeared after the failure of *Wolbachia* transmission to the particular female. The nature and mode of action of *f* remain unknown, although several speculations need testing (Rigaud 1997). Interestingly, an autosomal masculinizing gene *M* can restore a male phenotype in ZZ individuals rendered female by *f* but not those generated by *Wolbachia* (Rigaud & Juchault 1993). In accordance with theory, *M* is not found in populations with chromosomal sex determination and its frequency increases with the frequency of *f* (Juchault *et al.* 1992). Various experiments have shown that the different sex factors behave according to the following dominance sequence: $M > f/Wo > Z$. A further embellishment was suggested by Juchault and Mocquard (1993), who discovered that f sometimes develops a stable Mendelian pattern of inheritance in a given lineage through fixation on a male chromosome (Z). This effectively creates a new female-determining (W*) chromosome and provides a putative missing link for a cycle of changes in sex determination (Figure 8.3).

This example provides an insight into how the mode of sex determination may evolve due to sex ratio conflicts. It also emphasizes how the differences between sex chromosomes need not be great, and how nuclear and cytoplasmic sex factors may interact and possibly even interchange. Rigaud (1997) suggested that the recurrent appearance of feminizing cytoplasmic factors in isopods may drive frequent genetic changes and prevent the evolution of well-differentiated sex chromosomes (discussed by Bull 1983, his Chapter 17).

8.7 | Conclusions

Sex determination and sex ratios are intimately linked through the currency of sex ratio selection. Consequently, anybody interested in studying adaptive sex ratios would be ill-advised to ignore sex determination. In a few taxa, the inheritance of sex is understood well enough that sex ratio studies might reasonably proceed with the sex-determining system taken for granted. However, this applies to only a few species and there are also several, general, good reasons for also considering sex determination: (1) variation in sex determination occurs within some species (probably commoner than we yet realize), (2) variation in sex determination can occur between closely related species (common), (3) the existence of cytoplasmic sex factors (perhaps almost ubiquitous, Chapter 9) and (4) the inherent fascination of, and extra insight available from, integrated studies of sex ratio and sex determination.

Points 1–3 are essentially caveats but I wish to emphasize the fourth point. The case studies on Hymenoptera (section 8.5) and the woodlouse *A. vulgare* (section 8.6) illustrate how consideration of sex ratios and sex determination together can lead to a deeper and richer understanding of evolution than the study of either phenomenon in isolation. In addition, such examples show how sex determination and sex ratios can interact with other aspects of the biology of invertebrates, such as their social behaviour, mating systems and population dynamics (Cook & Crozier 1995).

Sex determination and sex ratios offer numerous opportunities to investigate cutting edge topics in evolution, such as genetic conflicts (Werren & Beukeboom 1998), using a wide range of approaches, whether in the office (theoretical models), laboratory (cytogenetics, classic genetics, molecular studies), or the field (geographical surveys, between-population variation). This enormous scope is best tackled with integrated studies, using a range of tools and an open mind.

Acknowledgements

I am grateful to Ian Hardy, Melanie Hatcher, Richard Stouthamer and an anonymous referee for valuable comments on the first version of this manuscript. My work is supported by a NERC Advanced Fellowship.

References

Atlan A, Mercot H, Landre C & Montchamp-Moreau C (1997) The *sex-ratio* trait in *Drosophila simulans*: geographical distribution of distortion and resistance. *Evolution*, **51**, 1886–1895.

Beukeboom LW (1995) Sex determination in Hymenoptera – a need for genetic and molecular studies. *Bioessays*, **17**, 813–817.

Beukeboom LW & Pijnacker LP (2001) Automictic parthenogenesis in the parasitoid wasp *Venturia canescens* revisited. *Genome*, **43**, 1–6.

Beukeboom LW, Ellers J & van Alphen JJM (2000) Absence of single-locus sex determination in the braconid wasps *Asobara tabida* and *Alysia manducator*. *Heredity*, **84**, 29–36.

Bopp D, Calhoun G, Horabin JI, Samuels M & Schedl P (1996) Sex-specific control of *sex-lethal* is a conserved mechanism for sex determination in the genus *Drosophila*. *Development*, **122**, 971–982.

Bridges CB (1922) The origin of variations in sexual and sex-limited characters. *American Naturalist*, **56**, 51–63.

Bull JJ (1983) *The Evolution of Sex Determining Mechanisms*. Menlo Park: Benjamin Cummings.

Bulnheim H-P (1967) Uber den einfluss der photoperiode auf die geschlechtsrealisation bei *Gammarus duebeni*. *Helgolander Wissenschaffen Meersunters*, **16**, 69–83.

Bulnheim H-P (1969) Zur analyse geschlechtes-bestimmender Faktoren bei *Gammarus duebeni* (Crustacea, Amphipoda). *Zoologische Anzeiger Supplement*, **32**, 244–260.

Butcher RDJ, Whitfield WGF & Hubbard SF (2000a) Single-locus complementary sex determination in *Diadegma chrysostictos* (Gmelin) (Hymenoptera : Ichneumonidae). *Journal of Heredity*, **91**, 104–111.

Butcher RDJ, Whitfield WGF & Hubbard SF (2000b) Complementary sex determination in the genus *Diadegma* (Hymenoptera : Ichneumonidae). *Journal of Evolutionary Biology*, **13**, 593–606.

Capillon C & Atlan A (1999) Evolution of driving X chromosomes and resistance factors in experimental populations of *Drosophila simulans*. *Evolution*, **53**, 506–517.

Carvalho AB, Sampaio MC, Varandas FR & Klaczko LB (1998) An experimental demonstration of Fisher's principle: evolution of sexual proportion by natural selection. *Genetics*, **148**, 719–731.

Charnov EL (1982) *The Theory of Sex Allocation*. Princeton, NJ: Princeton University Press.

Charnov EL & Bull JJ (1977) When is sex environmentally determined? *Nature* **266**, 828–830.

Charnov EL, Los-den Hartogh RL, Jones WT & van den Assem J (1981) Sex ratio evolution in a variable environment. *Nature*, **289**, 27–33.

Cline TW & Meyer BJ (1996) Vive la différence: males vs females in flies vs worms. *Annual Review of Genetics*, **30**, 637–702.

Cook JM (1993a) Sex determination in the Hymenoptera: a review of models and evidence. *Heredity*, **71**, 421–435.

Cook JM (1993b) Experimental tests of sex determination in *Goniozus nephantidis* (Hymenoptera, Bethylidae). *Heredity*, **71**, 130–137.

Cook JM (1993c) Inbred lines as reservoirs of sex alleles in parasitoid rearing programs. *Environmental Entomology*, **22**, 1213–1216.

Cook JM & Crozier RH (1995) Sex determination and population biology in the Hymenoptera. *Trends in Ecology and Evolution*, **10**, 281–285.

Cook JM, Rivero Lynch AP & Godfray HCJ (1994) Sex ratio and foundress number in the parasitoid wasp *Bracon hebetor*. *Animal Behaviour*, **47**, 687–696.

Cosmides ML & Tooby J (1981) Cytoplasmic inheritance and intra-genomic conflict. *Journal of Theoretical Biology*, **89**, 83–129.

Crozier RH (1971) Heterozygosity and sex determination in haplodiploidy. *American Naturalist*, **105**, 399–412.

Crozier RH (1977) Evolutionary genetics of the Hymenoptera. *Annual Review of Entomology*, **22**, 263–288.

Dobson SL & Tanouye MA (1998) Evidence for a genomic imprinting sex determination mechanism in *Nasonia vitripennis* (Hymenoptera; Chalcidoidea). *Genetics*, **149**, 233–242.

Dunn AM & Hatcher MJ (1997) Prevalence, transmission and intensity of infection by a microsporidian sex ratio distorter in natural *Gammarus duebeni* populations. *Parasitology*, **115**, 381–385.

Godfray HCJ (1990) The causes and consequences of constrained sex allocation in haplodiploid animals. *Journal of Evolutionary Biology*, **3**, 3–17.

Godfray HCJ (1994) *Parasitoids: Behavioural and Evolutionary Ecology*. Princeton, NJ: Princeton University Press.

Godfray HCJ & Cook JM (1997) Parasitoid mating systems. In: J Choe & BJ Crespi (eds) *The Evolution of Mating Systems in Insects and Arachnids*, pp 211–225. Cambridge: Cambridge University Press.

Gokhman VE & Quicke DLJ (1995) The last twenty years of parasitic Hymenoptera karyology: an update and phylogenetic implications. *Journal of Hymenoptera Research*, **4**, 41–63.

Graves JAM (1995) The evolution of mammalian sex chromosomes and the origin of sex-determining genes. *Philosophical Transactions of the Royal Society of London, Series B*, **350**, 305–311.

Griffiths AJF, Miller JH, Suzuki DT, Lewontin RC & Gelbart WM (2000) *An Introduction to Genetic Analysis*, 7th edn. New York: WH Freeman.

Hamilton WD (1967) Extraordinary sex ratios. *Science*, **156**, 477–488.

Hatcher MJ & Dunn AM (1995) Evolutionary consequences of cytoplasmically inherited feminizing factors. *Philosophical Transactions of the Royal Society of London, Series B*, **348**, 445–456.

Heimpel G, Antolin MF & Strand MR (1998) Diversity of sex-determining alleles in *Bracon hebetor*. *Heredity*, **82**, 282–291.

Hodgkin J (1988) Sex dimorphism and sex differentiation. In: WB Wood (ed) *The Nematode Caenorhabditis elegans*, pp 243–279. Cold Spring Harbor: Cold Spring Harbor Laboratory Press.

Hodgkin J (1990) Sex determination compared in *Drosophila* and *Caenorhabditis*. *Nature*, **344**, 721–729.

Horsfall WR & Anderson JF (1963) Thermally induced genital appendages on mosquitoes. *Science*, **141** 1183–1185.

Jaenike J (1996) Sex-ratio meiotic drive in the *Drosophila quinaria* group. *American Naturalist*, **148**, 237–254.

Johnston CM, Barnett M & Sharpe PT (1995) The molecular biology of temperature-dependent sex determination. *Philosophical Transactions of the Royal Society of London, Series B*, **350**, 297–303.

Juarez C & Banks JA (1998) Sex determination in plants. *Current Opinion in Plant Biology*, **1**, 68–72.

Juchault P & Legrand JJ (1976) Modification de la sex ratio dans les croisements entre différentes populations du Crustacé Oniscoide *Armadillidium vulgare* Latr. Notion de déterminisme polygénique et epigénétique du sexe. *Archives du Zoologie Expérimentale et Générale*, **117**, 81–93.

Juchault P & Mocquard JP (1993) Transfer of a parasitic sex factor to the nuclear genome of the host: an hypothesis on the evolution of sex-determining mechanisms in the terrestrial isopod *Armadillidium vulgare* Latr. *Journal of Evolutionary Biology*, **6**, 511–528.

Juchault P, Rigaud T & Mocquard JP (1992) Evolution of sex-determining mechanisms in a wild population of *Armadillidium vulgare* Latr. (Crustacea, Isopoda) – competition between 2 feminizing parasitic sex factors. *Heredity*, **69**, 382–390.

Juchault P, Rigaud T & Mocquard JP (1993) Evolution of sex determination and sex ratio variability in wild populations of *Armadillidium vulgare* Latr. (Crustacea, Isopoda) – a case-study in conflict-resolution. *Acta Oecologica*, **14**, 547–562.

Kerr RW (1970) Inheritance of DDT resistance in a laboratory colony of the house fly, *Musca domestica*. *Australian Journal of Biological Science*, **23**, 377–400.

Kukuk PF & May B (1990) Diploid males in a primitively eusocial bee *Lasioglossum* (*Dialictus*) *zephyrum* (Hymenoptera: Halicitidae). *Evolution*, **44**, 1522–1528.

Legrand JJ & Juchault P (1970) Modification experimentale de la proportion des sexes chez les Crustaces isopodes terrestre: induction de la thelygenie chez *Armadillidium vulgare* Latr. *Comptes Rendu de l'Academie des Sciences, Paris, Serie III*, **270**, 706–708.

Legrand JJ & Juchault P (1984) Nouvelles données sur la déterminisme genetique et epigenetique de la monogenie de la crustace isopodes terrestre *Armadillidium vulgare* Latr. *Genetics Selection Evolution*, **16**, 57–84.

Legrand JJ, Legrand-Hamelin E & Juchault P (1987) Sex determination in Crustacea. *Biological Reviews*, **62**, 439–470.

Leutert R (1975) Sex determination in *Bonellia*. In: R Reinboth (ed) *Intersexuality in the Animal Kingdom*, pp 84–90. Berlin: Springer-Verlag.

Luck RF, Stouthamer R & Nunney LP (1992) Sex determination and sex ratio patterns in parasitic Hymenoptera. In: DL Wrensch & MA Ebbert (eds) *Evolution and Diversity of Sex Ratios in Insects and Mites*, pp 442–476. New York: Chapman & Hall.

Lyttle TW (1977) Experimental population genetics of meiotic drive systems. I. Pseudo-Y chromosomal drive as a means of eliminating cage populations of *Drosophila melanogaster*. *Genetics*, **86**, 413–445.

Mainx F (1964) The genetics of *Megaselia scalaris* Loew (Phoridae). A new type of sex determination in Diptera. *American Naturalist*, **98**, 415–430.

Martin G, Legrand JJ & Juchault P (1973) Mise en évidence d'un micro-organisme intra cytoplasmique symbiote de l'Oniscoide terrestre *Armadillidium vulgare* Latr., dont la presence accompagne l'intersexualite ou la féminisation totale des males genetiques de la lignée thelygene. *Comptes Rendu de l'Academie des Sciences, Paris, Serie III*, **276**, 2313–2316.

McCabe J & Dunn AM (1997) Adaptive significance of environmental sex determination in an amphipod. *Journal of Evolutionary Biology*, **10**, 515–527.

Meise M, HilfikerKleiner D, Dubendorfer A, Brunner C, Nothiger R & Bopp D (1998) *Sex-lethal*, the master sex-determining gene in *Drosophila*, is not sex-specifically regulated in *Musca domestica*. *Development*, **125**, 1487–1494.

Milani R, Rubini PG & Franco MG (1967) Sex determination in the housefly. *Genetica Agraria*, **21**, 385–411.

Mittwoch U (1996) Sex-determining mechanisms in animals. *Trends in Ecology and Evolution*, **11**, 63–67.

Naito T & Suzuki H (1991) Sex determination in the sawfly *Athalia rosae ruficornis*: occurrence of triploid males. *Journal of Heredity*, **82**, 101–104.

Nöthiger R & Steinmann-Zwicky M (1985) A single principle for sex determination in insects. *Cold*

Spring Harbor Symposia on Quantitative Biology, **1**, 615–621.

Ode PJ, Antolin MF & Strand MR (1997) Brood-mate avoidance in the parasitic wasp *Bracon hebetor* Say. *Animal Behaviour*, **49**, 1239–1248.

Orzack SH & Parker ED (1986) Sex ratio control in a parasitic wasp *Nasonia vitripennis* I. Genetic variation in facultative sex ratio production. *Evolution*, **40**, 331–340.

Orzack SH & Parker ED (1990) Genetic variation for sex ratio traits within a natural population of a parasitic wasp. *Genetics*, **124**, 373–384.

Periquet G, Hedderwick MP, El Agoze M & Poirie M (1993) Sex determination in the hymenopteran *Diadromus pulchellus* (Ichneumonidae): validation of the one-locus multi-allele model. *Heredity*, **70**, 420–427.

Quicke DLJ (1997) *Parasitic Wasps*. London: Chapman & Hall.

Raymond CS, Shamu CE, Shen MM, Seifert KJ, Hirsch B, Hodgkin J & Zarkower D (1998) Evidence for evolutionary conservation of sex-determining genes. *Nature*, **391**, 691–694.

Rigaud T (1997) Inherited microorganisms and sex determination in arthropod hosts. In: SL O'Neill, AA Hoffman & JH Werren (eds) *Influential Passengers: Inherited Micro-organisms and Arthropod Reproduction*, pp 81–101. Oxford: Oxford University Press.

Rigaud T & Juchault P (1992) Genetic control of the vertical transmission of a cytoplasmic sex factor in *Armadillidium vulgare* Latr. (Crustacea, Oniscidea). *Heredity*, **68**, 47–52.

Rigaud T & Juchault P (1993) Conflict between feminizing sex-ratio distorters and an autosomal masculinizing gene in the terrestrial isopod *Armadillidium vulgare* Latr. *Genetics*, **133**, 247–252.

Rigaud T, Juchault P & Mocquard JP (1991) Experimental study of temperature effects on the sex-ratio of broods in terrestrial Crustacea *Armadillidium vulgare* Latr. possible implications in natural-populations. *Journal of Evolutionary Biology*, **4**, 603–617.

Rigaud T, Antoine D, Marcade I & Juchault P (1997) The effect of temperature on sex ratio in the isopod *Porcellionides pruinosus*: environmental sex determination or a by-product of cytoplasmic sex determination? *Evolutionary Ecology*, **11**, 205–215.

Rousset F, Bouchon D, Pintureau B, Juchault P & Solignac M (1992) *Wolbachia* endosymbionts responsible for various alterations of sexuality in arthropods. *Proceedings of the Royal Society of London, Series B*, **250**, 91–98.

Saccone G, Peluso I, Artiaco D, Giordano E, Bopp D & Polito LC (1998) The *Ceratitis capitata* homologue of the *Drosophila* sex-determining gene *sex-lethal* is structurally conserved, but not sex-specifically regulated. *Development*, **125**, 1495–1500.

Stouthamer R, Luck R & Hamilton WD (1990) Antibiotics cause parthenogenetic *Trichogramma* to revert to sex. *Proceedings of the National Academy of Sciences USA*, **87**, 2424–2427.

Stouthamer R, Luck R & Werren JH (1992) Genetics of sex determination and improvement of biological control using parasitoids. *Environmental Entomology*, **21**, 427–435.

Sulston J & Hodgkin J (1988) Methods. In: WB Wood (ed) *The Nematode* Caenorhabditis elegans, pp 587–601. Cold Spring Harbor: Cold Spring Harbor Laboratory Press.

Thompson PE & Bowen JS (1972) Interactions of differentiated primary sex factors in *Chironomus tentans*. *Genetics*, **70**, 491–493.

Ullerich FH (1984) Analysis of sex determination in the monogenic blowfly *Chrysomya rufifacies* by pole cell transplantation. *Molecular and General Genetics*, **193**, 479–487.

Vainola R (1998) A sex-linked locus (*Mpi*) in the opossum shrimp *Mysis relicta*: implications for early postglacial colonization history. *Heredity*, **81**, 621–629.

Watt PJ & Adams J (1994) Adaptive variation in sex determination in a crustacean, *Gammarus duebeni*. *Journal of Zoology*, **232**, 109–116.

Went DF & Camenzind R (1980) Sex determination in the dipteran insect *Heteropeza pygmaea*. *Genetica*, **52**, 373.

Werren JH (1987) The coevolution of autosomal and cytoplasmic sex-ratio factors. *Journal of Theoretical Biology*, **124**, 317–334.

Werren JH (1997) Biology of *Wolbachia*. *Annual Review of Entomology*, **42**, 587–609.

Werren JH & Beukeboom LW (1998) Sex determination, sex ratios, and genetic conflict. *Annual Review of Ecology and Systematics*, **29**, 233–261.

White MJD (1973) *Animal Cytology and Evolution*. Cambridge: Cambridge University Press.

Whiting PW (1939) Sex determination and reproductive economy in *Habrobracon*. *Genetics*, **24**, 110–111.

Whiting PW (1943) Multiple alleles in complementary sex determination of *Habrobracon*. *Genetics*, **28**, 365–382.

Whiting PW (1945) The evolution of male haploidy. *Quarterly Review of Biology*, **20**, 231–260.

Wilkins AS (1995) Moving up the hierarchy: a hypothesis on the evolution of genetic sex determination pathway. *Bioessays*, **17**, 71–77.

Wilson ACC, Sunnucks P & Hales DF (1997) Random loss of X-chromosome at male determination in an aphid, *Sitobion* near *fragariae*, detected using an X-linked polymorphic microsatellite marker. *Genetical Research, Cambridge*, **69**, 233–236.

Woyke J (1963) What happens to diploid drone larvae in a honeybee colony? *Journal of Apicultural Research*, **2**, 73–75.

Chapter 9

Sex ratio distorters and their detection

Richard Stouthamer, Gregory D.D. Hurst &
Johannes A.J. Breeuwer

9.1 | Summary

Sex ratio distorters (SRDs) are heritable elements
that modify the sex ratio of their host to promote their own transmission. In this chapter we
examine various theories relating to the evolutionary importance of SRDs, give an overview of
the various classes of SRDs and methods of how
to discover them in the field, and outline areas of current contention, and thus future work
in relation to their incidence and importance.
Sex ratio distorters include organelles, heritable bacteria and eukaryotes, B chromosomes and
meiotic-drive sex chromosomes. A high proportion of arthropod species that have been studied in detail harbour SRDs. They are important
in host evolution because they influence fundamental population dynamic processes, manipulate sex-determining mechanisms of their host
and may contribute to genetic isolation between
host populations. If SRDs are parasitic, selection may promote the spread of host genes to
prevent SRD action and transmission. Unusual
sex ratio phenotypes in the field may indicate
SRDs. This should be followed by genetic analysis of sex ratio phenotypes of isofemale lines.
If micro-organisms are suspected, they can be
identified molecularly with specific polymerase
chain reaction (PCR) primer pairs and sequence
analysis.

9.2 | Introduction

What are sex ratio distorters (SRDs)? SRDs are
broadly defined as those heritable elements that
modify the sex ratio of their host to promote
their own transmission, often at a cost to the
inclusive fitness of the individual bearing them.
In practice, these SRDs are recognized when
certain parents produce offspring sex ratios that
deviate from those produced by the rest of the
population.

The study of SRDs, like so many other aspects of sex ratio studies, is rooted in Hamilton's
(1967) publication 'Extraordinary sex ratios' and
the influential papers of Eberhard (1980) and
Cosmides and Tooby (1981). Sex ratio distorters
are examples of the general class of heritable elements called selfish genetic elements (Werren
et al. 1988). These are elements that enhance
their own transmission by manipulating their
host's reproduction, often at a cost to the nuclear genes of the host. Sex ratio distorters can
be located either on the nuclear chromosomes
(sex chromosome meiotic drive, supernumerary
B chromosomes) or in the cytoplasm of organisms (organelles and heritable micro-organisms,
including a range of bacteria and eukaryotes).
The number of cases where SRDs have been
discovered has increased substantially over the
last few years (L Hurst 1993). Most notably,

work on the symbiont proteobacterium *Wolbachia* has made it clear that heritable symbionts are common and cause several unusual sex ratio distortions (reviewed in O'Neill *et al.* 1997, Werren 1997, Stouthamer *et al.* 1999). Bacteria of this genus are known to cause the induction of parthenogenesis (Rousset *et al.* 1992, Stouthamer *et al.* 1993), male-killing (G Hurst *et al.* 1999a) and the conversion of genetic male individuals into functional females (Rousset *et al.* 1992).

Not only are SRDs very common amongst species, but within single species many different sex ratio distorting factors can be found. One of the most studied hymenopteran species, *Nasonia vitripennis*, harbours at least four different types of SRD elements: Werren and collaborators have discovered an unknown cytoplasmic factor causing all-female broods, a male-killing bacterium (Skinner 1985, Werren *et al.* 1986), a B chromosome causing all-male broods from females mated with a male carrying this factor (Werren *et al.* 1981, 1987), and finally *Wolbachia*, bacteria that enhance their own transmission through a process called cytoplasmic incompatibility which incidentally produces male-biased sex ratios (Breeuwer & Werren 1990). Additionally, a single sex ratio distorting phenotype may be employed by a number of different agents within a single population of the same species. For instance, in the ladybird *Adalia bipunctata*, there are at least three different bacteria that kill male embryos, all present in Moscow (Werren *et al.* 1994, G Hurst *et al.* 1999a,b, Majerus *et al.* 2000).

A high proportion of arthropod species that have been studied in detail, with the maintenance of different genetic lines of the species, appear to harbour some sex ratio distorters (G Hurst *et al.* 1997b), and some bear many. In this chapter we first examine various theories relating to the evolutionary importance of SRDs (section 9.3), give an overview of the various classes of sex ratio distorters that are currently known and methods of how to discover them in the field (section 9.5), and lastly outline areas of current contention, and thus future work in relation to their incidence and importance (section 9.6).

9.3 | The evolutionary importance of SRDs

Sex ratio distorting elements have been proposed as important in driving the evolution of their 'host' organism. In this sphere, most attention has focused on the fact that because SRDs are parasitic, selection promotes the spread of host genes that prevent SRD action and transmission. Genes preventing the action of drive are known, as are genes that prevent the transmission of cytoplasmic elements that distort the sex ratio (section 9.4).

The 'arms race' between SRD and host may be important in the evolution of host sex-determination systems. In certain populations of the pill woodlouse *Armadillidium vulgare*, the system of sex determination has altered from its ancestral state (female heterogamety) to one where sex is determined by the presence of *Wolbachia*. This itself did not require evolution in the host, merely the spread of the feminizing *Wolbachia*. However, the transmission of *Wolbachia* is now in part determined by the influence of nuclear genes (Rigaud & Juchault 1992). Heteromorphic sex chromosomes are no longer observed; rather sex is determined by a single nuclear locus determining bacterial transmission.

Coevolution between SRDs and their host may also contribute to genetic isolation between isolated populations. As discussed above, SRDs often produce selection for unlinked modifiers that repress their action. When these go to fixation within the population, the selfish phenotype is not exhibited in within-population crosses. However, the phenotype is seen in crosses involving individuals from different populations. In the case of meiotic drive in particular, it has been suggested that loss of repression of drive in the hybrid context may underlie (Haldane's rule (Haldane 1922)) the observation that sterility evolves more rapidly in the heterogametic sex (Frank 1991, L Hurst & Pomiankowski 1991). Whilst it is unrealistic to maintain that the loss of repression of selfish genetic elements is the unique cause of Haldane's rule, the question of whether they contribute to hybrid sterility remains (G Hurst & Schilthuizen 1998).

Aside from direct selection for repressors, selection will also favour individuals that choose mates that do not bear an SRD. This is a form of 'good genes' sexual selection, where it is the 'drive' of the SRD that maintains heritability for fitness, 'solving' the lek paradox that heritable variation diminishes as mate choice spreads. A potential example is the case of mate choice in stalk-eyed flies, where there is a female preference for males bearing meiotic drive suppressors (Wilkinson *et al.* 1998, but see Reinhold *et al.* 1999, Pomiankowski & Hurst 1999). A tantalizing possibility is also seen in the butterfly *Acraea encedon*, where females infected with male-killing *Wolbachia* are less likely to be mated than uninfected ones (Jiggins *et al.* 2000).

Selection to acquire fertilizing gametes that do not bear SRDs may also act upon female mating frequency. In a population bearing meiotic drive genes where drive occurs in males, there is selection upon females to mate multiply, as this reduces the probability of inheriting a driving element (Haig & Bergstrom 1995). The same principle can also apply to the acquisition of repressor genes. If eggs are laid in clutches, sperm mixing occurs, and the death of one individual in a clutch is partly compensated by the increased fitness of a sibling, then a benefit to multiple mating will arise from the decrease in the variance in repressor gene frequency between clutches that occurs with multiple mating (Zeh & Zeh 1996, 1997).

Aside from these direct selective pressures to prevent the action of SRDs and the exposure of progeny to SRDs, SRDs may have other 'indirect' evolutionary consequences for the host population. For instance, selection may act upon clutch size in a population bearing a male-killing bacterium (G Hurst & McVean 1998). Male-killing bacteria reduce clutch size, and, if the avoidance of sib-sib competition is a major feature in clutch size evolution, selection will favour an increase in mean clutch size in the population. Interestingly, selection for increased clutch size will produce an increase in the prevalence of the bacterium.

Additionally, it is clear that the effect of SRDs on population sex ratios may be important in evolution. SRDs have the ability to produce split sex ratios in populations of Hymenoptera, and may thus facilitate the evolution of eusociality (G Hurst 1997). More widely, much of the behavioural ecology of reproduction depends upon the population sex ratio, and strongly female-biased sex ratios may cause a degree of sexual role reversal. This is seen in *A. encedon*, where the high prevalence of a male-killing *Wolbachia* produces an extreme female-biased population sex ratio, high levels of female unmatedness in the field and unusual mating interactions (Jiggins *et al.* 2000).

9.4 | Mechanisms and population biology of SRDs

9.4.1 Nuclear genes
9.4.1.1 Meiotic drive systems
The first SRDs recorded were the cases of X chromosome meiotic drive in *Drosophila*. Highly female-biased broods in certain crosses were observed and, since embryonic and larval mortality were normal, a primary sex ratio bias was proposed (Gershenson 1928, Sturtevant & Dobzhansky 1936). The cause can be seen when spermiogenesis of the male involved in the cross is examined. Males producing only female progeny also produced sperm bundles bearing only half the normal quantity of sperm (Policansky & Ellison 1970). The implication is clear: Y-bearing sperm die, leaving X-bearing sperm only, leading to primary sex ratio bias. That this is an effect of the X chromosome can be seen from F1 and F2 crosses. The F1 daughters produced a normal sex ratio when crossed, ruling out a maternal factor. However, half of the sons of these females (i.e. F2 males) again produced all-female broods when crossed. These data indicate the involvement of the X chromosome of the original parental male, which gains preferential transmission over the Y in males.

Sex chromosome drive may also be associated with the Y chromosome, as has been found in the mosquito *Aedes aegypti* (Hickey & Craig 1966). Here, males bearing the driving Y produce all-male broods, and these sons in turn produce all-male broods.

The population biology of driving sex chromosomes was modelled by Hamilton (1967). The driving chromosome will invade if the fertility of the male is greater than 50% that of a normal male. This condition will be satisfied if males produce more sperm than is necessary to fertilize a given female, which is likely to be common.

Theory suggests that meiotically driving sex chromosomes can, in some circumstances, spread to fixation, causing the extinction of the population bearing them. Several factors may prevent this (Carvalho & Vaz 1999). The interaction may be 'naturally' balanced: equilibrium is reached without any further evolution. There are two potential mechanisms that promote 'natural' balance:

1. Female individuals homozygous for the driving X may be less fit than females bearing one or two wild-type X chromosomes. As the chromosome spreads, selection against the driving chromosome within females becomes stronger (i.e. the fitness of the driving X is frequency dependent), and may stabilize a polymorphism within the population.
2. As driving chromosomes spread, the population sex ratio changes, thus altering the pattern of mate competition. When a driving X chromosome increases in frequency, the population sex ratio will become female biased, and males' mating opportunities may increase. At some point, sperm production rates in males may limit male reproductive success, and at this point the driving and wild-type X have equal fitness.

Second, the system may not be naturally balanced, and the driving gene would spread to fixation were it not for further evolution of the host. However, repressors of drive, either on autosomes or on the Y chromosome, spread and prevent population extinction. Repressors will spread as drive is against the interests of unlinked genes (it overproduces the more common sex, and lowers host fertility). If repressors are costly (i.e. in the absence of drive, an individual bearing a repressor gene is less fit than one lacking it) then an equilibrium is reached at which the population is polymorphic for both the driving X chromosome and the repressor genes. Such equilibria are observed in the fruitfly *Drosophila mediopunctata*,

in which both autosomal and Y-linked repressors are known (Carvalho & Klazcko 1993, Carvalho et al. 1997), and in *D. quinaria* (Jaenike 1999). If repressors are cost-free, they will go to fixation, and may cause the loss of the drive phenotype within the population. This is observed in *D. simulans*: crosses between flies from the Seychelles show near-normal sex ratios (drive is repressed within the population), whereas when the X chromosome from the Seychelles is placed on the genetic background of the African mainland, males produce a female-biased sex ratio associated with drive (Merçot et al. 1995). The driving X has been introgressed from a population where drive is repressed (Seychelles) into a genetic background where there are no repressors (Africa).

9.4.1.2 B chromosomes

B chromosomes are extra chromosomes that are not duplicates of the normal complement of chromosomes (A chromosomes) and are not necessary for the survival of the organism in which they occur. Generally, B chromosomes carry few structural genes. For the present discussion, the most important feature of B chromosomes is that they display non-Mendelian inheritance. B chromosomes generally have little effect on sex ratios, with the notable exception of the so-called paternal sex ratio (PSR) chromosome found in some males of the parasitoid wasp *Nasonia vitripennis* (reviewed in Werren 1991). Hymenoptera have a haplodiploid sex-determination system, in which females arise from fertilized (diploid) eggs, while males arise from unfertilized (haploid) eggs (Chapter 8). When sperm carrying the PSR chromosome fertilizes an egg, the PSR chromosome somehow influences the behaviour of the paternal set of A chromosomes such that they do not participate successfully in the first mitotic division (Nur et al. 1988, Reed 1993). The only paternal chromosome that will persist in the zygote is the PSR chromosome. Such fertilized eggs do not develop as diploid females but become haploid males instead. These males carry the maternal complement of A chromosomes together with the paternal B chromosome.

Little is known about how the PSR chromosome accomplishes the selective destruction of the A complement of chromosomes. Werren and his collaborators have done extensive work on the DNA sequences located on this chromosome. Most of the DNA sequenced from the PSR chromosome belongs to a number of families of tandemly repeated DNA sequences (Eickbush *et al.* 1992, Beukeboom & Werren 1993, Reed *et al.* 1994). In addition a number of transposable elements have been found on this chromosome (McAllister 1995, McAllister & Werren 1997).

Population dynamic modelling of PSR in *Nasonia* populations predicts that in panmictic populations the PSR will spread to an equilibrium as long as the fertilization frequency of the eggs by females is larger than 50% (Skinner 1987). In sub-divided populations with a demic structure, such as that of *N. vitripennis*, the dynamics of the PSR are more complicated and it appears that the PSR can only maintain itself in the population due to the presence of another sex ratio distorting factor called maternal sex ratio (MSR) (Beukeboom & Werren 1992, Werren & Beukeboom 1993). This latter factor is maternally inherited and causes females to fertilize all of their eggs (section 9.4.2.3). An interesting feature of the mode of action of the PSR chromosome is that it seems particularly easy to cross species boundaries (Dobson & Tanouye 1998): in crosses between the different *Nasonia* species the PSR chromosome of *N. vitripennis* was easily introduced into the species *N. giraulti* and *N. longicornis.*

Another PSR-like factor was discovered by Hunter *et al.* (1993) in the parasitoid wasp *Encarsia pergandiella*. In field populations up to 39% of the males emerged from fertilized eggs because the paternal chromosomes were lost in the first mitotic division. Males emerging from fertilized eggs were the only ones that could pass on this trait again to their offspring. This pattern of paternal chromosome loss is reminiscent of the loss induced by the PSR chromosome of *Nasonia*. However, in a detailed cytogenetic study no evidence could be found for the presence of a B chromosome and the cause of the paternal chromosome loss in *E. pergandiella* remains to be determined.

9.4.2 Cytoplasmic genes

Sex ratio distorters generally gain a transmission advantage when they bias the sex ratio of the offspring in the direction of the transmitting sex. For cytoplasmically inherited factors this means producing more daughters, since only females can pass on cytoplasmically inherited SRDs (eggs contain cytoplasm, sperm effectively do not). Cytoplasmic genes include mitochondria and a variety of micro-organisms (bacteria and eukaryotes), these inherited micro-organisms being largely found in invertebrates. We here review the various manipulations observed, and their mechanistic basis.

9.4.2.1 Parthenogenesis-inducing Wolbachia bacteria

The spread of extreme female-biasing SRDs would lead to exclusively female populations. In those species where the production of daughters relies on fertilization, such SRDs would lead to the demise of the 'infected' population. Bacteria of the genus *Wolbachia* in parasitoid wasps have evolved a very effective method of sex ratio distortion by not only allowing females to produce daughters in the normal fashion, but by also causing unfertilized eggs to develop as females (Stouthamer *et al.* 1990, 1993, Stouthamer & Kazmer 1994). *Wolbachia*, in essence, doubles the chromosome number of unfertilized eggs, ensuring development as a female. To date approximately 50 species of parthenogenetic Hymenoptera have been found to be affected. Recently, *Wolbachia* have been reported in several parthenogenetic mite species of the genus *Bryobia* (A Weeks pers. comm.) and the parthenogenetic springtail *Folsomia candida* (Vanderkerckhove *et al.* 1999).

There are two types of populations of hosts infected with parthenogenesis-inducing (PI) *Wolbachia*: one where all the individuals of the species are infected and only female populations are known (fixed populations), and another where both infected and uninfected individuals coexist in a population (mixed populations) (Stouthamer 1997). The majority of all the species known to be infected have fixed populations. Among members of the parasitoid wasp genus *Trichogramma* there are many populations in which infected and uninfected wasps coexist.

Parthenogenesis in populations that are predominantly sexual can be difficult to determine, as it may require the collection of large numbers of females from the population, and the subsequent setting up of each female as an isofemale line. In *Trichogramma brevicapillum*, for instance, a total of 135 isofemale lines were started and only a single parthenogenetic line was found. How is coexistence of both infected and uninfected individuals in a population possible? It is unlikely that such relationships are naturally balanced. Rather, equilibrium is likely to be reached because of the invasion of other factors, such as PSR chromosomes or suppressor genes, that prevent the *Wolbachia* infection from going to fixation (Stouthamer 1997).

9.4.2.2 Feminization

Maternally transmitted cytoplasmic SRDs, which control sexual development in the embryo or juvenile by overriding the expression of nuclear genes, have been documented in a large number of crustacean species (Isopoda and Amphipoda). Feminizers discovered to date fall into two types: *Wolbachia* (in isopods) and protists (in amphipods). In isopods (as in Lepidoptera and birds), sex is normally determined by female heterogamety (females have ZW and males have ZZ sex chromosomes, Chapters 7 & 8) and an unbiased sex ratio is expected. However, females infected with *Wolbachia* produce only female offspring, irrespective of the sex chromosome constitution of the zygotes; thus ZZ zygotes develop into females instead of males (Rigaud 1997, Bouchon *et al.* 1998). In ZZ individuals the male phenotype is determined by the production of male hormones in the androgenic gland, whose differentiation is presumably suppressed by *Wolbachia* in spite of the ZZ sex chromosome and they develop into females (Rigaud 1997).

In amphipods, sex determination can be genetically controlled, based on photoperiod, or can depend on a combination of genetic and environmental factors. Infection with protists (microsporidia and haplosporidia) overrides the normal sex-determination mechanism and feminizes a proportion of the infected offspring (Bulnheim & Vávra 1968, Ginsberger-Vogel *et al.* 1980, Dunn *et al.* 1993, Terry *et al.* 1997). Sometimes, the infection results in the development of intersexes (Bulnheim 1965). The molecular mechanism is

unclear. Note that females, whether infected or not, still require sperm from males to fertilize and produce eggs. This has important consequences for the dynamics of feminizer and host, because sperm may become limiting at high infection frequencies. Access to mates within a local population may decrease the equilibrium frequency of the factor (Hatcher *et al.* 1999).

9.4.2.3 Maternal sex ratio

Maternal sex ratio (MSR) is a maternally inherited genetic factor found in *Nasonia vitripennis* that biases the primary sex ratio towards all-female production (Skinner 1982). In this species, females have control over fertilization of the egg and adjust their fertilization rate according to the number of ovipositing females in a host patch (deme size): female-biased sex ratios are produced in smaller demes, and approximately unbiased sex ratios produced in large demes (Werren 1983). MSR females, however, will always fertilize about 95–100% of their eggs, even in large demes. This results in all-female broods without mortality (in contrast to all-female broods caused by male-killers). MSR apparently overrides the normal control of fertilization, but details of the mechanism as well as the exact nature of the MSR factor are lacking. MSR is probably not a micro-organism because it is insensitive to antibiotics (R Stouthamer unpublished results). Sperm (and fertilization) are still required in order to produce daughters; unfertilized MSR females only produce males, which do not transmit the trait.

The dynamics of this trait depend upon the population mating structure (Werren 1987) and the presence of other distorters, such as PSR (Beukeboom & Werren 1992, Werren & Beukeboom 1993) and complex population sex ratio dynamics may result. So far MSR has only been described in a single wasp species. This is probably because it is likely to occur at low frequencies, as is the case in natural populations of *Nasonia*, and because other systems with comparable biology have not been examined in sufficient detail. MSR will be particularly difficult to find in species that adjust sex ratios according mating structure (Hamilton 1967), as normal females will also produce extremely female-biased

Table 9.1 | The taxonomic affiliation of male-killing bacteria, and the hosts they infect

Bacterium	Host	Reference
Spiroplasma (Mollicutes)	*Drosophila willistoni* (Diptera)	Hackett *et al.* 1986
	Adalia bipunctata (Coleoptera)	G Hurst *et al.* 1999b
Unnamed (Flavobacteria-Cytophaga-Bacteroides group)	*Coleomegilla maculata* (Coleoptera)	G Hurst *et al.* 1997b
	Adonia variegata (Coleoptera)	G Hurst *et al.* 1999c
Rickettsia (alpha group of proteobacteria)	*Adalia bipunctata* (Coleoptera)	Werren *et al.* 1994
Wolbachia (alpha group of proteobacteria)	*Adalia bipunctata* (Coleoptera)	G Hurst *et al.* 1999a
	Acraea encedon (Lepidoptera)	G Hurst *et al.* 1999a
Arsenophonus nasoniae (gamma group of proteobacteria)	*Nasonia vitripennis* (Hymenoptera)	Werren *et al.* 1986

sex ratios when they are the sole foundress of an offspring group. Thus the detection of MSR traits will only be possible when multiple foundresses are placed on host patches. If MSR is present, offspring sex ratios should remain female biased in contrast to more balanced sex ratios produced by groups of normal females.

9.4.2.4 Male-killing agents
There are a variety of arthropods in which female-biased secondary sex ratios are found, where the death of males is associated with sex-specific pathogenesis by an inherited parasite. These parasites are generally cytoplasmic, i.e. they live inside the cells of their host, and are maternally inherited. In all cases, the death of males can be seen as an adaptive strategy, the males being unable to pass the infection onto their progeny.

Male-killing micro-organisms can be broadly divided into two types (L Hurst 1991). First, there are agents that kill their host in an advanced stage of its life history. The death of males in these cases is accompanied by the release of the parasite into the environment. These 'late' male-killers are common in mosquitoes, and members of a variety of microsporidial genera (protists) have been found to produce sex ratio biases of this type: *Amblyospora*, *Thelohania*, *Parathelohania* (reviewed in L Hurst 1991). Second, there are agents that kill their host during embryogenesis or the first larval instar, 'early' male-killers. Here, the death of males is rarely accompanied by the spread of infective particles into the environment, the agents responsible being intracellular bacteria, with poor survival outside the host (except the case of *A. nasoniae* in *N. vitripennis*). Members of the genera *Spiroplasma*, *Rickettsia* and *Wolbachia* have been implicated, as have members of the gamma group of proteobacteria and members of the Flavobacteria (Table 9.1).

The explanation for the early male-killing trait lies in the effect that male host death has on the probability of survival of the remaining female hosts within the clutch. If, for instance, there is sibling egg consumption (the female hosts eat the dead male hosts), then the death of males will enhance the survival of the female hosts, which makes sense when viewed from the parasites' perspective: female and male siblings bear the same bacterium by descent, and the death of males reallocates resources from the male line (through which the bacterium cannot pass) to the female (through which it can). Male-killing is thus adaptive when there is sibling egg consumption, antagonistic interactions between siblings (Skinner 1985, L Hurst 1991), or deleterious effects of inbreeding on females (Werren 1987).

Late male-killers are in many senses classic infectiously transmitted parasites that are more benign in females, the host sex through which they can be transmitted vertically. Their life cycle is more complex, with spores of the microsporidia being released from male mosquito hosts into the water, where they infect a 'secondary' copepod host, which is subsequently killed, releasing spores that infect the mosquito (e.g. Sweeney et al. 1988). The probability of a spore released from a mosquito reentering a mosquito is likely to be related to the density of suitable intermediate hosts, and thus the density of mosquitos. Density dependence is therefore likely to be important in the dynamics of these agents.

Early male-killers will come to a stable polymorphic equilibrium if there is inefficient transmission of the parasite from mother to progeny. In this case, some uninfected types are generated each generation, preventing the parasite from increasing to fixation. If transmission is perfectly efficient, the parasite will increase to fixation with the extinction of the host population unless repressor genes invade (Uyenoyama & Feldman 1978), or there are group-level selective effects in operation.

9.4.2.5 Cytoplasmic male sterility
Cytoplasmic male sterility (CMS) is a distortion of sex allocation observed in dioecious angiosperms, in which male reproductive function is reduced such that little or no pollen is produced. Cytoplasmic male sterility is associated with mitochondrial mutants. The genetic basis underlying CMS is known in several cases and, although the genes involved differ between species, there appears to be a common role of recombination within the mitochondrial genome in creating the CMS mutant (Saumitou-Laprade et al. 1994). The plant mitochondrial genome is large (200–2500 kb), and

often present in the form of a variety of sub-circles, which combine to produce the complete mitochondrial DNA (mtDNA) molecule. Recombination is thus common, and this frequently produces rearrangements of gene order, duplications and chimeric genes (genes that form from the fusion of parts of two previously separate genes). Cytoplasmic male sterility is caused by a subset of these chimeric genes, a subset forming peptides incorporated in the inner mitochondrial membrane. Their phenotype is straightforward disruption of mitochondrial function, but the disruption is localized to the tapetum of anther tissue, where mitochondrial density within the plant is at its highest. Thus CMS mitochondria cause selective degeneration of male function.

Cytoplasmic male sterility types spread when reduction in male function (pollen production) enhances output through seed (Lewis 1941). This is analogous to male-killing in logic, maternally inherited genes here spreading if they enhance the production of the female gamete in a hermaphrodite, rather than the female progeny in a monoecious species. The CMS type will spread to fixation, causing population extinction unless either nuclear repressor genes spread, or there are group-level selective effects on CMS frequency. Both of these probably occur. Repressor genes are well known in several systems (Saumitou-Laprade et al. 1994), and indeed many cases of CMS are only seen in the hybrid context, the phenotype being completely repressed within the species (Frank 1989). Group-level selective effects are associated with the lowered frequency of pollination in populations where CMS is prevalent compared to populations where it is rare. The CMS cytotype can be maintained by a dynamic balance of local extinction of groups in which CMS is common (through lack of pollen) and recolonization from groups in which CMS is rare. There is evidence for such dynamics in some species, such as Thymus vulgaris (Mannicacci et al. 1996).

9.4.2.6 Cytoplasmically induced drive
The conversion of males to females and the killing of males are two stratagems used by cytoplasmic bacteria to increase their transmission

to the next generation. In Lepidoptera, and other groups where females are heterogametic, there is a third possibility. The cytoplasmic bacteria in females can promote the presence of the female-determining chromosome in the egg before fertilization: cytoplasmically induced drive. There is no compelling evidence of such a situation to date. G Hurst *et al.* (1997b) note that reexamination of Doncaster's (1913, 1914, 1922) work on the sex ratio trait of *Abraxus grossulariata* would be timely in this regard.

9.4.2.7 Cytoplasmic incompatiblity in haplodiploids

In a number of cases, *Wolbachia* cause cytoplasmic incompatibility (CI) between infected males and uninfected females. In eggs that are fertilized by sperm from infected males, the paternal chromosomes are fragmented, resulting in a haploid zygote (Breeuwer & Werren 1990, Reed & Werren 1995, Callaini *et al.* 1996). In haplodiploid species this skews the offspring sex ratio towards more males (as males normally develop from unfertilized eggs and females from fertilized eggs). Since CI only manifests itself in fertilized eggs, male offspring are not affected (Breeuwer & Werren 1990, Breeuwer 1997). The fate of fertilized eggs seems to depend upon the strength of CI and the chromosome system of the host. If paternal genome loss is complete, fertilized eggs can develop into males in haplodiploids (Breeuwer & Werren 1990, Breeuwer 1997, Vavre *et al.* 1999a,b, 2000), while in diplodiploids such embryos eventually die. If, however, paternal chromosome loss is incomplete and parts become incorporated into the nucleus of the zygote (Ryan & Saul 1968, Ryan *et al.* 1985, Reed & Werren 1995), this results in aneuploid zygotes, which are likely to be inviable in both haplodiploids and diplodiploids. This will be expressed in haplodiploids as female-biased mortality, in the presence of normal numbers of male offspring (Breeuwer 1997, Vavre *et al.* 1999a,b, 2000). The inclusion of paternal chromosome fragments may be more frequent in species with holokinetic chromosomes, such as mites, because microtubules can attach anywhere on the chromosomes, whereas in species with centromeric chromosomes, only centromeric frag-

ments can be included (Ryan *et al.* 1985, Breeuwer 1997). Both these effects will result in more male-biased broods. It should be noted that the sex ratio bias *per se* is not adaptive: a 'better' strategy for the *Wolbachia* is to kill the males.

9.5 | Detection and characterization of SRDs

9.5.1 Population sex ratio

Biased and unusual population sex ratios are often an indication of SRDs. As a starting point for the detection of sex ratio phenotypes one typically measures the number of males and females in natural populations or the offspring sex ratio from adult females collected in the field. Large differences in mean sex ratio between populations or between family sex ratios within populations are indicators that populations may harbour SRDs. However, sex ratio measurements in field populations are difficult to interpret because it is not possible to distinguish between the broods of individual females, and the sex ratio depends on mating structure (e.g. local mate competition) and population dynamics (e.g. sex-dependent dispersal and mortality). One major drawback of this method is that it can only detect SRDs occurring at high enough prevalence. Indeed, infection frequencies of male-killers and parthenogenesis inducing *Wolbachia* in a number of systems do not rise above 5%, which has only a marginal effect on the overall population sex ratio. Hence, population sex ratios should be taken as a first approach and interpreted with care.

9.5.2 Isofemale lines with phenotype

A frequent first step in a scheme particularly aimed at uncovering SRDs is to examine the sex ratio produced by individual females (Figure 9.1). In basis, many wild-collected females are confined individually with a single male, and the sex ratio produced recorded. Additional information is gained if mortality is scored, particularly failure of eggs to hatch. The loss of SRDs from culture lines without special maintenance also requires that variation is initially sought from

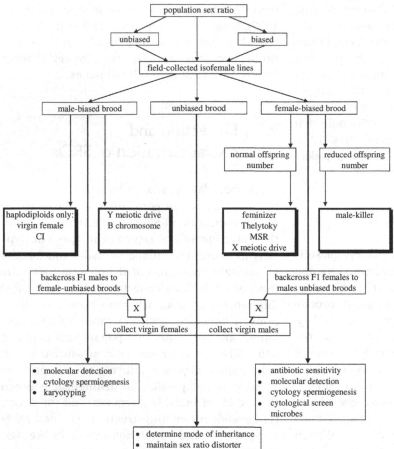

Fig 9.1 Flow diagram for the screening and identification of sex ratio distorters (SRDs). Initial screening involves the comparison of population sex ratios and setting up isofemale lines from field-collected females. Isofemale lines are subsequently classified according to the sex ratio and mortality rate of their broods. This determines the type of sex ratio distorter, which can be confirmed by a combination of molecular, cytological or antibiotic experiments. Single-sex broods will be lost if mates of the opposite sex are not available. An unbiased strain, free of sex ratio distorters, needs to be set up to provide the limiting sex to the biased isofemale lines to maintain the SRD. Further details of the procedure are in section 9.5.

wild-collected individuals, not from maintained laboratory cultures.

On finding heterogeneity in the sex ratio between lines, the inheritance pattern of any abnormal sex ratios is then analysed to determine the genetic basis of the trait (Figure 9.1 & sections 9.5.2.1–9.5.2.4). Some SRDs cause their female host to produce only one of the sexes and consequently the isofemale line including the SRD is likely to be lost. To circumvent this problem both virgin males and females from established and characterized cultures should be available during the period of isofemale screenings to be mated to members of single-sex broods. The following tests can be performed, as appropriate.

9.5.2.1 Female-biased sex ratio
Test for maternally inherited SRDs, and delineate their nature (microbial sex ratio distorters,

organelle mutants): on discovering a female-biased sex ratio in an isofemale line, the presence of maternally inherited agents can be tested by setting up the following crosses:

1. Between-line cross: daughter from biased brood × male from normal brood.

2. The reciprocal cross, if males are available: normal female × biased male.

3. Control experiment: both within-line crosses.

If the trait is maternally inherited only crosses (1) and (3) are expected to produce female-biased sex ratios. Some class (1) crosses may produce normal sex ratios. This may be due to unfaithful transmission of the microbe or to the presence of suppressors in the normal sex ratio line.

Careful recording of clutch sizes, egg hatch rate and progeny sex ratio should be undertaken to determine whether sex ratio bias is primary (e.g. feminization) or secondary (e.g. male-killing).

To examine whether the causative agent of a maternally inherited trait is a bacterium, adults of female-biased isofemale lines are typically exposed to antibiotics by mixing antibiotics (e.g. tetracycline, rifampin) in their diet (e.g. Stouthamer *et al.* 1990). Evidence for bacterial involvement derives from reversion of the female to producing a normal sex ratio, with an increase in the egg hatch-rate as well if the sex ratio bias is secondary. A single treatment of females may be sufficient to alter the resulting offspring sex ratio. Typically, however, treatment needs to be repeated for several generations for a permanent cure. In addition, a range of antibiotic concentrations should be tried out; too high a concentration can be as ineffective as too low a concentration if the host dies before cured offspring are produced.

A second method to indicate microbial involvement is to increase the rearing temperature. This method is most likely to cure larval stages. In many species, rearing at 28°C or higher will lead to loss of the infection (Stouthamer 1997). Note, however, that heat treatment can give false-negative results due to the insensitivity of certain microbial strains to elevated temperatures (Stouthamer 1997).

When antibiotics fail to cure a trait, it is then useful to examine the lines in question microscopically for the presence of micro-organisms. Egg or ovarian material is fixed, and then stains such as 4′6′-DAPI (diamidinophenolidone) may be used to detect bacterial and eukaryotic infections within the egg using fluorescence microscopy (Terry *et al.* 1997). An alternative, which often gives more confidence, is the use of transmission electron microscopy. Here, fixed eggs or ovaries are examined for micro-organisms under high magnification (Terry *et al.* 1999). If a trait is maternally inherited, and no micro-organisms are evident, then organelle involvement is suggested.

9.5.2.2 Female-biased sex ratio in male heterogametic species: tests for X-linked drive

X chromosome drive in a male heterogametic species is implicated if there is a female-biased primary sex ratio, and:

1. The female-biased sex ratio is associated with the male (i.e. crossing the male to other females also produces a biased sex ratio), and this male has disrupted spermiogenesis, as

seen in electron microscope sections of sperm bundles.

2. The trait is absent in the F1 daughters of the biased sex ratio crosses when these are crossed to males from 'normal' lines.

3. The trait reappears in half of the sons (F2 males) derived from the initial cross.

To fully confirm the association with the X chromosome, the scheme should be repeated from the F2 males showing the trait to the F4 generation, where again half the males should show the trait.

Diagnosis of meiotic drive can be complicated by the presence of resistance genes. Should the males used in mating to the F1 daughters in the above scheme bear resistance genes, then fewer than 50% of sons may exhibit the trait, even if all bear the driving chromosome. If resistance genes are present, then crosses to other populations may be helpful (section 9.5.4).

9.5.2.3 Male-biased sex ratio in male heterogametic species: tests for Y chromosome drive

Y chromosome drive is implicated if a male heterogametic species shows a primary sex ratio bias toward males. In addition:

1. The male-biased sex ratio is associated with the male in the initial cross (i.e. crossing the male to other females also produces a biased sex ratio).

2. Affected males have disrupted spermiogenesis, as seen in electron microscope sections of sperm bundles.

3. The trait is present in near 100% of F1 sons of the original cross.

9.5.2.4 Male-biased sex ratios in haplodiploid organisms

Male-biased sex ratios in haplodiploids may be due to either the male carrying driving B chromosomes, or the absence of a *Wolbachia* strain in the female, which is present in the male. All-male broods associated with B chromosomes will be inherited with high fidelity through the male line. If inheritance is not observed, then the initial distortion is more likely to be due to *Wolbachia* presence in the male, but absence in the female. If this is the case, male-biased

Table 9.2 | PCR primer pairs for molecular detection of known microbial sex ratio distorters (SRDs). Potential new bacterial SRDs may be picked up using a 16S eubacterial primer pair, but PCR products need to sequenced for further identification

Target DNA fragment			Fragment length	References
General eubacterial				
16S rDNA			≈1500	Lane 1991,
	27F	5'-AGA GTT TGA TCA TCA TGG CTC AG-3'		Weisberg et al. 1991
	1513R	5'-TAC GGT TAC CTT GTT TAC GAC TT-3'		
Wolbachia				
16S rDNA			≈930	O'Neill et al. 1992
	76F	5'-TTG TAG CCT GCT ATG GTA TAA CT-3'		
	1012R	5'-GAA TAG GTA TGA TTT TCA TGT-3'		
ftsZ			1050	Werren et al.1995
	130F	5'-GTT GTC GCA AAT ACC GAT GC-3'		
	1284R	5'-CTT AAG TAA GCT GGT ATA TC-3'		
	49IF	5'-GGA CCG GAT CCG TAT GCC GAT TGC AGA GCT TG-3'	730	Holden et al. 1993
	1262R	5'-TCG GTT CAC TTA TGA GTA TCC GCT TAA GCC AGG-3'		
wsp				
	81F	5'-TGG TCC AAT AAG TGA TGA AGA AAC-3'	590–632	Braig et al. 1998
	6.I9R	5'-AAA AAT TAA ACG CTA CTC CA-3'		
dnaA				
	74F	5'-GTC ATC TTG ATG GAG TGG AA-3'	480	Bourtzis et al. 1996
	553R	5'-GAT CTC GAC CGC CGA AGT TT-3'		
Rickettsia (typhi group)				
17 kDA antigen				
	Forward	5'-GCT CTT GCA ACT TCT ATG TT-3'	434	Williams et al. 1992,
	Reverse	5'-CAT TGT TCG TCA GGT TGG GCG-3'		G Hurst et al. 1996
Flavobacteria				
16S rDNA			762	G Hurst et al. 1999c
	202F	5'-ATT GTT AAA GTT CCG GCG-3'		
	1022R	5'-CTG TTT CCA GCT TAT TCG TAG TAC-3'		
Mollicutes (Mycoplasma, Spiroplasma, Acholeplasma, Ureaplasma)				
16s rDNA			429	Kuppeveld et al. 1992,
	MGSO	5'-TGC ACC ATC TGT CAC TCT GTT AAC CTC-3'		G Hurst et al. 1999a,b
	HaIn1	5'-GCT CAA CCC CTA ACC GC-3'		
Microsporidia				
18s rRNA			1100–15400	Baker et al. 1995
	Forward	5'-CAC CAG GTT GAT CTG CC-3'		
	Reverse	5'-TTA TGA TCC TGC TAA TGG TTC-3'		

sex ratios will not be inherited down the male line. Further, male-biased sex ratios will be apparent when males are crossed to antibiotic-treated females, and should disappear if the male is also subject to antibiotic treatment. It should be noted, however, that B chromosomes and *Wolbachia* may coexist, and inheritance studies therefore remain important in characterizing the cause of biases where they occur.

9.5.3 Mass screening for known agents on male/female templates

Sequence information from a variety of microbial SRDs is available. Several primer combinations have been designed that will amplify the DNA of specific candidate bacterial groups, or microsporidia in general, from DNA prepared from infected host individuals (Table 9.2). This may be particularly useful for initial diagnosis of the presence of a microbial SRD without recourse to breeding experiments. The prediction is straightforward: if an agent is an SRD, then amplification should only be successful from female hosts, not from males. Bouchon *et al.* (1998), for instance, used this approach to detect potential feminizing *Wolbachia* in isopod Crustacea. A bias towards presence of these bacteria in female specimens was seen as indicative of their role as a feminizing bacterium in these hosts, and used as the cue for later breeding work (as above). Mass screening may either involve a template derived from a series of individuals, or, more crudely, a template derived from pools of male and female individuals. In the latter case, it should be noted that SRDs that are not 100% efficient in their sex-distorting phenotype will be present in both female and male pools, potentially obscuring interpretation. In addition, care should be taken in all cases to avoid contaminating DNA from parasitoids within the tissue of the host; as an ideal, ovarian material or eggs should be used as a template, and eggs should be known to be unparasitized by egg parasitoids.

A drawback of this approach is that it will give false negatives if a new group of cytoplasmic SRDs is involved. Information may only be gained about the presence of microbial groups for which diagnostic PCRs have been developed.

More general primers could be developed that are specific for larger phylogenetic groups of the eubacteria, for example alpha or gamma proteobacteria, but a wider spectrum of detection also widens the potential for contaminating bacteria to obscure any pattern. In addition, the method may fail to detect the sex-biasing ability of microorganisms where host individuals sometimes escape micro-organism action (e.g. Rigaud *et al.* 1999).

9.5.4 Hybridization

One of the 'rules' of selfish genetic elements is that their effects are often manifested in hybrids. As can be seen from the discussions of the population biology of SRDs above, repressors can spread to fixation within a population, such that the distorting phenotype is no longer seen within the population. However, it will be seen in hybrids formed in between-population crosses, as these may lack resistance genes. Thus, meiotic drive can be manifest in between-population crosses when they are not observed within the populations in question (Merçot *et al.* 1995). Frank (1989) also notes that 30% of cases of CMS are observed in the hybrid context.

Manifestation of selfish genetic elements in the hybrid context also occurs for elements that kill individuals that do not bear them. Notably, CI-causing *Wolbachia* often spread to near fixation within a population/species, and the effects are seen on hybridization. In haplodiploid mites and insects, therefore, crosses between populations may show the characteristic male-biased sex ratio associated with CI in haplodiploid taxa. This has been observed in crosses between *N. vitripennis* and *N. giraulti*, where crosses in both directions give rise to male progeny only, by virtue of possession of different strains of *Wolbachia* (Breeuwer & Werren 1990). Indeed, after curing the wasps of their *Wolbachia* infection by antibiotics, they can interbreed across species and produce hybrids (i.e. diploid daughters) (Breeuwer & Werren 1990). These results rule out alternative reproductive isolation mechanisms, such as the inability to fertilize eggs using sperm of the other species. Nevertheless, hybrid breakdown occurs between these two *Nasonia* species, indicating that they have genetically diverged as well (Breeuwer & Werren 1995).

9.6 | Patterns of incidence

9.6.1 Meiotic drive

Sex chromosome meiotic drive is most commonly recorded within the Diptera, in which most examples are from *Drosophila*. An idea of the potential prevalence of meiotic drive in this group is provided by the work of Jaenike (1996). Over a period of time, nine species of mycophagous *Drosophila* were studied, with respect to answering questions in evolutionary ecology. Despite no explicit goal of seeking drive, Jaenike and coworkers have found drive to be present in five of the nine species (James & Jaenike 1990, Jaenike 1996). Outside the genus *Drosophila*, X chromosome meiotic drive has been found in sciarids and diopsids (Presgraves *et al.* 1997), and Y chromosome drive in mosquitoes (Wood & Newton 1991).

The incidence of meiotic drive within *Drosophila*, and in Diptera more widely, has led to the suggestion that sex chromosome drive may be common, thus making it an important evolutionary force (L Hurst & G Hurst 1996). However, the number of cases of drive outside the Diptera is perplexingly small. Apart from a case in lemmings (Fredga *et al.* 1976), no other confirmed case of sex chromosome drive exists. Indeed, two recent cases of female-biased sex ratios in the Lepidoptera (*Danaus chrysippus* and *A. encedon*) that had been thought to be cases of drive have been found to be associated with a secondary sex ratio bias rather than a primary one, and susceptible to antibiotics (Jiggins *et al.* 1998, Smith *et al.* 1998, Jiggins *et al.* 2000). These observations prompted Jiggins *et al.* (1999) to suggest that the Diptera may in fact be a hot-spot for the prevalence of sex chromosome drive. They cite the discovery of male-killing agents throughout the insects as a reason to tentatively reject the hypothesis of a study bias producing the discovery of drive only in dipteran insects: both are initially detected in the same way. If this conclusion is correct (and it certainly awaits further data), an explanation for the commonality of drive in the Diptera must be sought.

9.6.2 B chromosomes

The mode of action of the PSR chromosomes restricts them to occurring in species with a haplodiploid sex-determination system (Hymenoptera, Thysanoptera, some Coleoptera and some Acari). The mating structure of the population should be such that there is a relatively low frequency of brother-sister matings and the fertilization frequency should be more than 50%. In theory such populations should be rare, and female-biased sex ratios are generally associated with local mate competition situations, where most females are assumed to mate with their brothers. Yet for instance in some *Trichogramma* species the sib-mating frequencies are estimated to be around 54–65% (Stouthamer & Kazmer 1994) and the sex ratios are approximately 75% females; under such circumstances a PSR chromosome can invade the population (Stouthamer *et al.* 2001).

9.6.3 Parthenogenesis, *Wolbachia* and MSR-like factors

To date, the parthenogenesis inducing *Wolbachia* have only been detected in Hymenoptera and mites (recently, A Weeks showed that parthenogenesis in the mite genus *Bryobia* is caused by *Wolbachia*, see Charlat & Merçot 2000). Within the Hymenoptera *Wolbachia* occurs only in some families. Families that have a complementary sex-determination system (Chapter 8) seem to be free of parthenogenesis *Wolbachia* (Stouthamer & Kazmer 1994). An obvious reason for the lack of PI *Wolbachia* in species with CSD is that if parthenogenesis is induced by gamete duplication then the parthenogenetically produced individuals would become diploid males. We expect to find PI *Wolbachia* in other arthropod taxa that have the haplodiploid sex determination, such as thrips and some Coleoptera.

MSR-like factors have only been detected in a single species of parasitoid wasp but may be more common. Flanders (1943) reports a similar case where virgin females of the wasp species *Coccophagus lyciminia* produce male offspring only, whereas mated females produce almost exclusively female offspring. Little is known about the inheritance of this trait.

9.6.4 Male-killing agents

Late male-killers have, to date, only been recognized in mosquitoes. Here, the aquatic habitat acts as a means to disperse the microsporidia. However, there is no reason to suppose that microsporidia could not gain the same horizontal transmission in the terrestrial environment, and it is perhaps likely that late male-killers will therefore be found in terrestrial arthropods at some point.

Early male-killers are found in a range of insects. These differ in their basic sex-determination system (some are female-heterogametic, some male-heterogametic, some haplodiploid) suggesting that there are few hosts in which bacteria cannot evolve to kill males (G Hurst et al. 1997a). There are, however, factors associated with host ecology that make male-killers more likely to spread in certain taxa. Male-killing is advantageous to the parasite where the host has sibling egg cannibalism, antagonistic interactions between siblings, or inbreeding (G Hurst et al. 1997a, L Hurst 1991, Skinner 1985, Werren 1987). The greater the effect of male death on female host survivorship, the greater the chance of spread of a male-killer. Thus, groups like the coccinellid (ladybird) beetles, where sibling egg cannibalism is common, are most likely to bear male-killers. This appears to borne out by the data so far accumulated (G Hurst et al. 1997a).

9.6.5 Cytoplasmic male sterility

Cytoplasmic male sterility is common among angiosperm plants, with records from over 140 species spread across 20 families (Frank 1989). Around 50% of these cases are known only from interpopulation crosses or hybridizations, with CMS not being observed within the natural population. CMS is not known in animals, and is always a product of mitochondria and not chloroplasts in plants. Why is this? The absence of records in animals may be a result of a low level of study and awareness: hermaphrodite animals are little studied. However, there are mechanistic reasons to think it less likely to occur in animals. The CMS mutations in plants follow from the recombination between subcircles of a large mitochondrial genome. The animal mitochondrial genome is much smaller, subcircles are not commonly found, and although rearrangements in gene order are known, there are good mechanistic reasons to believe them to be much less common. Thus, mutation to male sterility probably occurs less commonly in animals than plants. The absence of CMS chloroplasts may also be accounted for easily. First, chloroplasts are not generally present in anther tissue. They are thus not in the physical position to alter male reproductive function. Second, whilst loss of mitochondrial function causes the death of the cell concerned, loss of chloroplast function does not. Thus, it is less easy to envisage mutation to male sterility in the chloroplast genome.

9.7 | Concluding remarks

It is clear that these sex ratio distorting elements are not merely rare curiosities but are very common. Some groups seem to bear a wider variety of these elements, notably species with haplodiploid sex determination. Others seem to bear a particular type of element more commonly than other groups, notably the incidence of male-killing bacteria in ladybirds.

Although, because of their use in biological control and for testing optimal sex ratio theories, the Hymenoptera have been studied more than some others, the commonality of elements they bear may represent a real 'extra' vulnerability of their genetic system to manipulation by SRDs. Where there is a plastic system of sex determination, it is more susceptible to corruption by SRDs. We predict from this that those species with environmental sex determination will play host to more agents of sex ratio distortion than their sister taxa with gene-inheritance-based sex determination. Additionally, ecological and biological differences explain the presence of so many male-killers in the ladybirds: the presence of sibling egg consumption in this group clearly favours the spread of male-killing strains.

It is clear that SRDs do interact subtly with the host's system of sex determination and may be the selective cause of a change in this system. What awaits assessment is the degree to which the diversity of systems of sex determination is a result of the conflict between hosts and SRDs.

On a microevolutionary level, we can begin to address this question by asking how commonly SRDs induce the spread of compensatory responses. Whilst some interactions may be 'naturally' balanced, it seems likely that the presence of resistance genes will be common as a number of the studies cited above already indicate. In all cases where the SRD has not gone to fixation we expect that compensatory responses, such as suppressor genes or the presence of SRDs of opposite effect, are present. If this is the case, their capacity to alter the sex-determination system of the host should not be underestimated. What requires delineation is the extent to which compensatory responses lead to fundamental changes in the sex-determination system of the host, as seen in A. vulgare.

A lack of compensatory responses by the nuclear genes may lead to the extinction of species that have been invaded by an extremely efficient SRD. One interesting question, about which we can only speculate, is the frequency with which SRDs cause extinction. There is no way for us to estimate this for most SRDs. Some insight may be gained from examining parthenogenesis-inducing Wolbachia, as these still allow the population to persist when the infection has gone to fixation. There are quite a large number of populations known that only consist of infected individuals (Stouthamer 1997). This suggests that the spread of SRDs to fixation is common and, by extension, that SRDs that rely on the sexual system of the host will commonly cause extinction. However, the parallel between parthenogenesis inducers and other SRDs is imperfect. SRDs that rely on host sexuality may induce much higher selection pressures on being able to produce male offspring than parthenogenesis inducers. At high prevalence, failure to mate usually results in the total loss of individual fitness, whereas this is not the case for PI Wolbachia. Thus, the frequency with which PI Wolbachia have gone to fixation may be an overestimate of the overall fixation and therefore extinction rate of species following invasion and fixation by other SRDs. It is clear, however, that such extremely efficient SRDs have a major impact on the population growth of many populations, as is illustrated by the male-killing Wolbachia in populations of Acraea (Jiggins et al. 1998).

A final question that is presently only the subject of speculation relates to the length of the association between SRDs and their host. In the case of Wolbachia it appears that the rate of horizontal transfer of the various Wolbachia is high and we can therefore conclude that the association between SRD and the host is not ancient. Little evidence exists for cospeciation between the SRD and its host. The generality of this conclusion needs to be tested. West et al. (1998) investigated the phylogeny of Wolbachia in two insect host-parasitoid communities: they concluded that there is no evidence for horizontal transfer between species, but a similar study by Vavre et al. (1999b) presents phylogenetic evidence for interspecies transfer. Recently, Huigens et al. (2000) provided conclusive experimental evidence for the horizontal transfer of Wolbachia within a species of Trichogramma.

Acknowledgements

We wish to thank Ian Hardy, Alison Dunn, Dave Parker and Paul Ode for comments on the manuscript. Greg Hurst wishes to acknowledge a BBSRC D. Phillips fellowship for personal funding.

References

Baker MD, Vossbrinck CR, Didier ES, Maddox JV & Shadduck JA (1995) Small subunit ribosomal DNA phylogeny of various microsporidia with emphasis on AIDS related forms. Journal of Eukaryotic Microbiology, 42, 564–570.

Beukeboom LW & Werren JH (1992) Population genetics of a parasitic chromosome: experimental analysis of PSR in subdivided populations. Evolution, 46, 1257–1268.

Beukeboom LW & Werren JH (1993) Deletion analysis of the selfish B-chromosome, Paternal sex ratio (PSR), in the parasitic wasp Nasonia vitripennis. Genetics, 133, 637–648.

Bouchon D, Rigaud T & Juchault P (1998) Evidence for widespread *Wolbachia* infection in isopod crustaceans: molecular identification and host feminization. *Proceedings of the Royal Society of London, series B*, **265**, 1081–1090.

Bourtzis K, Nirgianaki A, Markakis G & Savakis C (1996) *Wolbachia* infection and cytoplasmic incompatbility in *Drosophila* species. *Genetics*, **144**, 1063–1073.

Braig HR, Zhou WG, Dobson SL & O'Neill SL (1998) Cloning and characterization of a gene encoding the major surface protein of the bacterial endosymbiont *Wolbachia pipienties*. *Journal of Bacteriology*, **180**, 2373–2378.

Breeuwer JAJ (1997) *Wolbachia* and cytoplasmic incompatibility in the spider mites *Tetranychus urticae* and *T. turkestani*. *Heredity*, **79**, 41–47.

Breeuwer JAJ & Werren JH (1990) Microorganisms associated with chromosome destruction and reproductive isolation between two insect species. *Nature*, **346**, 558–560.

Breeuwer JAJ & Werren JH (1995) Hybrid breakdown between two haplodiploid species: the role of nuclear and cytoplasmic genes. *Evolution*, **49**, 705–717.

Bulnheim HP (1965) Untersuchungen über Intersexualität bei *Gammarus duebeni* (Crustacea, Amphipoda). *Helgoländer Wissenschaftliche Meeresuntersuchen*, **12**, 349–394.

Bulnheim HP & Vávra J (1968) Infection by the microsporidian *Octosporea effeminans* sp. n., and its sex determining influence in the amphipod *Gammarus duebeni*. *Journal of Parasitology*, **54**, 241–248.

Callaini G, Riparbelli MG, Giordano R & Dallai R (1996) Mitotic defects associated with cytoplasmic incompatibility in *Drosophila simulans*. *Journal of Invertebrate Pathology*, **67**, 55–64.

Carvalho AB & Klazcko LB (1993) Autosomal suppressors of sex ratio in *Drosophila mediopunctata*. *Heredity*, **71**, 546–551.

Carvalho AB & Vaz SC (1999) Are *Drosophila* SR drive chromosomes always balanced? *Heredity*, **83**, 221–228.

Carvalho AB, Vaz SC & Klazcko LB (1997) Polymorphism for Y-linked suppressors of *sex-ratio* in two natural populations of *Drosophila mediopunctata*. *Genetics*, **146**, 891–902.

Charlat S & Merçot H (2000) *Wolbachia* trends. *Trends in Ecology and Evolution*, **15**, 438–440.

Cosmides LM & Tooby J (1981) Cytoplasmatic inheritance and intragenomic conflict. *Journal of Theoretical Biology*, **89**, 83–129.

Dobson SL & Tanouye MA (1998) Interspecific movement of the paternal sex ratio chromosome. *Heredity*, **81**, 261–269.

Doncaster L (1913) On an inherited tendency to produce purely female families in *Abraxus glossurlariata*, and its relation to an abnormal chromosome number. *Journal of Genetics*, **3**, 1–10.

Doncaster L (1914) On the relations between chromosomes, sex limited transmission and sex determination in *Abraxux glossulariata*. *Journal of Genetics*, **4**, 1–21.

Doncaster L (1922) Further observations on chromosome and sex determination in *Abraxus glossulariata*. *Quarterly Journal of Microscopic Sciences*, **66**, 397–408.

Dunn AM, Adams J & Smith JE (1993) Transovarial transmission and sex ratio distortion by a microsporidian parasite in a shrimp. *Journal of Invertebrate Pathology*, **61**, 248–252.

Eberhard WG (1980) Evolutionary consequences of intracellular organelle competition. *Quarterly Review of Biology*, **55**, 231–249.

Eickbush DG, Eickbush TH & Werren JH (1992) Molecular characterization of repetitive DNA sequences from a B-chromosome. *Chromosoma*, **101**, 575–583.

Flanders SE (1943) The role of mating in the reproduction of parasitic Hymenoptera. *Journal of Economic Entomology*, **36**, 802–803.

Frank SA (1989) The evolutionary dynamics of cytoplasmic male sterility. *American Naturalist*, **133**, 345–376.

Frank SA (1991) Divergence of meiotic drive-suppression systems as an explanation for sex biased hybrid sterility and inviability. *Evolution*, **45**, 262–267.

Fredga K, Gropp A, Winking H & Frank F (1976) Fertile XX- and XY-type females in the wood lemming *Myopus schisticolor*. *Nature*, **261**, 225–227.

Gershenson S (1928) A new sex ratio abnormality in *Drosophila obscura*. *Genetics*, **13**, 488–507.

Ginsberger-Vogel T, Carre-Lecuyer MC & Fried-Montafier MC (1980) Transmission expérimentale de la thélygenie liée a l'intersexualité chez *Orchestia gammarellus* (Pallas): analyse des génotypes sexuels dans la descendance des femelles normales transformes en femelles thélygenes. *Archives de Zoologie Expérimentale et Générale*, **122**, 261–270.

Hackett KJ, Lynn DE, Williamson DL, Ginsberg AS & Whitcomb RF (1986) Cultivation of the *Drosophila* spiroplasma. *Science*, **232**, 1253–1255.

Haig D & Bergstrom CT (1995) Multiple mating, sperm competition and meiotic drive. *Journal of Evolutionary Biology*, **8**, 265–282.

Haldane JBS (1922) Sex ratio and unisexual sterility in hybrid animals. *Journal of Genetics*, **12**, 101–109.

Hamilton WD (1967) Extraordinary sex ratios. *Science*, **156**, 477–488.

Hatcher MJ, Taneyhill DE, Dunn AM & Tofts C (1999) Population dynamics under parasitic sex ratio distortion. *Theoretical Population Biology*, **56**, 11–28.

Hickey WA & Craig GBJ (1966) Genetic distortion of sex ratio in a mosquito, *Aedes aegypti*. *Genetics*, **53**, 1177–1196.

Holden PE, Brookfield JFY & Jones P (1993) Cloning and characterization of an *ftsZ* homologue from a bacterial symbiont of *Drosophila melanogaster*. *Molecular and General Genetics*, **240**, 213–220.

Huigens ME, Luck RF, Klaassen RHG, Maas MFPM, Timmermans MJTN & Stouthamer R (2000) Infectious parthenogenesis. *Nature*, **405**, 178–179.

Hunter MS, Nur U & Werren JH (1993) Origin of males by genome loss in an autoparasitoid wasp. *Heredity*, **70**, 162–171.

Hurst GDD (1997) *Wolbachia*, cytoplasmic incompatibility, and the evolution of eusociality. *Journal of Theoretical Biology*, **184**, 99–100.

Hurst GDD & McVean GAT (1998) Parasitic male-killing bacteria and the evolution of clutch size. *Ecological Entomology*, **23**, 350–353.

Hurst GDD & Schilthuizen M (1998) Selfish genetic elements and speciation. *Heredity*, **80**, 2–8.

Hurst GDD, Hammarton TC, Obrycki JJ, Majerus TMO, Walker LE, Bertrand D & Majerus MEN (1996) Male-killing bacterium in the fifth ladybird beetle, *Coleomegilla maculata* (Coleoptera: Coccinelidae). *Hereditary*, **77**, 177–185.

Hurst GDD, Hurst LD & Majerus MEN (1997a) Cytoplasmic sex ratio distorters. In: SL O'Neill, AA Hoffmann & JH Werren (eds) *Influential Passengers: Microbes and Invertebrate Reproduction*, pp 124–154. Oxford: Oxford University Press.

Hurst GDD, Hammarton TC, Bandi C, Majerus TMO, Bertrand D & Majerus MEN (1997b) The diversity of inherited parasites of insects – the male killing agent of the ladybird beetle *Coleomegilla maculata* is a member of the flavobacteria. *Genetical Research*, **70**, 1–6.

Hurst GDD, Jiggins FM, Schulenburg JHG von der, Bertrand D, West SA, Goriacheva II, Zakharov IA, Werren JH, Stouthamer R & Majerus MEN (1999a) Male-killing *Wolbachia* in two species of insect.
Proceedings of the Royal Society of London, series B, **266**, 735–740.

Hurst GDD, Schulenburg HG von der, Majerus TMO, Bertrand D, Zakharov IA, Baungaard J, Volkl W, Stouthamer R & Majerus MEN (1999b) Invasion of one insect species, *Adalia bipunctata*, by two different male-killing bacteria. *Insect Molecular Biology*, **8**, 133–139.

Hurst GDD, Bandi C, Sacchi L, Cochrane A, Bertrand D, Karaca I & Majerus MEN (1999c) *Adonia variegata* (Coleoptera: Coccinellidae) bears maternally inherited Flavobacteria that kill males only. *Parasitology*, **118**, 125–134.

Hurst LD (1991) The incidences and evolution of cytoplasmic male killers. *Proceedings of the Royal Society of London, series B*, **244**, 91–99.

Hurst LD (1993) The incidences, mechanisms and evolution of cytoplasmic sex ratio distorters in animals. *Biological Reviews*, **68**, 121–193.

Hurst LD & Hurst GDD (1996) Genomic revolutionaries rise up. *Nature*, **384**, 317–318.

Hurst LD & Pomiankowski AN (1991) Causes of sex ratio bias may account for unisexual sterility in hybrids: a new explanation for Haldane's rule and related phenomena. *Genetics*, **128**, 841–858.

Jaenike J (1996) Sex-ratio meiotic drive in the *Drosophila quinaria* group. *American Naturalist*, **148**, 237–254.

Jaenike J (1999) Suppression of sex-ratio meiotic drive and the maintenance of Y-chromosome polymorphism in *Drosophila*. *Evolution*, **53**, 164–174.

James AC & Jaenike J (1990) 'Sex ratio' meiotic drive in *Drosophila testacea*. *Genetics*, **126**, 651–656.

Jiggins FM, Hurst GDD & Majerus MEN (1998) Sex ratio distortion in *Acraea encedon* (Lepidoptera: Nymphalidae) is caused by a male-killing bacterium. *Heredity*, **81**, 87–91.

Jiggins FM, Hurst GDD & Majerus MEN (1999) How common are meiotically driving sex chromosomes? *American Naturalist*, **154**, 481–483.

Jiggins FM, Hurst GDD, Jiggins CD, Schulenburg JHG von der, Majerus MEN (2000) The butterfly *Danaus chrysippus* is infected by a male-killing *Spiroplasma* bacterium. *Parasitology*, **120**, 439–446.

Kuppeveld FJM van, Logt JTM van der, Angulo AF, Zoest MJ van, Quint WGV, Niesters HGM, Galama JMD & Melchers WJG (1992) Genus- and species-specific identification of mycoplasmas by 16S rRNA amplification. *Applied and Environmental Microbiology*, **58**, 2606–2615.

Lane DJ (1991) 16S/23S rRNA sequencing. In: E Stackebrandt & M Goodfellow (eds) *Nucleic Acid*

Techniques in Bacterial Systematics, pp 115–176. Chichester: John Wiley & Sons.

Lewis D (1941) Male sterility in natural populations of hermaphrodite plants: the equilibrium between females and hermaphrodites to be expected with different types of inheritance. *New Phytologist*, **40**, 56–63.

Majerus MEN, Schulenburg JHG von der & Zakharov IA (2000) Multiple causes of male-killing in a single sample of the two spot ladybird, *Adalia bipunctata* (Coleoptera: Coccinellidae) from Moscow. *Heredity*, **84**, 63–72.

Mannicacci D, Couvet D, Belhassen E, Gouyon PH & Atlan A (1996) Founder effects and sex-ratio in the gynodioecious *Thymus vulgaris* L. *Molecular Ecology*, **5**, 63–72.

McAllister BF (1995) Isolation and characterization of a retroelement from B-chromosome (PSR) in the parasitic wasp *Nasonia vitripennis*. *Insect Molecular Biology*, **4**, 253–262.

McAllister BF & Werren JH (1997) Hybrid origin of a B-chromosome (PSR) in the parasitic wasp *Nasonia vitripennis*. *Chromosoma*, **106**, 243–253.

Merçot II, Atlan A, Jacques M & Montchamp-Moreau C (1995) Sex-ratio distortion in *Drosophila simulans*: co-occurence of a meiotic driver and a supressor of drive. *Journal of Evolutionary Biology*, **8**, 283–300.

Nur U, Werren JH, Eickbush DG, Burke WD & Eickbush TH (1988) A selfish B-chromosome that enhances its transmission by eliminating the paternal genome. *Science*, **240**, 512–514.

O'Neill SL, Giordano R, Colbert AME, Karr TL & Robertson HM (1992) 16S rRNA phylogenetic analysis bacterial endosymbionts associated with cytoplasmic incompatibility in insects. *Proceedings of the National Academy of Sciences USA*, **89**, 2699–2702.

O'Neill SL, Hoffman AA & Werren JH (1997) *Influential Passengers: Inherited Microorganisms and Arthropod Reproduction*. Oxford: Oxford University Press.

Policansky D & Ellison J (1970) Sex ratio in *Drosophila pseudoobscura*: spermiogenic failure. *Science*, **169**, 888–889.

Pomiankowski A & Hurst LD (1999) Driving sexual preference. *Trends in Ecology and Evolution*, **14**, 425–426.

Presgraves DC, Severance E & Wilkinson GS (1997) Sex chromosome meiotic drive in stalk-eyed flies. *Genetics*, **147**, 1169–1180.

Reed KM (1993) Cytogenetic analysis of the paternal sex ratio chromosome of *Nasonia vitripennis*. *Genome*, **36**, 157–161.

Reed KM & Werren JH (1995) Induction of paternal genome loss by the paternal-sex-ratio chromosome and cytoplasmic incompatibility bacteria (*Wolbachia*): a comparative study of early embryonic events. *Molecular Reproduction and Development*, **40**, 408–418.

Reed KM, Beukeboom LW, Eickbush DG & Werren JH (1994) Junctions between repetitive DNAs on the PSR chromosome of *Nasonia vitripennis*: association of palindromes with recombination. *Journal of Molecular Evolution*, **38**, 352–362.

Reinhold K, Engquist L, Misof B & Kurtz J (1999) Meiotic drive and the evolution of female choice. *Proceedings of the Royal Society of London, series B*, **266**, 1341–1345.

Rigaud T (1997) Inherited microorganisms and sex determination of arthropod hosts. In: SL O'Neill, AA Hoffman & JH Werren (eds) *Influential Passengers: Inherited Microorganism and Arthropod Reproduction*, pp 81–102. Oxford: Oxford University Press.

Rigaud T & Juchault P (1992) Genetic control of the vertical transmission of a cytoplasmic sex factor in *Armadillidium vulgare*. *Heredity*, **68**, 47–52.

Rigaud T, Moreau J & Juchault P (1999) *Wolbachia* infection in the terrestrial isopod *Oniscus asellus*: sex ratio distortion and effect on fecundity. *Heredity*, **83**, 469–475.

Rousset F, Bouchon D, Pintureau B, Juchault P & Solignac M (1992) *Wolbachia* endosymbionts responsible for various alterations of sexuality in arthropods. *Proceedings of the Royal Society of London, series B*, **250**, 91–98.

Ryan SL & Saul GB (1968) Post-fertilization effect of incompatibility factors in *Mormoniella*. *Molecular and General Genetics*, **103**, 29–36.

Ryan SL, Saul GB & Conner GW (1985) Aberrant segregation of R-locus genes in male progeny from incompatible crosses in *Mormoniella vitripennis*. *Journal of Heredity*, **76**, 21–26.

Saumitou-Laprade P, Cuguen J & Vernet P (1994) Cytoplasmic male sterility in plants: molecular evidence and the nucleocytoplasmic conflict. *Trends in Ecology and Evolution*, **9**, 431–435.

Skinner SW (1982) Maternally-inherited sex ratio in the parasitoid wasp *Nasonia vitripennis*. *Science*, **215**, 1133–1134.

Skinner SW (1985) Son-killer: a third extra-chromosomal factor affecting sex ratios in the parasitoid wasp *Nasonia vitripennis*. *Genetics*, **109**, 745–754.

Skinner SW (1987) Paternal transmission of an extrachromosomal factor in a wasp. *Heredity*, **59**, 47–53.

Smith DAS, Gordon IJ, Depew LA & Owen DF (1998) Genetics of the butterfly *Danaus chrysippus* (L.) in a broad hybrid zone, with special reference to sex ratio, polymorphism and intragenomic conflict. *Biological Journal of the Linnean Society*, **65**, 1–40.

Stouthamer R (1997) *Wolbachia*-induced parthenogenesis. In: SL O'Neill, AA Hoffmann & JH Werren (eds) *Influential Passengers: Inherited Microorganisms and Arthropod Reproduction*, pp 102–124. Oxford: Oxford University Press.

Stouthamer R & Kazmer JD (1994) Cytogenetics of microbe-associated parthenogenesis and its consequences for gene flow in *Trichogramma* wasps. *Heredity*, **73**, 317–327.

Stouthamer R, Luck RF & Hamilton WD (1990) Antibiotics cause parthenogenetic *Trichogramma* to revert to sex. *Proceedings of the National Academy of Sciences USA*, **87**, 2424–2427.

Stouthamer R, Breeuwer JAJ, Luck RF & Werren JH (1993) Molecular identification of microorganisms associated with parthenogenesis. *Nature*, **361**, 66–68.

Stouthamer R, Breeuwer JAJ & Hurst GDD (1999) *Wolbachia pipientis*: microbial manipulator of arthropod reproduction. *Annual Review of Microbiology*, **53**, 71–102.

Stouthamer R, Tilborg M van, Jong H de, Nunney L & Luck RF (2001) Selfish element maintains sex in natural populations of a parasitoid wasp. *Proceedings of the Royal Society of London, series B*, **268**, 617–622.

Sturtevant AH & Dobzhansky T (1936) Geographical distribution and cytology of 'sex ratio' in *Drosophila pseudoobscura* and related species. *Genetics*, **21**, 473–490.

Sweeney AW, Graham MF & Hazard EI (1988) Life cycle of *Amblyospora dyxenoides* sp. nov. in the mosquito *Culex annulirostris* and the copepod *Mesocyclops albicans*. *Journal of Invertebrate Pathology*, **51**, 46–57.

Terry RS, Dunn AM & Smith JE (1997) Cellular distribution of feminizing microsporidian parasite: a strategy for transovarial transmissiom. *Parasitology*, **115**, 157–163.

Terry RS, Smith JE, Bouchon D, Rigaud T, Duncanson P, Sharpe RG & Dunn AM (1999) Ultrastructural characterisation and molecular taxonomic identification of *Nosema granulosis* n.sp., a transovariolly transmitted feminising (TTF)

microsporidium. *Journal of Eukaryotic Microbiology*, **46**, 492–499.

Uyenoyama MK & Feldman MW (1978) The genetics of sex ratio distortion by cytoplasmic infection under maternal and contagious transmission: an epidemiological study. *Theoretical Population Biology*, **14**, 471–497.

Vanderkerckhove TTM, Watteyne S, Willems A, Swings JG, Mertens J & Gillis M (1999) Phylogenetic analysis of the 16S rDNA of the cytoplasmic bacterium *Wolbachia* from the novel host *Folosomia candida* (Hexapoda, Collembola) and its implications for Wolbachial taxonomy. *FEMS Microbiology Letters*, **180**, 279–286.

Vavre F, Allemand R, Fleury F, Fouillet P & Bouletreau M (1999a) A new cytoplasmic incompatibility type due to *Wolbachia* in haplodiploid insects. *Annales de la Société Entomologique de France*, **35S**, 133–135.

Vavre F, Fleury F, Lepetit D, Fouillet P & Boulétreau M (1999b) Phylogenetic evidence for horizontal transmission of *Wolbachia* in host-parasitoid associations. *Molecular Biology and Evolution*, **12**, 1711–1723.

Vavre F, Fleury F, Varaldi J, Fouillet P & Bouletreau M (2000) Evidence for female mortality in *Wolbachia*-mediated cytoplasmic incompatibility in haplodiploid insects: epidemiologic and evolutionary consequences. *Evolution*, **54**, 191–200.

Weisburg WG, Barns SM, Pelletier DA & Lane DJ (1991) 16S ribosomal DNA amplification for phylogenetic study. *Journal of Bacteriology*, **173**, 697–703.

Werren JH (1983) Sex ratio evolution under local mate competition in a parasitic wasp. *Evolution*, **37**, 116–124.

Werren JH (1987) The coevolution of autosomal and cytoplasmic sex ratio factors. *Journal of Theoretical Biology*, **124**, 317–334.

Werren JH (1991) The paternal-sex-ratio chromsome of *Nasonia*. *American Naturalist*, **137**, 392–402.

Werren JH (1997) Biology of *Wolbachia*. *Annual Review of Entomology*, **42**, 587–609.

Werren JH & Beukeboom LW (1993) Population genetics of a parasitic chromosome: theoretical analysis of PSR in subdivided populations. *American Naturalist*, **142**, 224–241.

Werren JH, Skinner SW & Charnov EL (1981) Paternal inheritance of a daughterless sex ratio factor. *Nature*, **293**, 467.

Werren JH, Skinner SW & Huger AM (1986) Male-killing bacteria in a parasitic wasp. *Science*, **231**, 990–992.

Werren JH, Nur U & Eickbush D (1987) An extrachromosomal factor causing loss of paternal chromosomes. *Nature*, **327**, 75–76.

Werren JH, Nur U & Wu CI (1988) Selfish genetic elements. *Trends in Ecology and Evolution*, **3**, 297–302.

Werren JH, Hurst GDD, Zhang W, Breeuwer JAJ, Stouthamer R & Majerus MEN (1994) Rickettsial relative associated with male killing in the ladybird beetle (*Adalia bipunctata*). *Journal of Bacteriology*, **176**, 388–394.

Werren JH, Zhang W & Guo LR (1995) Evolution and phylogeny of *Wolbachia*: reproductive parasites of arthropods. *Proceedings of the Royal Society of London, series B*, **261**, 55–63.

West SA, Cook JM, Werren JH & Godfray HCJ (1998) *Wolbachia* in two insect host-parasitoid communities. *Molecular Ecology*, **7**, 1457–1465.

Wilkinson GS, Presgraves DC & Crymes L (1998) Male eye span in stalk eyed flies indicates genetic quality by meiotic drive suppression. *Nature*, **391**, 277–279.

Williams SG, Sacci JBJ, Schriefer ME, Anderson EM, Fujioka KK, Sorvillo FJ, Barr AR & Azad AF (1992) Typhus and typhus-like rickettsiae associated with opposums and their fleas in Los Angeles. *Journal Clinical Microbiology*, **30**, 1758–1762.

Wood RJ & Newton ME (1991) Sex-ratio distortion caused by meiotic drive in mosquitoes. *American Naturalist*, **137**, 379–391.

Zeh JA & Zeh DW (1996) The evolution of polyandry I: intragenomic conflict and genetic incompatibility. *Proceedings of the Royal Society of London, series B*, **263**, 1711–1717.

Zeh JA & Zeh DW (1997) The evolution of polyandry II: post-copulatory defences against genetic incompatibility. *Proceedings of the Royal Society of London, series B*, **264**, 69–75.

Part 4

Animal sex ratios under different life-histories

Chapter 10

Sex ratios of parasitic Hymenoptera with unusual life-histories

Paul J. Ode & Martha S. Hunter

10.1 | Summary

Hymenopteran parasitoids have proven to be exceptionally good organisms with which to test sex-allocation theory because their sex allocation is extremely labile. Sex allocation is frequently under maternal control, allowing mothers to adjust sex-allocation decisions in response to a variety of environmental conditions. Species with unusual life-histories can provide unique tests of sex-allocation theory either because maternal control over sex-allocation decisions is reduced relative to offspring control, or because the costs of producing male and female offspring differ. We consider three groups of parasitoids that are atypical in some aspect of their life-history: (1) polyembryonic species, which are unusual in that offspring may control the final sex ratio and clutch size; (2) heteronomous species, in which the sexes are generally placed in different hosts; and the balance between egg and host limitation may influence sex allocation; and (3) single-sex brood producing species, in which host versus egg limitation also influences sex ratios and which provide insight into the role of kin selection in the evolution of clutch size and sex ratios. We argue that these groups provide invaluable insights into sex-allocation behaviour by challenging the framework with which we interpret sex ratios of more typical parasitoids.

10.2 | 'Usual' life-histories, and the exceptions

Parasitic wasps have been favoured organisms for the study of sex ratios precisely because they are considered by many to have unusual life-histories. Indeed, observations of sex-allocation patterns in parasitic Hymenoptera have contributed greatly to the development of sex ratio theory (e.g. effects of local mate competition, Hamilton 1967; effects of spatial variation in resource quality, Charnov 1979, Charnov et al. 1981). Specifically, most parasitic wasps are haplodiploid, and ovipositing mothers can selectively fertilize eggs as these pass through the oviduct; males develop from unfertilized eggs and females from fertilized eggs. Parental control over the offspring sex ratio generates the potential to respond to environmental conditions that differentially affect the fitness of sons and daughters. Studies of parasitoid sex-allocation behaviour have thus provided some of the strongest tests of sex ratio theory.

In terms of feeding habits, the larvae of parasitic wasps (also referred to as parasitoids) share characteristics of both predators and true parasites. Parasitoids are similar to predators in that the host (another arthropod) is always killed; they are similar to true parasites in that one host is sufficient to complete development. Parasitic wasps may be classified functionally according to their life-history. They may be 'solitary', when only one offspring may develop per host, or 'gregarious', when more than one offspring generally develops per host. Endoparasitoids are species in which the offspring develop inside the host, while ectoparasitoid offspring feed from the outside of the host. Parasitoids can also be classified by the host stage that they attack and subsequently emerge from. For example, egg parasitoids are those which attack and whose progeny emerge from the eggs of their hosts; egg-larval parasitoids are those which attack host eggs but progeny emerge from the host larvae.

While parasitic wasps are tremendously diverse, most species share several common life-history features pertaining to sex ratios. In this chapter, we focus on three groups of parasitoids with atypical biologies with regards to their sex allocation.

1. In most parasitoids, aside from developmental mortality, the number and sex of offspring laid on a host are the same as the number and sex of emerging adults. The offspring eggs hatch and complete development using the resources provided by a single host. As a consequence, the ovipositing female is often in a position to assess the resources available to her offspring by measuring some aspect of host quality such as host size or age, and may make sex ratio and clutch size decisions accordingly. In polyembryonic species, however, ovipositing females may make only the most basic sex ratio adjustments by choosing to lay either a single egg of a particular sex or both a male and a female egg in a host. After hatching, the embryos proliferate, producing several to thousands of offspring from each egg. In mixed-sex broods, the resulting sex ratio is partly a result of competition between clonal cohorts of male and female offspring. In at least one species, clones compete through lethal combat; some female embryos develop as precocious larvae that preferentially kill male embryos over female embryos. In polyembryonic species therefore, control over sex allocation within a host resides largely with the progeny via the degree of proliferation and the behaviour of precocious larvae.

2. In the case of most solitary parasitoid species, each host can potentially be used for the development of either a daughter or a son. Therefore, time invested in host searching is always equal for both sons and daughters and the decision of which sex to lay is made after the host has been accepted for oviposition. In heteronomous aphelinid parasitoids, however, male and female eggs are laid in different host environments. Most commonly, female eggs are laid in hemipteran nymphs such as whiteflies and scale insects. Males develop as hyperparasitoids; male eggs are laid either in or on conspecific immature females or other primary parasitoids. In these species, females cannot choose what sex of egg to lay in a particular host. Rather, sex-allocation decisions may involve partitioning search time in the two host environments, or in some instances differentially accepting hosts of one type.

3. In most gregarious parasitoids, both males and females are routinely laid on the same host. Some parasitoids, however, produce only single-sex broods in which several individuals of one sex may be laid on a particular host. Like many gregarious parasitoids, these wasps must make sex-allocation and clutch-size decisions for each host. Unlike most gregarious parasitoids, however, adjustments to sex ratio occur between rather than within hosts.

In this chapter, we describe predictions and tests of sex-allocation behaviour in these exceptional groups. In so doing, we identify areas in which the theory is incomplete, or where experiments need to be done. First, however, we give a brief overview of sex-allocation theory especially as it pertains to most hymenopteran parasitoids.

10.3 | Parasitoid sex ratios

The field of sex-allocation research has a particularly rich history in terms of both theory and experiment. For a much more intensive treatment of this topic than we can present here we refer the reader to one of the many reviews available (e.g. Antolin 1993, King 1993, Godfray 1994, Hardy 1994, Godfray & Cook 1997). The vast majority of theoretical and empirical treatments have focused on two widespread sex-allocation patterns: the effect of population structure on the sex ratio and the role of host quality on the sex-allocation decisions of individual females.

10.3.1 Population mating structure and the sex ratio

Population sex ratios are typically affected by 'mating structure', i.e. how males and females obtain mates within the population. At one extreme is population-wide random mating. Fisher's (1930) explanation for the equal investment in sons and daughters in randomly mating populations is generally taken as the starting point for sex ratio theory. When the population sex ratio is 0.5 (proportion male), each son on average will mate with one female. The mother will realize equal gain in fitness through both her sons and daughters (assuming that sons and daughters are equally costly to produce). If, for some reason, the population sex ratio is biased towards one sex, selection will favour parents that overinvest in the less common sex until a population investment ratio of 0.5 is restored.

When the population is spatially structured so that relatives are more likely to interact with one another, biased sex ratios are expected. The best studied example of this, Hamilton's local mate competition (LMC) model, provides an appealing explanation for female-biased offspring sex ratios among parasitoids. LMC describes a mating system in which a variable number of females contribute offspring to a patch, which is frequently a single host or tightly clustered group of hosts. When only one female (foundress) contributes offspring to a patch, her best sex ratio strategy is to produce only enough sons to mate the daughters that are produced. Female-biased sex ratios reduce the level of competition between brothers, increase the number of mates for sons, and sib-mating results in a higher degree of relatedness between mother and offspring. As the number of foundresses increases, the optimal sex ratio increases, with an asymptote of approximately 0.5. Sons compete increasingly with the sons of other females for mates, resulting in more equal fitness returns to the mother through both sexes of offspring. While LMC has

been put forth as the most common explanation for female-biased parasitoid sex ratios, it is increasingly recognized that the mating systems of many parasitoid species fall in between the extremes of panmixis and complete local mating (e.g. Hardy 1994, Godfray & Cook 1997, Hardy *et al.* in press).

10.3.2 Host quality and sex ratio decisions

Parasitoid biologists have long been aware that ovipositing females tend to lay sons in the smallest hosts and daughters in the largest hosts (e.g. Chewyreuv 1913 (cited in Charnov 1982), Clausen 1939). This pattern has been well documented in a large number of solitary species (Charnov 1982, King 1993, Godfray 1994) and is explained by the Charnov host quality model (Charnov *et al.* 1981). In many solitary species, the size of an adult offspring is tightly correlated with the size of the host in which it developed. Furthermore, female body size is strongly correlated with various fitness measures such as fecundity and longevity. Male body size is correlated with mating ability. While both males and females are larger when they develop in larger hosts, the fitness returns for a female developing in a large versus a small host are likely to be greater than the fitness returns for a male. In other words, male fitness is probably less severely affected in a small host than is female fitness. Therefore, when an ovipositing female encounters a host, she should lay a son if the host is perceived to be small and a daughter if the host is perceived to be large. Furthermore, an ovipositing female should assess host size on a relative, rather than absolute, basis. A given host may be considered to be large if the majority of hosts are smaller; the same-sized host may be considered small if most other hosts are larger.

Both bodies of theory, concerning population structure effects and host quality effects on sex ratio, are reasonably well supported empirically. Nearly all of these empirical studies have been conducted with parasitoids possessing what we may call 'usual' life-histories. Specifically, the ovipositing mother controls the number and sex of each offspring that is deposited in or on a host. The search for and identification of hosts suitable for oviposition cannot be considered to represent investment in either male or female function because each host can be potentially used for either sex. The strongest advantage of examining sex allocation in species that diverge in one or more of these life-history characteristics is that unique insights can be gained in terms of our understanding of sex-allocation theory. For example, by studying sex allocation in species that do not exhibit maternal control over sex allocation, we can exclusively examine the role of offspring sex-allocation control issues in ways that are not possible in most 'typical' parasitoids. The taxa that we discuss here can be used to address one of two questions: what is the effect on sex-allocation patterns in the absence of maternal control and how is sex allocation affected when the sons and daughters are placed in different environments? We discuss the sex ratios of parasitoids with 'unusual' (even for parasitoids) life-histories and the implications for furthering our understanding of sex ratio biology.

10.4 | Parasitoids with unusual life-histories and the sex ratio

Here we discuss sex-allocation work with three different groups that are 'unusual' in one or more life-history traits: polyembryonic parasitoids, heteronomous parasitoids and single-sex brood producing parasitoids. Much of the following discussion follows from our own work on sex-allocation patterns within these groups of parasitoids. We identify some of the many unanswered questions in these groups.

10.4.1 Polyembryonic parasitoids
10.4.1.1 Definition, distribution and life-history
Most animals, including most parasitic wasps, are monoembryonic: a single embryo develops from each egg. Polyembryonic species are those in which more than one embryo develops from a single egg, giving rise to more than one genetically identical offspring. Two types of polyembryony are recognized (Strand & Grbić

1997a,b). Sporadic polyembryony (e.g. identical twinning) occurs in nearly all animal taxa as an unusual exception to monoembryonic development. Obligate polyembryony is comparatively rare, although widely distributed across six animal phyla (Craig et al. 1997). Within the class Insecta, obligate polyembryony occurs in certain species of the parasitic Strepsiptera and of four families of parasitic Hymenoptera: the Braconidae, Dryinidae, Encyrtidae and Platygasteridae (Ivanova-Kasas 1972).

By far the best studied polyembryonic wasp is the encyrtid *Copidosoma floridanum* (Ashmead) (see Strand & Grbić 1997a,b for reviews), an egg-larval parasitoid of plusiine moths (Lepidoptera: Noctuidae: Plusiinae). Necessarily, much of the remaining discussion of polyembyronics draws from our knowledge of this species. All polyembryonic wasps are endoparasitic and koinobionts (i.e. they allow their host to continue growth). All species apparently lay yolkless eggs. After oviposition, the polar bodies form an extraembryonic membrane that surrounds a group of embryonic cells creating the primary morula (Strand & Grbić 1997a). In *C. floridanum*, the primary morula divides repeatedly in synchrony with the host's moulting cycle to produce 'proliferating morulae' during the host's first, second, third and fourth (penultimate) larval stadia. In species of copidosomatine encyrtids morulae give rise to two larval castes: precocious larvae and reproductive larvae. Morphogenesis of *C. floridanum* morulae into precocious larvae is synchronized with the host's moulting cycle. Precocious larvae mediate intersexual competition for host resources (Grbić et al. 1992) and have been implicated in interspecies defence (Cruz 1981, Strand et al. 1990, Harvey et al. 2000) and possibly also in defence against nonsibling conspecifics (ICW Hardy, PJ Ode & ED Parker unpublished, *C. sosares*). Reproductive embryos undergo morphogenesis during the host's moult to the fifth (final) larval stadium. The reproductive larvae quickly consume the host and pupate within the remnant host cuticle. All precocious larvae die while the host is being consumed by the reproductive larvae. Approximately seven days later, adult wasps emerge virtually synchronously (over a 1- to 2-hour period).

10.4.1.2 Variation in clutch size and sex ratio

The number and sex ratio of offspring of polyembyronic wasps can be characterized at three different points in development (Ode & Strand 1995). Maternal clutch size and sex ratio refer to the number and sex of eggs laid by the mother (often referred to as the primary clutch size and sex ratio in other parasitoids). The primary clutch refers to the total number of offspring that develop during polyembryony (embryos, larvae, pupae and adults). Secondary clutch size and sex ratio refer to the number and sex of the emerging offspring. In *C. floridanum* (Strand 1989a) and *C. sosares* (PJ Ode unpublished), it is possible to determine both the number and the sex of eggs as they are oviposited by carefully observing the abdominal movements of an ovipositing female.

In polyembryonic species, maternal control over offspring clutch size and sex ratio is extremely limited compared to most parasitoid species. In species such as *C. floridanum*, ovipositing females control the number and sex of the eggs they lay per host: either one (male or female) or two (male and female) eggs (Strand 1989a) resulting in all-male, all-female and mixed broods respectively. Maternal decisions of how many and which sex egg to lay are influenced by host encounter rate (Hardy *et al.* 1993) and the distribution of host-egg ages (Ode & Strand 1995). However, the secondary clutch size and sex ratio of emerging polyembryonic offspring rarely reflect maternal oviposition decisions. Rather, the degree of polyembryonic divisions exhibited by male and female eggs as well as interference competition (mediated by the behaviour of precocious larvae in some species of copidosomatine encyrtids) within the host largely determine the clutch size and sex ratio of emerging adults. Across species of polyembryonic Hymenoptera, there is a wide range of clutch sizes that are produced from a single egg. In most species, fewer than 100 offspring arise from a single egg (Strand & Grbić 1997a). The copidosomatine encyrtids, on the other hand, include several species that routinely produce clutches exceeding 1000 offspring (e.g. Patterson 1919, Leiby 1922), as well as *C. floridanum*, where broods in excess of 3400 offspring developing from a single egg have been recorded (Ode & Strand 1995).

Even within a species, variation in both clutch size and sex ratio of polyembryonic wasps can be extreme. In *C. floridanum*, for instance, secondary clutch sizes vary from less than 500 to over 3400 and the secondary sex ratio in mixed-sex broods varies from less than 0.1% to over 99.6% male (Strand 1989b, Ode & Strand 1995). Extensive studies by Strand and colleagues have shown that much of this variation is due to two sexual asymmetries: differential production of precocious larvae and differential responses to host quality, specifically the age of the host egg at parasitism. Precocious larvae are found in moderate numbers in all-female and mixed-sex broods but are present in much lower numbers or even absent in all-male broods. Female precocious larvae have been shown to preferentially kill male reproductive embryos resulting frequently in female-biased offspring sex ratios in mixed-sex broods (Grbić *et al.* 1992). The age of the host eggs when parasitized by *C. floridanum* has a strong effect on both sex ratio (of mixed-sex broods) and clutch size (Ode & Strand 1995). While all-male clutch sizes are relatively unaffected by the age of the host egg, clutch sizes of all-female and mixed-sex broods are significantly smaller in older host eggs. Secondary sex ratios of mixed-sex broods laid in young host eggs are female biased; secondary sex ratios of broods laid in older host eggs are much closer to equality or even very male biased. The effect of host age on mixed-sex brood sex ratios appears to be mediated by the effect of host age on the number of male and female precocious larvae that are produced (Ode & Strand 1995). Female *C. floridanum* eggs produce fewer precocious larvae when laid in older host eggs, resulting in a greater survival of male embryos. When given a choice, ovipositing females prefer younger host eggs (Ode & Strand 1995). The female-biased secondary sex ratios of the majority of field-collected mixed-sex broods suggest that mostly young host eggs are encountered in the field (Ode & Strand 1995, PJ Ode & MR Strand unpublished).

10.4.1.3 Sibling rivalry and mating structure

Polyembryonic species provide an interesting test of sex ratio theory because control over secondary clutch size and sex ratio lies primarily with the offspring rather than the mother. This feature of polyembryonic species allows us unique insights into sex ratio theory, similar to studies concerning queen versus worker control of colony reproductive decisions, one of the fundamental conflicts in social hymenopteran species (Bourke & Franks 1995, Crozier & Pamilo 1996). In species such as *C. floridanum*, where the number and sex of eggs laid can be identified (Strand 1989a), the initial brood composition can be determined allowing us to study how such conflicts are resolved under a variety of initial brood conditions. Mothers are equally related to their sons and daughters, and when maternal control exists we expect to see sex allocation (across mixed-sex and single-sex broods) respond to mating structure and host quality in the ways outlined above (sections 10.3.1 & 10.3.2). If offspring control sex ratio in mixed-sex broods, the predicted sex ratio equilibria may differ substantially from the case of maternal control. While brothers and sisters value their own sex equally (relatedness of clonemates, $r = 1$), brothers are related to their sisters by $1/2$ whereas sisters are related to their brothers by $1/4$. Hence, males value their sisters more than sisters value their brothers. Therefore, under offspring control, daughters, largely through the action of their genetically identical precocious sisters, are predicted to exert greater control over the secondary sex ratio.

The degree to which conflict between a mother, her sons and her daughters exists depends largely on the mating structure of the population. When most matings are local and competition between brothers for mates is high, genetic conflict over the sex ratio is predicted to be low: mother, sons and daughters are all predicted to favour a highly female-biased sex ratio. The production of female-biased sex ratios reduces the amount of competition between brothers for mates and increases the number of sisters with whom brothers can mate. However, as mating opportunities for males away from the natal site increase, conflict over sex ratio arises. Several lines of evidence suggest that males can disperse in search of mates. Males are winged and readily disperse from both mixed-sex and all-male broods. Males of *C. bakeri* (Nadel 1987)

and *C. sosares* (PJ Ode, unpublished) have been observed in large swarms containing several hundred individuals. Finally, males live up to 14 days (under laboratory conditions) allowing sufficient time for them to disperse and find mates away from the natal site (Strand 1989a,b). Mothers and sons will prefer a much larger shift in sex ratio towards males than will daughters. Within mixed-sex broods, this conflict of interest appears to be resolved in favour of daughters through the siblicidal behaviour of female precocious larvae (Grbić *et al.* 1992, Ode & Strand 1995). Given that on average 60% of field-collected broods are mixed sex in *C. floridanum* (Ode & Strand 1995), daughter control of sex allocation is potentially an important phenomenon in this species. However, the mother may eliminate the potential for sibling rivalry by segregating her offspring by sex, laying male and female eggs in separate hosts. In many other copidosomatine species, mostly single-sex broods are produced (Patterson 1919, Leiby 1922, Doutt 1947, Alford 1976, Walter & Clarke 1992, Byers *et al.* 1993, PJ Ode & ICW Hardy unpublished, *C. sosares*), suggesting that the potential for sibling rivalry is reduced by maternal sex-allocation decisions. It is important to note, however, that the existence of precocious larvae is not known for all species of copidosomatine encyrtids. Comparative studies with species that either do not possess precocious larvae or do not exhibit the sexual asymmetries found in *C. floridanum* will provide further insight into the role of precocious larvae in the resolution of sex ratio conflicts.

The pattern of mixed-sex and single-sex brood production within a population has been used to make inferences regarding the population mating structure. In a field survey of a *Copidosoma* sp. attacking the noctuid *Chyrsodeixis argentifera*, Walter and Clarke (1992) found that most broods were single sex (63.89%) and most of these were all female (65.22%). Most mixed-sex broods were extremely female biased and the few sons present were presumed insufficient to ensure the mating of all the daughters present. They argued, given that offspring from single-sex broods must disperse to mate, that the female-biased adult sex ratios could not be explained by local mate competition. Hardy *et al.* (1993), however, pointed out that the reported bias is in the secondary sex ratio not the maternal sex ratio (in this species, all-female broods are generally larger than all-male broods), and that the maternal sex ratio was unlikely to have been significantly biased. Thus, female-biased secondary sex ratios could arise mechanistically under Fisherian sex allocation due to sexually differential clonal proliferation (in which case models assuming local mating would not apply). Alternatively, if neighbouring single-sex broods are offspring of the same mother, then local mate competition conditions may exist even though individuals must disperse from the hosts from which they emerged to find mates. Hardy *et al.* (1993) present evidence in *C. floridanum* that such circumstances exist. By manipulating the rate at which ovipositing females encountered host eggs, they altered the ratio of mixed-sex to single-sex broods. Mostly mixed-sex broods were produced when encounter rates were low; mostly single-sex broods (mostly all-female) were produced when encounter rates were high. In the field, low encounter rates are probably indicative of low host densities. Offspring from single-sex broods are not likely to find mates near the host from which they emerged. Mixed-sex broods are favoured under these circumstances because this ensures that progeny are mated. Likewise, high encounter rates are likely to reflect high host densities. Offspring developing in single-sex broods are likely to find mates from nearby hosts that have a high probability of being siblings.

An adaptive explanation for why few single-sex broods are all-male is the lower survivorship experienced by such broods. All-male broods produce fewer precocious larvae and produce them significantly later than all-female or mixed-sex broods (Grbić *et al.* 1992, Ode & Strand 1995). As a consequence, all-male broods are more susceptible to interspecific competitors, due to the inability to increase the production of precocious larvae as compared to all-female broods (Harvey *et al.* 2000, MR Strand pers. comm.). Furthermore, even in the absence of competitors, all-male broods produce smaller clutch sizes and show lower overall survivorship than do all-female or mixed-sex broods (Ode & Strand 1995). This hypothesis and local mate competition are two nonexclusive

explanations of this phenomenon; clearly either or both may be involved. Future work will show whether such explanations hold for why most single-sex broods are all-female in other polyembryonic species.

Perhaps the biggest hindrance to our understanding of sex allocation in polyembyronic species is the virtual lack of direct knowledge concerning the mating structure. However, recent field work on *Copidosoma sosares* using allozyme variation has begun to detect levels of population structure across a range of spatial scales (ICW Hardy, PJ Ode & ED Parker, unpublished). Such information is critical in determining the predicted strength of conflict over sex ratio as well as how such conflict will be resolved.

10.4.2 Heteronomous parasitoids

In one lineage of the hymenopteran family Aphelinidae, most species are heteronomous, meaning male and female offspring develop in different host environments (for reviews see Walter 1983, Hunter & Woolley 2001). Most commonly, female wasps develop as primary endoparasitoids of a hemipteran host such as a whitefly or armoured scale nymph (the *primary* host), while males develop as hyperparasitoids, either on conspecific females or on other primary parasitoid immatures (the *secondary* host). These wasps are called heteronomous hyperparasitoids, or more specifically autoparasitoids, when their secondary host range includes conspecific females (Walter 1983, Hunter & Woolley 2001). The sex-specific host relationships of autoparasitoids are entirely obligate; virgin females are highly reluctant to lay haploid male eggs in primary hosts and with a few exceptions involving sex ratio distorters (Zchori-Fein *et al.* 1992, Hunter *et al.* 1993, Hunter 1999), male eggs that are laid in the primary host do not develop (Gerling 1966, Williams 1972). Hyperparasitic female development is also very rare, and of the two reports of which we are aware (Nguyen & Sailer 1987, Hunter 1999), one involves infection with a parthenogenesis-inducing bacterium (Hunter 1999). Thus, in most instances, sexual autoparasitoid females search for hosts on plants, and deposit male and female eggs as they encounter hosts of the appropriate type.

It has long been noted that autoparasitoid sex ratios may be highly variable (Flanders 1942, Kuenzel 1975, Williams 1977) yet neither of the two models most commonly used to explain variable sex ratios in other parasitic wasps is likely to apply. The demic population structure necessary for local mate competition is unlikely to occur in autoparasitoid populations where the spatial distribution of primary and secondary hosts is often distinct (Donaldson & Walter 1991a,b) and males have been observed to move frequently among patches to mate (Kajita 1989). Instead, population structure is likely to vary spatially and temporally with changes in density and plant growth, the latter tending to stratify host stages (Noldus et al. 1987, Hunter 1993). The Charnov host quality model may only be applicable to the evolution of autoparasitism, in that it may be part of the explanation of why smaller secondary hosts are reserved for males. However, this model does not directly apply to the sex-allocation decisions of individuals in this group because the sex ratio is predetermined by the type of host encountered.

10.4.2.1 Host and egg limitation
The model that best explains autoparasitoid sex ratios is a variant of Fisher's theory. Autoparasitoid sex allocation is predicted to differ according to the degree of host and egg limitation experienced by individual wasps (Godfray & Waage 1990, Hunter & Godfray 1995). When hosts are scarce, wasps are said to be 'host limited'. When hosts are limiting, and evenly mixed in the environment, theory predicts that females should accept all suitable hosts encountered, and the sex ratio should then reflect the proportion of hosts for males (Godfray & Waage 1990). Interestingly, because adult females are simultaneously searching for both host types, and only one sex can be laid on a host, there is no predicted trade-off between the production of male and female offspring. Autoparasitoids foraging under these conditions may be unique among animals in their violation of this very basic, implicit assumption of Fisher's theory (Fisher 1930, Godfray & Hunter 1992). When hosts for male and female offspring are in different environments, the resulting sex ratios should be similar, but the assumptions more strictly Fisherian. This is likely to be the

case when the primary hosts are whiteflies that oviposit on the apical leaves of fast-growing plant hosts resulting in the vertical stratification of primary and secondary hosts (Hunter 1993). It may also occur when females develop on different hemipteran species than do the primary parasitoids used for the production of males (Walter 1983, Williams & Polaszek 1995). In these cases, there is a trade-off between searching for hosts for males and hosts for females, and wasps are predicted to invest equal time searching in each environment (Godfray & Waage 1990).

When wasps encounter more hosts than they have eggs to lay, they are said to be 'egg limited'. Egg-limited females should lay equal numbers of male and female eggs. Strictly egg-limited females should produce equal sex ratios regardless of the proportion of secondary hosts in the environment. Laboratory experiments with Encarsia tricolor have shown that offspring sex ratios were closest to equality when host densities were high and not limiting (Hunter & Godfray 1995). Further, when primary and secondary hosts were on separate leaves within the experimental arena, the number of observations of wasps on each leaf was similar even though the host densities differed, supporting the prediction that autoparasitoids invest equal amounts of time searching each host environment.

Evidence from the field is more difficult to interpret. In the field, autoparasitoid sex ratios are largely influenced by the proportion of hosts for males available (Donaldson & Walter 1991b, Hunter 1993, Bernal et al. 1998). This general finding has been used to support the contention that autoparasitoid sex ratios are not adaptive (Donaldson & Walter 1991b, Walter & Donaldson 1994), and highlights the fact that, in autoparasitoids, one would predict the same sex ratios from host-limited wasps that are behaving optimally as one would from wasps that have no adaptive control over sex ratios, but are constrained to accept hosts in proportion to their abundance. In field populations it is difficult to distinguish between these hypotheses. In a field population of Encarsia pergandiella, more male eggs were laid than would be predicted by the simple abundance of secondary hosts, possibly indicating some degree of egg limitation in the population. From these field data, however, no flexibility in sex allocation was demonstrated; the sex ratio did not move towards equality on plants with a higher total host density, as would be predicted by theory (Hunter 1993). It is important to note, however, that females need not show behavioural flexibility for sex-allocation patterns to be adaptive; if females are invariably host-limited in nature, there will be no selection for a conditional sex-allocation strategy. Furthermore, if host and egg limitation exist along a continuum rather than as a strict dichotomy (Hunter & Godfray 1995), and the environment of a given population is relatively constant, then a fixed degree of preference for the less common host type could be selected for. Almost by definition, secondary hosts are the less common host type, so a fixed preference for secondary hosts could be adaptive. Variation among species in their sex-allocation behaviour may in fact be useful in constructing testable predictions about what limits autoparasitoid populations' reproduction in the field (West & Rivero 2000).

10.4.2.2 Host selection and host feeding on primary hosts

While autoparasitoids are constrained to lay only one sex of egg on a particular host type, the preference for particular species of secondary hosts has been shown to influence sex ratios. In two laboratory experiments, exposure of wasps to the preferred heterospecific secondary hosts (in combination with whitefly primary hosts) led to more male-biased sex ratios than when the less-preferred conspecific secondary hosts were presented with the same whitefly primary hosts (Avilla *et al.* 1991, Williams 1991).

Host feeding may also influence sex ratios because primary hosts may be used for host feeding or oviposition, but pupal-stage secondary hosts are likely to be used only for oviposition. Host feeding of secondary hosts is much less likely, because of the physical difficulty of wasps getting haemolymph across the air space between the wasp pupa and the hemipteran cuticle enclosing it. The spatial arrangement of hosts has been shown to influence host feeding, and sex ratios in turn (Avilla *et al.* 1990, Hunter & Godfray 1995). *Encarsia tricolor* was more likely to feed

on (rather than oviposit in) primary hosts when primary and secondary hosts were presented together on the same leaf than when they were presented on separate leaves in the same arena. This result led to more male-biased sex ratios in the mixed patches (Hunter & Godfray 1995). These results indicate that in autoparasitoids one cannot easily extract sex allocation from other aspects of oviposition behaviour usually considered separately, such as host species selection and host-feeding behaviour.

10.4.2.3 Kin selection

Kin selection could play a role in autoparasitoid sex allocation if the fitness return of producing a male is discounted by the probability that the secondary host used to produce the male is a daughter or a sister (Colgan & Taylor 1981). Colgan and Taylor's (1981) consideration of this possibility led them to predict female-biased sex ratios where parasitism of relatives was common. There has been no test of this theory. It is insufficient as formulated because it makes no concession to variation in the relative abundance of primary and secondary hosts, but it could be incorporated into the more general Godfray and Waage (1990) model, if evidence of parasitism of relatives was found.

The extent of parasitism of relatives has not been documented for any autoparasitoid species, but the large colonial population structure of many hemipteran hosts and time delays due to autoparasitoid development may make encounters between adults and related immatures uncommon (Williams 1996). Indirect evidence of kin selection affecting sex ratios in nature would be provided by the finding that autoparasitoids discriminate between kin and nonrelatives. In the only experiment of this kind, however, female *E. tricolor* did not discriminate between pupal daughters and nonrelatives (Williams 1996), but relatives that may be more likely encountered, such as young larval daughters or immature sisters (Kajita 1989), should also be tested (Williams 1996). Lastly, an evolutionary response to selection on females to avoid kin may occur even if parasitism of kin occurs relatively infrequently. In some situations, for example when autoparasitoids colonize small isolated host patches, the probability of parasitism of kin is likely to rise. Selection in these environments may have been important in the evolution of a preference for heterospecific secondary hosts in some species (Williams 1991, 1996), especially if there is a constraint that prohibits the evolution of kin discrimination.

10.4.2.4 Other heteronomous parasitoids

Sex ratios studies to date have focused on what is likely to be the most prevalent heteronomous life-history, autoparasitism. Yet other intriguing life-histories exist in this group. In species called 'heterotrophic', females are produced as endoparasitoids of whitefly nymphs, while males are produced as primary parasitoids of Lepidoptera eggs (Beingolea 1959, Walter 1983, Hunter et al. 1996). The sex ratio predictions of Godfray and Waage (1990) apply to these animals. Whitefly nymphs and lepidopteran eggs may be separated or intermingled on the same leaves, as was found in soybean fields in Argentina where the heterotrophic species *Encarsia porteri* was collected (Hunter et al. 1996).

Indirect autoparasitoids produce males as hyperparasitoids, but may lay male eggs in unparasitized hemipterans in anticipation of later parasitism by a conspecific or heterospecific primary parasitoid. These ectoparasitic males lie in a quiescent state until the primary parasitoid approaches maturity, then hatch and consume it. Godfray and Waage (1990) predicted that while only unparasitized hosts are encountered, indirect autoparasitoids should lay equal numbers of male and female eggs. However, the sex ratio laid in unparasitized hosts should be increasingly female biased as the encounter rate with hosts previously parasitized by females (and suitable only for oviposition of male eggs) increases. This prediction has not been tested.

10.4.3 Single-sex brood producers

The vast majority of gregarious parasitoid species produce clutches containing both males and females (i.e. mixed-sex clutches). Species experiencing local mate competition are predicted to lay sufficient numbers of males per brood to maximize the number of daughters that are mated

(Hardy & Godfray 1990, Heimpel 1994). Randomly mating species may arrive at a 0.5 population sex ratio in a variety of ways, including some females producing single-sex broods composed of either all-daughters or all-sons. However, finite populations experience selection to reduce variance in offspring sex ratios by producing mixed-sex broods (Verner 1965, Taylor & Sauer 1980, Green et al. 1982, Chapter 5).

Gregarious species that routinely produce single-sex broods are quite rare. Aside from polyembryonic parasitoids, known species that produce single-sex broods (of both sexes) with a high frequency include several members of the eulophid genus *Achrysocharoides* (Askew & Ruse 1974, Bryan 1983, West et al. 1999), the eulophid *Eulophus larvarum* (Godfray & Shaw 1987), the braconid *Microbracon terebella* (Salt 1931), and the chrysidid *Argochrysis armilla* (Rosenheim 1993).

Species producing a preponderance of single-sex broods provide interesting tests of sex ratio theory because of the extreme differences in relatedness patterns that exist between all-male and all-female broods. When all broods are mixed-sex (with a sex ratio of 0.5) the mean relatedness of a male to his siblings $((0.5 + 0.5)/2 = 0.5)$ is the same as the mean relatedness of a female to her siblings $((0.25 + 0.75)/2 = 0.5)$. As long as offspring are unable to determine the sex of their siblings, males and females are predicted to exhibit equal levels of competitiveness. In single-sex broods, within-brood relatedness is 0.5 for all-male broods and 0.75 for all-female broods. Therefore, females in all-female broods are predicted to be less competitive than males in all-male broods. The relatedness patterns that exist in single-sex broods have yielded important insights into three areas of sex-allocation research: the role of sexual competitive asymmetries on clutch size, the effects of host versus egg limitation on sex ratios, and the evolutionary transition from a solitary to a gregarious life-history.

10.4.3.1 Single-sex brood production and clutch size

Competition between developing larvae frequently results in an overall decline in brood fitness. Consequently, a mother may be selected to reduce clutch size (relative to host size) thereby reducing the level of competition experienced by her developing offspring (Godfray & Parker 1991). Lower intrabrood relatedness (and a correspondingly higher incidence of sibling rivalry) should result in smaller clutch sizes. Ovipositing females are most likely to lay single-sex broods as a means of reducing the level of sibling competition when intersexual competition exceeds intrasexual competition (Godfray 1994). The greater relatedness that exists between sisters leads to the prediction that competition between sisters should be lower than that between brothers or between sister and brother. Therefore, assuming for the moment that only relatedness patterns influence clutch size optima, all-female broods should be larger than all-male broods (Godfray & Parker 1991). The handful of studies of single-sex brood-producing species have lent mixed support for this prediction. In the species of *Achrysocharoides* that produce single-sex broods, all-female broods are larger than all-male broods (Askew & Ruse 1974, Bryan 1980, 1983, West *et al.* 1999). On the other hand, all-male and all-female clutch sizes are not significantly different for *E. larvarum* (Godfray & Shaw 1987) or *Argochrysis armilla* (JA Rosenheim, pers. comm.) and in *Microbracon terebella* all-male clutches are larger than all-female (Salt 1931). Considering the equivocal support of the Godfray and Parker (1991) model using clutch size data from single-sex brood-producing species, it is important to recognize that other factors aside from relatedness patterns can account for differences in the optimal clutch sizes of all-male and all-female broods. In many species, a small female suffers a greater fitness penalty compared to a small male; therefore, all-male clutches are predicted to be larger than all-female clutches similar to the situation in *M. terebella* (Salt 1931). These predictions conflict, thus one would need to sort out the relative selective strengths of intersexual conflict and size-fitness relationships in order to make more refined testable predictions.

Single-sex broods of polyembryonic species provide an interesting comparison with single-sex brood-producing, monoembyronic species. Offspring of a single-sex polyembryonic brood are genetically identical. Within such broods there should be no sibling rivalry (i.e. no conflict of in-terest between clutch members) and clutch sizes are predicted to be very large (Godfray & Parker 1991). Indeed, the largest broods of any parasitoid species have been recorded from polyembryonic species (Clausen 1940, Ode & Strand 1995). However, the clutch sizes of many polyembryonic species, particularly nonencyrtids, are comparable to those of monoembryonic species (Clausen 1940, Ivanova-Kasas 1972). Also, because no sibling rivalry is predicted in single-sex polyembryonic broods, clutch size differences in all-male and all-female broods can be attributed to other factors, such as clutch size effects on individual adult male and female fitness measures. Similar to monoembyronic gregarious species, both male and female body sizes decrease with increasing clutch size in at least two polyembryonic species, *Copidosoma floridanum* (Ode & Strand 1995) and *C. sosares* (PJ Ode & ICW Hardy, unpublished). If, as noted above, individual females suffer a greater fitness penalty from being small than do males, female clutch sizes are predicted to be smaller than male clutch sizes. However, in the majority of studies, female clutch sizes are either equal to or greater than male clutch sizes (e.g. Patterson 1919, Ode & Strand 1995, PJ Ode & ICW Hardy unpublished). The apparent discrepancy between prediction and observation can be explained in part by noting that clutch size in at least one polyembryonic species is positively correlated with host weight. In *C. floridanum*, while all-female broods are larger than all-male broods, the final host weight for an all-female brood is on average larger than the host weight for an all-male brood, and the average female size is larger than the average male size (Ode & Strand 1995, PJ Ode & MR Strand unpublished).

10.4.3.2 Single-sex brood production under conditions of host versus egg limitation

Similar to polyembryonic and heteronomous species, the population sex ratio is further complicated by the issues of egg versus host limitation (West *et al.* 1999). When hosts are limiting, an equal number of male and female broods is predicted, since the limiting resource, search time, is then invested equally in male and female broods. If the optimal clutch size of all-female broods is larger than that of all-male broods, as is generally the case for *Achrysocharoides* (section 10.4.3.1), the population sex ratio should be female-biased. Alternatively, if eggs are limiting, females should seek to invest eggs equally in males and females, and the population sex ratio should then approach equality. In a survey of 19 populations of seven species of *Achrysocharoides*, West *et al.* (1999) showed that most contained equal numbers of male and female broods with an overall female-biased population sex ratio. This pattern suggests that hosts are limiting. Furthermore, as host density for *A. zwoelferi* was artificially increased in the field (i.e. increasing egg limitation), the proportion of female broods decreased (West *et al.* 1999).

10.4.3.3 Single-sex brood production and the evolution of clutch size

In solitary species, if more than one offspring is present, the additional eggs or larvae are eliminated via interference competition, for example physical attack, or physiological suppression (Clausen 1940, Salt 1961, Fisher 1971, Vinson & Iwantsch 1980, Strand 1986, Mayhew & van Alphen 1999). Larvae of gregarious species are rarely siblicidal; reduction in clutch size is typically the result of simple exploitation competition for host resources (Godfray 1994). Current phylogenetic data for the parasitic Hymenoptera indicate that gregarious development has arisen from solitary development on numerous occasions (Rosenheim 1993, Mayhew 1998). Gregarious development may arise from solitary species that routinely lay multiple egg clutches (more than one egg laid during a single host visit), a widespread behaviour among solitary parasitoids (Rosenheim & Hongkham 1996).

Despite the apparently large number of times that gregarious development has arisen, the conditions required for the spread of nonsiblicidal behaviour (allowing gregarious development to occur) are very stringent (Godfray 1987): nonsiblicidal behaviour will spread only when the per capita fitness of an individual sharing host resources exceeds that of an individual developing alone. The production of single-sex broods is predicted to relax these conditions by increasing the likelihood that two nonsiblicidal individuals are placed together in the same host (Rosenheim

1993). This can be understood intuitively by considering the within-brood relatedness for each of the three types of two-egg clutches: $r = 0.5$ for all-male broods, $r = 0.75$ for all-female broods, $r = 0.375$ for mixed-sex broods (a brother is related to his sister by 0.5 but a sister is related to her brother by 0.25). An overproduction of single-sex broods at the expense of producing fewer mixed-sex broods will elevate the average within-brood relatedness. In other words, the inclusive fitness cost of killing a brood mate increases as within-brood relatedness increases. In this way, single-sex broods make it more likely that nonsiblicidal behaviour will spread, allowing gregarious development to evolve.

The overall rarity of single-sex brood-producing species is most likely due to selection on individuals in finite populations to reduce the variance in their offspring sex ratios (Verner 1965, Taylor & Sauer 1980, Green *et al.* 1982, Chapter 5). Instead, single-sex brood production may be more common in the transition from solitary to gregarious development. As explained above, the oviposition of more than one egg per host and the production of single-sex broods are two important conditions allowing nonsiblicidal behaviour to spread in a population of siblicidal individuals. Studies of the adaptive function of such traits are important because they give an indication of the prevalence of such clutch-size and sex-allocation decisions in 'solitary' species and, thus, how likely they function as facilitators in this major life-history transition. Recent studies on the encyrtid wasp *Comperiella bifasciata*, an endoparasitoid of armoured scales, demonstrate the selective advantage of laying multiple eggs per host as well as the production of single-sex broods. *Comperiella bifasciata* is a solitary species that routinely lays more than one egg per host during a single host visit (Blumberg & Luck 1990, Rosenheim & Hongkham 1996, Ode & Rosenheim 1998). Multiple egg clutches increase the likelihood that at least one offspring escapes encapsulation by the host and successfully develops (Blumberg & Luck 1990, Ode & Rosenheim 1998). All-female broods were the most common brood type when more than one egg was laid per host (Ode & Rosenheim 1998). While both sexes experienced an increase in the likelihood that one individual will successfully develop, the benefit of multiple-egg clutches is clearly greatest for females. Single-sex brood production allows maternal control over which sex will emerge from a given host and also allows female offspring to avoid the competitive superiority of male brood mates. When both a son and a daughter are laid in a host, sons emerge nearly 90% of the time (Ode & Rosenheim 1998). In sum, given the high risk of encapsulation in a single-egg clutch and male competitive superiority in mixed-sex clutches, the production of a two-egg female clutch is necessary to ensure that one daughter will emerge; this is precisely the allocation pattern that facilitates the transition from solitary to gregarious development. Further studies on the sex-allocation decisions of other solitary species laying multiple egg clutches will indicate just how widespread these decisions are and the role of siblicide in the evolution of clutch size.

10.5 | Conclusions

'The chances for favorable serendipity [in research] are increased if one studies an animal that is not one of the common laboratory species. Atypical animals ... force one to use non-standard approaches and non-standard techniques, and even to think non-standard ideas.'

(Bartholomew 1982)

In the search for general principles, research often concentrates on a few model systems thought to represent a majority of animal species. An alternative approach is to embrace diversity, and use it as a tool to understand the interaction between adaptation and constraint in the evolution of life-history and behaviour. Hymenopteran parasitoids as a group represent tremendous diversity in almost every aspect of life-history. The species that are the focus of this chapter have life-histories that are almost unimaginably weird. Even more than other more typical species, these wasps force us to look beyond customary theoretical frameworks or to adapt existing theory to new settings. Sex-allocation theory for parasitoids generally tends to revolve around Fisher's theory,

local mate competition, and host quality/fitness relationships, yet in these unusual parasitoids, some of the assumptions of these theories are violated, and other ideas apply.

In polyembryonic parasitoids, mothers that lay a male and female egg within the same host relinquish further sex ratio control for that brood. Sex ratios at emergence are likely to be influenced by intersexual conflict between the precocial larvae. In the polyembryonic copidosomatine encyrtids, female precocious larvae, which are less related to their brothers than their brothers are to them, preferentially kill male embryos. Female-biased sex ratios are predicted to be favoured by both sons and daughters when local mating occurs, but conflict over sex ratio will increase when males have mating opportunities beyond the emergence site. An examination of mating structure in these populations would allow us to examine the extent of conflict between offspring and its resolution.

In the heteronomous parasitoids called autoparasitoids, and in parasitoids that produce single-sex broods, theory predicts that the degree of host or egg limitation will influence sex allocation. Host-limited autoparasitoids should parasitize hosts for male and female eggs as they encounter them, while egg-limited autoparasitoids should produce more equal sex ratios without regard to the relative abundance of the two host types. Similar predictions apply to single-sex brood producers; host-limited females should produce equal numbers of broods of males and females, while egg-limited females should lay equal numbers of male and female eggs. Sex ratios from the field in these two systems conform better to the predictions for host-limited populations than for egg-limited populations, a result which may be used as a hypothesis that can be tested with independent data (West & Rivero 2000). In autoparasitoids, a flexible sex-allocation behaviour has been demonstrated in one species, yet observations from other species suggest that egg limitation may not always result in a switch in sex-allocation behaviour. Independent observations of the degree of host or egg limitation in these autoparasitoid populations would help to distinguish between two explanations: that sex-allocation behaviour is constrained and that egg limitation is sufficiently rare that flexibility has not been selected for.

Acknowledgements

We thank Conrad Cloutier, Ian Hardy, Peter Mayhew and Michael Strand for their insightful comments on the manuscript. The authors were supported in part by an NSF-NATO Postdoctoral Fellowship (DGE–9633975) to PJO and a USDA NRI grant to MSH (98-35302-6904).

References

Alford DV (1976) Observations on *Litomastix aretas*, an encyrtid parasite of the strawberry tortrix moth. *Annals of Applied Biology*, **84**, 1–5.

Antolin MF (1993) Genetics of biased sex ratios in subdivided populations: models, assumptions, and evidence. In: D Futuyma & J Antonovics (eds) *Oxford Surveys in Evolutionary Biology*, volume 9, pp 239–281. Oxford: Oxford University Press.

Askew RR & Ruse JM (1974) Biology and taxonomy of species of the genus *Enaysma* Deluchhi (Hym., Eulophidae, Entedoninae) with special reference to the British fauna. *Transactions of the Royal Entomological Society of London*, **125**, 257–294.

Avila J, Argtigues M, Sarasúa MJ & Albajes R (1990) A review of the biological characteristics of *Encarsia tricolor* and their implications for biological control. *Bulletin O.I.L.B./S.R.O.P. (Organisation Internationale De Lutte Biologique/Section Regionale Ouest Palearctic)*, **13**, 14–18.

Avila J, Anadón J, Sarasúa MJ & Albajes R (1991) Egg allocation of the autoparasitoid *Encarsia tricolor* at different relative densities of the primary host (*Trialeurodes vaporariorum*) and two secondary hosts (*Encarsia formosa* and *E. tricolor*). *Entomologia Experimentalis et Applicata*, **59**, 219–227.

Bartholomew GA (1982) Scientific innovation and creativity: a zoologist's point of view. *American Zoologist*, **22**, 227–235.

Beingolea OD (1959) Nota sobre *Encarsia* spp. (Hymenop.: Aphelinidae) parásito de los huevos de *Anomis texana* Riley (Lepidop.: Noctuidae). *Revista Peruana de Entomologia Agraria*, **2**, 59–64.

Bernal JS, Luck RF & Morse JG (1998) Sex ratios in field populations of two parasitoids (Hymenoptera: Chalcidoidea) of *Coccus hesperidum* L. (Homoptera: Coccidae). *Oecologia*, **116**, 510–518.

Blumberg D & Luck RF (1990) Differences in the rates of superparasitism between two strains of *Comperiella bifasciata* (Howard) (Hymenoptera: Encyrtidae) parasitizing California Red Scale (Homoptera: Diaspididae): an adaptation to circumvent encapsulation? *Annals of the Entomological Society of America*, **83**, 591–597.

Bourke AFG & NR Franks (1995) *Social Evolution in Ants*. Princeton: Princeton University Press.

Bryan G (1980) Courtship behaviour, size differences between the sexes and oviposition in some *Achrysocharoides* species (Hym., Eulophidae). *Netherlands Journal of Zoology*, **30**, 611–621.

Bryan G (1983) Seasonal biological variation in some leaf-miner parasites in the genus *Achrysocharoides* (Hymenoptera, Eulophidae). *Ecological Entomology*, **8**, 259–270.

Byers JR, Yu DS & Jones JW (1993) Parasitism of the army cutworm, *Euxoa auxillaris* (Grt.) (Lepidoptera: Noctuidae), by *Copidosoma bakeri* (Howard) (Hymenoptera: Encyrtidae) and effect on crop damage. *Canadian Entomologist*, **125**, 329–335.

Charnov EL (1979) The genetical evolution of patterns of sexuality: Darwinian fitness. *American Naturalist*, **113**, 465–480.

Charnov EL (1982) *The Theory of Sex Allocation*. Princeton, NJ: Princeton University Press.

Charnov EL, los-den Hartogh RL, Jones WT & van den Assem J (1981) Sex ratio evolution in a variable environment. *Nature*, **289**, 27–33.

Clausen CP (1939) The effect of host size upon the sex ratio of hymenopterous parasites and its relation to methods of rearing and colonization. *Journal of the New York Entomological Society*, **47**, 1–9.

Clausen CP (1940) *Entomophagous Insects*. New York: McGraw-Hill.

Colgan P & Taylor P (1981) Sex-ratio in autoparasitic Hymenoptera. *American Naturalist*, **117**, 564–566.

Craig SF, Slobodkin LB, Wray GA & Biermann CH (1997) The 'paradox' of polyembryony: a review of the cases and a hypothesis for its evolution. *Evolutionary Ecology*, **11**, 127–143.

Crozier RH & Pamilo P (1996) *Evolution of Social Insect Colonies: Sex Allocation and Kin Selection*. Oxford: Oxford University Press.

Cruz YP (1981) A sterile defender morph in a polyembryonic hymenopterous parasite. *Nature*, **294**, 446–447.

Donaldson JS & Walter GH (1991a) Brood sex ratios of the solitary parasitoid wasp, *Coccophagus atratus*. *Ecological Entomology*, **16**, 25–33.

Donaldson JS & Walter GH (1991b) Host population structure affects field sex ratios of the heteronomous hyperparasitoid, *Coccophagus atratus*. *Ecological Entomology*, **16**, 35–44.

Doutt RL (1947) Polyembryony in *Copidosoma koehleri* Blanchard. *American Naturalist*, **81**, 435–453.

Fisher RA (1930) *The Genetical Theory of Natural Selection*. Oxford: Oxford University Press.

Fisher RC (1971) Aspects of the physiology of endoparasitic Hymenoptera. *Biological Reviews*, **46**, 243–278.

Flanders SE (1942) The sex-ratio in the Hymenoptera: a function of the environment. *Ecology*, **23**, 120–121.

Gerling D (1966) Studies with whitefly parasites of Southern California. I. *Encarsia pergandiella* Howard (Hymenoptera: Aphelinidae). *Canadian Entomologist*, **98**, 707–724.

Godfray HCJ (1987) The evolution of clutch size in parasitic wasps. *American Naturalist*, **129**, 221–233.

Godfray HCJ (1994) *Parasitoids: Behavioral and Evolutionary Ecology*. Princeton, NJ: Princeton University Press.

Godfray HCJ & JM Cook (1997) Mating systems of parasitoid wasps. In: JC Choe & BJ Crespi (eds) *The Evolution of Mating Systems in Insects and Arachnids*, pp. 211–225. Cambridge: Cambridge University Press.

Godfray HCJ & Hunter MS (1992) Sex ratios of heteronomous hyperparasitoids: adaptive or nonadaptive? *Ecological Entomology*, **17**, 89–90.

Godfray HCJ & Parker GA (1991) Clutch size, fecundity and parent-offspring conflict. *Philosophical Transactions of the Royal Society of London, series B*, **332**, 67–79.

Godfray HCJ & Shaw MR (1987) Seasonal variation in the reproductive strategy of the parasitic wasp *Eulophus larvarum* (Hymenoptera: Chalcidoidea: Eulophidae). *Ecological Entomology*, **12**, 251–256.

Godfray HCJ & Waage JK (1990) The evolution of highly skewed sex ratios in aphelinid wasps. *American Naturalist*, **136**, 715–721.

Grbić M, Ode PJ & Strand MR (1992) Sibling rivalry and brood sex ratios in polyembryonic wasps. *Nature*, **360**, 254–256.

Green RF, Gordh G & Hawkins BA (1982) Precise sex ratios in highly inbred parasitic wasps. *American Naturalist*, **120**, 653–665.

Hamilton WD (1967) Extraordinary sex ratios. *Science*, **156**, 477–488.

Hardy ICW (1994) Sex ratio and mating structure in the parasitoid Hymenoptera. *Oikos*, **69**, 3–20.

Hardy ICW & Godfray HCJ (1990) Estimating the frequency of constrained sex allocation in field populations of Hymenoptera. *Behaviour*, **114**, 137–147.

Hardy ICW, Ode PJ & Strand MR (1993) Factors influencing brood sex ratios in polyembryonic Hymenoptera. *Oecologia*, **93**, 343–348.

Hardy ICW, Ode PJ & Siva-Jothy M (in press) Mating systems. In: M Jervis (ed) *Insects as Natural Enemies. Practical Approaches to their Study and Evaluation*, 2nd edn. Dordrecht: Kluwer Academic Publishers.

Harvey JA, Corley LS & Strand MR (2000) Competition induces adaptive shifts in caste ratios of a polyembryonic wasp. *Nature*, **406**, 183–186.

Heimpel GE (1994) Virginity and the cost of insurance in highly inbred Hymenoptera. *Ecological Entomology*, **19**, 299–302.

Hunter MS (1993) Sex allocation in a field population of an autoparasitoid. *Oecologia*, **93**, 421–428.

Hunter MS (1999) The influence of parthenogenesis-inducing *Wolbachia* on the oviposition behavior and sex-specific developmental requirements of autoparasitoid wasps. *Journal of Evolutionary Biology*, **12**, 735–741.

Hunter MS & Godfray HCJ (1995) Ecological determinants of sex ratio in an autoparasitoid wasp. *Journal of Animal Ecology*, **64**, 95–106.

Hunter MS & Woolley JB (2001) Evolution and behavioral ecology of heteronomous aphelinid parasitoids. *Annual Review of Entomology*, **46**, 251–290.

Hunter MS, Nur U & Werren JH (1993) Origin of males by genome loss in an autoparasitoid wasp. *Heredity*, **70**, 162–171.

Hunter MS, Rose M & Polaszek A (1996) Divergent host relationships of males and females in the parasitoid *Encarsia porteri* (Hymenoptera: Aphelinidae). *Annals of the Entomological Society of America*, **89**, 667–675.

Ivanova-Kasas OM (1972) Polyembyrony in insects. In: SJ Counce & CH Waddington (eds) *Developmental Systems*, volume 2, *Insects*, pp 243–271. New York: Academic Press.

Kajita H (1989) Mating activity of the aphelinid wasp, *Encarsia* spp. in the field (Hymenoptera: Aphelinidae). *Applied Entomology and Zoology*, **24**, 313–315.

King BH (1993) Sex ratio manipulation by parasitoid wasps. In: DL Wrensch & MA Ebbert (eds) *Evolution and Diversity of Sex Ratio in Insects and Mites*, pp 418–441. New York: Chapman & Hall.

Kuenzel NT (1975) Population dynamics of protelean parasites (Hymenoptera: Aphelinidae) attacking a natural population of *Trialeurodes packardi* (Homoptera: Aleyrodidae) and new host records for two species. *Proceedings of the Entomological Society of Washington*, **79**, 400–404.

Leiby RW (1922) The polyembryonic development of *Copidosoma gelechiae* with notes on its biology. *Journal of Morphology*, **37**, 195–285.

Mayhew PJ (1998) The evolution of gregariousness in parasitoid wasps. *Proceedings of the Royal Society of London, series B*, **265**, 383–389.

Mayhew PJ & van Alphen JJM (1999) Gregarious development in alysiine parasitoids evolved through a reduction in larval aggression. *Animal Behaviour*, **58**, 131–141.

Nadel H (1987) Male swarms discovered in Chalcidoidea (Hymenoptera: Encyrtidae, Pteromalidae). *Pan-pacific Entomologist*, **63**, 242–246.

Nguyen R & Sailer RI (1987) Facultative hyperparasitism and sex determination of *Encarsia smithi* (Silvestri) (Hymenoptera: Aphelinidae). *Annals of the Entomological Society of America*, **80**, 713–719.

Noldus LPJJ, van Lenteren JC & Rumei X (1987) Movement of adult greenhouse whiteflies, *Trialeurodes vaporariorum*, and its relevance for the development of spatial distribution patterns. *Bulletin O.I.L.B/S.R.O.P*, **10**, 134–138.

Ode PJ & Rosenheim JA (1998) Sex allocation and the evolutionary transition between solitary and gregarious parasitoid development. *American Naturalist*, **152**, 757–761.

Ode PJ & Strand MR (1995) Progeny and sex allocation decisions of the polyembryonic wasp *Copidosoma floridanum*. *Journal of Animal Ecology*, **64**, 213–224.

Patterson JT (1919) Polyembryony and sex. *Journal of Heredity*, **10**, 344–352.

Rosenheim JA (1993) Single-sex broods and the evolution of nonsiblicidal parasitoid wasps. *American Naturalist*, **141**, 90–104.

Rosenheim JA & Hongkham D (1996) Clutch size in an obligately siblicidal parasitoid wasp. *Animal Behaviour*, **51**, 841–852.

Salt G (1931) Parasites of the wheat-stem sawfly, *Cephus pygmaeus*, Linnaeus, in England. *Bulletin of Entomological Research*, **22**, 479–545.

Salt G (1961) Competition among insect parasitoids. *Symposia of the Society for Experimental Biology*, **15**, 96–119.

Strand MR (1986) The physiological interactions of parasitoids with their hosts and their influence on reproductive strategies. In: J Waage & D Greathead (eds) *Insect Parasitoids*, pp 97–136. London: Academic Press.

Strand MR (1989a) Oviposition behavior and progeny allocation of the polyembryonic wasp *Copidosoma floridanum* (Hymenoptera: Encyrtidae). *Journal of Insect Behavior*, **2**, 355–369.

Strand MR (1989b) Clutch size, sex ratio and mating by the polyembryonic encyrtid Copidosoma floridanum (Hymenoptera: Encyrtidae). *Florida Entomologist*, **72**, 32–42.

Strand MR & Grbić M (1997a) The development and evolution of polyembryonic insects. *Current Topics in Developmental Biology*, **35**, 121–159.

Strand MR & Grbić M (1997b) The life history and development of polyembryonic parasitoids. In: NE Beckage (ed) *Parasites and Pathogens: Effects on Host Hormones and Behavior*, pp 37–56. New York: Chapman & Hall.

Strand MR, Johnson JA & Culin JD (1990) Intrinsic interspecific competition between the polyembryonic parasitoid *Copidosoma floridanum* and solitary endoparasitoid *Microplitis demolitor* in *Pseudoplusia includens*. *Entomologia Experimentalis et Applicata*, **55**, 275–284.

Taylor PD & Sauer A (1980) The selective advantage of sex-ratio homeostasis. *American Naturalist*, **116**, 305–310.

Verner J (1965) Selection for the sex ratio. *American Naturalist*, **99**, 419–422.

Vinson SB & Iwantsch GF (1980) Host suitability for insect parasitoids. *Annual Review of Entomology*, **25**, 397–419.

Walter GH (1983) 'Divergent male ontogenies' in Aphelinidae (Hymenoptera: Chalcidoidea): a simplified classification and a suggested evolutionary sequence. *Biological Journal of the Linnean Society*, **19**, 63–82.

Walter GH & Clarke AR (1992) Unisexual broods and sex ratios in a polyembryonic encyrtid parasitoid (*Copidosoma* sp.: Hymenoptera). *Oecologia*, **89**, 147–149.

Walter GH & Donaldson JS (1994) Heteronomous hyperparasitoids, sex ratios and adaptations. *Ecological Entomology*, **19**, 89–92.

West SA & Rivero A (2000) Using sex ratios to estimate what limits reproduction in parasitoids. *Ecology Letters*, **3**, 294–299.

West SA, Flanagan KE & Godfray HCJ (1999) Sex allocation and clutch size in parasitoid wasps that produce single-sex broods. *Animal Behaviour*, **57**, 265–275.

Williams JR (1972) The biology of *Physcus seminotus* Silv. and *P. subflavus* Annecke & Insley (Aphelinidae), parasites of the sugar-cane scale insect *Aulacaspis tegalensis* (Zhnt.) (Diaspididae). *Bulletin of Entomological Research*, **61**, 463–484.

Williams JR (1977) Some features of sex-linked hyperparasitism in Aphelinidae [Hymenoptera]. *Entomophaga*, **22**, 345–350.

Williams T (1991) Host selection and sex ratio in a heteronomous hyperparasitoid. *Ecological Entomology*, **16**, 377–386.

Williams T (1996) A test of kin recognition in a heteronomous hyperparasitoid. *Entomologia Experimentalis et Applicata*, **81**, 239–241.

Williams T & Polaszek A (1995) A re-examination of host relations in the Aphelinidae (Hymenoptera: Chalcidoidea). *Biological Journal of the Linnean Society*, **57**, 35–45.

Zchori-Fein E, Roush RT & Hunter MS (1992) Male production induced by antibiotic treatment in *Encarsia formosa* (Hymenoptera: Aphelinidae), an asexual species. *Experientia*, **48**, 102–105.

Chapter 11

Sex ratio control in arrhenotokous and pseudo-arrhenotokous mites

Maurice W. Sabelis, Cornelis J. Nagelkerke
& Johannes A. J. Breeuwer

11.1 | Summary

Some mite species are male-diploid while others are haplodiploid, with haploid males arising from unfertilized eggs (arrhenotoky) or from fertilized eggs via the elimination of the paternal genome at some stage before or during spermatogenesis (pseudo-arrhenotoky). Arrhenotoky confers the advantage to the female of controlling the offspring sex ratio by controlling the fertilization process. It is now well established that sex ratio control is also possible under pseudo-arrhenotoky. However, it is still not known how diplodiploidy affects the possibilities for sex ratio control.

Shifts in the offspring sex ratio have been demonstrated in relation to density, food availability and mating delays. This ability to control the sex ratio can be an adaptive trait when population mating structure varies. Theory based on single-generation mating groups predicts a female bias and can give qualitatively correct predictions, but in several cases sex ratios are more female biased than these predictions. It is argued that mites often show complex population mating structures, such as local multigeneration populations that are themselves subdivided into single-generation mating groups. This creates selection at various hierarchical levels and it is shown theoretically that these additional selection levels can create a stronger female bias in the offspring sex ratio. Before accepting this explanation, more critical tests are needed under different population mating structures.
Mites are ideal objects for such studies because their population mating structures vary greatly.

It is becoming increasingly clear that the role of endosymbionts, such as *Wolbachia*, has to be taken into consideration in order to understand mite sex ratios. Such endosymbionts are widespread in mites and can have various effects on sex ratios via male-killing, feminization, thelytoky and cytoplasmic incompatibility.

11.2 | Introduction

Mites exhibit great diversity in genetic systems (Norton *et al.* 1993). Apart from both sexes being diploid, there are various forms of male haploidy, differing in whether males arise from unfertilized eggs (arrhenotoky) or from fertilized eggs (pseudo-arrhenotoky) (Figure 11.1). Parthenogenetic production of females (thelytoky) occurs sporadically in several mite families, but some taxa exclusively consist of thelytokous forms (Norton *et al.* 1993). The genetic details underlying these various systems have been little explored, yet hold the promise of finding exciting new phenomena. For example, pseudo-arrhenotokous plant-inhabiting predatory mites appear to eliminate the paternal chromosome set before spermatogenesis (Helle *et al.* 1978, Hoy 1979), yet partially retain diploidy somatically (Perrot-Minnot *et al.* 2000); and earlier claims for the unique case of haploid parthenogenesis in a false spider mite (Pijnacker *et al.* 1981) have

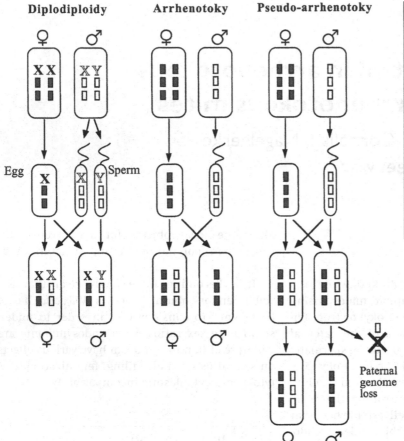

Diplodiploidy   **Arrhenotoky**   **Pseudo-arrhenotoky**

Egg

Sperm

Paternal
genome
loss

Fig 11.1 Three different genetic systems. Under diplodiploidy both sexes are diploid and arise from fertilized eggs. Arrhenotoky and pseudo-arrhenotoky share the property that males are effectively haploid as they transmit the maternal genome only. Under arrhenotoky males arise from unfertilized eggs and females from fertilized eggs, whereas under pseudo-arrhenotoky both sexes arise from fertilized eggs but males become haploid after inactivation and/or elimination of the paternal chromosome set (Sabelis & Nagelkerke 1998).

recently been corroborated (Weeks *et al.* 2001). Mites are therefore excellent for testing hypotheses on why taxa differ in genetic systems. Why do many species reproduce sexually and others by thelytoky? Why does the parthenogenetic production of males occur in some taxa and paternal genome loss in males in others? Whereas insects harbour much the same variety in genetic systems and prompt similar questions, mites are the ideal case for independently testing insect-inspired hypotheses on the evolution of genetic systems.

Assuming that diplodiploidy and sexual reproduction is the ancestral state, the central question is whether natural selection promotes male haploidy (arrhenotoky or pseudo-arrhenotoky), male rarity (spanandry) or even male absence (thelytoky). One answer is that females have 'the final say' over paternal inheritance and increase their genetic representation among the offspring by preventing males from

transmitting genes to their sons. If this were true, however, one wonders why these systems are not ubiquitous. Another potential answer is that male haploidy confers the advantage of sex ratio control to the mother. While this may be true under arrhenotoky, because females may control the access of sperm to unfertilized eggs, it is not at all clear why male haploidy is also manifested as pseudo-arrhenotoky; when both males and females arise from fertilized eggs, sex ratio control by egg fertilization is impossible. The exciting solution is that females of pseudo-arrhenotokous mites are able to control the ratio of females to males, just as arrhenotokous mites do (Sabelis 1985b, Sabelis & Scholman 1989, Nagelkerke & Sabelis 1991, Sabelis & Nagelkerke 1987, 1988, 1993, Nagelkerke & Sabelis 1998). They somehow exert control over the elimination of the paternally derived chromosome set. Thus, in principle there is no reason to infer that arrhenotoky confers a selective advantage over pseudo-arrhenotoky in terms of the ability to regulate the number of each sex produced.

Probably, the pivotal difference between the two types of male haploidy is manifested at very low population density. Suppose populations are occasionally severely reduced by predators, parasites, climatic catastrophes or lack of food. This may cause female survivors to become isolated. Under pseudo-arrhenotoky a virgin female would not be able to reproduce since fertilization is required to produce either sex. Under arrhenotoky, however, virgin females are able to produce sons first and mate with them, thus restoring the mother's capacity to reproduce (Sabelis & Nagelkerke 1987, 1988, 1993, Adamson & Ludwig 1993). Thelytokes also profit from being able to reproduce when isolated, but unlike arrhenotokes they have lost the possible advantages of sexual reproduction.

Thus, arrhenotoky is expected when females run a substantial risk of becoming isolated, e.g. due to severe population crashes or to single juveniles colonizing new food patches, whereas pseudo-arrhenotoky or diplodiploidy is expected when this risk is sufficiently low. Pseudo-arrhenotoky confers an advantage over diplodiploidy when females gain by controlling which sex to produce. However, if the risk of isolation is sufficiently low and the gain of sex ratio control sufficiently high, arrhenotoky and pseudo-arrhenotoky are equivalent unless there is a special advantage of initial (or exclusively somatic) retention of the paternal genome in sons (Sabelis & Nagelkerke 1988, 1993). The challenges in using mites as test objects for theories on the evolution of genetic systems are thus: (1) to elucidate the net advantages of retaining the paternal genome in sons (yet eliminate it before or during spermatogenesis), (2) to understand their life cycle in the field sufficiently to estimate the female's risk of becoming isolated, and (3) to assess the abilities and advantages of sex ratio control under realistic ecological conditions.

While meeting these challenges will provide new insight in the selective advantages to the individuals endowed with a given genetic system, it leaves unanswered how these systems evolved in the first place. We will first discuss the evolutionary origins in some detail and then embark on the three questions listed above. Finally, we will review how vulnerable each of the genetic systems is to parasitic manipulation. In particular, we will review what is known about parasites that manipulate the host's sex allocation

to their own advantage rather than that of their host.

11.3 | Evolutionary origins

Diplodiploidy prevails in ticks and (with some exceptions) in the oribatid mites, but co-occurs with arrhenotoky in several other higher taxa, such as the Prostigmata, the Astigmata and the Mesostigmata (Norton et al. 1993). Schrader and Hughes-Schrader (1931) suggested that arrhenotoky did not arise directly from a diplodiploid ancestor, but rather that it evolved through a series of intermediate steps involving pseudo-arrhenotokous systems. They recognized three character state changes in the evolution of arrhenotoky from a zygogenetic diplodiploid ancestor: (1) a switch in genetic system from ancestral diplodiploid zygogenesis to diplodiploid pseudo-arrhenotoky, (2) a switch in karyotype from diplodiploidy to haplodiploidy under pseudo-arrhenotoky and (3) a switch from haplodiploid pseudo-arrhenotoky to arrhenotoky. This evolutionary scenario involves ongoing switches from one to the other genetic system. Hence it becomes plausible when evolutionary intermediates are extant today. For mites the scenario holds because all known examples of pseudo-arrhenotoky occur in groups that also include arrhenotoky. Mesostigmatic mites are a striking example in that diplodiploidy (e.g. Parasitidae), pseudo-arrhenotoky (Phytoseiidae) and arrhenotoky (Macrochelidae, Ascidae, Laelapidae, Macronyssidae) all occur. In one arrhenotokous mesostigmatic family, the Laelapidae, two species have a heterochromatic chromosome arm, which may be interpreted as the vestige of the heterochromatized complement of a pseudo-arrhenotokous ancestor (de Jong et al. 1981). Cruickshank and Thomas (1999) mapped these genetic systems onto a molecular phylogeny using approximately 750 base pairs of 28rDNA and showed that arrhenotokous members of the mesostigmatic mites form a clade that arose from a pseudo-arrhenotokous ancestor, rather than from a diplodiploid one. This represents the first test and the first support for Schrader and Hughes-Schrader's (1931) hypothesis. Further tests with mites await further elucidation of genetic systems in some groups, as pointed out by Norton *et al.* (1993).

Mesostigmatic mites are thought to have a common ancestry with ixodid ticks, which have an XX-XO sex-determination system (Norton *et al.* 1993). How can haplodiploidy arise from such a sex-determining mechanism? Haig (1993a,b) developed a plausible evolutionary scenario for inactivation of paternal chromosomes in fertilized eggs: (1) meiotoic drive by the X chromosome in XO males causes female-biased sex ratios, (2) the maternal set of autosomes in males evolves effective sex linkage to exploit X drive and (3) genes expressed in mothers are selected to convert some of their XX daughters into sons. A similar scenario may well have led to the evolution of haplodiploidy.

Pseudo-arrhenotoky may be evolutionarily less stable than arrhenotoky. Bull (1983) pointed out that under pseudo-arrhenotoky females increase the transmission of maternal genes at the expense of the would-be father and that there is selection favouring genes that, when transmitted through sperm, avoid this elimination. Under arrhenotoky sperm has no access to sons, but under pseudo-arrhenotoky sperm penetrates eggs that will become female and eggs that will become male. Thus, any paternal mutant that avoids paternal genome loss in sons (and/or triggers the mothers to produce more daughters) is favoured. This renders pseudo-arrhenotoky more susceptible than arrhenotoky to revert to diplodiploidy. This argument may be especially important immediately after the origin of pseudo-arrhenotoky. However, this is not a satisfactory answer because it is equally valid to argue that under arrhenotoky there will be selection on paternal genes coding for any ploy to circumvent the mechanisms by which mothers prevent eggs from becoming fertilized.

Why do we still observe pseudo-arrhenotokous intermediates if they are unstable? The Phytoseiidae, a family of mainly plant-inhabiting predatory mites, are probably all pseudo-arrhenotokes, since all species share the property that females only reproduce after insemination and males of all species studied so far are haploid. The problem is that pseudo-arrhenotoky is not easily assessed. Arrhenotoky can be inferred

from the simple observation that uninseminated females produce sons only. Usually the unfertilized eggs are haploid, but cytological evidence is needed to rule out the possibility of diploid arrhenotoky (through fusion of the first two haploid cleavage nuclei). Thus, while providing more insight, cytological and genetic evidence is not a prerequisite for proving arrhenotoky. Under pseudo-arrhenotoky, uninseminated females do not lay eggs, but this also holds for diplodiploids and organisms reproducing by pseudogamy (sperm triggers reproduction without fertilization). Thus, it is more laborious to prove pseudo-arrhenotoky, because it has to be inferred from both genetic and cytological evidence, confirming that males develop from fertilized eggs, that they inactivate one set of chromosomes (by heterochromatization and/or elimination), that the genome lost or inactivated is of paternal origin and that the genome transmitted by fathers to their daughter's offspring is of maternal origin. Probably due to the considerable amount of work required, several potential cases of pseudo-arrhenotoky are as yet insufficiently investigated.

For plant-inhabiting predatory mites belonging to the family Phytoseiidae, there is cytological evidence of chromosome loss in males, as well as genetic evidence of males transmitting exclusively the maternal genome (Schulten 1985). Moreover, the joining of the two pronuclei has been observed in the first fertilized egg produced by a phytoseiid female (Toyoshima & Amano 1999a,b, Toyoshima *et al.* 2000). Since first eggs usually develop into a male, this is good evidence that males indeed start their life as diploid. Paternal genome loss occurs early in embryogenesis, but there is evidence that the paternal genes are somatically expressed in at least some tissues (Perrot-Minnot & Navajas 1995, Perrot-Minnot *et al.* 2000). Both cytological and genetic evidence for paternal genome loss is available for only one species, i.e. *Typhlodromus occidentalis* (Hoy 1979, Nelson-Rees *et al.* 1980, Roush & Hoy 1981). For two other species, *Phytoseiulus persimilis* and *Amblyseius bibens*, there is only genetic evidence for paternal genome loss (Helle *et al.* 1978). Paternal genome loss in males may well be widespread among the phytoseiid mites.

In all species studied so far, males are haploid, daughters and sons are produced by inseminated females only, and males transmit exclusively the maternal genome (Schulten 1985, Congdon & McMurtry 1988, Perrot-Minnot *et al.* 2000).

However, paternal genome loss in sons may have evolved more than once in mites. First, there are a few observations on chromosome loss during embryogenesis, such as in the parasitic mite *Dermanyssus gallinae* (Warren 1940). However, it is not yet known whether the elimination concerns paternally derived chromosomes, or whether it occurs in male embryos. Second, it is well known that in some families of mites males are haploid, while uninseminated females do not lay eggs (Nelson-Rees *et al.* 1980). This also applies to the parasitic mite *D. gallinae*. Other 'potential candidates for pseudo-arrhenotoky' are the Cunaxidae (Prostigmata), Brachypylina (Oribatida) and Histiostomatoidea (Astigmata) (Norton *et al.* 1993). Thus, it is premature to state that pseudo-arrhenotoky with currently three known origins among the invertebrates (fungal gnats, scale insects and phytoseiid mites) has evolved significantly less frequently than arrhenotoky with 11 known origins (Bull 1983).

11.4 | Selective advantages of arrhenotoky versus pseudo-arrhenotoky

According to Bull (1979, 1983), one selective advantage favouring male haploidy stems from a twofold representation of maternal genes in gametes of haploid sons in comparison to diploid sons of biparental origin. The probability of gene-identity-by-descent between grandmother and grandchild through uniparental sons is therefore double the probability of that through biparental sons. This twofold advantage of producing uniparental sons may overcome the potential lower fitness of these sons and may therefore be the key to our understanding of the evolution of male haploid systems. However, the twofold advantage applies equally well to both pseudo-arrhenotoky and arrhenotoky and thus the question remains as to why pseudo-arrhenotoky is found in some cases and arrhenotoky in others.

Following Bull (1983), a second advantage of arrhenotoky is that it allows the mother to control the sex of her offspring by influencing the fertilization of each egg. Hence, there is a flexible mechanism in arrhenotokous organisms that can change the sex ratio of the offspring in an adaptive way whenever investment in one sex is more profitable than investment in the other. However, this ability to control the sex ratio is now known to occur in pseudo-arrhenotokous phytoseiid mites as well (Sabelis & Nagelkerke 1988, 1993). How the females manage to induce paternal chromosome inactivation (heterochromatization) in eggs destined to become males is not known, but they appear to modify sex ratios with great precision (Nagelkerke & Sabelis 1998). Sex ratio control is therefore unlikely to provide an explanation for why one species reproduces by arrhenotokous parthenogenesis and another by pseudo-arrhenotoky. However, it still has to be shown that the reaction time of these two control mechanisms is the same. Hymenopterous parasitoids can decide upon offspring sex just before egg laying. Whether pseudo-arrhenotokous organisms have the same potential remains to be shown; their reaction time will largely depend on how long before egg laying the eggs are predisposed to undergo paternal genome loss.

The third advantage of arrhenotoky is that virgin females colonizing uninhabited sites can produce sons, whereas virgin females incapable of parthenogenesis cannot. If her sons mature before she ceases reproduction, they can mate with her so that she produces daughters and establishes a population. Arrhenotoky might evolve for this reason, although so would any other form of parthenogenesis. However, under diplodiploidy as well as pseudo-arrhenotoky eggs of either sex must be fertilized, so that virgin females cannot produce offspring unless they find a male produced by another inseminated female. Clearly, a low density of mates selects for arrhenotoky at the expense of pseudo-arrhenotoky. Thus, pseudo-arrhenotokous organisms are only expected when the chance of remaining unmated is virtually zero. For phytoseiid mites this may well be the case. They tend not to disperse aerially until local prey populations are virtually eliminated and local populations of phytoseiid predators are large, so that males are amply available. Another argument in support of a high mating chance is that females tend not to embark on long-distance dispersal until after mating; before mating, they stay near to the natal patch where population build-up ensures sufficient males to be around. Selection for arrhenotoky will thus be weak unless there are factors causing sudden crashes of phytoseiid populations. So far, however, there is little or no evidence that hyperpredators or pathogens cause such crashes in the field. Rain and wind can reduce population levels, but not catastrophically and it remains to be shown whether the extent of population reduction affects mate finding. In conclusion there seem to be good reasons to assume that under natural conditions phytoseiids have a low chance of remaining unmated.

It is interesting to note that the most important prey species of phytoseiids, i.e. tetranychid, tarsonemid and eriophyoid mites, all reproduce by arrhenotoky (Helle & Sabelis 1985, Lindquist et al. 1996). In their populations catastrophic crashes frequently occur, not because of weather-related factors, but because of pathogens and predators, especially acaropathogenic fungi and phytoseiid mites (Sabelis & van der Meer 1986, Sabelis et al. 1991, Pels & Sabelis 1999). Thus, if diploid eggs or juveniles of these prey species happen to escape from infection or predation, these individuals would greatly profit from the fact that they can eventually be inseminated by their own sons. This property can be considered as a preadaptation to factors, such as pesticide applications, that have become important in causing population crashes more recently in evolutionary history. However, under these conditions pseudo-arrhenotoky in phytoseiid mites represents a maladaptation. This provides a possible explanation for the fact that phytophagous mites, such as tetranychid mites, develop pesticide resistance at a faster rate than phytoseiid mites.

There must be advantages associated with initial retention of the paternal genome, large enough to outweigh the disadvantages at low densities. One advantage may accrue from the masking of errors in the maternal genome

during the period of paternal genome retention, but so far there is no available evidence to support this. Another advantage may accrue from double-strand repair of the maternal genome via copies of the equivalent part of the paternal genome. Errors in the maternal genome might be fatal, especially to the very early stages of embryonic development. The paternal genome is unlikely to contain the same errors and may therefore be suitable for reinstalling lost genetic information. Most interestingly, Nelson-Rees et al. (1980) observed that the pairing and opening of homologous chromosomes occurs in phytoseiid mites just before the paternal genome is eliminated. They stress that this is much like a diplotene event in meiosis. Meiotic repair is therefore a distinct possibility. However, using polymorphic codominant loci detected by direct amplification of length polymorphism (DALP), Perrot-Minnot et al. (2000) did not observe recombination between parental genomes prior to elimination of the paternal chromosomes. More work in this direction is needed to reach a definitive conclusion. One possible advantage of retaining the paternal genome is suggested by the experimental result that exposure of male phytoseiids to radiation causes reduced fertility in their sons (Helle et al. 1978). Since sons transmit their mother's genome only, one would not expect an influence of their father, unless (early or somatic) retention of the paternal genome in sons exerted a positive effect on their fitness.

Although retention of the paternal genome may have the above two advantages, at some stage during development this genome is eliminated. Two advantages of the elimination have been discussed here at length: by eliminating the paternal genome the mother increases genetic identity to her offspring and, if the elimination is somehow involved in sex determination, then it enables sex ratio control. But there may be several other advantages to paternal genome loss in sons. When realizing that the elimination never occurs immediately after zygote formation, but only after at least several mitotic divisions have taken place, it follows that the risk of fatal errors is spread over many cells. The malfunctioning of some cells may then be overcome by cell selection and replacement. In that case scarce nutrients needed for DNA synthesis can be reallocated. Moreover, because meiosis requires more time than mitosis (Lewis 1983), spermatogenesis may proceed faster. In addition, as suggested by Cavalier-Smith (1978, 1985), a smaller inventory of DNA may in turn imply a faster mitotic cycle time. A somewhat higher rate of development (relative to females) may be selectively advantageous in mate competition among males, especially under conditions of group-structured mating.

Exactly when the loss of the paternal chromosomes occurs during the life cycle may depend on the species under study, but in mites this has not been studied in sufficient detail. In some insect taxa, somatic tissues retain the paternal chromosome in an active or inactive state, whereas it is eliminated in the germline just before or during spermatogenesis. In other insect taxa, chromosome loss occurs early in embryogenesis, but not until after several mitotic divisions have taken place. Taken together, several questions have to be answered before we can understand the selective forces that have led to paternal genome loss in males: (1) why does the elimination concern the paternal genome, (2) why does it occur in males only, (3) why does the elimination take place at various moments between embryogenesis and spermatogenesis and (4) why do some organisms retain the paternal chromosome in somatic tissues while eliminating it from the germline? These questions are important challenges for future research.

11.5 | Population mating structure and the evolution of sex-allocation strategies

Mites exhibit a wide variety of population mating structures and there is much theory to show that these structures are important in determining the evolutionary stable sex ratio. Take plant-inhabiting mites as an example. Population-wide mixing is rather unlikely in these mites, because they are wingless and have restricted walking capacity. Moreover, local populations are probably founded by one or a few individuals, especially because long-distance dispersal by aerial means or phoresy (Sabelis & Dicke 1985) is extremely risky. Plant-inhabiting mites also have short generation times, which may result in more generations being spent on a single resource patch before they leave. As their population growth is fast compared to immigration and resources are limited, the resulting local populations develop internal relatedness. They cover several generations, yet last for a limited amount of time because of limited resources. Consequently, these local populations develop internal relatedness. This population structure observed in plant-inhabiting mites conforms to the assumptions underlying the so-called 'haystack' model of Maynard Smith (1964) (Figure 11.2), originally developed to understand the evolution of altruism but later extended to cover the evolution of sex allocation (Wilson & Colwell 1981, Sabelis & Nagelkerke 1993, Nagelkerke & Sabelis 1996). Low mobility within local populations may result in substructure within the haystacks and this may give rise to selection at the level of local mating groups within haystacks and at the level of haystacks within the population at large (Figure 11.2). The consequences of selection at more than one level have been explored theoretically (Frank 1986, 1987, Nagelkerke & Sabelis 1996), yet await definitive empirical tests to assess the role of multilevel selection. Mites are excellent test objects for these models.

Plant-inhabiting herbivorous mites are examples of 'haystack' population structures (Nagelkerke & Sabelis 1996). Herbivorous mites in the genus *Tetranychus* form high-density local populations covering one or more neighbouring host plants and these are in turn subdivided into mite colonies on different leaves. The mite colonies themselves may also show substructure because: (1) females oviposit where they feed, (2) they move very little, (3) their eggs consequently aggregate and (4) the emerging juveniles stay very near to where they developed. Consequently, there are likely to be local mating groups within colonies on leaves, although the borders of each group are vague. The local populations of herbivorous mites are initiated by few foundresses and may go through several generations before they crash, because of

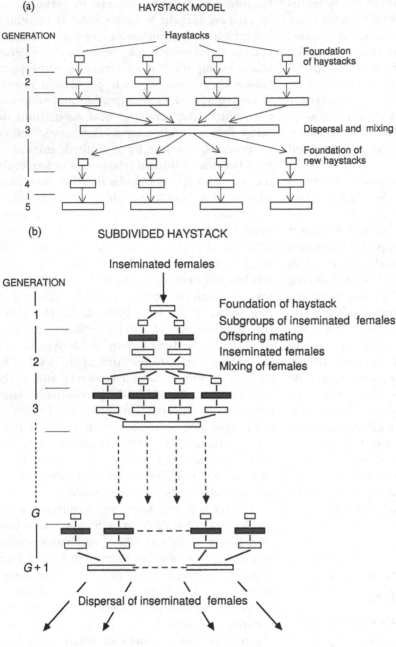

(a) HAYSTACK MODEL

GENERATION

Haystacks

1

2

3 — Dispersal and mixing

Foundation of haystacks

Foundation of new haystacks

4

5

(b) SUBDIVIDED HAYSTACK

Inseminated females

GENERATION

1 — Foundation of haystack
Subgroups of inseminated females
Offspring mating
Inseminated females
Mixing of females

2

3

G

G+1

Dispersal of inseminated females

Fig 11.2a,b Diagram of two multigeneration population structures implying selection at both the individual and the group level. Individual selection concerns the effect of a trait on the number of offspring of the actor, whereas group selection concerns the differential production of dispersers by groups of related individuals. (a) Haystacks. Multigeneration groups founded by a few individuals and expanding for a number of generations, after which population-wide dispersal takes place to start a new cycle of haystacks. (b) Example of a substructured haystack. Each haystack is started by a few females and lasts for G breeding generations, after which the inseminated females from the $(G + 1)$th generation disperse. In every generation inseminated females mix first and then found local mating groups.

If the haystack structure did not impose an additional selection level, then the females bias would only be due to the single-generation mating groups. However, when host plant overexploitation approaches, the spider mites become more mobile and hence homogeneously mix within the haystack, which in turn favours a less female-biased sex ratio. Thus, the strong female biases observed suggest an important role for selection under a haystack structure.

food exhaustion, pathogens or predators. Promising support for haystack models is provided by the extreme female bias found in broods of the spider mite *Oligonychus pratensis* (Stiefel & Margolies 1992) on a host plant nearing overexploitation. This stronger female bias in the last generation is exactly what is predicted by (subdivided) haystack models, provided the spider mites are capable of flexible sex allocation over generations (Nagelkerke & Sabelis 1996).

Another example of, possibly substructured, haystacks is phytoseiid predators in the genus *Phytoseiulus*, which prey on herbivorous mites in the genus *Tetranychus* and are specialized to overcome their prey's defences. One or a few inseminated females of these predatory mites invade local populations of the herbivorous mites. They consume prey mites and convert much of the ingested food (ca. 70%) into eggs which they deposit in the colonies of the herbivorous mites. In this way, they found a local population that expands rapidly for a number of generations until all prey is wiped out. Inseminated females then disperse to search for new local populations. Within infestations, the mobility of egg-laying females and of juveniles is low due to high prey density, and hence the local populations of predatory mites may be subdivided into local mating groups within colonies. Optimal sex allocation within such a substructured haystack will depend on the number of foundresses, the total number of generations, current generation number, the amount of local density dependence and the flexibility of sex allocation over generations (Bulmer & Taylor 1980, Wilson & Colwell 1981, van Tienderen & de Jong 1986, Frank 1986, 1987, Nagelkerke & Sabelis 1996).

Most interestingly, specialist phytoseiid predators of spider mites exhibit such a strong female bias that it cannot be explained by the existence of local single-generation mating groups alone. The addition of a multigeneration haystack structure to the level of single-generation mating groups has been shown to provide a potential explanation for this extreme bias (Nagelkerke & Sabelis 1996). However, many empirical questions need a more precise answer. For example, we need more information on the number of foundresses and their immigration pattern over time, on the migration of males between local mating groups within local populations and between local populations, on the dynamics of local mating groups (role of movement patterns, mate choice, etc.) and on the ability to adjust sex allocation to current conditions. All this information is necessary to determine the relative importance of selection at the haystack level (local population) compared to the level of the substructure (local mating group) (Nagelkerke & Sabelis 1996). There is also much need to extend the theory by considering: (1) the full feedback loop in predator–prey interactions and the resulting dynamics of invasions into local populations (instead of a fixed number of foundresses invading simultaneously), (2) overlapping generations, (3) the migration of males within and between local populations and (4) dynamic changes in mating structure.

The population mating structure of specialist phytoseiid mites feeding on herbivorous mites

in the genus *Tetranychus* is just one extreme in a continuum of possibilities. Generalist phytoseiid predators are probably near the other extreme. These have more stable local populations, with continuous migration between them, and they seem to exhibit more local mobility within local populations (e.g. McMurtry & Croft 1997). These population structures, termed 'islands' in the group selection literature, would in theory give rise to much lower female bias. Whereas such a lower female bias is indeed observed in generalist phytoseiid mites (Sabelis 1985b, Sabelis & Nagelkerke 1987, 1988, 1993, Sabelis & Janssen 1993, Nagelkerke & Sabelis 1996), the biological details of oviposition behaviour and mate choice may crucially matter as to whether 'haystacks' or 'islands' are substructured into local mating groups or not. For example, the generalist phytoseiid predator *Iphiseius degenerans* feeds on pollen and flower thrips in flowers, but lays its eggs in hair tufts near veins on leaves. Females develop and lay one egg at a time so that they have to commute between flowers and leaves. This would normally lead to mixing of the eggs within local populations, but females tend to return to the same oviposition sites and to lay their eggs next to each other (Faraji *et al.* 2000). This mechanism of kin recognition leads to clusters of strongly related eggs and this would normally promote sib-mating once matured. However, the individuals emerging from eggs in a cluster tend not to stay together during their development, possibly because of the risk of cannibalism, and this may create a tendency towards random mating. This example nicely illustrates that one has to study the biological details quite precisely before deciding on which population structure prevails in a given species.

Apart from the above examples of tetranychid and phytoseiid mites, there are many other mite systems that exhibit various interesting kinds of population structure. Some parasitic mites in the genus *Pyemotidae* exhibit physogastry, a phenomenon whereby the body increases tremendously in size due to the numerous offspring developing inside. Typically one descendant develops into a male that mates with all of his sisters either inside their mother's body or once they emerge from it (Kaliszewski & Wrensch 1993). Usually, many of these parasites are found

Table 11.1 | Methods for assessing the two sexes in offspring of mites and notes on their suitability

Methods	Reference
Morphological evaluation of sex differences in adult phase	Helle & Sabelis 1985
Possible bias due to sex differential mortality	
Morphological evaluation of sex differences in nymphal phase[1]	Wrensch & Johnston 1983
Less dependent on sex differential mortality	
Sex morphology of juveniles not known for all mite taxa	
Counting number of chromosomes after aceto-orceïn treatment	Helle & Sabelis 1985
In haplodiploids males have n and females $2n$ chromosomes,	
but assessment using orceïn treatment is quite laborious	
Weighing eggs using an electrobalance	MW Sabelis, unpubl.[2]
Egg size or weight differs between sexes,	
but overlap in size/weight distributions is considerable	

[1] Developed for deutonymphs of Parasitidae (Acari, Mesostigmata) by Wrensch & Johnston (1983).

[2] In *Phytoseiulus persimilis* eggs destined to become males weigh 4.32 μg (SD = 0.59), whereas eggs destined to develop into females weigh 4.97 μg (SD = 0.44).

together on their host (ant nests, small mammals) and one may wonder to what extent vagrant males may steal potential mates from broods of females other than their mother. The local population of parasitic mites may well represent another level of population structure relevant to understanding sex-allocation strategies. Other potential examples of haystack-like groups may be found among the feather mites that live in the shafts of bird feathers (Kethley 1971), various tarsonomid mites phoretic on insects (Kaliszewski & Wrensch 1993), vagrant and gall-forming eriophyoid mites that feed on plants (Sabelis & Bruin 1996), hemisarcoptid mites phoretic on coccinellid beetles (Izraylevich & Gerson 1995, 1996), spider mites that create family nests within a web (Saito 1990a,b, 1994, 1995, 2000, Saito & Sahara 1999, Mori *et al.* 1999, Saito *et al.* 2000a,b), *Varroa* mites that parasitize bee larvae (Boot *et al.* 1995), laelapid predatory mites that live in the soil (Ruf 1993, Lesna & Sabelis 1999), antennophorid mites that parasitize ants (Franks *et al.* 1991) and flower mites that are vectored by hummingbirds (Wilson & Colwell 1981).

11.6 | Sex ratio control

Using various technical procedures (Table 11.1), sex ratio control has been investigated in detail in tetranychid mites (Sabelis 1985a, Wrensch 1993, Young *et al.* 1986) and phytoseiid mites (Sabelis 1985b, Sabelis & Nagelkerke 1987, 1988, 1993, Nagelkerke & Sabelis 1998). For both mite taxa, the sex ratio usually shifts from a female bias to equality when the number of foundresses (= density) increases, and also when food availability decreases. This is expected from sex-allocation theory because both conditions promote population mixing. However, the sex ratio response to delayed first insemination depends strongly on the mite taxa under consideration, and in particular on the genetic system. In tetranychid mites, mating delay causes unusually female-biased sex ratios after insemination (Krainacker & Carey 1990), whereas in phytoseiid mites it causes unusually male-biased sex ratios in the offspring produced after mating (Momen 1994). Note that the arrhenotokous tetranychids continue to produce sons before mating, whereas the pseuo-arrhenotokous phytoseiid mites cannot reproduce before mating. Possibly, the delay of mating induces arrhenotokes to overproduce daughters because sons produced before mating will soon mature, whereas it induces pseudo-arrhenotokes to overproduce because mating delay signals male scarcity. Taken together, all these shifts in offspring sex ratios in response to external and internal conditions demonstrate an ability to control sex allocation. To date a range of techniques has been applied.

It is not easy to analyse whether the sex ratios produced in response to environmental conditions (prey density, predator density) are near-optimal decisions. Mites often have viscous populations with vague borders between egg clutches and resulting mating groups (Nagelkerke & Sabelis 1998). This is a result of a combination of limited mobility, resources that are less concentrated than a host is to the parasitoid and oviposition that occurs sequentially and at intervals instead of laying one batch of eggs at a time. The exact degree of mixing is therefore difficult to assess. Since sex-allocation theory generally assumes the existence of distinct clutches and mating groups, there is need for a theory that abandons the clutch concept and takes local mating group dynamics in space and time into account.

The absence of clearly demarcated mating groups also calls for unconventional methods for investigating the precision of sex allocation (Nagelkerke & Sabelis 1998). Standard methods are all clutch based and measured precision depends on the scale (e.g. clutch size or time window) of analysis. A start has been made to remedy this by analysing a range of scales along a continuum and by searching for scale-independent measures of precision (Nagelkerke & Sabelis 1998). Phytoseiid mites apparently produce very precise sex ratios over longer scales, maybe maximally so over their whole lifetime; but the ability of control over short time scales is still not clear. However, rather regular son-daughter sequences in clutches of phytoseiid mites have been observed (Toyoshima & Amano 1998).

Observations on phytoseiid mites have prompted the development of game theoretical models that combine the optimization of clutch size and sex allocation (Nagelkerke 1994, Nagelkerke et al. 1996), two problems that until then had been considered separately. The question of whether a female should deposit eggs where a clutch of another female is already in place (so-called 'double oviposition') turned out to be strongly influenced by sex-allocation decisions, depending, among others, on reproductive asynchrony between clutches and male migration between patches. When 'double oviposition' occurs, sex-allocation strategies can have a large influence on clutch size and hence on the number of patches visited by a female. The reason for this is that by adding a relatively small but male-biased clutch to the female-biased clutch of another, a female can obtain greater fitness per egg (via her son's mating with the daughters of the unrelated female) and these gains have to balanced against the costs of travelling to other sites. Sex-allocation strategies can therefore be an important driving force to distribute eggs over more patches. Predatory mite females indeed disperse their eggs over many patches even when there is an ample supply of prey. Integration of sex-allocation and egg-distribution games into one modelling framework has therefore provided new insights into the ecology of phytoseiid mites.

Mites have also inspired research to extend sex-allocation theory beyond clutch size approximation by continuous variables: eggs of phytoseiid mites are large relative to female body weight (ca. 20%) and phytoseiid females lay only a few (one to five) eggs per day. This calls for a discrete approximation of clutch sizes. When analysing sex-allocation behaviour it is therefore preferable to construct models with integer-sized clutches. Such models have been developed for the case of local mate competition, more than one foundress and sex allocation that has nonzero variance in sex allocation (Nagelkerke & Sabelis 1996). The predictions differed from those of standard continuous models; for instance, polymorphic sex-allocation strategies were possible. Discrete models have also been applied to investigate when phytoseiid females should add an egg (probably a son) to the egg clutch of another female and when they are expected to avoid this. It appeared that egg size was critical to the predictions emerging from the model. Thus, despite sex-allocation models becoming more cumbersome by the inclusion of the assumption of discrete eggs, this seems worthwhile in terms of the new insights gained (a message also emphasized by Greeff 1997, discussing combined clutch size and sex-allocation models for fig wasps).

Mate choice and kin recognition may play an important role in shaping the mating structure of mite populations. Within multigeneration groups relatives may meet and have the opportunity to mate with each other or share oviposition sites. Because optimal sex allocation (and clutch size) may depend on relatedness, the question arises as to whether mites adjust their sex-allocation and oviposition behavior to their interactants. Females of herbivorous mites in the genus *Tetranychus* indeed adjust their progeny sex ratio to the relatedness with neighbouring females (Roeder *et al.* 1996). Hence, they can evidently discriminate between those who are kin and those who are not. Kin recognition has also been found in phytoseiid mites, albeit exclusively in relation to where they oviposit (Faraji *et al.* 2000) or to cannibalism (P Schausberger, pers. comm.). Also there is good evidence from tetranychid mites and laelapid mites that mating is not random when given a choice (Gotoh *et al.* 1993, Lesna & Sabelis 1999). Mate choice in two selected prey preference lines in a laelapid predatory mite can even switch from assortative to disassortative depending on whether the hybrids between the lines are reproductively inferior or superior (Lesna & Sabelis 1999). Kin recognition and mate choice will obviously alter the population mating structure, but their role in sex-allocation strategies in mites is yet to be demonstrated.

11.7 | Sex ratio distorters

Control over sex allocation and genetic systems of sex determination are often seriously compromised by heritable micro-organisms transmitted from females to progeny (Chapter 9). These micro-organisms are maintained because they manipulate host reproduction, promoting vertical transmission via the increased production of daughters at the expense of sons. Obviously, the presence of such heritable sex ratio distorters (SRDs) will have great impact on the evolution of the host and reciprocal interactions between hosts and microbial SRDs. Information on SRDs in mites is limited to medically or agriculturally important species (van der Geest *et al.* 2000). They provide an ideal opportunity to assess how

they influence the success of various genetic systems, and to investigate whether the host is an innocent bystander or evolves in response to the SRD.

Currently four classes of microbe-associated sex ratio distortions are known: male-killing, feminization, thelytoky and cytoplasmic incompatibility (Chapter 9). The most notable SRD is the symbiont bacterium *Wolbachia*, which is involved in all these aberrant sex ratio traits. Recent molecular surveys for *Wolbachia* indicate that many mites are infected (Breeuwer & Jacobs 1996, Johanowicz & Hoy 1996, 1998a,b, 1999) and can cause female mortality (Breeuwer 1997, Vala *et al.* 2000). We expect that examination of intraspecific variation and the inheritance of sex ratio in mites will yield many more examples.

So far, male-killing bacteria have not been recorded in mites. However, close relatives of male-killing bacteria in insects, such as *Spiroplasma* and *Rickettsia*, have been found in ticks (Hastriter *et al.* 1987, Hurst *et al.* 1997), and in mites the conditions for the spread of male-killing strains are met (Hurst *et al.* 1997, Hurst & Jiggins 2000). Thus, male-killing bacteria are likely to occur in mites. Two examples merit particular examination. First, infection with *Orientia tsutsugamushi* is associated with the production of all-female broods in the trombiculid mite, *Leptotrombidium fletcheri* (Roberts *et al.* 1977, Takahashi *et al.* 1997, Guanghua *et al.* 1999). This may well be due to male-killing bacteria, but parthenogenesis and feminization cannot be excluded. The second case is the association of *Spiroplasma ixodetis* with its tick host *Ixodes pacificus* (Hurst & Jiggins 2000), which needs to be assessed for male-killing or more generally for sex ratio distortion. Interestingly, most rickettsia are serious vertebrate pathogens that are vectored by ticks. Although a large body of medically oriented literature exists on the transmission dynamics of these pathogens, it is not known whether they can cause SRD. If so, this will be of great interest to the epidemiology of rickettsial diseases.

Until recently, parthenogenesis-inducing (PI) *Wolbachia* have only been found in Hymenoptera, with many species of parasitic wasps showing this phenomenon. However, AR Weeks recently found *Wolbachia* to cause parthenogenesis in several mite species of the genus *Bryobia* (see Charlat & Merçot 2000, Weeks & Breeuwer 2001). This genus consists of approximately 100 nominal species and although precise data on their reproductive mode are scarce, thelytokous reproduction predominates (Norton *et al.* 1993, Palmer & Norton 1990, 1992). A few sexual haplodiploid species are known, which are thought to represent the ancestral genetic system. Possibly, *Wolbachia* have played, or still do play, a role in the diversification of this genus. Preliminary data using microsatellite markers indicate that the diploidy of infected *Bryobia* eggs is not restored by gamete duplication (Weeks & Breeuwer 2001), as has been found in *Trichogramma* wasps (Chapter 9). Clearly, the *Bryobia* case shows that interactions between PI *Wolbachia* and their host can have different genetic consequences.

Recently, a feminizing microbe in a female-haploid mite species of the genus *Brevipalpus* was discovered (Weeks *et al.* 2001). This is an exciting discovery for several reasons. First, the responsible bacterium is unrelated to *Wolbachia*, suggesting that other bacteria than *Wolbachia* can feminize as well. Second, it is the first case of feminization in mites; the only other two known examples are in a number of isopods and a microsporidian amphipod. Finally, *Brevipalpus* species are the only reported example among the higher arthropods in which thelytokous females are haploid (Pijnacker *et al.* 1980, 1981). They are essentially apomictic. Thus, there is no possibility for preserving genetic variation through heterozygosity. Furthermore, the haploid chromosome number is extremely low, $n = 2$. This in itself reduces the amount of genetic variation the genome can contain and limits the number of linkage groups (Crozier 1985, Wrensch 1993). Elimination of the microbes results in the production of haploid males that are uninfected. Because females are haploid, the infection apparently results in the feminization of haploid eggs. Nevertheless, thelytokous *Brevipalpus* mites are highly variable, occurring on many different host plant species, and are known to rapidly develop resistance to acaricides (Weeks & Breeuwer 2001). This is in contrast to current dogma on the evolutionary potential of asexual species: adaptation requires genetic variation, but thelytoky restricts the generation of genetic variation.

As in *Bryobia*, sexual species in the genus *Brevipalpus* are haplodiploid. Similarly in insects, the majority of SRDs are found in haplodiploid genetic systems (Chapter 9). This suggests that haplodiploid species may be more vulnerable to invasion by SRDs. These two examples also indicate that thelytoky in many more mites is likely to be associated with heritable micro-organisms. Additional surveys for SRDs and genetic systems of their host are needed to test that idea. Oribatid mites form an interesting group: more than 10% of the species is thelytokous and large families are entirely thelytokous (Palmer & Norton 1990, 1992, Norton *et al.* 1993). A specific survey for *Wolbachia* in a limited number of oribatid species did not yield positive results (Perrot-Minnot & Norton 1997). This does not mean that parthenogenesis has a genetic basis in these mites. Other microbial species can do the trick as well, as is shown by the recent discovery of microbes, unrelated to *Wolbachia*, that feminize *Brevipalpus* mites.

The final SRD trait is cytoplasmic incompatibility (CI), the most common effect that is associated with *Wolbachia* bacteria. This is the phenomenon whereby infected males become reproductively incompatible with uninfected females or with females harbouring different microbial strains. The paternal chromosomes in fertilized eggs of incompatible crosses are eliminated. The fate of the resulting embryos is variable and depends, among others, upon the genetic system of the host. In diplodiploids and also pseudo-arrhenotokous mites most embryos die (Johanowicz & Hoy 1998a,b). In arrhenotokous mites the effect varies: fertilized eggs may develop into normal haploid males (no mortality), die (female mortality), result in F1 females that produce offspring which suffer from hybrid breakdown or even develop into normal females (e.g. Breeuwer 1997, Gotoh *et al.* 1999, Vala *et al.* 2000). Current introgression experiments on the two-spotted spider mite, *Tetranychus urticae*, indicate that part of the variation can be attributed to genetic differences between hosts (F Vala, pers. comm.). Temperature may be another factor affecting the expression of CI in mites (van Opijnen & Breeuwer 1999).

11.8 | Future perspectives

To understand the evolution of various genetic systems in mites, several aspects need further investigation. Clearly, we still lack an answer to the question of why the paternal genome is initially retained in males of pseudo-arrhenotokes. Testing the role of recombinational repair should be a prime objective for future research. Another issue is to elucidate how pseudo-arrhenotokes manage to manipulate the sex ratio of their offspring. Even for some arrhenotokes this is not at all clear. For example, in spider mites females have a spermatheca that opens into the oviduct, so all conditions for the control of fertilization seem to be fulfilled. However, the sperm cells move through the wall of the spermatheca into the haemocoel, where they (actively?) move to eggs in the ovarium (Helle & Sabelis 1985). How can a spider mite control this mode of fertilization? Finally, interesting new research is waiting to be done with acarid mites that are usually diplodiploid, yet express all the features typical of local mate competition (Ignatowicz 1986, Radwan 1993a,b). They exhibit strong male aggression under certain conditions and some species (*Rhizoglyphus*) produce two male morphs, one of which is specially suited to fight against other males ('Captain Hook' males) (Radwan 1993a,b, 1995, Radwan & Bogacz 2000, Radwan *et al.* 2000). Do these expressions of male aggression indicate strong local mate competition and do females of these diplodiploid species control the sex ratio somehow? These are all interesting questions that may shed new light on sex ratio control and the selective advantages of genetic systems.

The development of new theory on sex allocation is another important issue for future research. Many plant-inhabiting arthropods have haystack-like population structures where natural selection acts on the level of individuals and on groups of related individuals. The evolution of sex allocation and dispersal in these systems depends on the exact spatial structure of the population, and in turn influences this structure. The latter effect of the 'feedback loop' is commonly ignored in currently available models and may well be pivotal to the development of the theory. Until now the evolution of sex allocation and dispersal in metapopulations have only been treated separately. In addition, the local and global population dynamics are generally considered to be invariant in the models published to date. An integrated analysis is an essential next step because: (1) optimal sex allocation depends on dispersal behaviour and the payoff of dispersal in turn depends on sex allocation (Nagelkerke 1994), (2) dispersal and sex allocation influence the interaction with local resources (prey for a predator, plants for a herbivore) and (3) metapopulation structure depends on the interaction between (global) immigration pressure and (local) relatedness. Disentangling the interrelationship of evolution and population dynamics by a combination of population modelling on the one hand and experiments with small arthropods (herbivorous and predatory mites) to validate assumptions and predictions on the other hand seems a promising next step for future research. However, no prediction on sex allocation can withstand scrutiny if it is not shown how it withstands manipulation by SRDs. Clearly, demonstrating evolutionarily stable sex-allocation strategies within a given population is one thing, but demonstrating that it can resist parasites that promote their transmission by manipulating the offspring sex ratio of their host is another.

The finding of *Wolbachia*-induced cytoplasmic incompatibility (CI) in mites may have several interesting implications. First, incompatible crosses give rise to altered brood sex ratios and this may have an impact on sex allocation of other individuals in the population. Second, one may wonder whether the host is an innocent bystander or profits from *Wolbachia*-induced CI, e.g. by retention of its coadapted genome (Vala *et al.* 2000). Third, *Wolbachia*-induced CI will create selection to avoid mating with incompatible partners (Lesna & Sabelis 1999) and thereby reinforces genetic isolation and possibly sympatric host race formation and speciation (Vala *et al.* 2000).

Acknowledgements

We wish to thank Ian Hardy and two anonymous referees for comments on the manuscript.

References

Adamson M & Ludwig D (1993) Oedipal mating as a factor in sex allocation in haplodiploids. *Philosophical Transactions of the Royal Society, London, series B*, **341**, 195–202.

Boot WJ, van Baalen M & Sabelis MW (1995) Why do *Varroa* mites invade worker brood cells of the honey bee despite lower reproductive success? *Behavioral Ecology and Sociobiology*, **36**, 283–289.

Breeuwer JAJ (1997) *Wolbachia* and cytoplasmic incompatibility in the spider mites *Tetranychus urticae* and *T. turkestani. Heredity*, **79**, 41–47.

Breeuwer JAJ & Jacobs G (1996) *Wolbachia*: intracellular manipulators of mite reproduction. *Experimental and Applied Acarology*, **20**, 421–434.

Bull JJ (1979) An advantage for the evolution of male haploidy and systems with similar genetic transmission. *Heredity*, **43**, 361–381.

Bull JJ (1983) *The Evolution of Sex-determining Mechanisms.* Menlo Park, Canada: Benjamin Cummings.

Bulmer MG & Taylor PD (1980) Sex ratio under the haystack model. *Journal of Theoretical Biology*, **86**, 83–89.

Cavalier-Smith T (1978) Nuclear volume control by nucleoskeletal DNA, selection for cell volume and cell growth rate, and the solution of the DNA C-value paradox. *Journal of Cell Science*, **34**, 247–278.

Cavalier-Smith T (1985) Cell volume and the evolution of eukaryote genome size. In: T Cavalier-Smith (ed) *The Evolution of Genome Size*, pp 105–184. Chichester, UK: John Wiley.

Charlat S & Merçot H (2000) *Wolbachia* trends. *Trends in Ecology and Evolution*, **15**, 438–439.

Congdon BD & McMurtry JA (1988) Morphological evidence establishing the loss of paternal chromosomes in males of predatory phytoseiid mites, genus *Euseius. Entomologia Experimentalis et Applicata*, **48**, 95–96.

Crozier RH (1985) Adaptive consequences of male haploidy. In: W Helle & MW Sabelis (eds) *Spider Mites: Their Biology, Natural Enemies and Control*, pp 201–222. World Crop Pests volume 1B. Amsterdam: Elsevier.

Cruickshank RH & Thomas RH (1999) Evolution of haplodiploidy in dermanyssine mites (Acari: Mesostigmata). *Evolution*, **53**, 1796–1803.

Faraji F, Janssen A, van Rijn PCJ & Sabelis MW (2000) Kin recognition by the predatory mite *Iphiseius degenerans*: discrimination among own, conspecific and heterospecific eggs. *Ecological Entomology*, **25**, 147–155.

Frank SA (1986) Hierarchical selection theory and sex ratios. I. General solutions for structured populations. *Theoretical Population Biology*, **29**, 312–342.

Frank SA (1987) Demography and sex ratio in social spiders. *Evolution*, **41**, 1267–1281.

Franks NR, Healey KJ & Byrom L (1991) Studies on the relationship between the ant ectoparasite *Antennophorus grandis* (Acarina: Antennophoridae) and its host *Lasius flavus* (Hymenoptera: Formicidae). *Journal of Zoology, London*, **225**, 59–70.

Geest van der LPS, Elliot SL, Breeuwer JAJ & Beerling EAM (2000) Diseases of mites. *Experimental and Applied Acarology*, **24**, 497–560.

Gotoh T, Bruin J, Sabelis MW & Menken SBJ (1993) Host race formation in *Tetranychus urticae*: genetic differentiation, host plant preference, and mate choice in a tomato and a cucumber strain. *Entomologia Experimentalis et Applicata*, **68**, 171–178.

Gotoh T, Sugasawa J & Nagata T (1999) Reproductive compatibility of the two-spotted spider mite (*Tetranychus urticae*) infected with *Wolbachia. Entomological Science*, **2**, 289–295.

Greeff JM (1997) Offspring allocation in externally ovipositing fig wasps with varying clutch size and sex ratio. *Behavioral Ecology*, **8**, 500–505.

Guanghua W, Moahua XYuL, & Mingrong B (1999) Studies on biting and transovarial transmission of *Rickettsia tsutsugamushi* in *Leptotrombidium* (L.) *goahuense* Tsai and Chow (Trombiculidae). In: Proceedings of the XI International Acaralogy Congress, pp 447–449.

Haig D (1993a) The evolution of unusual chromosomal systems in coccoids – extraordinary sex-ratios revisited. *Journal of Evolutionary Biology*, **6**, 69–77.

Haig D (1993b) The evolution of unusual chromosomal systems in sciarid flies – intragenomic conflict and the sex-ratio. *Journal of Evolutionary Biology*, **6**, 249–261.

Hastriter MW, Kelly DJ, Chan TC, Phang OW & Lewis GE (1987) Evaluation of *Leptotrombidium* (*Leptotrombidium*) *fletcheri* (Acari: Trombiculidae) as a potential vector of *Ehrlichia sennetsu. Journal of Medical Entomology*, **24**, 542–546.

Helle W & Sabelis MW (1985) *Spider Mites: Their Biology, Natural Enemies and Control*. World Crop Pest Series, volume 1A, B. Amsterdam: Elsevier.

Helle W, Bolland HR, van Arendonk R, de Boer R, Schulten GGM & Russell VM (1978) Genetic evidence for biparental males in haplo-diploid predator mites (Acarina: Phytoseiidae). *Genetica*, **49**, 165–171.

Hoy MA (1979) Parahaploidy of the 'arrhenotokous' predator *Metaseiulus occidentalis* (Nesbitt) (Acari: Phytoseiidae) demonstrated by X-irradiation of males. *Entomologia Experimentalis et Applicata*, **26**, 97–104.

Hurst GGD & Jiggins FM (2000) Male-killing bacteria in insects: mechanisms, incidence, and implications. *Emerging Infectious Diseases*, **6**, 329–336.

Hurst GGD, Hurst LD & Majerus MEN (1997) Cytoplasmic sex ratio distorters. In: SL O'Neill, AA Hoffmann & JH Werren (eds) *Influential Passengers: Microbes and Invertebrate Reproduction*, pp 124–154. Oxford, UK: Oxford University Press.

Ignatowicz S (1986) Sex ratio in the acarid mites (Acaridae: Acaroidea). *Roczniki Nauk Rolniczych, Seria E*, **16**, 101–110.

Izraylevich S & Gerson U (1995) Sex ratio of *Hemisarcoptes coccophagus*, a mite parasitic on insects: density-dependent processes. *Oikos*, **74**, 439–446.

Izraylevich S & Gerson U (1996) Sex allocation by a mite parasitic on insects: local mate competition, host quality and operational sex ratio. *Oecologia*, **108**, 676–682.

Johanowicz DL & Hoy MA (1996) *Wolbachia* in a predator-prey system: 16S ribosomal DNA analysis of two phytoseiids (Acari: Phytoseiidae) and their prey (Acari: Tetranychidae). *Annals of the Entomological Society of America*, **89**, 435–441.

Johanowicz DL & Hoy MA (1998a) The manipulation of arthropod reproduction by *Wolbachia* endosymbionts. *Florida Entomologist*, **81**, 310–317.

Johanowicz DL & Hoy MA (1998b) Experimental induction and termination of non-reciprocal reproductive incompatibilities in a parahaploid mite. *Entomologia Experimentalis et Applicata*, **87**, 51–58.

Johanowicz DL & Hoy MA (1999) *Wolbachia* infection dynamics in experimental laboratory populations of *Metaseiulus occidentalis*. *Entomologia Experimentalis et Applicata*, **93**, 259–268.

Jong JH de, Lobbes PV & Bolland HR (1981) Karyotypes and sex determination in two species of laelapid mites (Acari, Gamasida). *Genetica*, **55**, 187–190.

Kaliszewski M & Wrensch DL (1993) Evolution of sex determination and sex ratio within the mite cohort Tarsonemina (Acari: Heterostigmata). In: DL Wrensch & MA Ebbert (eds) *Evolution and Diversity of Sex Ratio in Insects and Mites*, pp 192–213. New York: Chapman & Hall.

Kethley J (1971) Population regulation in quill mites (Acarina: Syringophilidae). *Ecology*, **52**, 1113–1118.

Krainacker DA & Carey JR (1990) Effect of age at first mating on primary sex-ratio of the two-spotted spider-mite. *Experimental and Applied Acarology*, **9**, 169–175.

Lesna I & Sabelis MW (1999) Diet-dependent female choice for males with 'good genes'. *Nature*, **401**, 581–584.

Lewis WMJ (1983) Interruption of synthesis as a cost of sex in small organisms. *American Naturalist*, **121**, 825–834.

Lindquist EE, Sabelis MW & Bruin J (eds) (1996) *Eriophyoid Mites – Their Biology, Natural Enemies and Control*. World Crop Pest Series, volume 6. Amsterdam: Elsevier.

Maynard Smith J (1964) Group selection and kin selection. *Nature*, **201**, 1145–1147.

McMurtry JA & Croft BA (1997) Life-styles of phytoseiid mites and their role in biological control. *Annual Review of Entomology*, **42**, 291–321.

Momen FM (1994) Fertilization and starvation affecting reproduction in *Amblyseius barkeri* (Hughes) (Acari, Phytoseiidae). *Anzeiger für Schädlingskunde, Pflanzenschutz und Umweltschutz*, **67**, 130–132.

Mori K, Saito Y & Sakagami T (1999) Effects of the nest web and female attendance on survival of young in the subsocial spider mite *Schizotetranychus longus* (Acari: Tetranychidae). *Experimental and Applied Acarology*, **235**, 411–418.

Nagelkerke CJ (1994) Simultaneous optimization of egg distribution and sex allocation in a patch-structured population. *American Naturalist*, **144**, 262–284.

Nagelkerke CJ & Sabelis MW (1991) Precise sex-ratio control in the pseudo-arrhenotokous phytoseiid mite, *Typhlodromus occidentalis* Nesbitt. In: R Schuster & PW Murphy (eds) *The Acari: Reproduction, Development and Life-History Strategies*, pp 193–208. London: Chapman & Hall.

Nagelkerke CJ & Sabelis MW (1996) Hierarchical levels of spatial structure and their consequences for the evolution of sex allocation in mites and other arthropods. *American Naturalist*, **148**, 16–39.

Nagelkerke CJ & Sabelis MW (1998) Precise control of sex allocation in pseudo-arrhenotokous phytoseiid mites. *Journal of Evolutionary Biology*, **11**, 649–684.

Nagelkerke CJ, van Baalen M & Sabelis MW (1996) When should a female avoid adding eggs to the clutch of another female? A simultaneous oviposition and sex allocation game. *Evolutionary Ecology*, **10**, 475–497.

Nelson-Rees WA, Hoy MA & Roush RT (1980) Heterochromatinization, chromatin elimination and haploidization in the parahaploid mite *Metaseiulus occidentalis* (Nesbitt) (Acarina: Phytoseiidae). *Chromosoma* (Berl.), **77**, 263–276.

Norton RA, Kethley JB, Johnston DE & O'Connor BM (1993) Phylogenetic perspectives on genetic systems and reproductive modes of mites. In: DL Wrensch & MA Ebbert (eds) *Evolution and Diversity of Sex Ratio in Haplodiploid Insects and Mites*, pp 8–99. New York: Chapman & Hall.

Opijnen T van & Breeuwer JAJ (1999) High temperatures eliminate *Wolbachia*, a cytoplasmic incompatibility inducing endosymbiont, from the two-spotted spider mite. *Experimental and Applied Acarology*, **23**, 871–881.

Palmer SC & Norton RA (1990) Further experimental proof of thelytokous parthenogenesis in oribatid mites (Acari, Oribatida, Desmonomata). *Experimental and Applied Acarology*, **8**, 149–159.

Palmer SC & Norton RA (1992) Genetic diversity in thelytocous oribatid mites (Acari, Acariformes, Desmonomata). *Biochemical Systematics and Ecology*, **20**, 219–231.

Pels B & Sabelis MW (1999) Local dynamics, overexploitation and predator dispersal in an acarine predator-prey system. *Oikos*, **86**, 573–583.

Perrot-Minnot MJ & Navajas M (1995) Biparental inheritance of RAPD markers in males of the pseudo-arrhenotokous mite *Typhlodromus pyri*. *Genome*, **38**, 838–844.

Perrot-Minnot MJ & Norton RA (1997) Obligate thelytoky in oribatid mites: no evidence for *Wolbachia* inducement. *Canadian Entomologist*, **129**, 691–698.

Perrot-Minnot MJ, Lagnel J, Migeon A & Navajas M (2000) Tracking paternal genes with DALP markers in a pseudoarrhenotokous reproductive system: biparental transmission but haplodiploid-like inheritance in the mite *Neoseiulus californicus*. *Heredity*, **84**, 702–709.

Pijnacker LP, Ferwerda MA, Bolland HR & Helle W (1980) Haploid female parthenogenesis in the false spider mite *Brevipalpus obovatus* (Acari: Tenuipalpidae). *Genetica*, **51**, 211–214.

Pijnacker LP, Ferwerda MA & Helle W (1981) Cytological investigations of the female and male reproductive system of the parthenogenetic privet mite *Brevipalus obovatus* Donnadieu (Phytoptipalpidae, Acari). *Acarologia*, **22**, 157–163.

Radwan J (1993a) Kin recognition in the acarid mite, *Caloglyphus berlesei* – negative evidence. *Animal Behaviour*, **45**, 200–202.

Radwan J (1993b) The adaptive significance of male polymorphism in the acarid mite *Caloglyphus berlesei*. *Behavioral Ecology and Sociobiology*, **33**, 201–208.

Radwan J (1995) Male morph determination in 2 species of acarid mites. *Heredity*, **74**, 669–673.

Radwan J & Bogacz I (2000) Comparison of life-history traits of the two male morphs of the bulb mite, *Rhizoglyphus robini*. *Experimental and Applied Acarology*, **24**, 115–121.

Radwan J, Czyz M, Konior M & Kolodziejczyk M (2000) Aggressiveness in two male morphs of the bulb mite *Rhizoglyphus robini*. *Ethology*, **106**, 53–62.

Roberts LW, Rapmund G & Cadigan FC (1977) Sex ratios in *Rickettsia tsutsugamushi*-infected and noninfected colonies of *Leptotrombidium* (Acari: Trombiculidae). *Journal of Medical Entomology*, **14**, 89–92.

Roeder C, Harmsen R & Mouldey S (1996) The effects of relatedness on progeny sex ratio in spider mites. *Journal of Evolutionary Biology*, **9**, 143–151.

Roush RT & Hoy MA (1981) Genetic improvement of *Metaseiulus occidentalis*: selection with methomyl, dimethoate and carbaryl and genetic analysis of carbaryl resistance. *Journal of Economic Entomology*, **74**, 138–141.

Ruf A (1993) Die Morphologische Variabilität und Fortpflanzungsbiologie der Raubmilbe *Hypoaspis aculeifer* (Canestrini 1883) (Mesostigmata: Laelapidae). Bremen: Universität Bremen.

Sabelis MW (1985a) Reproductive Strategies. In: W Helle & MW Sabelis (eds) *Spider Mites: Their Biology, Natural Enemies and Control*, pp 265–278. World Crop Pest Series, volume 1A. Amsterdam: Elsevier.

Sabelis MW (1985b) Sex Allocation. In: W Helle & MW Sabelis (eds) *Spider Mites: Their Biology, Natural Enemies and Control*, pp 83–94. World Crop Pest Series, volume 1B. Amsterdam: Elsevier.

Sabelis MW & Bruin J (1996) Evolutionary ecology: life history patterns, food plant choice and dispersal. In: EE Lindquist, MW Sabelis & J Bruin (eds) *Eriophyoid Mites: Their Biology, Natural Enemies and Control*, pp 329–366. Amsterdam: Elsevier.

Sabelis MW & Dicke M (1985) Long-range dispersal and searching behaviour. In: W Helle & MW Sabelis (eds) *Spider Mites: Their Biology, Natural Enemies and*

Control, pp 141–160. World Crop Pest Series, volume 1B. Amsterdam: Elsevier.

Sabelis MW & Janssen A (1993) Evolution of life-history patterns in the Phytoseiidae. In: MA Houck (ed) *Mites. Ecological and Evolutionary Analyses of Life-History Patterns*, pp 70–98. New York: Chapman & Hall.

Sabelis MW & Nagelkerke CJ (1987) Sex allocation strategies of pseudo-arrhenotokous phytoseiid mites. *Netherlands Journal of Zoology*, **37**, 117–136.

Sabelis MW & Nagelkerke CJ (1988) Evolution of pseudo-arrhenotoky. *Experimental and Applied Acarology*, **4**, 301–318.

Sabelis MW & Nagelkerke CJ (1993) Sex allocation and pseudoarrhenotoky in phytoseiid mites. In: DL Wrensch & MA Ebbert (eds) *Evolution and Diversity of Sex Ratio in Haplodiploid Insects and Mites*, pp 512–541. New York: Chapman & Hall.

Sabelis MW & Scholman M (1989) Sex ratio control in a pseudo-arrhenotokous phytoseiid mite. In: GP Channabasavanna & CA Viraktamath (eds) *Progress in Acarology*, p 267. VII International Acarology Congress, 1986. Oxford and I.B.H., New Delhi, India.

Sabelis MW & van der Meer J (1986) Local dynamics of the interaction between predatory mites and two-spotted spider mites. In: JAJ Metz & O Diekmann (eds) *Dynamics of Physiologically Structured Populations*, pp 322–344. Lecture Notes in Biomathematics, **68**. Berlin: Springer-Verlag.

Sabelis MW, Diekmann O & Jansen VAA (1991) Metapopulation persistence despite local extinction: predator-prey patch models of the Lotka-Volterra type. *Biological Journal of the Linnean Society*, **42**, 267–283.

Saito Y (1990a) Factors determining harem ownership in a subsocial spider-mite (Acari, Tetranychidae). *Journal of Ethology*, **8**, 37–43.

Saito Y (1990b) Harem and non-harem type mating systems in 2 species of subsocial spider-mites (Acari, Tetranychidae). *Researches on Population Ecology*, **32**, 263–278.

Saito Y (1994) Do males of *Schizotetranychus miscanthi* (Acari, Tetranychidae) recognize kin in male competition. *Journal of Ethology*, **12**, 15–17.

Saito Y (1995) Clinal variation in male-to-male antagonism and weaponry in a subsocial mite. *Evolution*, **49**, 413–417.

Saito Y (2000) Do kin selection and intra-sexual selection operate in spider mites? *Experimental and Applied Acarology*, **24**, 351–363.

Saito Y & Sahara K (1999) Two clinal trends in male-male aggressiveness in a subsocial spider mite (*Schizotetranychus miscanthi*). *Behavioral Ecology and Sociobiology*, **46**, 25–29.

Saito Y, Mori K, Chittenden AR & Sato Y (2000a) Correspondence of male-to-male aggression to spatial distribution of individuals in field populations of a subsocial spider mite. *Journal of Ethology*, **18**, 79–83.

Saito Y, Sahara K & Mori K (2000b) Inbreeding depression by recessive deleterious genes affecting female fecundity of a haplo-diploid mite. *Journal of Evolutionary Biology*, **13**, 668–678.

Schrader F & Hughes-Schrader S (1931) Haploidy in Metazoa. *Quarterly Review of Biology*, **6**, 411–438.

Schulten GGM (1985) Pseudo-arrhenotoky. In: W Helle & MW Sabelis (eds) *Spider Mites, their Biology, Natural Enemies and Control*, pp 67–72. World Crop Pest Series, volume 1B Amsterdam: Elsevier.

Stiefel VL & Margolies DC (1992) Do components of colonization–dispersal cycles affect the offspring sex ratios of Banks grass mites (*Oligonychus pratensis*)? *Entomologia Experimentalis et Applicata*, **64**, 161–166.

Takahashi M, Urakami H, Yoshida Y, Furuya Y, Misumi H, Hori E, Kawamura A & Tanaka H (1997) Occurrence of high ratio of males after introduction of minocycline in a colony of *Leptotrombidium fletcheri* infected with *Orientia tsutsugamushi*. *European Journal of Epidemiology*, **13**, 19–23.

Tienderen PH van & de Jong G (1986) Sex ratio under the haystack model: polymorphism may occur. *Journal of Theoretical Biology*, **122**, 69–81.

Toyoshima S & Amano H (1998) Effect of prey density on sex ratio of two predacious mites, *Phytoseiulus persimilis* Athias-Henriot and *Amblyseius womersleyi* (Acari: Phytoseiidae). *Experimental and Applied Acarology*, **22**, 709–723.

Toyoshima S & Amano H (1999a) Cytological evidence of pseudo-arrhenotoky in two phytoseiid mites, *Phytoseiulus persimilis* Athias-Henriot and *Amblyseius womersleyi* Schicha. *Journal of the Acarological Society of Japan*, **8**, 135–142.

Toyoshima S & Amano H (1999b) Comparison of development and reproduction in offspring produced by females of *Phytoseiulus persimilis* Athias-Henriot (Acari: Phytoseiidae) under two prey conditions. *Applied Entomology and Zoology*, **34**, 285–292.

Toyoshima S, Nakamura M, Nagahama Y & Amano H (2000) Process of egg formation in the female body cavity and fertilization in male eggs of *Phytoseiulus persimilis* (Acari: Phytoseiidae). *Experimental and Applied Acarology*, **24**, 441–451.

Vala F, Breeuwer JAJ & Sabelis MW (2000) *Wolbachia*-induced 'hybrid breakdown' in the two-spotted spider mite *Tetranychus urticae* Koch. *Proceedings of The Royal Society of London, series B*, **267**, 1931–1937.

Warren E (1940) On the genital system of *Dermanyssus gallinae* (De Geer), and several other Gamasidae. *Annals Natal Museum*, **9**, 409–459.

Weeks AR & Breeuwer JAJ (2001) *Wolbachia* induced parthenogenesis in a genus of phytophagous mites. *Proceedings of the Royal Society, London, series B*, **268**, 2245–2251.

Weeks AR, Marec F & Breeuwer JAJ (2001) A mite species that consists entirely of haploid females. *Science*, **292**, 2497–2482.

Wilson DS & Colwell RK (1981) Evolution of sex ratio in structured demes. *Evolution*, **35**, 882–897.

Wrensch DL (1993) Evolutionary flexibility through haploid males or how chance favors the prepared genome. In: DL Wrensch & MA Ebbert (eds) *Evolution and Diversity of Sex Ratio in Insects and Mites*, pp 118–149. New York: Chapman & Hall.

Wrensch DL & Johnston DE (1983) Sexual dimorphism in deutonymphs of mites of the family Parasitidae (Acari: Mesostigmata). *Annals of the Entomological Society of America*, **76**, 473–474.

Young SSY, Wrensch DL & Kongchuensin M (1986) Control of sex ratio by female spider mites. *Entomologia Experimentalis et Applicata*, **40**, 53–60.

Chapter 12

Aphid sex ratios

William A. Foster

12.1 | Summary

Aphids produce males and females by a partheno-genetic process that gives the mother proximate control of sex allocation. Sex is an infrequent, usually annual, punctuation in a sequence of asexual generations. The aphid genome, initiated in the fertilized egg, is replicated in a sequence of bodies that make up the aphid clone and which can be thought of as a disaggregated hermaphrodite: selection acts at the clonal level to produce an optimal allocation in sperm and ova. Two crucial factors influencing sex allocation are the degree of within-clone mating and the timing by the clone of investment in males and mating females. Extreme sex ratios are very common in aphids, to an extent that is probably unique amongst diploid organisms. The aphids, unshackled from the constraints imposed by meiosis on sex determination, therefore provide an excellent opportunity for those interested in the evolutionary biology of sex allocation.

12.2 | Introduction

Aphids are of special importance to evolutionary biologists because they are a diplodiploid group in which the mother clearly has proximate control of the sex of her offspring. They thus provide a genetic system other than haplodiploidy in which the default allocation ratio is not 0.5 (proportion investment in males, i.e. males/(males+females)). My aim in this chapter is to review current knowledge of aphid sex ratios, to provide a brief, accessible account of the relevant biology of these animals, and to highlight aphid groups and specific ideas that would be especially fruitful to study. Aphid sex ratios were reviewed by Moran (1993) and other useful accounts are provided by Ward and Wellings (1994a,b) and Dixon (1998). The aphid classification used follows Nieto Nafria *et al.* (1998).

12.3 | Aphid life cycles

The aphid genome, having been assembled in the fertilized egg, is replicated by parthenogenesis, passing through a sequence of individuals, which together constitute the aphid clone. These individuals may become very numerous, very widely dispersed and adopt radically diverse phenotypes, but the clonal genotype is not disrupted, because there is no recombination (Blackman 1987), although it is subject to mutation (e.g. Sunnucks *et al.* 1998, Loxdale *et al.* 2002). Eventually, some or all of the clone members produce males and mating females, usually in a single annual generation, and a new set of aphid genomes is produced. Parthenogenetic and mating females are radically different, and these differences are apparent very early in development (Blackman 1987). In mating females, a small number of yolky eggs develop slowly within the aphid (only a single one in the Eriosomatinae), which also eventually has accessory glands and a sperm receptacle. In the parthenogenetic female morphs, relatively large numbers of embryos are continually developing within the mother from well before her own birth.

This sequence of asexual generations annually punctuated by a sexual generation (cyclical parthenogenesis) is almost certainly a derived feature shared by the Aphidoidea and of ancient origin (Heie 1980, 1987). Aphids now show such an astonishing diversity of life cycles that almost any generalization about them will have important exceptions; for example there are many species with populations that no longer have a sexual phase, species with parthenogenetic clones that occasionally produce males (e.g. Blackman 1972, Rispe *et al.* 1999) and species where the foundress can give rise directly to the sexual generation (e.g. Dixon 1972, Strathdee *et al.* 1993a).

An aphid clone can be thought of as a disaggregated hermaphrodite, with selection acting at the level of the clone to produce an optimal allocation of resources in sperm and ova (e.g. Ward & Wellings 1994a,b). Unlike the cells of a conventional hermaphrodite, the individuals of an aphid clone are mobile and independent, and each is able to generate a lineage that produces both males and females. In some aphid populations only some of the lineages produce sexuals, others may persist parthenogenetically. But in most of these cases, the mothers that produce sexuals produce no other offspring, so by concentrating on the sexual-producing lineages, we can ignore the complications of facultative parthenogenesis (Lively 1987, Moran 1993). In some aphids mothers can produce both sexuals and parthenogens, which makes investment ratios very difficult to assess; these cases are largely ignored here.

Two crucial factors influence sex allocation in aphids. The first is the degree of within-clone mating. This is largely determined by the details of the life cycle of a particular species; for example, whether the clone alternates between two hosts, a primary or 'winter' host (where mating occurs) and a secondary or 'summer' host (where successive, entirely parthenogenetic, generations occur), and whether both the males

and sexual females are winged. Clones with individuals that disperse little will approximate to selfing hermaphrodites and are predicted to have female-biased sex ratios.

The second crucial factor is the timing of the investment by the clone in males and sexual females. In many aphid species, the sexuals are produced by a single individual, the sexupara: the allocation ratio is then simply the fraction of her resources she invests in males relative to females. In others, the focal investing female produces males and a special morph, the gynopara, which parthenogenetically produces the mating females: the focal female is thus the mother of the males but the grandmother of the sexual females, which are produced one generation later. As pointed out by Ward and Wellings (1994a,b), the critical allocation ratio is that of males to gynoparae, since resources invested in the latter are committed to future ova. The extent of the asynchrony between gynopara and male production will have significant effects on the expected allocation ratio, as modelled by Ward and Wellings (1994b).

Three representative life cycles are shown in Figure 12.1. Most aphids (about 90%) do not host-alternate, but live on one or a few species of a particular plant genus (Eastop 1973, Dixon 1998). In many of these species, the sexuals are produced in the autumn by a special morph, the sexupara; the males may be winged and the females wingless (Figure 12.1a) or the sexuals may be wingless (Figure 12.1b). A minority of aphids host-alternate, and there are two distinct patterns: a winged parthenogenetic female (the sexupara) may fly back to the primary host and give birth to the sexuals (usually both wingless) (Figure 12.1b), or the return journey may be made by the winged males and the winged mother (the gynopara) of the mating females (Figure 12.1c). This difference is crucially important to sex allocation: in the first, eriosomatine, pattern (Figure 12.1b), the sexuals are produced close together, greatly increasing the likelihood of within-clone mating, whereas in the second, aphidine, pattern (Figure 12.1c), the males and the mothers of the mating females migrate independently back to the primary host, so that within-clone mating is much less likely. There are many other

(a)

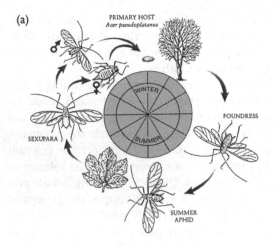

(b)

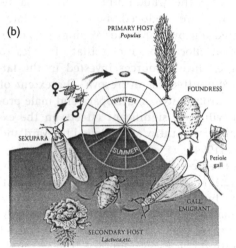

(c)

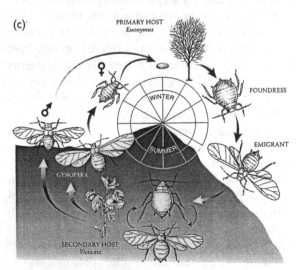

Fig 12.1a–c Representative aphid life cycles. (a) Non host-alternating aphid. *Drepanosiphum platanoides* (Schrank) [Drepanosiphoninae] on *Acer pseudoplatanus*. Note that a winged sexupara gives birth to winged males and wingless mating females. (b) Host-alternating aphid. *Pemphigus bursarius* (L.) [Pemphiginae] which alternates between *Populus nigra* (primary host) and lettuce (secondary host). Note that a winged sexupara, migrating from the secondary host, gives birth to wingless males and wingless mating females on the bark of the primary host. Populations can overwinter asexually near the lettuce roots in the soil. (c) Host-alternating aphid. *Aphis fabae* Scopoli [Aphidinae], which alternates between *Euonymus europaeus* (primary host) and beans, docks (secondary host). Note that the forms that fly back to the primary host are the gynoparae (mother of the mating females) and the males. Modified from Figures 4 and 6 in Blackman and Eastop (1994), by permission of CAB International. Drawing by J. Rodford.

variations on these basic patterns (e.g. Lampel 1968, Blackman & Eastop 1994).

12.4 | Proximate mechanisms relevant to sex allocation

The cues used by the clone to prompt the switch to the production of sexuals are, generally speaking, environmental factors that signal the onset of autumn. These include increased night-length, sometimes modulated by temperature, and possibly also crowding and host-plant condition (e.g. Marcovitch 1924, Lees 1973, Ward *et al.* 1984, Kawada 1987). For root-dwelling pemphigines, which include some of the main subjects of sex ratio studies, there is no effect of photoperiod: low temperatures and crowding appear to be the critical factors (Judge 1968, Moran *et al.* 1993). There may in addition be maternal effects, so-called 'interval-timers', that inhibit the young clone from producing sexuals in the long nights of spring (e.g. Bonnemaison 1951, Lees 1966).

In aphids, sex determination does not involve a meiotic process; therefore, in sharp contrast to the vast majority of other diploid organisms, there is no proximate enforcement of a sex ratio at conception that is close to 0.5. Male and female sexuals are always produced parthenogenetically within the clone: fusion of gametes is not involved. The sex of a mother's offspring therefore

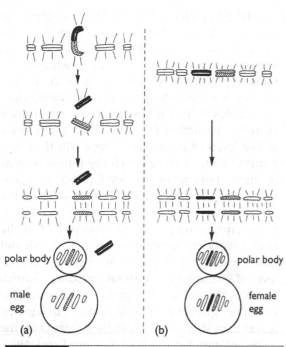

polar body

male egg

(a)

polar body

female egg

(b)

Fig 12.2a,b Diagram of the behaviour of X chromosomes during the maturation divisions of (a) male and (b) female parthenogenetic eggs. One X chromosome is shown in black, the other stippled. After Orlando (1974): Figure 3.5. in Blackman (1987), by permission of Elsevier.

depends neither on information carried in the sperm or the eggs (as in most diploids) nor on whether the eggs are fertilized (as in haplodiploids). In aphids, sex determination is XO/XX (male/female) and is determined by the behaviour of the X chromosome during the maturation of the oocytes. In the development of male oocytes, one of the X chromosomes is completely lost and an XO individual is produced (Orlando 1974, Blackman & Hales 1986) (Figure 12.2). Males produce only X-carrying sperm, so that all offspring of the mated females are XX and therefore are themselves female (e.g. Blackman 1987). These females are the founders of all the new clones that will begin their life in the following season.

Although it is clear that sex determination occurs within the mother, the precise physiological mechanisms have not been fully established. Environmental cues act via the endocrine system, in particular via the levels of juvenile hormone, to regulate the production of males and mating females inside the mother (Hales & Mittler 1983, 1987, Hardie *et al.* 1985). Although it seems

reasonable to infer that the sex ratio is sensitive to selection, there may be significant constraints in how this occurs, as has been emphasized by Moran (1993). For example, there may be developmental costs in the switch from producing mating females to males (e.g. Searle & Mittler 1981), and the lag between the sex determination and birth of an aphid may limit the precision with which the sex ratio can be matched to the environment.

12.5 | Major patterns of sex allocation

Basic sex-allocation theory tells us that at equilibrium the genetic return to the controlling party per unit of energy expended is equal for the two sexes (Fisher 1930, Charnov 1982). In most animal populations, this results in equal allocation in the two sexes (so-called Fisherian systems). That this is, in fact, only a special case of Fisher's theory (Bourke & Franks 1995) is tellingly illustrated by the aphids, where, to date, there are no detailed examples of Fisherian 0.5 allocation ratios.

There are a number of distinct ecological contexts for aphids that might markedly influence the fitness return to the clone of the two sexes and lead to highly unequal allocation ratios. I will now consider each of these.

12.5.1 Population subdivision
12.5.1.1 Local mate competition
Biased sex ratios are frequently the result of interactions between relatives in spatially structured populations. By far the most common and best studied type of interaction is local mate competition (LMC), in which relatives (usually male siblings) compete for access to (usually female) siblings (Hamilton 1967). In many aphid species, the sexuals are produced in a context that is remarkably close to the scenario envisaged by Hamilton in his original formulation of LMC. In eriosomatines, for example, the winged sexuparae (equivalent to Hamilton's 'foundresses') fly to the primary host and give birth to males and mating females, which are wingless, do not feed,

and mate in their fifth instar in a 'patch' close to where they were born (Foster & Benton 1992). Fecundity is easy to measure, since each mating female lays a single egg. If there is only one sexupara in the patch, the mating system is equivalent to self-fertilization. This type of life cycle characterizes the Eriosomatinae and Hormaphidinae, but sexuparae producing wingless males and mating females occur sporadically throughout most aphid taxa.

These aphids therefore provide an excellent model diploid system, complementing the much better studied haplodiploid parasitoids and fig-pollinating wasps, in which to investigate and develop ideas about LMC. In eriosomatines, the investment ratio ranges in different species from being very female biased to slightly male biased (e.g. Kurosu & Aoki 1991), in broad agreement with LMC predictions. In the earliest study, Yamaguchi (1985) described an allocation ratio of 0.27 (incorrectly reported by her as 0.37) in the eriosomatine *Prociphilus oriens*. She significantly extended LMC theory by providing the first detailed account of a situation in which there is variation in the fecundity of females in a patch.

The basic idea underlying this situation is know as the constant male hypothesis (CMH); since it was developed initially in aphids, it is worth considering in more detail (see Figure 12.3, based on Frank 1987b). Under LMC, the return on

investing in males is initially high but decreases rapidly with further investment, whereas the returns on investing in females are linear. But the average returns on both sexes, following Fisher, must be the same at equilibrium. If the rate of returns per investment is lower for males than for females at high levels of investment, it must therefore be higher for males than for females at low levels of investment. There will thus be a critical point, C, above which the rate of returns for males falls below that for females. An aphid (sexupara) should produce the sex yielding the greater return, and she should therefore follow the composite (upper) line in Figure 12.3. If she has a level of investment lower than C, she should invest exclusively in males. If she has a higher level of investment, she should produce C-worth of males (the 'constant' number of males) and the rest of her investment should be made in females. A sexupara should never produce more than the 'constant' number of males (see also Frank 1985, 1987a,b).

Yamaguchi's data, based on the dissection of the embryos in 176 sexuparae, provided excellent support for her model, but there are difficulties in interpreting the data because of a severe practical problem faced by the sexuparae. Because their offspring are at an advanced stage of development before the sexuparae leave the secondary hosts, these aphids are committed to a particular investment ratio long before they arrive in the 'patch' on the primary host, and therefore in the absence of any information about patch quality for their offspring. This is in striking contrast to female parasitoids, which can determine offspring gender at more or less the last moment before egg laying. CMH predicts that within a patch there should be constant investment in males (above the critical threshold), whatever the sexupara's level of parental investment; however, between patches, the number of males produced by a particular female should be linearly related to the total parental investment by females in her patch.

Kindlmann and Dixon (1989) suggested a solution to this problem. They posit a negative feedback relationship between aphid density (and hence numbers of sexuparae in the patch) and the size and fecundity of the sexuparae: when

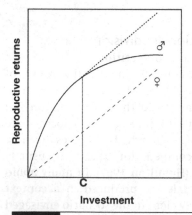

Fig 12.3 The constant male hypothesis. Curves that describe the rate of fitness return with increasing investment in females and males under local mate competition. Up to C, aphids should invest in males only. Above C, aphids should invest in females only (upper dotted line). From Frank (1987b).

density is high, the small sexuparae produce a high proportion of males; when density is low, the large sexuparae produce a high proportion of females. This remains an LMC model, but does not require advanced forecasting by the sexuparae: the constant number of males is fixed for each species. There is, however, no compelling evidence that size variation in eriosomatines is predominantly determined by density, as opposed to food quality or temperature for example (Ward & Wellings 1994a). The model might just work for a non-host-alternating species, such as *Pemphigus spyrothecae*, but it is extremely difficult to see how it could ever work for host-alternating species. The overall population of sexuparae arriving at the primary host need not be related to the local density experienced by the aphids on the secondary hosts (between which there is little summer dispersal): the same overall population could result from either many small colonies or a few dense ones. The degree of crowding on the secondary host will be greater for some aphids than for others, but all these sexuparae will experience a comparable density on the pool of primary hosts.

Stubblefield and Seger (1990) developed a model of LMC with variable fecundity that is somewhat more realistic than Yamaguchi's and Frank's, and in which it is assumed that the sexupara's sex ratio is determined before she arrives at the patch and that she cannot thereafter modify it: each sexupara knows her own fecundity and has an expectation of patch size (n, number of foundresses) but does not know the sex ratio or fecundity of the other sexuparae. This model is similar to CMH, except that foundresses (sexuparae) with more than the threshold level of resources are predicted to invest most, *but not all*, of the excess in daughters.

Foster and Benton (1992), in a study of *Pemphigus spyrothecae* (Figure 12.4), were able to test some of the assumptions and predictions of these models. Most sexuparae (147/149) produced two males and between two and six (mean 5.1) females, in accordance with predictions of LMC and CMH models. The average investment in males was 0.135. The average number of sexuparae forming a group on the bark surface where mating could occur was 2.48, but it is

Fig 12.4 Drawing of a sexupara of *Pemphigus spyrothecae* and the two males and six mating females she has given birth to on the bark of *Populus nigra*. The length of the wingless female is approximately 0.8 mm. Drawn by J. Rodford.

clear that the variance can be very high. The adult males were quite mobile, could mate with up to 14 females, and neither they nor the females were observed to avoid sib-mating. The overall investment in sons by the sexuparae was 0.135 (investment estimated on a wet-weight basis): this is consistently more female biased than is predicted by Stubblefield and Seger's model. The most likely explanation for this, according to Foster and Benton (1992), is that the model does not take into account the fact that the sexuparae arrive asynchronously at the patch. In this situation, sexuparae which arrive later will be contributing to a patch that contains males that matured earlier and are ready to mate with their daughters; these sexuparae would therefore be selected to invest less in sons, as has been modelled by Nunney and Luck (1988). The biological observations on *P. spyrothecae* are qualitatively consistent with this idea: males can fertilize all their sisters within one hour; it is highly likely that the sexuparae in a patch arrive at wider intervals than this; and the males can fertilize up to 14 females.

Other observations on sex ratios in eriosomatines and aphids with similar life-histories (e.g. Hormaphidinae) generally confirm the female-biased sex ratios and constant number of males predicted by the above models. Kurosu and Aoki (1991) reported numerical investment ratios of between 0.2 and 0.56 in nine eriosomatine and hormaphidine species, but in all these the males were much smaller than the females, indicating that the investment ratio was markedly female biased. They observed that in several of the species the variation in the number of males produced was very high. This occurred only in those species that no longer had a reliable connection with the primary host: in some areas, since there were no trees for them to fly to and lay eggs on, the sexuparae were an evolutionary dead-end. They suggest that the sex-allocation ratio is under strong stabilizing selection in the obligate host-alternators, but in the other species without a primary host 'bad genes' accumulate and the sex ratio drifts away from the optimum. It is surprising that selection has not acted to eliminate the sexuparae altogether in these species. Moran (1993) summarizes data on four eriosomatines. *Eriosoma crataegi* and *Pemphigus betae* both had female-biased investment ratios, and approximately constant investment in males. *Prociphilus corrugatans* and *Prociphilus fraxinifolii* had investment ratios close to 0.5, and Moran suggests that this is because they outbreed: the sexuparae of *P. corrugatans* occur in very large aggregations and both species have a large variance in sex ratio, often producing unisex broods. The investment ratio has also been shown to be female biased in *Eriosoma ulmi* (Janiszewska-Cichocka 1969) and a constant investment in males is noted in two unnamed *Eriosoma* species by NJB Parker & AFG Dixon (unpublished; cited in Kindlmann & Dixon 1989).

Local mate competition is almost certainly the reason for the female-biased sex ratio recorded in the high-arctic aphidine *Acyrthosiphum svalbardicum* (Strathdee *et al.* 1993a). The sexuals in this species (the females are slightly larger than the males, Strathdee *et al.* 1993b) are produced either directly from the foundress or from the second-generation parthenogenetic females. All the morphs are wingless and restricted to patches of the host plant. The sex ratio produced by the foundress (0.30–0.35) is less female biased than that produced by the second generation of viviparous females (0.24–0.25); this may be related to differences in patch size, but data are not available. It is also possible that, in many other aphid species with wingless sexuals, the males produced within a single clone compete for a limited number of females, principally their clone mates, and these species might be expected to have very female-biased investment ratios.

12.5.1.2 Local resource competition

In aphid species in which the males disperse but the females remain on the maternal patch, we might expect to find local resource competition (LRC), as suggested by Moran (1993). The conditions for LRC occur in some non-host-alternating Aphidinae, Calaphidinae and Pterocommatinae. Here females from within one clone might compete with each other for resources, but the males could disperse. Here, in a kind of mirror image of LMC, male-biased sex ratios would be selected for and one might observe the outcome of a constant female hypothesis: small mothers would invest only in females but would produce mainly males above a certain level of female production. This suggestion requires formal modelling: there are currently no experimental data available with which to test the predictions.

12.5.2 Variation in environmental conditions

A second set of factors that might affect the fitness return to the clone of the two sexes is the environmental conditions in which they develop. Even in the absence of subdivisions within the population, there may be circumstances when it is advantageous for the clone to invest in one sex rather than the other. Two scenarios may be relevant in aphids. First, in non-host-alternating species with winged males and wingless females, investment in sons may provide an escape from deteriorating local conditions (Moran 1993). There is certainly evidence that investment can become more male biased in more stressful conditions (e.g. *Masonaphis maxima* (Gilbert 1980); *Uroleucon gravicorne* (Moran 1993)). This could perhaps be thought of as a gender-specific subset of winged/wingless polyphenism in aphids generally.

Another trend is for larger and more fecund mothers to produce more female-biased broods (e.g. *Sitobion avenae* (Newton & Dixon 1987); *Masonaphis maxima* (Gilbert 1980); *Uroleucon gravicorne* (Moran 1993)). Can this be interpreted as condition-dependent sex allocation, equivalent to parasitoids laying mainly females in larger hosts and mainly males in smaller hosts (e.g. Charnov *et al.* 1981, Godfray 1994)? We probably need to be cautious in interpreting this trend. For example, while there is clear evidence that more fecund females produce more female-biased broods (e.g. Foster & Benton 1992, Moran 1993), this is almost certainly because of LMC (section 12.5.1.1). The critical difficulty, as in parasitoid wasps, is in obtaining field evidence for the relationship between size and fitness in the two sexes. For example, in the eriosomatines, where each female only produces one egg, it is difficult to see how increasing the investment in any particular female can yield impressive fitness benefits, although it remains possible that egg size is related to female fitness. Since there is some evidence for mate-guarding in *Pemphigus spyrothecae* (Foster & Benton 1992) and other aphids (see Figure 5.6 in Dixon 1998), it might be argued that, if there were no LMC, larger sexuparae should invest more in males, according to a Trivers and Willard (1973) type model.

The aphid *Tamalia coweni* is a possible candidate for a condition-dependent sex-allocation mechanism (Miller & Avilés 2000). The species is not host-alternating and, rather unusually for aphids, both the males and the sexual females are winged. Outbreeding is almost certain and there is probably no scope for demographic effects (section 12.5.3) on sex allocation. There is a female bias in both larger broods and in early season galls. Miller and Avilés suggest that this is because aphids in better condition invest in the more rewarding sex. This is plausible, but there is in fact no evidence that sexuals from larger broods, or earlier produced galls, are larger, or that size has a more significant effect on female than on male fitness. These ideas are worth investigating further.

12.5.3 Demographic effects

The life cycle of host-alternating aphidines (less than 10% of aphid species) has a distinctive feature that, in terms of sex allocation, sets them apart from all other aphids. The focal female (on the secondary host) invests in winged males and winged gynopara, both of which fly back to the primary host, where the gynopara parthenogenetically produces the wingless sexual females.

The relevant allocation ratio is that of males to gynoparae (e.g. Rispe *et al.* 1999). There are two critical features in this system: the delay between the production of the males and (one generation later) the mating females; and the deadline, for oviposition, of leaf-fall on the primary host (Ward & Wellings 1994a,b). A game-theoretic model of sex allocation in these aphids was developed by Ward and Wellings (1994a,b), and the model is also discussed by Dixon (1998). It is difficult to evaluate this model fully, since no mathematical details have ever been published, but its main predictions are intuitively clear.

The clones of these aphidines should switch to sex as late as possible in the season, i.e. one 'gynopara + ovipara' lifetime before leaf-fall. This is analogous to the switch to the production of reproductives in annual social insect colonies, which should occur approximately one worker lifetime before the end of the season (Oster & Wilson 1978), although this is rarely observed owing to the intense reproductive conflicts within most social insect colonies (Bourke & Franks 1995). It is also clear that gynoparae should be produced before males: late-produced gynoparae are in danger of overshooting the deadline and early-produced males might die before finding a mate. The exact allocation ratio depends on the development time of the sexual females relative to the total time between the cueing of sex and leaf-fall, and the mortality rates of the asexual clonal mothers, the gynoparae and the males. The most critical factors are the development time of the egg-laying females and the mortality of the asexual mothers (of the males and the gynoparae). For example, if development time is zero, investment in males and egg-layers should be unbiased; but if the development time of the egg-layers is long, the males must wait longer for their mates to mature, the optimal switch to male production becomes later and the investment in gynoparae increases.

This is clearly an intriguing model. It would be useful to have a detailed version available, so its general validity could be tested more explicitly. Its predictions are consistent with what little is known about sex allocation in these aphidines. Sex allocation is much more difficult to study in these aphids than in eriosomatines; for example, aphids live longer in the laboratory, artificially extending the period of male production, and males may be more mobile than gynoparae, skewing their sampling frequency in the field (Ward & Wellings 1994b). But it seems clear that clones do invest in gynoparae first and then males, which is the exact opposite of what happens in the eriosomatines. This has been shown for example in *Rhopalosiphum padi* (Ward *et al.* 1984, Hullé *et al.* 1999), *Myzus persicae* (Hales *et al.* 1989), *Dysaphis plantaginea* (Bonnemaison 1951) and *Hyalopterus pruni* (Smith 1936). Suction trap data indicate that the investment ratios of aphids migrating between the primary and secondary hosts are strongly female biased (e.g. Wiktelius 1987), as predicted by all the models with realistic values of the lag in egg production. However, Newton and Dixon (1987) reported an allocation of 0.5 in laboratory studies on *Sitobion avenae* (a non-host-alternating, but gynopara-producing, aphidine); they interpreted this as a Fisherian investment ratio, but Ward and Wellings (1994b) point out that this is in fact consistent with their demographic model and only fortuitously Fisherian. They also suggest that the male-biased investment ratios seen in some other aphidines are an artefact of laboratory study (e.g. Hales *et al.* 1989, Guldemond & Tigges 1991). Clearly, more detailed observations and measurements need to be taken on the relevant variables in the field.

Recent observations on the host-alternating aphid *Rhopalosiphum padi* may be relevant to this model. Rispe *et al.* (1999) demonstrated a striking bimodal distribution of sex allocation in this species: in many populations, clones produced either only females (gynoparae) or mainly males. The overall population allocation ratio, from their laboratory studies, was a Fisherian 0.5, although it is not certain that this would also be found in field conditions. A similar pattern has been observed in the cyclically parthenogenetic cladoceran *Daphnia* (Innes & Dunbrack 1993). We understand neither the proximate mechanism nor the adaptive significance of this sex ratio polymorphism.

12.5.4 Cytoplasmic sex ratio distorters

Given the widespread occurrence of sex ratio distorters in insects (e.g. Hurst *et al.* 1997, Chapter 9), it would perhaps be surprising if they did not also occur in aphids. There are a range of secondary symbionts in aphids, selection on which might favour sex ratio distortion. The most likely role for selfish genetic elements would be to suppress sex altogether: permanently asexual lineages have evolved in aphids many times, but there is as yet no evidence for the involvement of selfish endosymbionts. *Wolbachia* has been looked for, but not found, by research workers in more than one laboratory (N Moran, per. comm.).

12.6 | Future work

Aphids provide an underexplored yet potentially highly fruitful arena for sex ratio research, which should provide genuine insights both into the evolutionary biology of sex allocation and into the life-histories of these fascinating animals. The crucial barrier to progress, which compromises all studies of aphid evolutionary biology, is our ignorance of the clonal structure of aphid populations in the wild. Without this information, we can only make guesses about the extent of clonal inbreeding, based on our limited knowledge of aphid life cycles and vagility. However, technology for clonal delineation is rapidly becoming available (e.g. Hales *et al.* 1997, Simon *et al.* 1999) and the application of this to the sexual phase in aphid life cycles should transform our understanding of sex allocation. If aphids were themselves able to discriminate non-clonemates from clonemates, this could have far-reaching effects on their optimal sex ratios. The bulk of evidence, for example for soldier aphids (e.g. Foster 1990, Aoki *et al.* 1991, Stern & Foster 1996, Shibao 1999) and for duelling aphids (Carlin *et al.* 1994), suggests that aphids cannot distinguish kin, but I am aware of only one study on kin discrimination in mating (*Pemphigus spyrothecae*; Foster & Benton 1992). This work needs to be extended to other systems, in particular other, preferably non-eriosomatine, non-host-alternating species, where kin discrimination might be advantageous.

The pemphigines and hormaphidines are probably the easiest aphids in which to study sex allocation. Investment ratios can be measured simply by dissecting the sexuparae, and the males can be readily distinguished because they are smaller. The sexuals, which do not feed, can be kept between glass slides and their entire life-history and mating behaviour observed under the microscope. This needs to be related to observations on patch size and mating behaviour in the wild. The general model ('self-knowledge with avoidance of sib-mating') of Stubblefield and Seger (1990) seems to be appropriate for these insects.

It would also be useful to develop and test the Ward and Wellings (1994a,b) demographic model, by measuring the critical relevant variables in field populations of common and easily available host-alternating aphidines. We also need to establish whether the aphid bacterial symbionts play any *Wolbachia*-like role in sex allocation: the obvious candidate aphids here are eriosomatines, or possibly hormaphidines. *Wolbachia* or other sex ratio distorters should be explicitly tested for in aphids more generally. Finally, if we can find out more about the detailed developmental biology of the sexual morphs and of sex determination generally, this will enable us to understand the constraints that are operating on the focal, sex-determining mothers within the clone.

Acknowledgements

I am very grateful to Nancy Moran, Tony Dixon, Andrew Bourke, Ian Hardy, Don Miller and David Stern for their comments on previous versions of the manuscript.

References

Aoki S, Kurosu U & Stern DL (1991) Aphid soldiers discriminate between soldiers and non-soldiers, rather than between kin and non-kin, in *Ceratoglyphina bambusae*. *Animal Behaviour*, **42**, 865–866.

Blackman RL (1972) The inheritance of life cycle differences in *Myzus persicae* (Sulz) (Hem., Aphididae). *Bulletin of Entomological Research*, **62**, 281–294.

Blackman RL (1987) Reproduction, cytogenetics and development. In: AR Minks & P Harrewijn (eds) *Aphids: their Biology, Natural Enemies and Control*, volume 2A, pp 163–195. Amsterdam: Elsevier.

Blackman RL & Eastop VF (1994) *Aphids on the World's Trees*. London: The Natural History Museum.

Blackman RL & Hales DF (1986) Behaviour of the X chromosomes during growth and maturation of parthenogenetic eggs of *Amphorophora tuberculata* (Homoptera: Aphididae), in relation to sex determination. *Chromosoma (Berlin)*, **94**, 59–64.

Bonnemaison L (1951) Contribution à l'étude des facteurs provoquant l'apparition des formes ailées et sexuées chez les Aphidinae. *Annales des Épiphyties*, **2**, 263–280.

Bourke AFG & Franks NR (1995) *Social Evolution in Ants*. Princeton, NJ: Princeton University Press.

Carlin NF, Gladstein DS, Berry AJ & Pierce NE (1994) Absence of kin discrimination in a soldier-producing aphid, *Ceratovacuna japonica* (Hemiptera: Pemphigidae: Cerataphidini). *Journal of the New York Entomological Society*, **102**, 287–298.

Charnov EL (1982) *The Theory of Sex Allocation*. Princeton, NJ: Princeton University Press.

Charnov EL, Los-de Hartogh RL, Jones WT & van den Assen J (1981) Sex ratio evolution in a variable environment. *Nature*, **289**, 27–33.

Dixon AFG (1972) The 'Interval Timer', photoperiod and temperature in the seasonal development of parthenogenetic and sexual morphs in the Lime aphid, *Eucallipterus tiliae* L. *Oecologia (Berlin)*, **9**, 301–310.

Dixon AFG (1998) *Aphid Ecology*, 2nd edn. London: Chapman & Hall.

Eastop VF (1973) Deductions from the present day host plants of aphids and related insects. In: HF van Emden (ed) *Insect/Plant Relationships. Symposia of the Royal Entomological Society of London*, volume 6, pp 157–178. Oxford: Blackwells.

Fisher RA (1930) *The Genetical Theory of Natural Selection*. Oxford: Clarendon Press.

Foster WA (1990) Experimental evidence for effective and altruistic colony defence against natural predators by soldiers of the gall-forming aphid *Pemphigus spyrothecae* (Hemiptera: Pemphigidae). *Behavioral Ecology and Sociobiology*, **27**, 421–430.

Foster WA & Benton TG (1992) Sex ratio, local mate competition and mating behaviour in the aphid *Pemphigus spyrothecae*. *Behavioral Ecology and Sociobiology*, **30**, 297–307.

Frank SA (1985) Hierarchical selection theory and sex ratios. II. On applying the theory, and a test with fig wasps. *Evolution*, **39**, 949–964.

Frank SA (1987a) Individuals and population sex ratio allocation patterns. *Theoretical Population Biology*, **31**, 47–74.

Frank SA (1987b) Variable sex ratios among colonies of ants. *Behavioural Ecology and Sociobiology*, **20**, 195–201.

Gilbert N (1980) Comparative dynamics of a single-host aphid. *Journal of Animal Ecology*, **49**, 351–369.

Godfray HCJ (1994) *Parasitoids: Behavioral and Evolutionary Ecology*. Princeton, NJ: Princeton University Press.

Guldemond JA & Tigges WT (1991) Production of sexuals and sex ratio in *Cryptomyzus* species in relation to dispersal and host-alternation (Homoptera: Aphidinea: Aphididae). *Entomologia Generalis*, **16**, 257–264.

Hales DF & Mittler TE (1983) Precocene causes male determination in the aphid *Myzus persicae*. *Journal of Insect Physiology*, **29**, 819–823.

Hales DF & Mittler TE (1987) Chromosomal sex determination in aphids controlled by juvenile hormone. *Genome*, **29**, 107–109.

Hales DF, Wellings PW & Parkes RA (1989) Investment in gynoparae and males by *Myzus persicae* (Sulzer). *Functional Ecology*, **3**, 727–734.

Hales DF, Tomiuk J, Wöhrmann K & Sunnucks P (1997) Evolutionary and genetic aspects of aphid biology: a review. *European Journal of Entomology*, **94**, 1–155.

Hamilton WD (1967) Extraordinary sex ratios. *Science*, **156**, 477–488.

Hardie J, Baker FC, Jamieson GC, Lees AD & Schooley DA (1985) The identification of an aphid juvenile hormone and its titre in relation to photoperiod. *Physiological Entomology*, **10**, 297–302.

Heie OE (1980) The Aphidoidea (Hemiptera) of Fennoscandia and Denmark. *Fauna Entomologica Scandinavica*, **9**, 1–236.

Heie OE (1987) Palaeontology and phylogeny. In: AR Minks & P Harrewijn (eds) *Aphids: their Biology, Natural Enemies and Control*, volume 2A, pp 367–391. Amsterdam: Elsevier.

Hullé M, Maurice D, Rispe C & Simon J-C (1999) Clonal variability in sequences of morph production during the transition from parthenogenetic to sexual reproduction in the aphid *Rhopalosiphum padi* (Sternorrhyncha: Aphididae). *European Journal of Entomology*, **96**, 125–134.

Hurst GDD, Hurst LD & Majerus MEN (1997) Cytoplasmic sex ratio distorters. In: SL O'Neill, AA Hoffmann & JH Werren (eds) *Influential Passengers: Inherited Microorganisms and Arthropod Reproduction*, pp 125–154. Oxford: Oxford University Press.

Innes DJ & Dunbrack RL (1993) Sex allocation variation in *Daphnia pulex*. *Journal of Evolutionary Biology*, **6**, 559–575.

Janiszewska-Cichocka E (1969) Zur Morphologie und Biologie der Ulmenblattlaus, *Eriosoma ulmi* (Linnaeus, 1758) (Homoptera, Pemphigidae). *Annales Zoologici*, **27**, 205–221.

Judge FD (1968) Polymorphism in a subterranean aphid, *Pemphigus bursarius*. 1. Factors affecting the development of sexuparae. *Annals of the Entomological Society of America*, **61**, 819–827.

Kawada K (1987) Polymorphism and morph determination. In: AR Minks & P Harrewijn (eds) *Aphids: their Biology, Natural Enemies and Control*, volume 2A, pp 255–268. Amsterdam: Elsevier.

Kindlmann P & Dixon AFG (1989) Role of population density in determining the sex ratios of species that show local mate competition, with aphids as a model group. *Functional Ecology*, **3**, 311–314.

Kurosu U & Aoki S (1991) Why are aphid galls so rare? *Evolutionary Theory*, **10**, 85–99.

Lampel G (1968) *Die Biologie des Blattlaus-Generationswechsels, mit Besonderer Berücksichtugung Terminologischer Aspekte*. Jena: Gustav Fisher Verlag.

Lees AD (1966) The control of polymorphism in aphids. *Advances in Insect Physiology*, **3**, 207–277.

Lees AD (1973) Photoperiodic time measurement in the aphid *Megoura viciae*. *Journal of Insect Physiology*, **19**, 2279–2316.

Lively CM (1987) Facultative parthenogenesis and sex-ratio evolution. *Evolutionary Ecology*, **1**, 197–200.

Loxdale HD, Lushai G, Brookes CP & Allen JA (2002) Intraclonal genetic variation and its potential for evolutionary change. *Biological Reviews* (in press).

Marcovitch S (1924) The migration of Aphididae and the appearance of the sexual forms as affected by the relative length of daily light exposure. *Journal of Agricultural Research*, **27**, 513–522.

Miller DG III & Avilés L (2000) Sex ratio and brood size in a monophagous outcrossing gall aphid, *Tamalia coweni* (Homoptera: Aphididae). *Evolutionary Ecology Research*, **2**, 745–759.

Moran NA (1993) Evolution of sex ratio variation in aphids. In: DL Wrensch & MA Ebbert (eds) *Evolution and Diversity of Sex Ratio in Insects and Mites*, pp 346–368. New York: Chapman & Hall.

Moran NA, Seminoff J & Johnstone L (1993) Induction

of winged sexuparae in root-inhabiting colonies of the aphid *Pemphigus betae*. *Physiological Ecology*, **18**, 296–302.

Newton C & Dixon AFG (1987) Cost of sex in aphids: size of males at birth and the primary sex ratio in *Sitobion avenae* F. *Functional Ecology*, **1**, 321–326.

Nieto Nafria JM, Mier Durante MP & Remaudière G (1998) Les noms des taxa du group-famille chez les Aphididae (Hemiptera). *Revue Francais Entomologie (N.S.)*, **19**, 77–92.

Nunney L & Luck RF (1988) Factors influencing the optimum sex ratio in a structured population. *Theoretical Population Biology* **33**, 1–30.

Orlando E (1974) Sex determination in *Megoura viciae* (Homoptera, Aphididae). *Monitore Zoologico Italiano (N.S.)*, **8**, 61–70.

Oster GF & Wilson EO (1978) *Caste and Ecology in the Social Insects*. Princeton, NJ: Princeton University Press.

Rispe C, Bonhomme, J & Simon J-C (1999) Extreme life-cycle and sex ratio variation among sexually produced clones of the aphid *Rhopalosiphum padi* (Homoptera: Aphididae). *Oikos*, **86**, 254–264.

Searle JB & Mittler TE (1981) Embryogenesis and the production of males by apterous viviparae of the green peach potato aphid *Myzus persicae* in relation to photoperiod. *Journal of Insect Physiology*, **27**, 145–153.

Shibao H (1999) Lack of kin discrimination in the eusocial aphid *Pseudoregma bambucicola* (Homoptera: Aphididae). *Journal of Ethology*, **17**, 17–24.

Simon J-C, Baumann S, Sunnucks P, Hebert PDN, Pierre J-S, Le Gallic J, Dedryver C-A (1999) Reproductive mode and population genetic structure of the cereal aphid *Sitobion avenae* using phenotypic and microsatellite markers. *Molecular Ecology*, **8**, 531–545.

Smith LM (1936) Biology of the mealy plum aphid, *Hyalopterus pruni* (Geoffroy). *Hilgardia*, **10**, 167–209.

Stern DL & Foster WA (1996) The evolution of soldiers in aphids. *Biological Reviews*, **71**, 27–79.

Strathdee AJ, Bale JS, Block WC, Webb NR, Hodkinson ID & Coulson SJ (1993a) Extreme adaptive life-cycle in a high arctic aphid. *Acyrthosiphon svalbardicum*. *Ecological Entomology*, **18**, 254–258.

Strathdee AJ, Bale JS, Block WC, Webb NR, Hodkinson ID & Coulson SJ (1993b) Identification of three previously unknown morphs of *Acyrthosiphon svalbardicum* Heikinheimo (Hemiptera: Aphididae) on Spitsbergen. *Entomologica Scandinavica*, **24**, 43–47.

Stubblefield JW & Seger J (1990) Local mate competition with variable fecundity: dependence of offspring sex ratios on information utilization and mode of male production. *Behavioral Ecology*, **1**, 68–80.

Sunnucks P, Chisholm D, Turak E & Hales DF (1998) Evolution of an ecological trait in parthenogenetic *Sitobion* aphids. *Heredity*, **81**, 638–647.

Trivers RL & Willard DE (1973) Natural selection of parental ability to vary the sex ratio of offspring. *Science*, **179**, 90–92.

Ward SA & Wellings PW (1994a) Clonal ontogeny and aphids' sex ratio. In: SR Leather, AD Watt, NJ Mills & KFA Walters (eds) *Individuals, Populations and Patterns in Ecology*, pp 397–408. Andover: Intercept.

Ward SA & Wellings PW (1994b) Deadlines and delays as factors in aphid sex allocation. *European Journal of Entomology*, **91**, 29–36.

Ward SA, Leather SR & Dixon AFG (1984) Temperature prediction and the timing of sex in aphids. *Oecologia (Berlin)*, **62**, 230–233.

Wiktelius S (1987) The role of grassland in the yearly life-cycle of *Rhopalosiphum padi* (Homoptera: Aphididae) in Sweden. *Annals of Applied Biology*, **110**, 9–15.

Yamaguchi Y (1985) Sex ratios of an aphid subject to local mate competition with variable maternal condition. *Nature*, **318**, 460–462.

Chapter 13

Sex ratios in birds and mammals: can the hypotheses be disentangled?

Andrew Cockburn, Sarah Legge & Michael C. Double

13.1 | Summary

Birds and mammals are sometimes viewed as black sheep, notorious for their failure to pay attention to the elegance of adaptive sex ratio theory to the same extent as other animals. Here we show that this accusation is unfair. A number of hypotheses contribute to our understanding of why birds and mammals might alter the sex ratio of their young. Most importantly, the long-term studies that have addressed sex ratios have inevitably come to the conclusion that sex allocation is subject to multiple influences. This suggests that where the constraint of Mendelian sex allocation is overcome, a suite of adaptive and nonadaptive influences will combine to affect the sex ratio. As a consequence, predictions of strong bias resulting from the tenets of a single hypothesis are unlikely to be met. Here we review profitable approaches to untangling this complexity.

13.2 | Introduction

As will be evident from earlier chapters in this book, the study of sex allocation has proved one of the most triumphant areas of evolutionary theory, with explicit theoretical predictions often anticipating parental investment in male and female offspring with great precision. However, comparable progress in understanding sex allocation in birds and mammals has proved much more difficult. Several obstacles have been recognized. First, sex determination is almost ubiquitously associated with chromosome heterogamety, constraining the mechanisms that might be used to alter sex ratios (Williams 1979, 1992). Fisher's original prediction of equal investment in male and female offspring leads to similar predictions for the sex ratio to those that arise from a model of random segregation of sex chromosomes at meiosis, except where offspring impose obviously different costs on the parents (Williams 1979). Second, literature reports of sex ratios are too unevenly distributed across species to comment with confidence on the frequency of sex ratio control in any taxon. This could arise from a disinclination to publish results where sex ratios do not depart from equal investment, because of the ambiguous meaning of balanced sex ratios (Festa-Bianchet 1996). Third, the primary and secondary sex ratios of many species cannot easily be determined. Fourth, the mechanisms of sex allocation in almost all species are poorly understood (Krackow 1995a,b, 1999). Fifth, the sex-allocation patterns that have been reported in the literature are famed for their inconsistency (e.g. Clutton-Brock & Iason 1986, Hewison & Gaillard 1999, Radford & Blakey 2000).

Despite these serious difficulties, there are theoretical and practical reasons why sex ratio control in birds and mammals should be of unusual interest. First, birds and mammals are homeothermic, and suffer its necessary corollary: rearing a brood involves very heavy investment in a comparatively small number of young. Indeed, many birds and mammals only rear a single offspring at a time, so the choice between son and daughter is particularly stark. Second, birds and mammals often have complicated, highly structured societies, negating the assumption of panmixia (population-wide random mating) that lies at the heart of models that predict equal investment in sons and daughters (Fisher 1930). Third, the life cycle of birds and mammals involves complicated interactions between overlapping generations to an extent that makes theory much more difficult. Fourth, empirical studies of birds and mammals have sometimes revealed such strong patterns in sex allocation that it is clear that the tyranny of Mendelian assortment of sex chromosomes can be overcome in some cases. Last, a window into hitherto intractable problems has been provided by the advent of new molecular techniques that allow gender to be determined very early in development (Griffiths et al. 1998).

These considerations prompt this review, in which we develop an optimistic prognosis for future research on avian and mammalian sex ratios. Because this topic has been reviewed from a variety of perspectives quite recently, we have narrowed our focus relative to previous reviews (Clutton-Brock 1986, Clutton-Brock & Iason 1986, Frank 1990, Clutton-Brock 1991, Gowaty 1993, Hardy 1997, Sheldon 1998). First, our main aim is to consider factors influencing the relative number of male and female offspring (sex ratio) rather than the resources devoted to male and female offspring (sex-biased investment), except where the latter imposes selection on the former. Second, we hope to illustrate profitable approaches and potential pitfalls in the study of sex ratios. We do this by considering some examples at considerable length, rather than attempting an exhaustive tabulation of recent studies.

13.3 | Is the sex ratio constrained?

The extraordinary sex ratios noted in many (mainly invertebrate) animals appear to be the exception in both birds and mammals (Williams 1979, Clutton-Brock 1986, Clutton-Brock & Iason 1986, Cockburn 1990, Krackow 1995b). Although several apparently exciting patterns have been

recorded in birds and mammals, they often do not withstand the accumulation of further data (Palmer 2000). For example, initial studies of great tits, Parus major, suggested that the father's body size (Kölliker et al. 1999) and the mother's reproductive state (Lessells et al. 1996) influence sex allocation, but these results were not sustained in a larger dataset (Radford & Blakey 2000). Indeed, some authors have argued that sex-allocation research is also plagued by publication biases, so that the cases of bias reported in the literature represent an even smaller subset of available data (Festa-Bianchet 1996). Our lack of understanding of the physiological basis of sex allocation has also led to scepticism about the extent of adaptive manipulation of the sex ratio (Krackow 1995b, 1997a, 1999).

Several hypotheses could explain the rarity of extraordinary biases:

A. Selection for sex ratio manipulation is weak or absent. Sex ratios are determined largely by random Mendelian segregation of sex chromosomes, which gives a sex ratio of 0.5 (proportion male).

B. Selective pressures favouring biased ratios are present, but they are insufficient to overcome the constraints imposed by the Mendelian segregation of sex chromosomes. In these circumstances it is better to bias patterns of investment in male and female offspring rather than the sex ratio.

C. Selection favours equal investment in males and females. Because males and females are equally costly, such equal investment is easily achieved by the Mendelian segregation of sex chromosomes.

D. Selection primarily leads to facultative adjustment around a sex ratio close to parity, and as a consequence bias is not expected.

E. Once sex ratio control has arisen, the social complexity of vertebrate societies is likely to pose a suite of differing selection pressures. Once there is no mechanistic constraint, all these pressures will be free to act, potentially operating in different directions, with the net result that (large) bias is rare.

Testing hypotheses of absolute constraint is extremely difficult. Ultimately, we believe insights might arise from comparing the frequency of biased ratios in the different groups of birds and mammals. The nature of sex determination differs between eutherians, marsupials and birds, and hence the levels of constraint may differ. In birds, females are heterogametic (WZ), while males are homogametic (ZZ). In mammals, the converse is true; males are heterogametic (XY) and females are homogametic (XX). In mammals the presence of two copies of X-linked genes in females and one in males is balanced by inactivation of one of the X chromosomes (dosage compensation: Heard *et al.* 1997). However, marsupials and eutherians have different forms of dosage compensation. In eutherians, each individual is a mosaic of tissues that differ in whether the maternal or paternal X has been inactivated. By contrast, in marsupials all inactivated X-linked genes are paternally derived, and inactivation is incomplete and less stable (Graves 1996). These differences have the potential to influence the role of sex-linked genes in social evolution, and hence to influence the sex ratio (Haig 2000).

Another fundamental difference among these groups is that the constraints of flight have prevented birds from producing all their brood at once. This imposes an order effect within the brood that can be much more pronounced than occurs in mammals, particularly when incubation commences before the clutch is complete, causing the early hatching of some nestlings which subsequently enjoy a substantial size advantage over their nest mates. Nearly all repeatable biases in the sex ratio of birds are associated with changes in the sequence with which young are produced (Krackow 1999).

Unfortunately, the quality and availability of data currently preclude comparison between the major groups. In birds and marsupials, the young hatch into a nest or attach into a pouch at a very early stage of development, allowing direct intervention by the parents in the brood, and direct observation by students of sex ratios. In eutherians, the situation is often much less tractable, as young are often concealed in nests until they are well developed. It is not possible to infer primary sex ratios from a sample of older offspring, because mortality can be sex-biased (Fiala 1980). Until recently, the availability of avian data has also been constrained because the sexes usually

do not differ until late in development. It may therefore be unsurprising that biased sex ratios are recorded most frequently from marsupials (Cockburn 1990). However, the problem of determining the sex of morphologically indistinguishable young and embryos has been resolved by genetic techniques that allow the sexing of day-old bird nestlings and embryos recovered from eggs (Griffiths *et al.* 1998, Fridolfsson & Ellegren 1999), as well as mammalian embryos (Valdivia *et al.* 1993). Appropriate comparisons may become possible in the future.

The best evidence that constraints are not absolute comes from the extraordinary sex ratios recorded from some birds and mammals. For example, biases reported for the wood lemming *Myopus schisticolor* (Kalela & Oksala 1966), marsupial antechinuses *Antechinus swainsonii* and *A. agilis* (Cockburn *et al.* 1985a, Davison & Ward 1998), and Seychelles warblers *Acrocephalus sechellensis* (Komdeur *et al.* 1997) approach those seen in extreme invertebrate cases. While we examine these cases in detail below, we also highlight numerous other cases that lead us to suggest that the intrinsic complexity of vertebrate societies may also be of great importance in constraining the extent of bias (Hypothesis E above). Once control over the sex ratio has been attained, numerous selective pressures will affect most species simultaneously or sequentially. This complexity will hinder the development of extreme sex ratios that can result from a single selective pressure. In order to understand these arguments, it is first necessary to review the ways that sex ratio could respond to selection.

13.4 | Adaptive models of sex allocation

There are five main classes of models purporting to provide adaptive explanations of sex ratio patterns in birds and mammals (Box 13.1). In this section, we review each of these in turn, highlighting some of the difficulties in testing those hypotheses, and showing that they cannot be considered independently of each other. In the following section, we use case studies to examine how the hypotheses interact in well-studied populations. We conclude by

Box 13.1 | A classification of adaptive hypotheses that have been applied to explain sex ratio in bird and mammal societies

1. Sex allocation reflects the frequency-dependent advantage enjoyed by the rare sex (Fisherian hypotheses).

    1.1 **Fisher's hypothesis.** Parents converge on an evolutionarily stable sex ratio where the investment in males and females is approximately equal (section 13.4.1).

    1.2 **Homeostasis hypothesis.** Parents respond to a low number of one sex by producing that rare sex (section 13.4.1).

2. Parents in good condition invest most in the sex that derives the greater increase in reproductive value from a given level of investment (Trivers-Willard hypotheses).

    2.1 **Narrow sense Trivers-Willard hypothesis.** In polygynous societies, males have a higher variance in reproductive success. Increased maternal investment will have a larger effect on the return from sons from mothers, so mothers in good condition produce sons and mothers in poor condition produce daughters (section 13.4.2).

    2.2 **Advantaged daughter hypothesis.** In societies based on female groups, daughters may acquire a rank close to that of their mothers. High-ranked females produce daughters, while low-ranked females produce sons, which usually disperse (section 13.5.2).

    2.3 **Attractiveness hypothesis.** If females can enhance the fitness of their offspring by mating with an attractive partner by virtue of genes for viability and attractiveness, they overproduce sons if they can mate with an attractive male, but produce daughters when paired to an unattractive male (section 13.5.4).

3. In societies where one sex disperses earlier or further than the other, investment reflects the costs or benefits derived from related individuals of the philopatric sex remaining closely associated for a long period of time.

    3.1 **Local mate competition hypothesis.** In environments where a number of individuals colonize a habitat and their progeny mate amongst themselves, sex ratios should be more biased towards daughters as competition among related males to mate with their sisters increases (which occurs when the number of colonizers decreases) (section 13.4.3).

    3.2 **Local resource competition hypothesis.** Where one sex remains philopatric, competition for resources between related members of that sex leads to selection for overproduction on the dispersive sex, which do not compete among themselves or with their mother (section 13.4.4).

    3.3 **Local resource enhancement (repayment) hypothesis.** Where the presence of philopatric descendants enhances the fitness of breeders, there may be selection for overproduction of the philopatric sex (section 13.4.4).

4. Where resources for parental investment are limited, parents can manipulate sex ratios to lessen the risk of reproductive failure and/or increase the prospects that they will survive to reproduce again.

    4.1 **Cost of reproduction hypothesis.** Females in poor condition are reluctant to invest in the sex that imposes greater demands on resources, to minimize the risk of failure or brood reduction, or to reduce costs in terms of future reproductive success (section 13.4.5).

4.2 **Male exploitation hypothesis.** Females exploit differential provisioning of the sexes by males to reduce the costs of parental care (section 13.4.5).

5. Where the competitive interaction between siblings is influenced by their gender, females alter the sex ratio or sex sequence of their brood to maximize productivity.

5.1 **Brood reduction hypothesis.** Females manipulate the sex ratio of their offspring to influence the probability of brood reduction (section 13.5).

examining a number of empirical results that do not seem to conform to any of the existing theories.

13.4.1 Fisher's hypothesis

Fisher (1930) developed the famous argument that parental expenditure on sons and daughters should be equal. Bull and Charnov (1988) provide a thoughtful analysis of the implications, assumptions and empirical basis of Fisher's ideas. In particular, they point out that one of Fisher's motivations was to explain the prevalence of sex ratios of 0.5 in nature. Assuming that sons and daughters are equally costly, this hypothesis could be taken to mean that there should be a sex ratio of 0.5 in most birds and mammals (Hypothesis C above). There are a number of problems with this argument. First, measuring the relative costs of sons and daughters is profoundly difficult, involving the integration of several interacting fitness currencies, and has probably never been satisfactorily accomplished. Moreover, the empirical prevalence of sex ratios of 0.5 could also arise because of the random Mendelian assortment of sex chromosomes (Hypothesis B above) or, because Mendelian assortment achieves the Fisherian aim, there is little selection to deviate from chromosomal sex determination (Hypothesis A above). A sex ratio of 0.5 therefore provides no evidence either way.

Williams (1979) discussed these difficulties, and suggested that good evidence for adaptive control would come from sub-binomial variance around a sex ratio of 0.5. For example, if there was strong selection for producing equal numbers of males and females, mothers producing dizygotic twins should have a son and a daughter, rather than two sons or two daughters. However, this suggestion is controversial, and various hypotheses suggest that sub-binomial or super-binomial variance could arise under Fisher's model, and also as an epiphenomenon of other processes (Kolman 1960, Fiala 1981, Frank 1990, Nagelkerke 1996, Krackow 1997a). This issue is explored in Chapter 5.

A specialized variant of the Fisher hypothesis is the suggestion that we would expect selection to favour production of the rare sex whenever it was in short supply (**homeostasis hypothesis**). Trivers (1985) reviews some data suggesting that these effects may be important in human populations. Creel and Creel (1997) suggested that the legal hunting of male lions *Panthera leo* in the Selous region of Tanzania had led to selection for the production of male cubs. In hunted areas, 66–81% of juveniles are male. However, some explicit tests reject the homeostasis hypothesis (Bensch *et al.* 1999), and its theoretical basis remains questionable (Leigh 1970, Kumm *et al.* 1994, West & Godfray 1997). It therefore remains the case that adaptive moderation of the sex ratio is most likely to be detected when the sex ratio is predicted to vary from 0.5 in response to environmental or social pressures. Balanced sex ratios provide negligible insights into adaptive sex ratio control (Bull & Charnov 1988).

13.4.2 The Trivers-Willard hypothesis and its extensions

Trivers and Willard (1973) developed the most influential hypothesis for mammalian sex ratios. They pointed out that mammals are commonly polygynous, so that one male can monopolize the reproduction of many females, with the obvious corollary that many other males will fail to reproduce. Hence the variance in reproductive success among males will be much greater than the variance among females, which will have no trouble in gaining a mate. Females in good condition should be able to produce higher quality offspring than females in poor condition. There is excellent evidence from a variety of birds and mammals that perturbing early development has important long-term effects on fitness, despite the possibility that some retardation in development can be compensated by catch-up growth (reviewed by Lindström 1999). Therefore, a female that can invest more heavily than other females could potentially gain a greater advantage from investing in sons, because that additional investment increases her potential to produce a high-quality breeding male, and hence reap a substantial harvest of grandchildren. By contrast, a female in poor condition is likely to produce a weaker son, which may fail to reproduce. She should therefore produce daughters, which increase her probability of having at least some grandchildren.

It is easy to see that this hypothesis can be generalized, as has been done on numerous occasions (Charnov 1982). Leimar (1996) provides a particularly insightful extension of the theory relevant to birds and mammals, emphasizing reproductive value. He used state-dependent life-history theory to show that if the impact of a given unit of parental investment enhances the reproductive value of one offspring sex more than the other, females most able to invest should produce the sex that responds more, and lower quality females should produce the less responsive sex.

There has been enormous empirical interest in Trivers-Willard effects. However, there is often a tendency to assume that results supporting one assumption in one population can easily be generalized to a second population where another assumption has been verified. Furthermore, there has been an unfortunate tendency to concentrate on a subset of the many steps in the chain of logic, and assume that the remainder must also be true. For examples of the shortcomings of such assumptions, see Hewison and Gaillard's (1999) analysis of the effect of maternal investment on offspring fitness in ungulates, and Krackow's (1997b) careful experimental studies of sex ratios in mice.

For example, what can we conclude if females in poor condition produce daughters? Although sex differences in susceptibility to early conditions could have a number of explanations, current evidence supports the view that the larger sex may require more resources for normal development, because of the large size and faster growth that result from sexual selection. Males are the larger sex in most birds and mammals, and suffer higher mortality rates under conditions of food restriction (Widdowson 1976, Clutton-Brock et al. 1985, Sheldon et al. 1998). In species with reversed sex dimorphism, the converse is true (Torres & Drummond 1997). Biased sex ratios among mothers in poor condition could therefore result from either differential mortality or adaptive sex ratio bias, and hence provide no direct evidence for Trivers-Willard logic (differential mortality hypothesis: Clutton-Brock 1991).

The most difficult problem is demonstrating that a given unit of investment has a different impact on the reproductive value of sons relative to daughters (Hewison & Gaillard 1999). This is most complicated where the effects on fitness of sons and daughters can operate through different aspects of early development. For example, in red deer *Cervus elaphus*, birth weight affects subsequent male fitness, but not the fitness of females. However, fitness in female deer is affected by population density and cold spring temperatures in the year of birth, but male fitness is not (Albon et al. 1987, Kruuk et al. 1999b).

A further problem arises when there is more than one young in the brood, as a trade-off between the number and size of the young in the brood can alter the appropriate pattern of investment in males and females. A female could produce a large brood because she is in good condition, but her offspring may be of smaller size because of competition for resources (Williams 1979, Frank 1990). There is some evidence to support the idea that polytocous mammals (producing more than one young per litter) are sensitive to these trade-offs, although devising an appropriate set of predictions can be difficult (Cassinello & Gomendio 1996). In one remarkable case, Gosling (1986) showed that in coypu *Myocastor coypus*, females in good condition aborted small litters that were predominantly female, retaining large female-biased litters and small litters containing males. These females were able to conceive another litter quickly, and hence produced numerous small daughters or a few large sons.

It is less clear how birds can produce small females and large males or vice versa. One possibility is to exploit the natural tendency of egg size to decline through the laying sequence, so the larger sex is produced by early eggs and the smaller sex through later eggs (e.g. Bednarz & Hayden 1991, Olsen & Cockburn 1991). However, some birds can maintain sex dimorphism in eggs regardless of laying sequence (Mead *et al.* 1987, Cordero *et al.* 2000), and the pattern is inconsistent with the prediction of the Trivers-Willard model in some others (Weatherhead 1985). We discuss an extremely pronounced tendency for gender to change with laying sequence in section 13.5.4.

A related pattern that occurs in some birds is a seasonal trend in the sex ratio. These are most commonly recorded from raptors, which show reverse size dimorphism. However, despite the strength of the trends in many species, they are inconsistent in direction, with kestrels changing from male-bias to female-bias, and other raptors changing in the opposite direction (Dijkstra *et al.* 1990, Olsen & Cockburn 1991, Zijlstra *et al.* 1992, Daan *et al.* 1996, Smallwood & Smallwood 1998). This trend has been related to whether being fledged early increases the probability of reproducing in the following year. In small species such as kestrels, females are able to breed in the following season if they are fledged early in the season, while the converse may be true for larger species (Daan *et al.* 1996, Smallwood & Smallwood 1998). Theoretical analysis supports this logic (Pen *et al.* 1999).

Before discussing the most famous example supporting the Trivers-Willard hypothesis (section 13.5.1), it is necessary to introduce the alternative hypotheses.

13.4.3 Local mate competition

Hamilton (1967) proposed that when females (termed 'foundresses') colonize a discrete and ephemeral resource patch and their offspring then mate among themselves, competition among males for mates (local mate competition) leads to selection for female-biased sex ratios when one or a small number of foundresses contribute offspring to the mating group. Local mate competition profoundly influences the sex ratio of many animals (e.g. Chapters 10, 11, 19 & 20), but its relevance to birds and mammals seems limited, as they typically occur in quite stable populations. However, several authors have suggested that the isolation of rodent populations into small demes that episodically colonize new patches of habitat could have led to local mate competition (Maynard Smith & Stenseth 1978, Stenseth 1978, Carothers 1980, Bull & Bulmer 1981, Werren & Hatcher 2000).

This suggestion is partly motivated by the remarkable sex-determining system found in some rodents, including some species of lemming, famed for their dramatic population fluctuations. In the wood lemming *Myopus schisticolor*, collared lemming *Dicrostonyx torquatus* and some species of *Akodon* an X chromosome (X*) has evolved that suppresses the male-determining effect of the Y chromosome, so X*Y individuals are female, and the population sex ratio is typically heavily female-biased (Kalela & Oksala 1966, Fredga 1988, Espinosa & Vitullo 1996). In wood lemmings, X*Y females produce only daughters because a double disjunction during mitotic anaphase in the foetal ovary causes the exclusive production of X* eggs. They therefore avoid any fertility penalty associated with the production of YY zygotes. By contrast, in collared lemmings X*Y females produce both X*- and Y-carrying eggs, so they should suffer a substantial fertility cost because of the death of YY individuals. However, the females compensate for this disadvantage by a higher ovulation rate so their fertility is not reduced relative to XX females (Gileva *et al.* 1982). At least in wood lemmings, this X* effect is associated with a deletion in the region Xp21–23 of the X chromosome (Liu *et al.* 1998), an area also associated with the occasional occurrence of XY females in humans (Wachtel 1998). The pattern in rodents is extremely interesting, because it represents the only well understood sex-ratio-modification system in any bird or mammal, and it has evolved convergently several times, suggesting that it may reflect common evolutionary pressures affecting populations of small rodents.

Sadly, it is not clear why such a system has evolved. Notably, studies of population dynamics and habitat use by wood lemmings do not provide evidence for spatial clustering and inbreeding (Gileva & Fedorov 1991, Bondrup-Nielsen et al. 1993, Ims et al. 1993, Stenseth & Ims 1993, Eskelinen 1997), undermining the local mate competition hypothesis.

Local mate competition models have also been criticized by Bulmer (1988) on the basis of work by Gileva (1987), who found evidence for an autosomal gene that leads to preferential fertilization by Y sperm. Bulmer argues that the extent of segregation distortion in Y sperm is consistent with expectations under random mating, as selection for a female-biased sex ratio under local mate competition would eliminate pressure for segregation distortion favouring the production of Y sperm. Furthermore, habitual inbreeding typically eliminates female-biased sex ratios in laboratory colonies for reasons that remain controversial (Jarrell 1995, Gileva 1998). Bulmer (1988) concludes that the X* chromosome indicates the evolution of a selfish element that has not yet been eliminated by the evolution of an appropriate repressor (sex ratio distorters are discussed in Chapter 9).

Benenson (1993) has elaborated models of Bengtsson (1977) to show that X* can be evolutionarily stable and resistant to a Y suppressor provided that it confers a substantial increase in fertility. Field observations support the pattern of fertility required for evolutionary stability in wood lemmings (Bondrup-Nielsen et al. 1993) and tend to do so in collared lemmings (Gileva et al. 1982). However, at least initially, it is likely that populations would have had to pass through a crisis in which fertility was drastically reduced because of the production of YY embryos (Hurst et al. 1996, McVean & Hurst 1996). Evidence for such a crisis comes from the different methods of fertility restoration used by wood and collared lemmings, and in other rodents with X* chromosomes (Fredga 1988, Espinosa & Vitullo 1996). One possible advantage of producing strongly biased litters is that the embryos would not be subject to the intrauterine androgenization that occurs when females occupy sites adjacent to a male embryo, promoting sociability (Bondrup-Nielsen et al. 1993, Espinosa & Vitullo 1996),

though this possibility has not been addressed empirically.

We are thus left disappointed. Detailed unravelling of the mode of sex ratio distortion has not yet led to a clear understanding of the basis of the convergent evolution of the X* chromosome in these rodents. A number of recent reviewers have argued that an understanding of the mechanism will advance adaptive models of sex allocation, or perhaps lead to a rejection of adaptive models (Krackow 1995b, 1997a, Oddie 1998, Krackow 1999). That view is not supported by this example.

13.4.4 Local resource competition and enhancement

In many species, one sex is philopatric, settling near its birthplace. The other sex tends to disperse earlier and further from the birthplace. In mammals, males are usually the dispersive sex, while in birds females are more likely to disperse (Greenwood 1980, Pusey 1987, Wolff 1994, Clarke et al. 1997). The sex that disperses appears to represent an interplay between selection for inbreeding avoidance and a tendency of the sex that gains most from resource defence to remain philopatric (Greenwood 1980, Cockburn et al. 1985b, Pusey 1987, Wolff 1994, Wolff & Plissner 1998). Such philopatry has a profound effect on the intensity of interactions between same-sex relatives. Competition will be most pronounced and prolonged among the philopatric sex. Clark (1978) thus proposed that female mammals should invest more heavily in the dispersive sex, as, unlike the philopatric sex, the return from investment is not devalued by competition for local resources (local resource competition hypothesis).

Clark (1978) also realized that if the presence of relatives enhances reproductive success, the converse would be true, and therefore parents should overproduce the philopatric sex (local resource enhancement hypothesis). African wild dogs, Lycaon pictus, whose males assist the troop with rearing young, overproduce male young if they are giving birth for the first time (Malcolm & Marten 1982, Creel et al. 1998). An identical effect has been reported for red cockaded woodpeckers, Picoides borealis, breeding on a territory for the first time (Gowaty & Lennartz 1985). These results led to the formal modelling of the repayment hypothesis (Emlen *et al.* 1986, Lessells & Avery 1987). These models explicitly predict sex ratios on the basis of measurable parameters to a greater extent than any other model of mammalian and avian sex ratio, and have prompted a number of empirical tests.

Unlike the Trivers-Willard hypothesis, in their simple form local resource competition and enhancement are predicted to lead to biased sex ratios at the level of the population. Not unsurprisingly, biased population sex ratios have been cited as evidence for the operation of these effects (e.g. Cockburn 1990, Gowaty 1993). Conversely, unbiased sex ratios have been cited as evidence of the failure of the repayment models to capture the complexity of these interactions (Koenig & Walters 1999).

There are a number of important reasons why predictions of population-wide sex ratio bias need not be fulfilled. For example, consider sex allocation in the Seychelles warbler, which provides one of the most dramatic examples of avian sex ratio bias. In this species most help is provided by females, and helpers direct care mainly to their relatives (Komdeur 1994a). Helpers enhance productivity on good territories, but this effect diminishes and is reversed as the number of helpers increases (Komdeur 1994b). On poor territories, helpers always depress productivity. As expected, females without helpers on good territories produce daughters (helpers), while females with larger numbers of helpers or those on poor territories produce sons (dispersers) (Komdeur 1996, 1998, Komdeur *et al.* 1997). However, the population sex ratio shows no bias, in contrast to the predictions of the repayment model. Pen and Weissing (2000) have recently shown that if parents facultatively adjust the sex ratio according to the benefit of having helpers, evolutionarily stable strategy models make no clear predictions about the population sex ratio. This is particularly true if parents are able to increase the clutch size when they have helpers, and when the benefit of having helpers declines with increasing helper number. This result is similar to one obtained for the sex ratio when gender is determined by environmental conditions (Frank & Swingland 1988).

A second problem arises because the direction of evolution is assumed: dispersal influences sex allocation. However, the converse could be true. The extent of dispersal could in turn be influenced by the pattern of sex allocation (Perrin & Mazalov 2000). The implications for the population sex ratio are unclear, as this effect has not been formally modelled.

Nonetheless, a number of studies of mammals provide compelling evidence for local resource competition effects at the level of the family or subpopulation. In both rodents and primates, the presence of a single additional female in a social group or habitat patch is sufficient to cause females to increase their production of sons (Perret 1990, 1996, Aars *et al.* 1995). We will return to the complexities of this hypothesis below (section 13.5.2), but here reiterate the importance of measuring the lifetime fitness consequences for male and female offspring (e.g. Komdeur 1998).

13.4.5 Females should adjust the sex ratio according to their ability to withstand the costs of reproduction

An alternative argument encapsulates some of the premises of the Trivers-Willard and the local resource competition hypotheses, but from a completely different perspective. Females in poor condition might be prevented from expressing certain sex ratio options because they are unable to bear the costs of doing so. This could be either because reproduction is likely to fail or be impaired, or because there is an impact on the survival and/or subsequent reproduction of the mother (cost of reproduction hypothesis: Myers 1978, Gomendio *et al.* 1990, Wiebe & Bortolotti 1992). This idea differs from Trivers-Willard logic because the impact is on the reproductive value of the mother rather than that of her offspring (Bensch 1999). It is also conceptually different from the local resource competition model as the emphasis is on the fitness of the mother, rather than on competition among female relatives, including siblings.

There are numerous examples suggesting that females may suffer immediate or deferred life-history costs according to the sex of the offspring they rear. For example, young rhesus macaques, *Macaca mulatta*, suffer subsequent fertility penalties if they rear a son rather than a daughter (Bercovitch & Berard 1993). In bighorn sheep, *Ovis canadensis*, lambs born the year following the weaning of a son had lower survival than lambs born after a daughter, particularly at high population density (Berube *et al.* 1996). The year after weaning a son, ewes were more likely to have a daughter than a son, while ewes that had previously weaned a daughter had similar numbers of sons and daughters. Festa-Bianchet and Jorgenson (1998) provide direct evidence that bighorn sheep females may value their own survival and reproduction over their current offspring's development and survival. A population increase reduced the mass of lambs dramatically, but had only a modest effect on the mass of mothers.

Female southern elephant seals, *Microunga leonina*, only attempt to breed when they weigh more than 300 kg. However, between 300 and 380 kg they always produce female pups (Arnbom *et al.* 1994). Female size varies dramatically (up to a tonne), but there is no relationship between female size and sex ratio for females above 380 kg. Elephant seals fast during lactation, and must suckle their young from a fixed body reserve (Le Boeuf *et al.* 1989). Small females use as much as 85% of their body reserves during weaning while large females may use as little as 45% (Fedak *et al.* 1996). When the size of the calf is controlled, females do not suckle male and female calves differently (Arnbom *et al.* 1997). However, because male calves are larger than females regardless of maternal birth weight, it may be too risky for small females to attempt to rear males, or males may die during gestation.

In at least three species of polygynous birds, primary females produce broods biased towards sons while secondary females produce broods biased towards daughters (Patterson & Emlen 1980, Nishiumi 1998, Westerdahl *et al.* 2000). In these species the male usually assists only the primary female with nest provisioning. Although the pattern could therefore be compatible with Trivers-Willard explanations (Nishiumi *et al.* 1996), Westerdahl *et al.* (2000) found no difference in the fitness of male young raised

in primary and secondary nests of great reed warblers, *Acrocephalus arundinaceus*. The preponderance of daughters in nests of secondary females could reflect greater susceptibility of male nestlings to starvation, or a reduction in total investment in order to increase prospects for future reproduction. However, the bias towards sons could also be because males provide more food to male-biased broods (Westerdahl *et al.* 2000). This male exploitation hypothesis is potentially applicable in any circumstances where males provide different resources to male and female offspring.

The similarity between the sex ratio trend predicted by these hypotheses and the Trivers-Willard hypothesis means that it is difficult to distinguish between them. Clear support for the Trivers-Willard hypothesis depends on showing that a given unit of investment *by high-quality females* affects the reproductive value of their sons and daughters differently. As we commented above, this has rarely been achieved, so we believe that support for the Trivers-Willard hypothesis is much less strong than is often implied.

It is also clear that the Trivers-Willard and cost of reproduction hypotheses are difficult to distinguish from the differential mortality hypothesis. This distinction has been achieved in studies of lesser black-backed gulls, *Larus fuscus*. Male chicks are larger than females, and more susceptible to mortality when food is in short supply (Nager *et al.* 2000). In an impressive experiment, Nager *et al.* (1999) were able to manipulate laying patterns in nests by removing eggs, thus inducing females to lay extended clutches. The size of eggs declined during the laying sequence because of a decline in maternal condition (eggs remained a constant size when mothers were provided with supplementary food). Clutches became progressively more female biased in broods laid by mothers that received no food supplementation, while females with food supplementation produced a slight surplus of females across all brood sizes. This effect occurred over and above any mortality of male embryos, so the cost of reproduction effect occurs in addition to any differential mortality effect. Gulls seem to be manipulating the sex ratio to minimize incapacity of their young or themselves.

13.4.6 Certain brood combinations should be favoured because they prevent fatal combinations of offspring

The final suite of hypotheses deal not with interactions between the parents and their brood, but with the interactions that occur within broods. If certain combinations of same-sexed or different-sexed offspring exacerbate conflict, they could be selected against because they lead to brood reduction, or selected for because they achieve brood reduction efficiently.

Bortolotti (1986) found that bald eagles, *Haliaeetus leucocephalus*, almost never produced broods when the male chick preceded the female chick in the hatching sequence (MF), while all other possibilities (MM, FM & FF) were equally common. This has no effect on the population sex ratio, as the missing combination by definition has a sex ratio of 0.5. However, a simple pattern of this sort leads to a 0.33 ratio among first eggs, and a 0.66 ratio among second eggs. As in other raptors, males are small compared to females. Bortolotti predicted that the MF combination was avoided because it would be particularly prone to conflict. Subsequent data collected over 17 years suggested that MF was much more common in years when food availability was poor, suggesting that in these conditions brood reduction was being actively facilitated (Dzus *et al.* 1996). Several other authors provide data supporting the occurrence of biased sex sequences in reverse sex-dimorphic raptorial birds. Frustratingly, these data are usually presented without attention to the specific combinations, so they cannot be compared directly to the bald eagle data. The diversity of results is bewildering. For example, in Harris's hawks, *Parabuteo unicinctus*, there is a male bias in first chicks (0.69: Bednarz & Hayden 1991); and in blue-footed boobies, *Sula nebouxii*, there is a male bias in second chicks (0.67: Torres & Drummond 1999).

The complexity of these order effects is best demonstrated by laughing kookaburras, *Dacelo novaeguineae* – which are cooperative breeders where a socially and genetically monogamous pair are assisted by both male and female helpers (Legge & Cockburn 2000). The pattern of help is complicated, because the presence of male helpers has no effect on fledging success, and

female 'helpers' may even depress productivity (Legge 2000a). The lack of an effect of supplementary feeding is surprising, as brood reduction is very common and results from siblicide (Legge 2000b). Groups with female helpers produce male-biased broods, which is consistent with local resource competition, as female helpers reduce productivity (Legge *et al.* 2001). However, in this species, brood sex ratios are determined primarily by the sequence of eggs within the brood. Clutch size is usually three. Most first-hatched young are male, while the second egg is usually female; however, the second egg bias is conditional. Females without helpers almost invariably produce daughters, and females that produce a daughter first never produce a son second. Because females are larger than males in this reverse dimorphic species, it may pay the female to produce a large daughter in the second position as a bulwark against siblicide leading to the loss of the second and third chicks.

In a related example, consider the unusual influences on sex allocation in spotted hyaenas, *Crocuta crocuta*. These animals live in large clans dominated by females (East *et al.* 1993). Sibling aggression is intense, and dominance can be established in fights as early as the first day of life (Smale *et al.* 1995). In free-living populations, these sibling interactions are hidden from observation, and the sex ratio can only be inferred once young emerge from burrows. Frank *et al.* (1991) noted that by the time of emergence same-sexed twins were rare, and hypothesized that because individuals compete for rank within their own sex, aggression should be more intense and hence always fatal. In another population, Hofer and East (1997) suggested that FF combinations are most likely to fail. Later work on the clan studied by Frank suggested that the disadvantage suffered by same-sex sibs is confined to periods of intense competition when resources are scarce (Smale *et al.* 1995, 1999). After a clan fission reduced local density, litters containing two daughters often prospered (see also Golla *et al.* 1999). It is not known whether the apparent disadvantage suffered by same-sexed siblings ever leads to manipulation of the sex ratio at birth, but the nature of these interactions confirms that siblicide could influence sex ratios in mammals as well as in birds.

13.5 | When several factors interact

The greatest difficulties of interpretation arise when several selection pressures operate on the same population. In these cases, it is also clear that the pattern of bias at the level of the population is unpredictable (Frank 1987, 1990).

13.5.1 Red deer: a classic study gets complicated

The most compelling and complete data in support of the Trivers-Willard argument come from work by Clutton-Brock and colleagues (Clutton-Brock *et al.* 1984, 1986). They used long-term data on patterns of reproductive success by red deer on the Scottish island of Rum to demonstrate that maternal dominance rank influenced the lifetime reproductive success of male progeny much more markedly than the success of daughters. Sons outperformed daughters of high-ranking mothers, while for low-ranking mothers the converse was true. High-ranking mothers produced more sons than daughters, while low-ranking mothers tended to produce more daughters.

However, Kruuk *et al.* (1999a) recently re-examined the situation on Rum, where the density has been increasing on some parts of the island for many years because culling has been prevented. The tendency of high-ranking females to produce male-biased sex ratios disappeared as the density increased, and also when winter rainfall was high, both conditions that make reproduction more difficult for females. A plausible explanation for the difference is that a different influence on sex ratio comes into play. Because male foetuses are more vulnerable to poor conditions, the mortality of male foetuses may obscure the general tendency of high-ranking females to produce sons. Kruuk *et al.* (1999a) point out that there has never been support for the Trivers-Willard hypothesis from high-density populations of ungulates, suggesting that these interactions might be general. There are some surprising aspects of these data. High-ranking females tend to conceive each year, regardless of conditions, and hence suffer mortality of their sons, whereas low-ranking females often skip a year

of reproduction (Cockburn 1999). However, perhaps the most salutary aspect of this study is that sex ratio biases would probably not have been detected had the study commenced in the 1990s.

13.5.2 Local resource competition and Trivers-Willard effects in primates

A conditional role for the Trivers-Willard effect in primates has also been suggested. One of the most widely studied problems is the relationship between maternal rank and sex ratio. Many primates live in large groups. As group size increases, competition for resources between female relatives will increase. Male dispersal is the norm in many primate societies. Such considerations led Clark (1978) to propose the local resource competition hypothesis. Johnson (1988) showed that male bias in the birth sex ratios of different primate species in zoos increased with the intensity of reproductive competition within female groups in the wild. Furthermore, in spider monkeys where females disperse the sex ratio is biased towards females (McFarland Symington 1987). Early empirical studies suggested that sex allocation in primates was conditional and strongly related to maternal rank. However, contrary to the original formulation of the Trivers-Willard hypothesis, high-ranking females were reported to bias the sex ratio towards daughters (Simpson & Simpson 1982). Silk (1983) proposed that this result could arise under local resource competition because females might use social harassment to reduce the recruitment of daughters into the group, and this may lead to differential loss of the daughters of low-ranking females.

It was soon realized that this result could also arise because of the way that females inherit rank. Females usually acquire a rank just below that of their mother, so, as suggested by Leimar (1996), Trivers-Willard logic predicts that high-ranking females should benefit greatly from producing daughters, while females of lower rank should benefit more from producing sons: the advantaged daughter hypothesis (Altmann *et al.* 1988, Altmann & Altmann 1991).

However, further inspection of primate datasets made it clear that while high-ranked females produce daughters in some populations, in others there is no effect and in some populations high-ranking females overproduce sons

(Hiraiwa-Hasegawa 1993). All three outcomes have been reported for rhesus macaques. Some authors have dismissed this variation as indicative of sampling artefacts, as even in long-term studies of impressive breadth sample sizes tend to be modest. However, comparative analysis shows that in populations of cercopithecine primates (baboons, macaques, mangabeys and other African monkeys with cheek pouches) with good growth rates, where local competition is presumed to be low, high-ranking mothers favour males (van Schaik & Hrdy 1991). In populations where the growth rate is low, and hence competition is intense, high-ranking females overproduce daughters. These data are consistent with the operation of a Trivers-Willard advantage of transferring good condition to sons while density is low, but an overwhelming importance of local resource competition at high density.

This argument is remarkably similar to that proposed by Kruuk et al. (1999a) to explain long-term variation in red deer sex ratios (section 13.5.1). Hiraiwa-Hasegawa (1993) was critical of these arguments because of uncertainty over whether it had ever been demonstrated that the fitness of sons would be differentially enhanced by primate mothers, but increasingly all the key assumptions have been demonstrated (Dittus 1998, Bercovitch et al. 2000). The convergence between results from long-term studies of primate societies and ungulates confirms the importance of long-term data in sex ratio studies, and also suggests a profitable avenue for achieving synthesis of results from different data.

13.5.3 Local resource competition and Trivers-Willard effects in antechinuses

The simplest population structure known from birds and mammals occurs in those marsupials where all males die abruptly after mating (Lee & Cockburn 1985). Some populations show extreme biases in the sex ratio (Cockburn et al. 1985a, Dickman 1988), and some bias must occur in the primary sex ratio (Davison & Ward 1998), possibly via differential sperm assortment (Cockburn 1990). The best studied species is Antechinus agilis (hitherto A. stuartii), a small carnivorous marsupial that inhabits wet forests in south-eastern Australia. The young of this species can be sam-

pled at an early stage because young are born weighing only 16 mg, and spend the first few weeks of their life obligatorily attached to the teat. Two explanations have been proffered for strong interpopulation differences in sex ratios. First, local resource competition could be important. Sex ratios are most female biased as mothers approach complete semelparity (reproducing only once in a lifetime: Cockburn et al. 1985a, Davison & Ward 1998), supporting the view that mothers adjust their sex ratio in response to likely mother-daughter competition, as all males disperse and daughters are highly philopatric (Cockburn et al. 1985b). Second, Trivers-Willard effects are implicated by an increased production of males when food is supplemented (Dickman 1988).

In the hope of disentangling the two hypotheses, Cockburn (1994) studied individual variation in a population where it transpired that the secondary sex ratio among primiparous females was neither biased nor showed any evidence of departure from binomial expectation. However, frequent infanticide led to dramatic departures from binomial expectation by the time of weaning. Mothers with female-biased litters killed daughters regardless of their own underlying condition, so the number of daughters weaned was typically only two (from an initial brood containing an average of four daughters). Mothers always weaned at least one daughter. By contrast, mothers typically weaned all their sons or none of them. Infanticide against sons was strongly related to known predictors of maternal performance. In particular, older females (that exhibited senescence) were much more likely to kill their sons. Infanticide against daughters is most consistent with the local resource competition hypothesis, but infanticide against sons is more difficult to interpret. While it may indicate that the Trivers-Willard effect applies, it could also act as prophylaxis against reproductive failure. Survivorship of mothers declines strongly with the number of sons in their litter, and this effect is most pronounced for older mothers.

This example provides a further case where extremely strong sex ratio effects occurred in response to local resource competition, but without biasing sex ratios at the level of the population.

13.5.4 Zebra finches: a case study in confusion

Species that are easily reared in captivity seem best suited to rapid advances in our understanding of the mechanism of sex ratio control. Considerable attention has been paid to sex allocation in zebra finches, *Poephila guttata*. The first evidence of maternal manipulation of the sex ratio came from Burley's provocative work that exploited the mother's ability to enhance or diminish the attractiveness of zebra finches by using different coloured leg bands (Burley 1981, 1986). Females pair preferentially with males with the attractive leg bands. Females paired to attractive males are more likely to invest in sons, whereas those with less attractive mates invest in daughters. These results were exciting because they extended Trivers-Willard logic in two ways. First, they potentially removed polygyny as a prerequisite for Trivers-Willard effects. Second, they suggested that the attractiveness of the partner could motivate females to invest differently in their young (attractiveness hypothesis).

Subsequently, Gil *et al.* (1999) showed that females paired to attractive males deposit more testosterone in eggs, although testosterone concentrations tend to decline with hatch order. In other bird species, developing young exposed to testosterone grow faster, beg more vigorously and acquire higher dominance status. Gil *et al.* (1999) speculate that females do not always provide additional testosterone to their young, either because: (1) the females themselves suffer a cost associated with greater aggressive interactions with neighbours, and only high-quality chicks are worth this investment or (2) only high-quality chicks can cope with the additional investment. This is a version of the differential allocation hypothesis (Burley 1988, Sheldon 2000), which argues that because investment by iteroparous females (those that breed several times in their lifetime) in any brood should be balanced against future investment, females should ordinarily be conservative, but should invest more heavily if the current offspring are of unusually high quality.

There are considerable difficulties in extending these models to zebra finches. First, Kilner (1998) showed that within zebra finch broods, there is extreme female bias in early laid and hatched chicks, while late laid and hatched chicks are exclusively male (this result is analysed in greater detail in Chapter 3). If Kilner's results are generally true, and they represent one of the strongest sex ratio patterns thus far reported from birds, it is primarily the females that would be the recipients of additional testosterone, as they are produced earliest in the sequence. This is directly contrary to Burley's observation that females with attractive mates will primarily invest in sons of presumed high quality. Furthermore, it is not immediately obvious why female chicks should benefit more from testosterone, or be more resilient to it, compared to male chicks.

Indeed, female zebra finches with abundant food produce more daughters for a given hatch rank (Clotfelter 1996), and female chicks are much more sensitive to mortality when food is restricted (Bradbury & Blakey 1998, Kilner 1998), or where broods are experimentally enlarged (De Kogel 1997). Kilner (1998) argues that this reflects the greater sensitivity of female reproductive success to condition, as zebra finches breed opportunistically in response to rare rainfall events. Females strip their body reserves in order to breed, and their eventual fecundity is related to their weight at fledging. If this is true, according to Trivers-Willard logic females in good condition should produce daughters. Once again, these observations run directly counter to Burley's view that male attractiveness drives sex allocation in this species. Indeed, attractive males make poor parents, which may exacerbate the problems that females have in rearing young (Burley 1988). Unfortunately, neither of these exactly opposite predictions concerning the effect of additional female investment on the lifetime success of their offspring has been tested rigorously in the field, and the natural history of the species conspires against the likelihood that this will be achieved (Zann 1996). Nor have any of the aviary studies of zebra finches yet been designed to investigate both possibilities.

Despite the unsatisfactory nature of current results from zebra finches, the attractiveness hypothesis has been investigated in a number of other species, with mixed results. Females with attractive partners produce more males in collared flycatchers, *Ficedula albicollis* (Ellegren *et al.* 1996, but see Sheldon & Ellegren 1996), and blue tits, *Parus caeruleus* (Kempenaers *et al.* 1997, Sheldon *et al.* 1999), but not in great reed warblers, *Acrocephalus arundinaceus* (Westerdahl *et al.* 1997), red-winged blackbirds, *Agelaius phoeniceus* (Westneat *et al.* 1995), or barn swallows, *Hirundo rustica* (Saino *et al.* 1999). These data are too few and too new to allow any generalizations to be drawn, but future studies should concentrate on tests that allow alternative hypotheses to be evaluated.

13.6 | Unexplained empirical results: artefact or insight

This chapter has placed sex ratio research within an explicit theoretical context. However, some patterns, though dramatic and extremely interesting, sit uncomfortably within any of the existing models. As we have seen, these include sex ratio modification in lemmings (section 13.4.3) and the relationship between hatching sequence and sex in zebra finches (section 13.5.4). In an extreme example, Heinsohn *et al.* (1997) reported that captive eclectus parrots, *Eclectus roratus*, can produce extremely long runs of individuals of one sex. For example, a female at Chester Zoo produced 30 sons before producing a single daughter. Another produced 20 sons before fledging 13 daughters in a row. As the name suggests, eclectus parrots have a singular biology, combining cooperative breeding with reverse sex dichromatism that is displayed conspicuously shortly after hatching. While these traits suggest that the sex ratio is malleable, and that the sexes may be under unusual selection pressures, the most likely explanation for the long runs is that a facultative sex-allocation mechanism has been tipped dramatically in a single direction by the unusual conditions of captivity. It may also be that new theory will be required to interpret these troublesome patterns (see a recent approach by Mesterton-Gibbons & Hardy 2001).

13.7 | Concluding advice

In summary, we suggest the following guides to successful pursuit of research on sex ratios in the complex selective milieu provided by mammal and bird societies:

- Biased sex ratios show that sex ratio modification is occurring, but tell us almost nothing about the underlying mechanistic or adaptive cause (sections 13.4.1, 13.4.2, 13.4.4, 13.5).
- Some key empirical patterns do not sit comfortably within the framework provided by available theory. Do not assume that all results have to be fitted into the straightjacket of theory (section 13.6). All results, including negative ones, are valuable in dissecting hypotheses of constraint (section 13.3).
- Knowledge of mechanism will be immensely useful, particularly in the design of manipulative experiments, but it does not necessarily lead to an understanding of the adaptive basis of sex allocation (sections 13.4.2 and 13.4.3).
- The literature is littered with studies that are described as being 'consistent with' the predictions of one hypothesis or another. The real distinction between the predictions of hypotheses such as the Trivers-Willard model, the cost of reproduction model and the differential mortality model are slight, and require focus on the key predictions, not just one step in the chain of logic (section 13.4.5).
- As is often the case, cleverly designed experiments are likely to afford far greater insights than correlative studies (all sections). However, it is important that the experiments deal with the issues elaborated in this section, and that sample sizes are adequate.
- Most important, detailed long-term analysis of sex ratios has inevitably uncovered complexity rather than concluding that sex ratios vary in response to a single factor (section 13.5). Resist the temptation to assume that a pattern present in one year will be present the next, or that you have uncovered a system responding to a single factor.

Acknowledgements

Our work on sex ratios has been improved by insights from David Green, Rob Heinsohn, Sarah Hrdy, Chris Johnson and Penny Olsen, and supported by the Australian Research Council.

Graeme Buchanan, Ian Hardy, Becky Kilner, Sven Krackow, John Lazarus, Marty Leonard, Penny Olsen and Ben Sheldon provided helpful comments on the manuscript.

References

Aars J, Andreassen HP & Ims RA (1995) Root voles: litter sex ratio variation in fragmented habitat. *Journal of Animal Ecology*, **64**, 459–472.

Albon SD, Clutton-Brock TH & Guinness FE (1987) Early development and population dynamics in red deer. II. Density-independent effects and cohort variation. *Journal of Animal Ecology*, **56**, 69–81.

Altmann M & Altmann J (1991) Models of status-correlated bias in offspring sex ratio. *American Naturalist*, **137**, 556–566.

Altmann J, Hausfater G & Altmann SA (1988) Determinants of reproductive success in savannah baboons. In: TH Clutton-Brock (ed) *Reproductive Success*, pp 403–418. Chicago, IL: University of Chicago Press.

Arnbom T, Fedak MA & Rothery P (1994) Offspring sex ratio in relation to female size in southern elephant seals, *Mirounga leonina*. *Behavioral Ecology and Sociobiology*, **35**, 373–378.

Arnbom T, Fedak MA & Boyd IL (1997) Factors affecting maternal expenditure in southern elephant seals during lactation. *Ecology*, **78**, 471–483.

Bednarz JC & Hayden TJ (1991) Skewed brood sex ratio and sex-biased hatching sequence in Harris's hawks. *American Naturalist*, **137**, 116–132.

Benenson EA (1993) On the maintenance of the unique sex determination system in lemmings. *Oikos*, **41**, 211–218.

Bengtsson BO (1977) Evolution of the sex ratio in the wood lemming, *Myopus schisticolor*. In: TM Fenchel & FB Christiansen (eds) *Measuring Selection in Natural Populations*, pp 333–343. Berlin: Springer-Verlag.

Bensch S (1999) Sex allocation in relation to parental quality. In: NJ Adams & RH Slotow (eds) *Proceedings of the 22nd International Ornithogical Congress, Durban*, pp 451–466. Johannesburg: BirdLife South Africa.

Bensch S, Westerdahl H, Hansson B & Hasselquist D (1999) Do females adjust the sex of their offspring in relation to the breeding sex ratio? *Journal of Evolutionary Biology*, **12**, 1104–1109.

Bercovitch FB & Berard JD (1993) Life history costs and consequences of rapid reproductive maturation in female rhesus macaques. *Behavioral Ecology and Sociobiology*, **32**, 103–109.

Bercovitch FB, Widdig A & Nurnberg P (2000) Maternal investment in rhesus macaques (*Macaca mulatta*): reproductive costs and consequences of raising sons. *Behavioral Ecology and Sociobiology*, **48**, 1–11.

Berube CH, Festa-Bianchet M & Jorgenson JT (1996) Reproductive costs of sons and daughters in rocky mountain bighorn sheep. *Behavioral Ecology*, **7**, 60–68.

Bondrup-Nielsen S, Ims RA, Fredriksson R & Fredga K (1993) Demography of the wood lemming (*Myopus schisticolor*). In: NC Stenseth & RA Ims (eds) *The Biology of Lemmings. Linnean Society Symposium Series No. 15.* pp 493–507. London: Academic Press.

Bortolotti GR (1986) Influence of sibling competition on nestling sex ratios of sexually dimorphic birds. *American Naturalist*, **127**, 495–507.

Bradbury RR & Blakey JK (1998) Diet, maternal condition, and offspring sex ratio in the zebra finch, *Poephila guttata*. *Proceedings of the Royal Society of London, series B*, **265**, 895–899.

Bull JJ & Bulmer MG (1981) The evolution of XY females in mammals. *Heredity*, **47**, 347–365.

Bull JJ & Charnov EL (1988) How fundamental are Fisherian sex ratios? *Oxford Surveys in Evolutionary Biology*, **5**, 96–135.

Bulmer M (1988) Sex ratio evolution in lemmings. *Heredity*, **61**, 231–233.

Burley N (1981) Sex ratio manipulation and selection for attractiveness. *Science*, **211**, 721–722.

Burley N (1986) Sex-ratio manipulation in color-banded populations of zebra finches. *Evolution*, **40**, 1191–1206.

Burley N (1988) The differential allocation hypothesis: an experimental test. *American Naturalist*, **132**, 611–628.

Carothers AD (1980) Population dynamics and the evolution of sex determination in lemmings. *Genetical Research*, **36**, 199–209.

Cassinello J & Gomendio M (1996) Adaptive variation in litter size and sex ratio at birth in a sexually dimorphic ungulate. *Proceedings of the Royal Society of London, series B*, **263**, 1461–1466.

Charnov EL (1982) *The Theory of Sex Allocation*. Princeton, NJ: Princeton University Press.

Clark AB (1978) Sex ratio and local resource competition in a prosimian primate. *Science*, **201**, 163–165.

Clarke AL, Saether BE & Roskaft E (1997) Sex biases in avian dispersal: a reappraisal. *Oikos*, **79**, 429–438.

Clotfelter ED (1996) Mechanisms of facultative sex-ratio variation in zebra finches (*Taeniopygia guttata*). *Auk*, **113**, 441–449.

Clutton-Brock TH (1986) Sex ratio variation in birds. *Ibis*, **128**, 317–329.

Clutton-Brock TH (1991) *The Evolution of Parental Care*. Princeton, NJ: Princeton University Press.

Clutton-Brock TH & Iason GR (1986) Sex ratio variation in mammals. *Quarterly Review of Biology*, **61**, 339–374.

Clutton-Brock TH, Albon SD & Guinness FE (1984) Maternal dominance, breeding success and birth sex ratios in red deer. *Nature*, **308**, 358–360.

Clutton-Brock TH, Albon SD & Guinness FE (1985) Parental investment and sex differences in juvenile mortality in birds and mammals. *Nature*, **313**, 131–133.

Clutton-Brock TH, Albon SD & Guinness FE (1986) Great expectations: maternal dominance, sex ratios and offspring reproductive success in red deer. *Animal Behaviour*, **34**, 460–471.

Cockburn A (1990) Sex ratio variation in marsupials. *Australian Journal of Zoology*, **37**, 467–479.

Cockburn A (1994) Adaptive sex allocation by brood reduction in antechinuses. *Behavioral Ecology and Sociobiology*, **35**, 53–62.

Cockburn A (1999) Deer destiny determined by density. *Nature*, **399**, 407–408.

Cockburn A, Scott MP & Dickman CR (1985a) Sex ratio and intrasexual kin competition in mammals. *Oecologia (Berlin)*, **66**, 427–429.

Cockburn A, Scott MP & Scotts DJ (1985b) Inbreeding avoidance and male-biased natal dispersal in *Antechinus* spp. (Marsupialia: Dasyuridae). *Animal Behaviour*, **33**, 908–915.

Cordero PJ, Griffith SC, Aparicio JM & Parkin DT (2000) Sexual dimorphism in house sparrow eggs. *Behavioral Ecology and Sociobiology*, **48**, 353–357.

Creel S & Creel NM (1997) Lion density and population structure in the Selous game reserve: evaluation of hunting quotas and offtake. *African Journal of Ecology*, **35**, 83–93.

Creel S, Creel NM & Monfort SL (1998) Birth order, estrogens and sex-ratio adaptation in African wild dogs (*Lycaon pictus*). *Animal Reproduction Science*, **53**, 315–320.

Daan S, Dijkstra C & Weissing FJ (1996) An evolutionary explanation for seasonal trends in avian sex ratios. *Behavioral Ecology*, **7**, 426–430.

Davison MJ & Ward SJ (1998) Prenatal bias in sex ratios in a marsupial, *Antechinus agilis*. *Proceedings of the Royal Society of London, series B*, **265**, 2095–2099.

De Kogel CH (1997) Long-term effects of brood size manipulation on morphological development and sex-specific mortality of offspring. *Journal of Animal Ecology*, **66**, 167–178.

Dickman CR (1988) Sex-ratio variation in response to interspecific competition. *American Naturalist*, **132**, 289–297.

Dijkstra C, Daan S & Buker JB (1990) Adaptive seasonal variation in the sex ratio of kestrel broods. *Functional Ecology*, **4**, 143–147.

Dittus WPJ (1998) Birth sex ratios in toque macaques and other mammals: integrating the effects of maternal condition and competition. *Behavioral Ecology and Sociobiology*, **44**, 149–160.

Dzus EH, Bortolotti GR & Gerrard JM (1996) Does sex-biased hatching order in bald eagles vary with food resources. *Ecoscience*, **3**, 252–258.

East ML, Hofer H & Wickler W (1993) The erect penis is a flag of submission in a female-dominated society: greetings in Serengeti spotted hyenas. *Behavioral Ecology and Sociobiology*, **33**, 355–370.

Ellegren H, Gustafsson L & Sheldon BC (1996) Sex ratio adjustment in relation to paternal attractiveness in a wild bird population. *Proceedings of the National Academy of Sciences USA*, **93**, 11723–11728.

Emlen ST, Emlen JM & Levin SA (1986) Sex-ratio selection in species with helpers-at-the-nest. *American Naturalist*, **127**, 1–8.

Eskelinen O (1997) On the population fluctuations and structure of the wood lemming *Myopus schisticolor*. *Zeitschrift für Saugetierkunde*, **62**, 293–302.

Espinosa MB & Vitullo AD (1996) Offspring sex-ratio and reproductive performance in heterogametic females of the South American field mouse *Akodon azarae*. *Hereditas*, **124**, 57–62.

Fedak MA, Arnbom T & Boyd IL (1996) The relation between the size of southern elephant seal mothers, the growth of their pups, and the use of maternal energy, fat and protein during lactation. *Physiological Zoology*, **69**, 887–911.

Festa-Bianchet M (1996) Offspring sex ratio studies of mammals: does publication depend upon the quality of the research or the direction of the results? *Ecoscience*, **3**, 42–44.

Festa-Bianchet M & Jorgenson JT (1998) Selfish mothers: reproductive expenditure and resource availability in bighorn ewes. *Behavioral Ecology*, **9**, 144–150.

Fiala KL (1980) On estimating the primary sex ratio from incomplete data. *American Naturalist*, **115**, 442–444.

Fiala KL (1981) Sex ratio constancy in the red-winged blackbird. *Evolution*, **35**, 898–910.

Fisher RA (1930) *The Genetical Theory of Natural Selection.* Oxford: Clarendon.

Frank LG, Glickman SE & Licht P (1991) Fatal sibling aggression, precocial development, and androgens in neonatal spotted hyenas. *Science,* **252,** 702–704.

Frank SA (1987) Individual and population sex allocation patterns. *Theoretical Population Biology,* **31,** 47–74.

Frank SA (1990) Sex allocation theory for birds and mammals. *Annual Review of Ecology and Systematics,* **21,** 13–55.

Frank SA & Swingland IR (1988) Sex ratio under conditional sex expression. *Journal of Theoretical Biology,* **135,** 415–418.

Fredga K (1988) Aberrant chromosomal sex-determining mechanism in mammals, with special reference to species with XY females. *Philosophical Transactions of the Royal Society of London, series B,* **322,** 83–95.

Fridolfsson AK & Ellegren H (1999) A simple and universal method for molecular sexing of non-ratite birds. *Journal of Avian Biology,* **30,** 116–121.

Gil D, Graves J, Hazon N & Wells A (1999) Male attractiveness and differential testosterone investment in zebra finch eggs. *Science,* **286,** 126–128.

Gileva EA (1987) Meiotic drive in the sex chromosome system of the varying lemming, *Dicrostonyx torquatus* Pall. (Rodentia, Microtinae). *Heredity,* **59,** 383–389.

Gileva EA (1998) Inbreeding and sex ratio in two captive colonies of *Dicrostonyx torquatus* Pall., 1779: a reply to Jarrell, G.H. *Hereditas,* **128,** 185–188.

Gileva EA & Fedorov VB (1991) Sex ratio, XY females and absence of inbreeding in a population of the wood lemming *Myopus schisticolor* Lilleborg, 1844. *Heredity,* **66,** 351–355.

Gileva EA, Benenson IE, Konopistseva LA, Puchkov VF & Makarenets IA (1982) XO females in the varying lemming, *Dicrostonyx torquatus:* reproductive performance and its evolutionary significance. *Evolution,* **36,** 601–609.

Golla W, Hofer H & East ML (1999) Within-litter sibling aggression in spotted hyaenas: effect of maternal nursing, sex and age. *Animal Behaviour,* **58,** 715–726.

Gomendio M, Clutton-Brock TH, Albon SD, Guinness FE & Simpson MJ (1990) Mammalian sex ratios and variation in costs of rearing sons and daughters. *Nature,* **343,** 261–263.

Gosling LM (1986) Selective abortion of entire litters in the coypu: adaptive control of offspring production in relation to quality and sex. *American Naturalist,* **127,** 772–795.

Gowaty PA (1993) Differential dispersal, local resource competition, and sex ratio variation in birds. *American Naturalist,* **141,** 263–280.

Gowaty PA & Lennartz MR (1985) Sex ratios of nestling and fledgling red-cockaded woodpeckers (*Picoides borealis*) favor males. *American Naturalist,* **126,** 347–357.

Graves JAM (1996) Mammals that break the rules: genetics of marsupials and monotremes. *Annual Review of Genetics,* **30,** 233–260.

Greenwood PJ (1980) Mating systems, philopatry and dispersal in birds and mammals. *Animal Behaviour,* **28,** 1140–1162.

Griffiths R, Double MC, Orr K & Dawson RJG (1998) A DNA test to sex most birds. *Molecular Ecology,* **7,** 1071–1075.

Haig D (2000) Genomic imprinting, sex-biased dispersal, and social behaviour. *Annals of the New York Academy of Sciences,* **907,** 149–163.

Hamilton WD (1967) Extraordinary sex ratios. *Science,* **1565,** 477–478.

Hardy ICW (1997) Possible factors influencing vertebrate sex ratios: an introductory overview. *Applied Animal Behaviour Science,* **51,** 217–241.

Heard E, Clerc P & Avner P (1997) X-chromosome inactivation in mammals. *Annual Review of Genetics,* **31,** 571–610.

Heinsohn R, Legge S & Barry S (1997) Extreme bias in sex allocation in eclectus parrots. *Proceedings of the Royal Society of London, series B,* **264,** 1325–1329.

Hewison AJM & Gaillard JM (1999) Successful sons or advantaged daughters? The Trivers-Willard model and sex-biased maternal investment in ungulates. *Trends in Ecology and Evolution,* **14,** 229–234.

Hiraiwa-Hasegawa M (1993) Skewed birth sex ratios in primates: should high-ranking mothers have daughters or sons. *Trends in Ecology and Evolution,* **8,** 395–400.

Hofer H & East ML (1997) Skewed offspring sex ratios and sex composition of twin litters in Serengeti spotted hyaenas (*Crocuta crocuta*) are a consequence of siblicide. *Applied Animal Behaviour Science,* **51,** 307–316.

Hurst LD, Atlan A & Bengtsson BO (1996) Genetic conflicts. *Quarterly Review of Biology,* **71,** 317–364.

Ims RA, Bondrup-Nielsen S, Fredriksson R & Fredga K (1993) Habitat use and spatial distribution of the wood lemming (*Myopus schisticolor*). In: NC Stenseth & RA Ims (eds) *The Biology of Lemmings. Linnean Society Symposium Series No. 15,* pp 509–518. London: Academic Press.

Jarrell GH (1995) A male-biased natal sex-ratio in inbred collared lemmings, *Dicrostonyx groenlandicus*. *Hereditas*, **123**, 31–37.

Johnson CN (1988) Dispersal and the sex ratio at birth in primates. *Nature*, **332**, 726–728.

Kalela O & Oksala T (1966) Sex ratio in the wood lemming, *Myopus schisticolor* (Lillijeb.), in nature and in captivity. *Annales Universitatus Turkuensis, Series AI*, **37**, 1–24.

Kempenaers B, Verheyren GR & Dhondt AA (1997) Extrapair paternity in the blue tit (*Parus caeruleus*): female choice, male characteristics, and offspring quality. *Behavioral Ecology*, **8**, 481–492.

Kilner R (1998) Primary and secondary sex ratio manipulation by zebra finches. *Animal Behaviour*, **56**, 155–164.

Koenig WD & Walters JR (1999) Sex-ratio selection in species with helpers at the nest: the repayment model revisited. *American Naturalist*, **153**, 124–130.

Kölliker M, Heeb P, Werner I, Mateman AC, Lessells CM & Richner H (1999) Offspring sex ratio is related to male body size in the great tit (*Parus major*). *Behavioral Ecology*, **10**, 68–72.

Kolman W (1960) The mechanism of natural selection on the sex ratio. *American Naturalist*, **94**, 373–377.

Komdeur J (1994a) The effect of kinship on helping in the cooperative breeding Seychelles warbler (*Acrocephalus sechellensis*). *Proceedings of the Royal Society of London, series B*, **256**, 47–52.

Komdeur J (1994b) Experimental evidence for helping and hindering by previous offspring in the cooperative-breeding Seychelles warbler *Acrocephalus sechellensis*. *Behavioral Ecology and Sociobiology*, **34**, 175–186.

Komdeur J (1996) Facultative sex ratio bias in the offspring of Seychelles warblers. *Proceedings of the Royal Society of London, series B*, **263**, 661–666.

Komdeur J (1998) Long-term fitness benefits of egg sex modification by the Seychelles warbler. *Ecology Letters*, **1**, 56–62.

Komdeur J, Daan S, Tinbergen J & Mateman C (1997) Extreme adaptive modification in sex ratio of the Seychelles warblers eggs. *Nature*, **385**, 522–525.

Krackow S (1995a) The developmental asynchrony hypothesis for sex ratio manipulation. *Journal of Theoretical Biology*, **176**, 273–280.

Krackow S (1995b) Potential mechanisms for sex ratio adjustment in mammals and birds. *Biological Reviews*, **70**, 225–241.

Krackow S (1997a) Further evaluation of the developmental asynchrony hypothesis of sex ratio variation. *Applied Animal Behaviour Science*, **51**, 243–250.

Krackow S (1997b) Maternal investment, sex-differential prospects, and the sex ratio in wild house mice. *Behavioral Ecology and Sociobiology*, **41**, 435–443.

Krackow S (1999) Avian sex ratio distortions: the myth of maternal control. In: NJ Adams & RH Slotow (eds) *Proceedings of the 22nd International Ornithogical Congress, Durban*, pp 425–433. Johannesburg: BirdLife South Africa.

Kruuk LEB, Clutton-Brock TH, Albon SD, Pemberton JM & Guinness FE (1999a) Population density affects sex ratio variation in red deer. *Nature*, **399**, 459–461.

Kruuk LEB, Clutton-Brock TH, Rose KE & Guinness FE (1999b) Early determinants of lifetime reproductive success differ between the sexes in red deer. *Proceedings of the Royal Society of London, series B*, **266**, 1655–1661.

Kumm J, Laland KN & Feldman MW (1994) Gene-culture coevolution and sex ratios: the effects of infanticide, sex selective abortion, sex selection, and sex-biased parental investment on the evolution of sex ratios. *Journal of Theoretical Biology*, **186**, 213–221.

Le Boeuf BJ, Condit R & Reiter J (1989) Parental investment and the secondary sex ratio in northern elephant seals. *Behavioral Ecology and Sociobiology*, **25**, 109–117.

Lee AK & Cockburn A (1985) *Evolutionary Ecology of Marsupials*. Cambridge: Cambridge University Press.

Legge S (2000a) The effect of helpers on reproductive success in the laughing kookaburra. *Journal of Animal Ecology*, **69**, 714–724.

Legge S (2000b) Siblicide in the cooperatively breeding laughing kookaburra (*Dacelo novaeguineae*). *Behavioral Ecology and Sociobiology*, **48**, 293–302.

Legge S & Cockburn A (2000) Social and mating system of cooperatively breeding laughing kookaburras (*Dacelo novaeguineae*). *Behavioral Ecology and Sociobiology*, **47**, 220–229.

Legge S, Heinsohn R, Double MC, Griffiths R & Cockburn A (2001) Complex sex allocation in the laughing kookaburra. *Behavioral Ecology*, **12**, 524–533.

Leigh EGJ (1970) Sex ratio and differential mortality between the sexes. *American Naturalist*, **104**, 205–210.

Leimar O (1996) Life-history analysis of the Trivers and Willard sex-ratio problem. *Behavioral Ecology*, **7**, 316–325.

Lessells CM & Avery MI (1987) Sex-ratio selection in species with helpers at the nest: some extensions of the repayment model. *American Naturalist*, **129**, 610–620.

Lessells CM, Mateman AC & Visser J (1996) Great tit hatchling sex ratios. *Journal of Avian Biology*, **27**, 135–142.

Lindström J (1999) Early development and fitness in birds and mammals. *Trends in Ecology and Evolution*, **14**, 343–348.

Liu WS, Eriksson L & Fredga K (1998) XY sex reversal in the wood lemming is associated with deletion of Xp(21–23) as revealed by chromosome microdissection and fluorescence in situ hybridization. *Chromosome Research*, **6**, 379–383.

Malcolm JR & Marten K (1982) Natural selection and the communal rearing of pups in African wild dogs (*Lycaon pictus*). *Behavioral Ecology and Sociobiology*, **10**, 1–13.

Maynard Smith J & Stenseth NC (1978) On the evolutionary stability of female biased sex ratio in the wood lemming (*Myopus schisticolor*): the effect of inbreeding. *Heredity*, **41**, 205–214.

McFarland Symington M (1987) Sex ratio and maternal rank in wild spider monkeys: when daughters disperse. *Behavioral Ecology and Sociobiology*, **20**, 421–425.

McVean G & Hurst LD (1996) Genetic conflicts and the paradox of sex determination: three paths to the evolution of female intersexuality in a mammal. *Journal of Theoretical Biology*, **179**, 199–211.

Mead PS, Morton ML & Fish BE (1987) Sexual dimorphism in egg size and implications regarding facultative manipulation of sex in mountain white-crowned sparrows. *Condor*, **89**, 789–803.

Mesterton-Gibbons M & Hardy ICW (2001) A polymorphic effect of sexually differential production costs when one parent controls the sex ratio. *Proceedings of the Royal Society of London, series B*, **268**, 1429–1431.

Myers JH (1978) Sex ratio adjustment under food stress: maximization of quality or numbers of offspring. *American Naturalist*, **112**, 381–388.

Nagelkerke CJ (1996) Discrete clutch sizes, local mate competition, and the evolution of precise sex allocation. *Theoretical Population Biology*, **49**, 314–343.

Nager RG, Monaghan P, Griffiths R, Houston DC & Dawson R (1999) Experimental demonstration that offspring sex ratio varies with maternal condition. *Proceedings of the National Academy of Sciences USA*, **96**, 570–573.

Nager RG, Monaghan P, Houston DC & Genovart M (2000) Parental condition, brood sex ratio and differential young survival: an experimental study in gulls (*Larus fuscus*). *Behavioral Ecology and Sociobiology*, **48**, 452–457.

Nishiumi I (1998) Brood sex ratio is dependent on female mating status in polygynous great reed warblers. *Behavioral Ecology and Sociobiology*, **44**, 9–14.

Nishiumi I, Yamagishi S, Maekawa H & Shimoda C (1996) Paternal expenditure is related to brood sex ratio in polygynous great reed warblers. *Behavioral Ecology and Sociobiology*, **39**, 211–217.

Oddie K (1998) Sex discrimination before birth. *Trends in Ecology and Evolution*, **13**, 130–131.

Olsen PD & Cockburn A (1991) Female-biased sex allocation in peregrine falcons and other raptors. *Behavioral Ecology and Sociobiology*, **28**, 417–423.

Palmer AR (2000) Quasi-replication and the contract of error: lessons from sex ratios, heritabilities and fluctuating asymmetry. *Annual Review of Ecology and Systematics*, **31**, 441–480.

Patterson CB & Emlen JM (1980) Variation in nestling sex ratios in the yellow-headed blackbird. *American Naturalist*, **115**, 743–747.

Pen I & Weissing FJ (2000) Sex-ratio optimization with helpers at the nest. *Proceedings of the Royal Society of London, series B*, **267**, 539–543.

Pen I, Weissing FJ & Daan S (1999) Seasonal sex ratio trend in the European kestrel: an evolutionarily stable strategy analysis. *American Naturalist*, **153**, 384–397.

Perret M (1990) Influence of social factors on sex ratio at birth, maternal investment and young survival in a prosimian primate. *Behavioral Ecology and Sociobiology*, **27**, 447–454.

Perret M (1996) Manipulation of sex ratio at birth by urinary cues in a prosimian primate. *Behavioral Ecology and Sociobiology*, **38**, 259–266.

Perrin N & Mazalov V (2000) Local competition, inbreeding, and the evolution of sex-biased dispersal. *American Naturalist*, **155**, 116–127.

Pusey AE (1987) Sex-biased dispersal and inbreeding avoidance in birds and mammals. *Trends in Ecology and Evolution*, **2**, 295–299.

Radford AN & Blakey JK (2000) Is variation in brood sex ratios adaptive in the great tit (*Parus major*)? *Behavioral Ecology*, **11**, 294–298.

Saino N, Ellegren H & Møller AP (1999) No evidence for adjustment of sex allocation in relation to paternal ornamentation and paternity in barn swallows. *Molecular Ecology*, **8**, 399–406.

Sheldon BC (1998) Recent studies of avian sex ratios. *Heredity*, **80**, 397–402.

Sheldon BC (2000) Differential allocation: tests, mechanisms and implications. *Trends in Ecology and Evolution*, **15**, 397–402.

Sheldon BC & Ellegren H (1996) Offspring sex and paternity in the collared flycatcher. *Proceedings of the Royal Society of London, series B*, **263**, 1017–1021.

Sheldon BC, Merilä J, Lindgren G & Ellegren H (1998) Gender and environmental sensitivity in nestling collared flycatchers. *Ecology*, **79**, 1939–1948.

Sheldon BC, Andersson S, Griffith SC, Ornborg J & Sendecka J (1999) Ultraviolet colour variation influences blue tit sex ratios. *Nature*, **402**, 874–877.

Silk JB (1983) Local resource competition and facultative adjustment of sex ratios in relation to competitive abilities. *American Naturalist*, **121**, 56–66.

Simpson MJA & Simpson AE (1982) Birth sex ratios and social rank in rhesus monkey mothers. *Nature*, **300**, 440–441.

Smale L, Holekamp KE, Weldele M, Frank LG & Glickman SE (1995) Competition and cooperation between litter-mates in the spotted hyaena, *Crocuta crocuta*. *Animal Behaviour*, **50**, 671–682.

Smale L, Holekamp KE & White PA (1999) Siblicide revisited in the spotted hyaena: does it conform to obligate or facultative models? *Animal Behaviour*, **58**, 545–551.

Smallwood PD & Smallwood JA (1998) Seasonal shifts in sex ratios of fledgling American kestrels (*Falco sparverius paulus*): the early bird hypothesis. *Evolutionary Ecology*, **12**, 839–853.

Stenseth NC (1978) Is the female biased sex ratio in wood lemming *Myopus schisticolor* maintained by cyclic inbreeding. *Oikos*, **30**, 83–89.

Stenseth NC & Ims RA (1993) Population biology of lemmings with skewed sex ratio – an introduction. In: NC Stenseth & RA Ims (eds) *The Biology of Lemmings. Linnean Society Symposium Series No. 15*, pp 449–463. London: Academic Press.

Torres R & Drummond H (1997) Female-biased mortality in nestlings of a bird with size dimorphism. *Journal of Animal Ecology*, **66**, 859–865.

Torres R & Drummond H (1999) Variably male-biased sex ratio in a marine bird with females larger than males. *Oecologia*, **118**, 16–22.

Trivers R (1985) *Social Evolution*. Menlo Park: Benjamin/Cummings.

Trivers RL & Willard DE (1973) Natural selection of parental ability to vary the sex ratio of offspring. *Science*, **179**, 90–92.

Valdivia RPA, Kunieda T, Azuma S & Toyoda Y (1993) PCR sexing and development rate differences in preimplantation mouse embryos fertilized and cultured *in vitro*. *Molecular Reproduction and Development*, **35**, 121–126.

van Schaik CP & Hrdy SB (1991) Intensity of local resource competition shapes the relationship between maternal rank and sex ratios at birth in cercopithecine primates. *American Naturalist*, **138**, 1555–1562.

Wachtel SS (1998) X-linked sex-reversing genes. *Cytogenetics and Cell Genetics*, **80**, 222–225.

Weatherhead PJ (1985) Sex ratios of red-winged blackbirds by egg size and laying sequence. *Auk*, **102**, 298–304.

Werren JH & Hatcher MJ (2000) Maternal-zygotic gene conflict over sex determination: effects of inbreeding. *Genetics*, **155**, 1469–1479.

West SA & Godfray HCJ (1997) Sex ratio strategies after perturbation of the stable age distribution. *Journal of Theoretical Biology*, **186**, 213–221.

Westerdahl H, Bensch S, Hansson B, Hasselquist D & von Schantz T (1997) Sex ratio variation among broods of great reed warblers *Acrocephalus arundinaceus*. *Molecular Ecology*, **6**, 543–548.

Westerdahl H, Bensch S, Hansson B, Hasselquist D & von Schantz T (2000) Brood sex ratios, female harem status and resources for nestling provisioning in the great reed warbler (*Acrocephalus arundinaceus*). *Behavioral Ecology and Sociobiology*, **47**, 312–318.

Westneat DF, Clark AB & Rambo KC (1995) Within brood patterns of paternity and paternal behavior in red-winged blackbirds. *Behavioral Ecology and Sociobiology*, **37**, 349–356.

Widdowson EM (1976) The response of the sexes to nutritional stress. *Proceedings of the Nutrition Society*, **35**, 175–180.

Wiebe KL & Bortolotti GR (1992) Facultative sex ratio manipulation in American kestrels. *Behavioral Ecology and Sociobiology*, **30**, 379–386.

Williams GC (1979) The question of adaptive sex-ratio in outcrossed vertebrates. *Proceedings of the Royal Society of London, series B*, **205**, 567–580.

Williams GC (1992) *Natural Selection: Domains, Levels, and Challenges*. Oxford: Oxford University Press.

Wolff JO (1994) More on juvenile dispersal in mammals. *Oikos*, **71**, 349–352.

Wolff JO & Plissner JH (1998) Sex biases in avian natal dispersal: an extension of the mammalian model. *Oikos*, **83**, 327–330.

Zann RA (1996) *The Zebra Finch: A Synthesis of Field and Laboratory Studies*. Oxford: Oxford University Press.

Zijlstra M, Daan S & Bruinenberg-Rinsma J (1992) Seasonal variation in the sex ratio of marsh harrier *Circus aeruginosus* broods. *Functional Ecology*, **6**, 553–559.

Chapter 14

Human sex ratios: adaptations and mechanisms, problems and prospects

John Lazarus

14.1 Summary

This chapter considers the factors influencing population birth sex ratios and variations of the sex ratio within populations, together with adaptive interpretations of these effects and the mechanisms underlying them. Adaptations and mechanisms for birth sex ratios are poorly understood, whereas variation in the sexual biasing of postnatal investment is more clearly adaptive. Social and demographic factors influencing birth sex ratios and postnatal investment include: parental status (the Trivers-Willard effect), father's age, birth order and local resource enhancement and competition. Complex effects are expected since many causal factors may interact at different times during offspring dependency. Possible mechanisms for control of the birth sex ratio include: the proportion of X and Y sperm in ejaculates; the relative success of X and Y sperm in fertilization, as determined by hormonal changes over the menstrual cycle interacting with the time of insemination and coital frequency; and embryonic mortality. Prospects for new work emerge from the following analyses: mechanisms for the Trivers-Willard effect (X and Y sperm proportions in ejaculates, coital frequency, embryonic mortality, family size and paternal age; predicting adaptation from proposed mechanisms and ancestral states; relationships between status, reproductive success and its variance; measuring status more realistically; and marginal return as the correct measure for predicting investment.

14.2 Introduction

The peculiar fascination with the human sex ratio at birth has a number of sources. Parents are faced by its unpredictability at the level of the individual birth, frustratingly at the mercy of their own biology for one of the most important events in their lives. Humans have consequently sought ways to influence offspring sex for personal or economic reasons, though with little success (Grant 1998). The male bias at birth and changes in population sex ratios over time have also aroused interest. For example, the birth sex ratio has become more female-biased in Europe and North America in recent decades, possibly because of exposure of the male reproductive system to environmental toxins (James 2000). From classical times there has been speculation that the time of insemination in the menstrual cycle influences the sex ratio, and modern studies support and extend this idea (section 14.6.1).

Scientists have identified numerous influences on the human sex ratio and sought to understand the mechanisms underlying them. At the same time evolutionary hypotheses concerning the birth sex ratio of populations, and of variation within populations, have been tested on human groups. Following birth, the long period of childhood dependency provides many opportunities for variation in the parental treatment of sons and daughters, and these have been studied by anthropologists and tested against evolutionary theories of differential parental investment in the two sexes. The present state of knowledge for these two aspects of sex ratio study, mechanism and adaptation, is very incomplete. In this chapter I examine work in these areas critically, explore the implications of mechanisms for questions of adaptation and discuss some problems requiring further work. As a result of this analysis prospects for new avenues of research emerge.

Certain medical and social concerns require further knowledge of sex ratio mechanisms and evolution. The selection of children by sex (Grant 1998), which promises the avoidance of sex-linked diseases, calls for a far greater understanding of mechanisms (section 14.6), and of the consequences for sex ratios at the population level (sections 14.3.1 & 14.5). Understanding the sex-biasing of postnatal investment (section 14.4) may help to predict and prevent child neglect (Cronk 1991).

14.2.1 Terminology

Human sex ratios are conventionally reported as the number of males per 100 females, and this convention is adopted here; a 'higher' sex ratio is therefore more male-biased. The sex ratio at conception is termed the 'primary sex ratio', and that at birth the 'secondary sex ratio'. I use the term 'modern', applied to human populations, to refer to both historical and contemporary societies, in contrast to the ancestral hunter-gatherer phase of human evolution.

14.3 | Birth sex ratios

14.3.1 Population birth sex ratios

Sex ratios at the population level are understood primarily in terms of Fisher's (1930, 1958) argument that since, for an entire generation, total reproductive value for males must equal that for females, 'the sex ratio will so adjust itself, under the influence of natural selection, that the total parental expenditure incurred in respect of children of each sex, shall be equal' (Fisher 1958, p 159). Fisher's theory therefore predicts not conception or birth sex ratios, but the total investment in offspring up to the end of parental

investment. Fisher's conclusion assumes a linear relationship between parental investment and its effect on the reproductive value of offspring. If this assumption is relaxed, and the marginal return on additional investment in the two sexes differs, then equal allocation is no longer expected (Charnov 1979a, Maynard Smith 1980, Frank 1990).

These theories provide the framework for understanding why the average sex ratio at birth is above parity, with values for human populations averaging 105–106 (proportion of males = 0.512–0.514) (Visaria 1967). Although a full answer is elusive, the theory summarized above indicates that a departure from 100 in the primary or secondary sex ratio is not in itself a cause for surprise (and is found in other mammals, Clutton-Brock & Albon 1982). Fisher (1958) explained the bias in terms of a greater male mortality during the period of parental care. If birth sex ratio was equal this mortality differential would result in total investment at the end of the investment period being lower for boys than for girls. Consequently the sex ratio at birth (or conception) would have to be male-biased for overall investment to be the same for each sex, as Fisher predicted (although control of the sex ratio would have to be facultative for this departure from parity to evolve, Werren & Charnov 1978). Since a nonlinear return on investment renders the equal-allocation prediction incorrect (see above), the validity of Fisher's argument for the male-biased sex ratio is uncertain. A nonlinear return would itself predict a male-biased sex ratio given different investment-return functions for each sex; for example, a linear return for sons and a diminishing returns function for daughters (Frank 1990), or an additional frequency-dependent component of fitness in males (Maynard Smith 1980). One theory for adaptive sex ratio variation at the family level (Trivers & Willard 1973: section 14.3.2.1) is based on differing investment-return functions for sons and daughters, so that between-family selection may result in allocation bias at the population level (Frank 1987, 1990). Finally, a male-biased sex ratio would be selected if males were more altruistic to kin than females (Trivers & Willard 1973,

Trivers 1985, p 297); a 'local resource enhancement' effect (section 14.4.2).

According to Fisher's theory, selection maintains sex allocation at equality by favouring shifts in the opposite direction to any current population bias (Werren & Charnov 1978). A study of preindustrial Finnish parishes provides supporting evidence for this (Lummaa *et al.* 1998). The Fisherian nature of the relationship between a bias in the *adult* sex ratio and a compensating shift in *birth* sex ratio is questionable, since the two generations involved are unlikely to form part of the same breeding population. However, the autocorrelation between population birth sex ratios in successive years (Graffelman & Hoekstra 2000: section 14.3.2.3) diminishes the problem of the lag in the shift in sex ratio in the breeding population and makes a Fisherian mechanism more plausible. A similar analysis for many twentieth century countries does not support the Fisherian prediction (James 2000). However, since any effect of adult sex ratio must work proximately through parental perceptions, the more local analysis of the Finnish parishes seems more appropriate (Williams 1979, James 2000). Similarly, there could be a Fisherian explanation for the increase in the birth sex ratio during war time (James 1971, Graffelman & Hoekstra 2000), since local adult sex ratios are low at such times (Trivers 1985, p 288, James 1995, 2000; see Grant 1990 and section 14.6.1 for possible mechanisms).

14.3.2 Variation in the birth sex ratio

Over the last century a great many influences on the birth sex ratio have been examined, including demographic, environmental, psychological and anatomical factors (Teitelbaum 1970). Despite all this work we lack a unified understanding of influences on the secondary sex ratio for three reasons. Methodologically, there has been a shortage of studies that control for confounding influences (Teitelbaum 1970, Moore & Gledhill 1988). Conceptually, understanding variation in the secondary sex ratio lacked an evolutionary basis until the 1970s. Before this, explanatory theories were generally statements of putative prox-

imate mechanism (Teitelbaum & Mantel 1971, section 14.6.4). Finally, we are largely ignorant of the mechanisms mediating influences on the sex ratio (section 14.6). Reviews of factors influencing the birth sex ratio are provided by Teitelbaum (1972), James (1987a), Chahnazarian (1988) and Grant (1998). Biases away from the average tend to be small.

14.3.2.1 Status: The Trivers-Willard effect

Trivers and Willard (1973) proposed that parental manipulation of offspring sex ratio as a function of the parents' social status (rank, condition) will be favoured by selection if status enhances reproductive success (RS) differentially for sons and daughters. For example, if high status in parents confers greater RS on sons than daughters, and low status has the opposite effect, then high-status parents are predicted to invest relatively more in sons, and low-status parents more in daughters (Figure 14.1a). The hypothesis was reviewed for mammals by Hrdy (1987) and for nonhuman primates by Brown (2001). Its application to humans relies on the following sources of reproductive differentials as a function of offspring sex and parental status.

Source A: In many cultures high-status individuals are predicted to favour sons over daughters (and low-status individuals vice versa) due to the combination of two influences. First, status can be culturally inherited from parents through parental investment, including the inheritance of wealth and property (Hrdy & Judge 1993, Buss 1999). Second, status increases RS more in sons than it does in daughters. This is because women seek mates of high status (in terms, for example, of earning capacity and industriousness) more than men seek the same trait in women, for the benefit that status bestows on them and their offspring (Trivers & Willard 1973, Buss 1989, 1992, Ellis 1992). Consequently, a man's RS is expected to increase with status more steeply than a woman's as a result of his greater opportunities of: (1) gaining a more fertile mate (which men prefer, Buss 1992), (2) gaining one earlier, (3) in polygynous

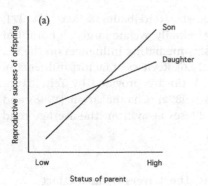 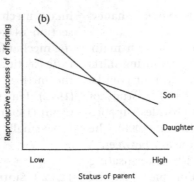

Fig 14.1 The Trivers-Willard effect. (a) Reproductive success of offspring increases with parental status. (b) Reproductive success of offspring decreases with parental status. In (a) status has a greater effect on the reproductive success of sons than it does on daughters, but in (b) this inequality is reversed.

societies gaining more wives, and perhaps (4) having greater access to sexual partners outside marriage (Pérusse 1993). The following sources do not rely on female choice (although female choice may exploit and strengthen male status advantages that follow from them).

Source B: Men are stronger and more aggressive than women and tend to breed in their place of birth, while women migrate to marry. This male philopatry, or patrilocality (which is by no means universal) facilitates kin-based alliances among males and erodes female autonomy since women lose the support of close kin when they marry. These features have traditionally given men greater 'resource-procuring and resource-holding potential' than women (Hrdy & Judge 1993, p 21) and this male advantage would tend to produce a Trivers-Willard (TW) effect if its size increased with status; for example, if male–male competition increased the variance in resources between males with the highest and lowest resource holding power (RHP). Then low-status (low RHP) men may gain greater fitness through daughters than sons.

Source C: Inheritance passes to sons rather than daughters in a majority of human cultures (Hrdy & Judge 1993). In such cases, high-status parents would gain greater fitness through their sons than their daughters as long as inheritance increased a child's RS. However, it is also necessary that low-status men with a small inheritance are outreproduced by their sisters, and this could be realized by hypergyny, i.e. women marrying

up the socioeconomic scale (e.g. Trivers & Willard 1973, Boone 1986, Bereczkei & Dunbar 1997). Hypergyny increases the relative benefit of daughters to low-status parents and strengthens the TW effect (section 14.4.1). However, it only obviously does so where RS increases with status; where status and fertility are inversely related (section 14.7.2.1) hypergyny may weaken the TW effect (Myers 1978).

Source D: Male status might bring relatively greater gains in various aspects of life, outside the sexual sphere, that increase RS (e.g. lowered risk-taking: Waldron 1983, Wilson & Daly 1997).

Source E: The emphasis above has been on paternal investment as the source of reproductive differentials. However, maternal investment (sometimes consequent on her partner's status) is also important, most obviously in terms of nutritional support during pregnancy and lactation. Mothers who are well-fed and in good health would therefore be expected to favour sons (and vice versa for poor-condition mothers) if the nutritional and health status of offspring had an impact on their status as adults.

Trivers and Willard's model has been extended in a number of ways. First, it has been formalized, producing further predictions yet to be tested (Frank 1987, Charnov 1979b, Anderson & Crawford 1993, Joshi 2000). Second, the hypothesis should be framed in terms of the reproductive value of sons and daughters, rather than their RS (Leimar 1996). This is because if quality is passed on from, say, mother to offspring, 'it is not correct to count all grandoffspring as being of equal value. For a high quality mother, a daughter's offspring would tend to be of higher than average quality, whereas a son's offspring might be of only average quality' (Leimar 1996, p 316), since the daughter would pass on her high quality to *her* offspring. While this insight is important in predicting that high-quality mothers might sometimes prefer daughters rather than sons, its relevance to humans is uncertain since it assumes offspring quality to be influenced by maternal but not paternal quality, and maternal and paternal quality to be independent. If instead offspring quality was influenced only by paternal quality, the prediction that high-quality fathers would produce more sons would be *strengthened*. In practice, male and female quality would be correlated if high-status men have partners who are highly fertile (see above).

There is an alternative to the TW hypothesis that could account for a positive relationship between status and sex ratio. If males are more costly to raise, and the sex difference in cost is greater for lower status parents, then such parents could be selected to produce daughters in the absence of any effect of parental status on offspring fitness (cost of reproduction hypothesis, Myers 1978, Oksanen 1981, Chapter 13). Distinguishing between TW and cost of reproduction hypotheses requires measuring the differential impact of investment in sons and daughters, by high- and low-status parents, on offspring and on parental reproductive value, respectively (Chapter 13).

Finally, there is a theory of female sexual strategy that, like TW, predicts individual differences in offspring sex ratio (Gangestad & Simpson 1990). Two female mating strategies, maintained by frequency-dependent selection, are proposed. 'Restricted' women seek commitment in a sexual partner and benefit through paternal investment, while 'unrestricted' women eschew commitment but benefit through genetic qualities passed on to sons who, in turn, attract unrestricted women. As predicted, unrestricted women have a higher offspring sex ratio than restricted women, as measured by occupational and personality correlates of these strategies, and by number of premarital partners.

14.3.2.2 Evidence for the Trivers-Willard effect

The Trivers-Willard hypothesis has been tested by examining the relationship between birth sex ratio and status. Out of 54 analyses I know of (e.g. Teitelbaum 1970, Teitelbaum & Mantel 1971, Trivers & Willard 1973, James 1987a, Chahnazarian 1988, Mackey 1993, Bereczkei & Dunbar 1997), 26 (48%) support the hypothesis. Measures of status include headman status, social class, education, church rank, wealth and entry in *Who's Who*. Where comparisons are made between high- and low-status *populations* sex ratio differences might be due to genetic differences (e.g. Gypsy *versus* Hungarian: Bereczkei & Dunbar 1997; Mukogodo *versus* global average: Cronk 1989; see section 14.3.2.3 for racial effects). Possible support for the TW hypothesis also comes from evidence that contemporary New Zealand mothers with a more dominant or emotionally independent personality are more likely to bear sons (Grant 1990, 1992, 1994). However, the relationship between these personality traits and societal measures of status is unknown.

So, about half the tests of the hypothesis support it. This summary is no more than a crude indication of the balance of evidence since it is not exhaustive; some results are based on small samples (sample sizes range from hundreds to millions) and therefore have low power (given the small effects expected, Moore & Gledhill 1988); and many studies fail to control for confounding variables such as paternal age and birth order (section 14.3.2.3). In addition, the influence of 'local factors' (sections 14.4.1 & 14.4.2) has been neglected, and issues concerning novel environments (section 14.7.1) and status (section 14.7.2) have consequences for the kind of studies expected to show the effect. Finally, concerns of publication bias have been expressed, nonsignificant findings perhaps being less likely to find their way into the literature (Maynard Smith 1983, Festa-Bianchet 1996).

It is important to note that the interaction between status and a son-daughter difference in RS, on which the TW effect relies (Figure 14.1), has rarely been demonstrated in tests for the effect (for an exception see Borgerhoff Mulder 1998a).

14.3.2.3 Demography: parental age, birth order and race

The sex ratio has been shown to decrease with paternal age, maternal age and birth order. However, since all three variables are intercorrelated it is important to control for two of them in a multivariate analysis when testing for an independent effect of the third. When this is done the maternal age effect generally disappears, while the other effects remain. Significant inverse relationships between sex ratio and these variables were found: for maternal age in 1/16 studies (6%); for paternal age in 8/13 studies (62%) and for birth order in 13/16 studies (81%); in none was it significantly positive (multivariate analyses, James & Rostron 1985, Chahnazarian 1988, Jacobsen *et al*. 1999; not an exhaustive review). Although a meta-analysis (Rosenthal 1991) is required to determine whether the evidence overall supports a paternal age effect, there is sufficient evidence to justify exploring a potentially important consequence of this effect for the TW effect (section 14.6.4). In Jacobsen *et al*.'s (1999) study of all Danish births in 1980–1993, for example, the sex ratio dropped from 106.6 for fathers aged 13–24 to 104.1 for fathers aged >39.

There are racial differences in birth sex ratio, ratios being highest in Orientals, lowest in Black Africans, and intermediate in Whites, although parental age and birth order are not always controlled for (Visaria 1967, Teitelbaum 1970, Teitelbaum & Mantel 1971, James 1984, 1985, Ruder 1986). There is some evidence that the effect of race acts through the father, rather than the mother (Ruder 1986).

14.4 | Sexual biasing in postconception parental investment

In sex allocation there is a distinction between *producing* offspring of a certain sex ratio (the primary sex ratio decision) and *investing* in offspring once they have been produced. Birth is in some respects an arbitrary stage in life at which to focus on the sex ratio, but valuable in that the means for parental manipulation of offspring change radically at this point. The previous section dealt with birth rather than conception sex ratios for the practical reason that human primary sex ratios are not accurately known (Creasy 1977, Waldron 1983, Stinson 1985). I now consider variations in investment before and after birth, but discuss embryonic mortality later as a mechanism of birth sex ratio biasing (section 14.6.2). Postnatally there is a distinction between parental decisions to alter the sex ratio by infanticide or neglect, and to maintain the sex ratio but modify the relative reproductive value of sons and daughters by differential investment (Anderson & Crawford 1993).

14.4.1 The Trivers-Willard effect

The TW effect may be manifested in many ways after conception (Cronk 1991), high-status parents being predicted to invest more in sons and low-status parents more in daughters (although a problem with this formulation of the TW effect for investment following conception is considered in section 14.7.3.) A number of studies have tested the TW hypothesis, and are reviewed by Hrdy (1987), Cronk (1991) and Hrdy and Judge (1993).

Two components of prenatal investment have been examined for a TW effect: birth weight and interbirth interval, greater weight and longer intervals being assumed to indicate greater investment. In the contemporary US, with income and presence/absence of co-resident male as status measures, birth weight and birth interval following the next-to-youngest child were found not to follow the TW prediction, while the birth interval preceding the youngest child did, for both status measures (Gaulin & Robbins 1991). For example, for women living without a coresident male (assumed low status), the birth interval between her two youngest children was almost two years greater if the younger child was a daughter rather than a son. It seems curious that a birth interval should be correlated with the sex of the *subsequent* child but the relationship might be a consequence of low coital frequency in women living without a man since low coital frequency may increase the likelihood of a female birth (section 14.6.1), and would be associated with long interbirth intervals because of the lowered chance of conception.

The greater sensitivity of male prenatal growth to stressors and nutritional enrichment (Stinson 1985), and the greater vulnerability of infant males, can be understood as TW adaptations, providing greater opportunity for modifying male morbidity and mortality as a function of status (Wells 2000). The increase in the secondary sex ratio under improved environmental conditions (Trivers 1985, p 297) may also be a consequence. Such an effect is evident in the contemporary US, where the sex ratio of infant mortality decreases as educational level rises (Abernethy & Yip 1990). The effect is probably due to differential resourcing rather than active neglect, but in other populations patterns of neglect and infanticide often fit the TW prediction. If a child of a certain sex has very poor prospects it may be adaptive to kill, abandon or neglect it to save the costs of investment. However, care is necessary in interpreting differential mortality since it may not be a result of parental investment (Clutton-Brock 1991, Cronk 1991, Borgerhoff Mulder 1998a). Additionally, differential allocation to sons and daughters may be a consequence of offspring rather than parent behaviour (Clutton-Brock 1991, p 221), and parent–offspring conflict (Trivers 1974) will sometimes be implicated in investment outcomes.

The pioneering study by Dickemann (1979) shows that female-biased infanticide in high caste groups in precolonial North India fits the TW prediction. In these stratified societies high caste men are reproductively successful, marrying lower caste women (hypergyny) who provide a dowry and reward their parents with enhanced lineage survival through grandchildren. High caste women, in contrast, have poor marriage prospects. Similar patterns are found in Imperial China and mediaeval Europe, where convent life replaced infanticide as the fate of many noble-born women (Dickemann 1979, Boone 1986). In the mediaeval Portuguese elite the proportion of sons and daughters who ever married, and their reproductive success, show a cross-over

interaction (as in Figure 14.1a) between the effects of sex and parental status (Boone 1986). Importantly for the TW effect, this is one of the few studies presenting data on the RS of sons and daughters as a function of parental status. A finding of *lower* female child mortality in a high-status group in eighteenth and nineteenth century Germany (Voland *et al.* 1991) disappeared when the dataset was enlarged (Voland *et al.* 1997).

In Gabbra pastoralists, and agropastoralist Kipsigis, both Kenyan groups, the numbers of sons and daughters show no TW bias even though the relationship between wealth and RS is that shown in Figure 14.1a, so that TW effects through neglect or infanticide might be predicted. For the Kipsigis, however, a TW effect was evident for education, with the advantage enjoyed by boys in length of education narrowing in poorer families (Borgerhoff Mulder 1998a). Differential inheritance patterns are suggested as an alternative method of the sexual biasing of investment in the Gabbra (Mace 1996). In the Mukogodo, an impoverished low-status Kenyan pastoralist group, daughters receive closer attention and have better growth performance, sons suffer higher childhood mortality and, compared to higher status non-Mukogodo, there is less of a male bias in medical care (Cronk 2000). The Hungarian Gypsies provide another example of female-biased investment of various kinds in a low-status group (Bereczkei & Dunbar 1997).

A contemporary study of parental investment in US adolescents used family income and educational level as status measures, and a wide range of investment measures. Controlling for number of sibs, maternal age, family residence pattern and race, the results did not support the TW prediction. In fact higher status parents were relatively more involved in their *daughters'* schooling and cultural activities (Freese & Powell 1999). Bequests in contemporary Canada show a TW pattern, with the wealthiest favouring sons and the least wealthy favouring daughters (Smith *et al.* 1987; see Hrdy & Judge 1993 for further examples).

There is thus partial support for a TW interpretation of postnatal investment patterns. As we shall see in the next section this mixed picture might be resolved by considering other effects that interact with TW.

14.4.2 Interaction between local factors, Trivers-Willard and other effects

By cooperating or competing with its parent or other kin, an offspring can enhance or depress its parents' inclusive fitness (Cronk 1990, Borgerhoff Mulder 1998a). A parent is selected to favour the sex of offspring that better enhances its fitness and neglect the sex that reduces it (but see section 14.7.3). These 'local resource enhancement' (LRE) and 'local resource competition' (LRC) processes (Silk 1984) have been reviewed for human postnatal investment by Sieff (1990) and Cronk (1991); I will refer to these collectively as 'local factors'.

To take an example of LRE from Sieff (1990), a first-born daughter can provide help to her mother by caring for later-born siblings. This may be the reason why first-born daughters in the Punjab are spared the disproportionately high mortality of later-born daughters (see also Trivers 1985, p 296 for a cross-cultural prediction). As an LRC example, a Gabbra father controls competition among his sons for inheritance of the camel herd by favouring the first-born. Concentrating wealth in this way, rather than sharing it equally, may maximize the father's number of grandchildren (Mace 1996, see also Borgerhoff Mulder 1998a).

Depending on the direction of sex biasing due to local factors, these processes and the TW effect may act in concert or in opposition and, considered together, may help to explain otherwise discrepant datasets (Van Schaik & Hrdy 1991). For six populations in eighteenth and nineteenth century Germany, the magnitude of the TW effect on infant mortality declined as resource competition intensified (Voland *et al.* 1997). This means that as resources became scarcer higher-ranked families benefited relatively less from sons, perhaps because only a single son could inherit (Voland *et al.* 1991). However, this interaction between TW and LRC does not explain population density effects in the Kipsigis (Borgerhoff Mulder 1998a).

Further aspects of social life must be considered for a more realistic appraisal of TW and

local factor predictions (Cronk 1990, Borgerhoff Mulder 1998a). Thus, the economic value of children in many African groups may explain the absence of differential childhood mortality in the Gabbra and Kipsigis (an LRE effect, but one that may not show a sexual bias; section 14.4.1), whilst opportunities for hypergyny help to explain the favouring of daughters in the Kipsigis, Mukogodo and Hungarian Gypsies (Borgerhoff Mulder 1998a). An idea only recently assimilated into studies of post-natal investment (Borgerhoff Mulder 1998a) is the impact on the TW effect of whether mother's or father's quality influences offspring fitness (Leimar 1996, section 14.3.2.1). It must also be said that studies do not always control for potentially confounding factors such as parental age and birth order, which are known to influence the secondary sex ratio (section 14.3.2.3) and are likely to influence postnatal investment too.

14.5 Birth sex ratios and postconception investment: summing up

Our present understanding of the sexual biasing of postnatal investment is an increasingly complex one of parental responses to TW effects, local factors and other social influences, broadly comprehensible within an adaptive framework. In contrast, understanding of the secondary sex ratio in this manner is less advanced, due to the smaller magnitude of the effects and the relative neglect of local factors and social context. In compensation, birth sex ratio studies have paid more attention to parental age and birth order effects, and there is a lesson here for the study of postnatal investment. Indeed there may be data on these effects in the sociological, psychological and anthropological literatures awaiting attention in the sex ratio arena.

The importance of the timing of investment has only recently been appreciated. Impacts of investment are expected to vary with offspring age, sometimes favouring different forms of care in sons and daughters (Borgerhoff Mulder 1998a). These age-dependent effects will vary with parental age too, and with costs, constraints and opportunities for investment. Adaptive control of sex allocation is expected to result in *selective* enhancement and inhibition of offspring prospects; we should not expect every aspect of investment to show, say, a TW or a local factor effect. In particular, the moment of birth has largely divided thinking on parental investment. We should try now to understand the pattern of investment from conception ·to independence in a single framework.

The analysis of investment needs to take account of other complexities too. First, parents must make many investment decisions, and the optimal solution for each will not be independent; Rosenheim *et al.* (1996) provide a possible model for this problem. Allocation at one stage may be compromised by the need to reserve resources for later investment. Second, optimal allocation between sons and daughters at each stage depends on the functional relationship between investment and its consequence for the fitness of each sex, and for each child in the family (Davis & Todd 1999, section 14.7.3). This will depend, among other things, on whether the investment is shared or unshared (Lazarus & Inglis 1986). Third, evolutionarily stable allocations depend on differential offspring survival, the cost of each sex and frequency-dependent effects on fitness (Maynard Smith 1980). Fourth, allocation patterns within families have sex ratio impacts at the population level through cultural and genetic routes (Nordborg 1992, Laland *et al.* 1995, section 14.3.1).

14.6 | Birth sex ratio mechanisms

'convincing evidence of a physiological mechanism of sex ratio adjustment is completely lacking'

(Krackow 1995a, p 274).

Individuals might adjust the birth sex ratio through influences on the primary sex ratio or on mortality *in utero*. I consider only evidence from human studies, although for the primary sex ratio, at least, this largely conforms with the evidence for mammals more generally (Krackow 1995b). After discussing hypotheses for control of the sex ratio I consider how they might explain

the TW effect and the demographic influences already described (section 14.3.2). In discussing primary sex ratio mechanisms the sex ratio data come inevitably from births rather than conceptions, so further sex biasing during embryonic life is possible (sections 14.6.2 & 14.6.4).

14.6.1 Primary sex ratio

In principle, the primary sex ratio could be adjusted either by differential production of X- and Y-bearing sperm, or by their differential success in fertilization. Although hypotheses for sex ratio variation have focused on the latter mechanism, there is some evidence for control of the sex ratio as a function of the relative proportion of X and Y sperm in ejaculates. First, the proportion of X or Y sperm is greater in men who have fathered only daughters or only sons, respectively, but not in all studies (Dmowski *et al.* 1979, Bibbins *et al.* 1988, Irving *et al.* 1999). In one study, men with only daughters had 67.9% X sperm in their ejaculate, compared to 50.7% in men with both sons and daughters (Bibbins *et al.* 1988). Second, a reduction of sexual abstinence from ≥14 days to ≤48 hours before taking a sperm sample was associated with a significant increase in the proportion of Y sperm detectable in the sample from 37.2% to 43.5% (Schwinger *et al.* 1976). A just nonsignificant finding of the same type was found when the longer period of abstinence was only 7–10 days (Hilsenrath *et al.* 1997), which may be consistent with the first study, given the smaller difference between the two groups in the period of abstinence. These findings on variation in the proportions of X and Y sperm in ejaculates require further confirmation. Another way in which sexual abstinence might influence the sex ratio is that the fertilizing capacity of sperm might decline with their age more rapidly for Y than for X sperm (Hilsenrath *et al.* 1997).

There are currently three hypotheses for the control of the sex ratio by differential success of X and Y sperm in fertilization, all involving the surge in the gonadotropin, luteinizing hormone (LH), which occurs in the middle of the menstrual cycle, around the time of ovulation. All hypotheses predict that the sex ratio is responsive to the frequency of coitus and its timing in relation to the menstrual cycle. Demographic and TW influences on the sex ratio might then be explained in terms of these adjustments (sections 14.6.3 & 14.6.4). For none of these hypotheses have all the putative physiological processes been established.

James (1980, 1986, 1987b) proposed that a high level of maternal LH at the time of conception is associated with an increased likelihood of a female birth, and that high levels of maternal oestrogen and testosterone, and paternal testosterone, have an opposite, but weaker, effect. Evidence relevant to the theory is discussed by James (1986, 1987b, 1990, 1992, 1996), and includes the lowered sex ratio following induction of ovulation with gonadotropin (James 1980) and the positive association between a female's android body fat distribution (a testosterone indicator) and offspring sex ratio (Manning et al. 1996, Singh & Zambarano 1997, but see also Manning et al. 1999). (Men could consequently influence offspring sex ratio through body shape preference in sexual partners.)

According to this hypothesis, conceptions closer to ovulation, when LH peaks, should have a low sex ratio, while conceptions both early and late in the fertile period of the cycle, when LH levels are low, should have a high sex ratio (i.e. a U-shaped pattern of sex ratio through the cycle). The interpretation of studies examining this relationship has been controversial (e.g. James 1994, Martin 1994, Bernstein 1995). Some have found the predicted U-shaped pattern of sex ratios while others have not. While in sum (James 1999) the studies lend support to James' prediction, Martin's (1995, 1997a) hypothesis, to be discussed below, is able to predict which studies do and do not show the U-shaped pattern.

James (1971, 1997b) argued further that an increased coital frequency would lead to insemination earlier in the cycle, which in turn would lead to earlier conception and a consequently higher sex ratio. Modelling of the relevant processes supports James' view, a coital frequency of once per day resulting in fertilization 95 minutes earlier in the life of the ovum, on average, than a coital frequency of once per 10 days (Roberts 1978).

The following evidence supports James' prediction that coital frequency increases the sex ratio:

1. The early months of marriage are characterized by high coital rates (James 1981, 1983) and (in a different population) high sex ratios. For first postnuptial births in Australia, 1908–1967, the sex ratio for conceptions in the first month of marriage was 109.4, falling immediately to 106 in the following month and remaining around this value for all later births (Renkonen 1970).

2. Birth sex ratios in combatant countries are high during and just after wars (e.g. James 1971, Graffelman & Hoekstra 2000). James (1971) proposed that this was due to a high coital frequency in the short home leaves of servicemen, and shortly after demobilization.

3. The lower sex ratios of non-co-resident polygynous (87.3) compared to monogamous (114.1) women in seven East African groups is explained by the assumed lower coital frequency in women who must share their husbands with co-wives resident elsewhere (Whiting 1993, Borgerhoff Mulder 1994).

The second hypothesis (Roberts 1978, Martin 1995, 1997a) is based on the greater motility of Y compared to X sperm in cervical mucus and the increase in penetrability of the cervical mucus in the 1- to 2-day period prior to ovulation, followed by a decrease in penetrability following ovulation (Katz 1991). It is proposed that this pattern of mucus penetrability, which is under hormonal control, gives an advantage to Y sperm early in the fertile part of the cycle, which declines at ovulation and rises again following ovulation. Correspondingly, the probability of a male conception should be high for inseminations early in the fertile period, decline for inseminations closer to ovulation and increase for inseminations thereafter (as James also predicted). Martin (1997a) also employs evidence that mucus penetrability is reduced by the seminal debris left in the cervix by prior inseminations. He argues that a greater coital rate will consequently increase the Y sperm advantage and flatten the U-shaped relationship between time of insemination and sex ratio (see also James 1997a, Martin 1997b).

Martin (1995, 1997a) tested this hypothesis against the studies of the relationship between sex ratio and time of insemination in the menstrual cycle already described. He reconciles the conflicting results with evidence that those showing the U-shaped pattern have low coital frequencies, while those not showing the pattern have high coital frequencies, as his hypothesis would predict. (A further study not included in Martin's (1997a) analysis, nor in James' (1999) review (WHO 1984) fits neither Martin's nor James' hypothesis.)

Krackow's (1995a) developmental asynchrony hypothesis combines the assumed U-shaped effect of timing of insemination on the sex ratio, and a sex difference in the timing of blastocyst implantation, with reduced implantation success for blastocysts whose development is out of phase with the state of readiness of the uterus. Maternal control of embryonic sex and survival are a function of timing of insemination and of uterine readiness for implantation.

14.6.2 Embryonic mortality and the secondary sex ratio

Following fertilization and implantation, a sex difference in embryonic mortality is the remaining potential source of variation in the secondary sex ratio. Earlier conclusions that male embryos have a greater risk of mortality have given way to a more complex picture. For mortality in the first six months the current evidence is inconclusive but does not suggest a male disadvantage. For the last three months of pregnancy the greater male mortality in the period 1930–1960 in Europe and North America has been replaced by an unbiased risk in recent decades. This is due to improvements in obstetric care and maternal health, which have caused a greater reduction in the causes of mortality that impact more on males (Waldron 1983, Stinson 1985). Variation in the secondary sex ratio due to embryonic mortality is therefore a cultural variable and it may be possible to infer the ancestral state from the known causes of mortality.

14.6.3 Mechanisms for demographic effects

Given the proposed role of coital frequency, the decline in sex ratio with paternal, but not with maternal, age (section 14.3.2.3) would be consistent with coital rate declining more rapidly with husband's than with wife's age, but the evidence is mixed (James 1974, 1983, Udry *et al.* 1982, Rao & Demaris 1995). Explaining the decline in

sex ratio with birth order in terms of coital frequency is also problematic since the relationship between coital frequency and birth order is not straightforward (James 1974, Rao & Demaris 1995), probably because of two opposing processes. First, a higher coital frequency may lead to a larger family size; second, coital frequency may decline with birth order because of the various ways in which children inhibit sexual activity.

Whether variation in coital rate could produce the magnitude of effect required to explain the influence of demographic variables on the sex ratio requires comparison with a model such as Roberts' (1978). James (1994) argues that coital rate is only a weak determinant of the sex ratio, and Roberts (1978) concludes that the birth order data cannot be explained fully by variation in coital frequency.

14.6.4 Mechanisms for the Trivers-Willard effect

By what psychological-physiological mechanisms might the TW effect work? If females control sex allocation via the primary sex ratio then the effect must come about by some aspect of status acting on female physiology. There are three possibilities. First, aspects of the male's status influence the sex ratio via cognitive-physiological pathways in his female partner (including coital frequency: section 14.6.1). Second, the sex ratio is modified via cognitive-physiological correlates in women of their own status, or some other trait, such as attractiveness or personality (see section 14.3.2.2 for work by Grant). If father's, rather than mother's, status is more important for the TW effect (section 14.3.2.1), the woman's own trait would need to correlate reliably (during evolution of the TW effect, but not necessarily in a novel environment: section 14.7.1) with the status of her partner. Finally, both sex ratio and this female trait might be conditionally expressed, pleiotropic effects of the same genes in females. The relevant psychological and physiological influences might include nutrition and stressors (Gaulin & Robbins 1991) and the social dimensions of identity salience, group identification and group ideology, which have been identified by social psychologists as factors determining feelings of relative deprivation (Smith *et al.* 1999, section 14.7.2.3). Their impact on the sex ratio is likely to work via hormonal pathways (section 14.6.1).

Males might also play a role in sex allocation, by varying the proportion of X and Y sperm in the ejaculate in two ways (section 14.6.1). First, the ejaculates of higher-status men might contain a higher proportion of Y sperm. Second, a positive correlation between status and coital rate might produce the TW effect since there is evidence that coital rate correlates with the proportion of Y sperm in ejaculates. A high coital rate would also increase the sex ratio according to the theories implicating timing of insemination (section 14.6.1). The relationship between coital rate and status in contemporary societies is not consistent, though sometimes positive (e.g. James 1974, Pérusse 1993, Wellings et al. 1994, Rao & Demaris 1995, Weinberg et al. 1997). It may have been positive in early humans, and historically where status correlates positively with family size (section 14.7.2.1). Finally, if status correlates with testosterone level (Kemper 1990) and high-status men had a higher level of testosterone in the seminal fluid, this would raise the sex ratio, according to James' (1986, 1987c) theory.

Differential embryonic mortality may also contribute to the TW effect. Greater embryonic mortality under adverse environmental conditions (Wells 2000), and the advantage to male embryos of recent improvements in obstetric care and maternal health (section 14.6.2), may partially reflect status differences.

Trivers and Willard (1973) were not the first to propose a relationship between status and the birth sex ratio, although they were the first to place this relationship in an evolutionary context. An increase in sex ratio with socioeconomic status had previously been predicted in two ways (Teitelbaum & Mantel 1971). The first followed from a combination of an assumed male-biased embryonic mortality (section 14.6.2), and an assumption that such mortality was inversely related to socioeconomic status. The second implicated birth control and the assumption that its use increased with status (see Potts 1997). A greater proportion of sons in high-status families would then occur if couples practising birth control were more likely to stop having children

after the birth of a boy (e.g. to inherit property: Grant 1998, pp 154–155) than after the birth of a girl, or if they had smaller families, since sex ratio and birth order are negatively related (section 14.3.2.3). The latter relationship could explain the TW effect in those postdemographic transition societies in which there is an inverse relationship between status and family size (section 14.7.2.1).

These various mechanisms can be considered as the selected (evolved) means by which the TW effect is implemented, or as nonselected means, depending on one's view of the evolved mechanism, an issue I take up in the following section.

14.6.5 Mechanism and adaptation

it is not yet possible to tell whether any given sex ratio bias is adaptive or the consequence of some physiological constraint . . . the problem should best be tackled by examining the proximate mechanisms underlying sex ratio variations which are likely to be far fewer than the number of variables with which sex ratio has been found to vary

(James 1993, p 8).

It is not yet clear which, if any, of the influences on the birth sex ratio (section 14.3.2), or the proposed mechanisms underlying them (sections 14.6.1 & 14.6.2), are adaptations. For example, the influences of paternal age and birth order may be a consequence of physiological and psychological constraints (section 14.6.3), such as coital frequency. Alternatively, coital frequency may be an adaptive mechanism, and the birth order effect would be adaptive where early investment or primogeniture (inheritance to the first born) was more beneficial via sons. Another possibility is that the birth order effect is a nonadaptive epiphenomenon of an inverse relationship between coital frequency and birth order, which is itself an adaptation to control family size. This would fit with later offspring being of less value when family size approaches its optimal value asymptotically with respect to parental fitness.

Coital frequency might also be an adaptation for optimizing offspring production as a function of father's age, with the offspring sex ratio being a nonadaptive consequence. The adaptiveness of paternal age effects on the sex ratio will depend on the differential value of parental investment in sons and daughters, and their differential costs and benefits, as a function of father's age.

Locating the mechanism that was selected for the implementation of effects on the sex ratio in early humans will help us to predict which populations should show the effect. For example, if coital frequency was the selected mechanism for the TW effect then only those populations in which status and coital frequency were positively related would be expected to show the effect. Again, if coital frequency predicts family size then a coital frequency mechanism would rule out the TW effect in societies with a negative status/family size relationship. Knowledge of mechanisms therefore narrows the range of possible predictions concerning the variables influencing the sex ratio, and consequently the populations in which certain effects are predicted; I take up a particular example in the following section.

The picture would be more complicated if modern individuals consciously influence the birth sex ratio of their offspring as a function of status or other variables in an adaptive way (by deciding when to stop having children or by sex-selective abortion). Such actions might be a consequence of a general evolved psychological mechanism designed to make adaptive reproductive decisions. Its existence would complicate the task of isolating an evolved mechanism designed specifically to implement TW or some other effect on the sex ratio (see also section 14.7.2.3).

14.6.5.1 Inferring adaptation in modern populations from ancestral mechanisms: paternal age as an example

It is instructive to explore the distinction between selected and nonselected mechanisms for one possible TW mechanism: paternal age. The decline in sex ratio with paternal age (section 14.3.2.3) would produce a TW effect if higher status men had children when younger, on average, than lower status men. This would occur if higher status men took sexual partners when

younger, as long as conception was not delayed longer in such women. For this to be a selected mechanism, it must have evolved in the environment of early humans (section 14.7.1), or earlier. In early humans an early mating advantage for high-status men, together with a high adult mortality that prevented a much longer period of offspring production for such men, is a possible scenario for the effect. In modern populations the common increase in status with age acts against this effect but it would work if a *combination* of status and youth provided the greatest male advantage in sexual activity; the data differ on this point (e.g. Mealey 1985, Kenrick & Keefe 1992, p 84).

Let us explore the consequences for TW of a decline in the sex ratio with paternal age in a modern population, assuming that this relationship held in ancestral populations, and that higher status men do indeed father children when younger on average in the same modern population. In addition to the particular interest in paternal age effects, I offer this analysis as an example of a method for deriving predictions for adaptive behaviour in modern populations as a function of proposed mechanisms and ancestral states.

In Scenario 1 (Figure 14.2) this 'high status–young father' relationship is assumed to have held in the ancestral environment as well. The TW effect could then have been implemented (or strengthened) by the joint existence of the 'high status–young father' relationship and the effect of paternal age on the sex ratio, *status having no direct effect on the sex-ratio-determining mechanism*. The paternal age effect could then have been selected for in the ancestral environment, precisely because it provided a mechanism for implementing TW, and might continue in the modern environment (in which the TW effect might no longer be adaptive), since selection would not have had time to act against it, or because it was a constraint. In Scenario 2 (Figure 14.2) it is assumed that: (1) there was no relationship between age at fatherhood and status in ancestral populations; (2) there was no direct TW status–birth sex ratio relationship in ancestral populations either; but (3) the 'high status–young father' relationship does hold in the modern population. This scenario could result in an apparent but spurious

Scenario 1

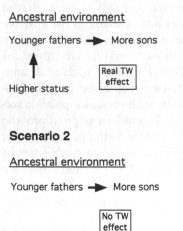

Scenario 2

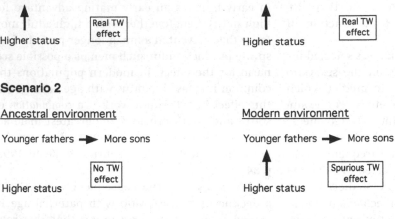

Fig 14.2 Possible influences of the 'paternal age effect' (sex ratio declines with father's age) on the Trivers-Willard (TW) effect. Arrows indicate causal relationships. In Scenario 1 a Trivers-Willard effect is implemented through the paternal age effect, which evolved in the ancestral environment. In Scenario 2 a Trivers-Willard effect appears to be at work in a modern population but is merely a spurious consequence of independent correlations of father's age with sex ratio and with status. See text for further explanation.

TW effect in the modern population because of the independent correlation of both status and sex ratio with age at fatherhood. By 'spurious' I mean that there had been no selective history for the effect.

If the 'high status–young father–more sons' causal chain (Scenario 1) was the only mechanism selected for implementing the TW effect in the ancestral environment, then the effect would be expected only in modern populations in which this causal chain still existed. However, even where this causal chain did exist in a modern environment a TW effect could, alternatively, be spurious (Scenario 2) or the result of some other mechanism. Path analysis (Mueller 1996) could be used to discover whether there is a direct causal relationship between status and sex ratio, or whether a TW effect relies on the common correlation of these two variables with paternal age. If in a modern population there is no relationship between status and age at fatherhood, then the lack of a TW effect in that population could be due to the breaking of that ancestral relationship (Scenario 1, but with no modern link between 'Higher status' and 'Younger fathers'). The absence of the TW effect in a modern population is therefore not necessarily evidence against the effect having evolved (see also section 14.7.1). Finally, if higher status men father children when *older*, on average, then the

paternal age effect runs counter to the TW effect and should be controlled for.

Evidence on the relationship between status and age at fatherhood in modern populations is therefore of interest. Different kinds of relationship between these and other variables, in conjunction with a given ancestral state, lead to different conclusions concerning the TW hypothesis.

The possible adaptiveness of the paternal age effect for TW raises the question of the origin of the effect itself. There are three possibilities. First, it evolved as a means of implementing or strengthening the TW effect, as just discussed. Second, it is due to physiological constraints (including, perhaps, a declining coital rate with age, Martin 1994). Third, the effect has an adaptive advantage in some other sphere, such as sons being of greater benefit to younger fathers, or daughters to older fathers.

14.7 | Problems in human sex ratio research

14.7.1 Novel environments: should behavioural adaptations be expected in modern humans?

A major issue in applying evolutionary hypotheses to humans is whether behavioural adaptations that evolved early in the evolution of our species remain adaptive in modern populations whose environment differs from the Pleistocene 'environment of evolutionary adaptedness' (EEA, Foley 1996) of early humans. This issue has been debated for more than a decade (e.g. Symons 1989, Turke 1990, Daly & Wilson 1999, 2000, Smith 2000, Smith *et al.* 2000).

While some scientists have sought evidence for adaptation in the behaviour of modern humans, others take Symons' (1989, p 143) view that, 'Darwin's theory of natural selection sheds light on human affairs only to the extent that it sheds light on phenotypic design, and design is usually manifested at the psychological rather than the behavioral level'. This seems to confuse mechanism with design. In the EEA we would expect both the psychological mechanism *and* its behavioural consequences to show evidence of adaptive design. The problem comes in interpreting the adaptiveness of behaviour in post-Pleistocene environments, since an evolved psychological mechanism may fail to generate adaptive behaviour if its environmental inputs differ from those it was selected to work with. The debate has partly been about the specificity of such evolved mechanisms (Turke 1990); at one extreme mechanisms designed for inputs unique to the EEA should not be expected necessarily to produce adaptive outcomes in modern populations, while general purpose mechanisms (including those involving conscious decision-making) can make adaptive use of a variety of novel environmental cues.

Views also differ concerning the utility of measuring reproductive success (RS) in modern populations as a means of understanding adaptation. Evolutionary psychologists, who believe that adaptive behaviour is controlled by specific cognitive modules evolved during the EEA, see little value in measuring reproductive consequences in modern populations. Darwinian anthropologists, on the other hand, argue that the novelty of modern environments, defined in terms of the range of cues that our psychological mechanisms can adaptively deal with, is an empirical issue. Consequently, they do measure RS, and attempt to distinguish between adaptive and maladaptive behaviour in modern populations (Turke 1990, Kaplan & Lancaster 2000).

This debate impacts on the analysis of the adaptive nature of sex ratio variation in modern populations. In particular, the TW hypothesis was predicated on the assumption that status or condition enhances RS. Does this hold for modern human populations, and if not what are the implications for the hypothesis in such populations?

14.7.2 Problems in the Trivers-Willard effect

14.7.2.1 The relationship between status and reproductive success

In traditional societies, and in Western societies before the demographic transition, status and wealth in men commonly predict RS. In developing and technologically advanced societies, however, the picture is mixed, with many societies exhibiting a negative relationship between wealth and fertility (family size) (e.g. Freedman & Thornton 1982, Vining 1986, Turke 1990, Pérusse 1993, Potts 1997, but see Low 2000). The demographic transition is the change from high mortality and fertility to low mortality and fertility that occurred, in Europe, in the eighteenth and nineteenth centuries, and is probably due to health improvements and fertility regulation (Potts 1997, Borgerhoff Mulder 1998b).

The relationship between status and fertility also depends on the scale at which the population is studied. For example, a negative relationship in Germany between fertility and income measured at the level of occupational groupings conceals a *positive* relationship between these variables *within* such groupings (Potts & Selman 1979, p 207). Fertility is also associated with relative, rather than absolute, measures of status within a reference group (Turke 1990, Mace 2000), as shown by the influence of changes in income over time (Freedman & Thornton 1982) and of comparison with parental income as an indicator of economic aspiration (Easterlin 1980). There are many reasons for these effects, including the fact that status is itself a function of the reference group (section 14.7.2.3).

The negative relationship between status and fertility in the modern world may be a nonadaptive response to a novel environment. Alternatively, by limiting family size, high-status parents might maximize fitness in the longer term by concentrating their greater available resources in fewer offspring. For high-status parents to produce more grandchildren, but fewer children, it is straightforward to show that the cost of an increase in family size (in terms of reduced family size in the following generation) would have to be greater for low- than for high-status families. The status/fitness issue is an active and unresolved one, with theoretical modelling of reproductive decisions and application to demographic datasets (e.g. Kaplan *et al.* 1995, Cronk *et al.* 2000).

If fitness declines with status, what happens to the TW prediction? As originally framed, the hypothesis does not assume that parents of higher status enjoy greater fitness, but it does assume that their *offspring* do. Then if children inherit the status of their parents, as in many modern societies (Buss 1999), or if fitness is correlated in successive generations, a correlation between fitness and status in the same individual would follow as an inferred assumption of the hypothesis. However, a positive relationship between status and fitness, although assumed by Trivers and Willard, is not *necessary* for the predicted effect of status on sexual allocation. As explained in the next section, for the TW effect to work what *is* necessary is that parental status interacts with offspring fitness such that the sons of high-status parents are more successful than the daughters, and this relationship is reversed for low-status parents.

14.7.2.2 Reproductive success and reproductive success variance

Figure 14.1a is a familiar representation of the TW effect showing the interaction between status and sex in determining offspring RS, with a positive status/RS correlation. Figure 14.1b shows the same interaction, necessary for the TW effect, accompanied by a negative correlation between parental status and offspring RS. It should be evident that this negative correlation does nothing to alter the TW prediction; it is the relative success of sons and daughters that produces the effect.

If offspring RS does decline with status then conditions for the TW effect occur where the RS/status gradient is steeper for daughters than it

is for sons, which means that the variance in RS due to status differences is greater for females than for males (Figure 14.1b). If this sex difference in variance translates into the same effect at the population level, which is plausible, then the normally greater variance in RS found in males will be reversed. It would be interesting to test this prediction.

It is sometimes stated that the TW effect requires that the sex of offspring benefiting more from having high-status parents (commonly males) has a greater variance in RS (e.g. Hrdy 1987, Anderson & Crawford 1993, Hrdy & Judge 1993). The claim may have come about by confusing a greater *between-status category* variance in RS for males (which *is* one of Trivers and Willard's assumptions) with the common finding of a greater *population* variance in male compared to female RS. However, if variance *within* status categories was higher in females than males, then the population variance could similarly be greater for females, even though between-status variance was greater for males. These conditions may be exceptional, or even nonexistent. I simply wish to point out that the assumption of greater population variance in male than in female RS is not a necessary condition for the TW effect. And as the preceding paragraph argues, the opposite may sometimes be expected.

14.7.2.3 The nature of status

Measuring status in tests of TW raises important questions that have received rather little attention in tests for the effect. How do status measures map on to the sources of reproductive differentials on which the TW hypothesis rests? What is the relevant reference group for measuring status? Is status perceived in absolute or relative terms? From whose perspective should status be measured? By what strategies do women select mates in terms of status, and what are their implications for the mapping of status on to sex ratio? What kind of status differentials existed in early humans?

Since notions of status and status striving are human universals (Brown 1991, Buss 1999), it is likely that selection favoured a general trait of striving for resources, and traits, such as intelligence, that make the most of material resources and social opportunities (Cummins

1998). In the TW context, however, the definition of status has to be tailored to the hypothesis and must take into account the perspective from which status is being viewed. First are those traits sought by women in a long-term mate (*Source A*, section 14.3.2.1). In contemporary societies the traits women most value, compared to those sought by men, are earning potential, ambition and industriousness. However, among the most highly valued mate characteristics, for both sexes, are intelligence, kindness and understanding (Buss 1989, 1992). Tests of TW have not measured status in terms of these last three traits, instead employing measures such as socioeconomic index, church rank, income and other material resources. Furthermore, tests have not measured female mate preferences directly, although there are good reasons for expecting such preferences to match the kind of objective measures described above.

Of course, intelligence, ambition and industriousness may all correlate with material resources, so that each might stand proxy for resources as status measures (although accidents of inheritance can disengage the causal relationship between these traits and material success). However, it is less obvious that intelligence would correlate with ambition and industriousness, or that kindness and understanding would correlate with any of these.

The other sources of male status attributes relevant to TW (*Sources B–D*, section 14.3.2.1) do not *rely* on female choice, but nevertheless include some of the attributes sought by women in mates: possession of resources and traits responsible for success in their procurement; and possibly aspects of life unconcerned with reproduction, such as risk-taking. Finally, maternal health (*Source E*) has received little attention as a potential status measure in TW studies. Which status traits (if any) show a stronger TW effect may therefore indicate whether, say, female choice, inheritance effects or male–male competition has been more important in establishing the effect (section 14.3.2.1). The clearest prediction is that only if female choice has been the predominant influence in status determination will the traits of kindness and understanding produce a TW effect. In addition, a prediction from one putative source of TW is that the effect

will be stronger in patrilocal societies (*Source B*, section 14.3.2.1).

Different sources of reproductive differentials for TW imply that status should be measured from differing viewpoints. As already argued, *Source A* differentials rely on female choice. *Sources B* and *C* rely on male resource holding power and other attributes that are objectively measurable but, if male–male competition is involved, may also depend on male perceptions. Finally, *Source C*, reliant on inheritance, should be measured by parental perceptions of children's status. This discussion of viewpoints does not deny that purely objective measures of status may correlate highly with the requisite subjective measures, but this is an empirical question.

Evidence from economics and social psychology shows that individuals, in part, evaluate status in relative rather than absolute terms, and do so 'locally', within a 'reference group'; a firm, for example, or those with whom they have face-to-face interaction (Frank 1985, Smith *et al.* 1999). Such local groups are smaller, sometimes very much smaller, than the populations studied in tests of TW and, although the importance of local status has been appreciated by some workers (Turke 1990, James 1996), studies have typically used measures whose values carry the same absolute status meaning across the whole population under study (for an exception see Freese & Powell 1999). It is therefore plausible that some individuals rated as low status in the population are perceived as relatively high status within the smaller community of people with whom they interact, and vice versa. As an example, consider the case of women listed in American *Who's Who* in 'traditional occupations'; those in which women constituted more than 15% of the work force (e.g. teaching, nursing, service organizations, the arts). Such women might be considered of high status due to their achievements, compared to the general population. Alternatively, within a smaller social circle they may be viewed as low status compared with the wives of the kind of men who gain entry into *Who's Who*, whose occupations provided higher incomes than their own (Mackey 1993).

If status is best measured at the local level, and is multidimensional, then the nature of the group within which it should be measured will vary from individual to individual, and with the attribute being measured. An individual's 'overall status' could then be measured as the average of a number of separate status measures, weighted by their perceived importance (e.g. Pérusse 1993, p 272). On the other hand, to the extent that the influences of status on the sex ratio work through simple physiological mechanisms, such as nutritional or health status, absolute status measures may reveal TW effects, even in cross-national comparisons (Mackey 1993, section 14.3.2.2). Such tests, although crude, may reveal evolved mechanisms.

Finding an appropriate measure of status is clearly important for testing the TW effect. But the mapping of status onto the sex ratio will also be influenced by the mate-selection strategy employed by women, once male status has been assessed. For example, women may accept the first male that exceeds a fixed threshold status value (Ellis 1992), assess a sample to set this threshold (Todd & Miller 1999) or use relative rather than absolute criteria. These different strategies are likely to produce different mappings, which will also vary with the distribution of male attributes in the population.

Any adaptive relationship between status and sex ratio in humans must have evolved, at the latest, in the immediate-return economies of Pleistocene hunter-gatherers, in which there were no accrued resources to inherit. If present-day hunter-gatherers are a good guide (Kelly 1995), early human societies were relatively egalitarian, although status differentials in terms of hunting, competition and social skills would have been likely, providing routes for the TW effect. Relationships between the sexes (Foley 1996) would probably have provided minimal conditions for *Sources A* and *B* of the reproductive differentials required for the TW effect (section 14.3.2.1), with mate selection for traits akin to the industriousness, intelligence and kindness seen in contemporary societies, and the material benefits they would bring. The extent to which modern measures of status will be recognized by evolved assessment processes will depend on the specificity of these processes (sections 14.6.5 & 14.6.5.1). Material evidence of status differences

first appears in the Late Mesolithic (approximately 8000 years ago), as shown by the relative richness of grave goods in burial sites (Constandse-Westermann & Newell 1989).

14.7.3 Predicting investment: current offspring value or marginal return?

How should optimal investment be calculated? As a function of the current reproductive value of the offspring ('current value') or of the increase in that reproductive value as a consequence of the investment ('marginal return')? Selection works on the consequences for fitness of the feature being selected, so 'marginal return' is the answer. Consider how this applies to the TW hypothesis. For decisions about the primary sex ratio the familiar representation of the problem in Figure 14.1a works well, with the ordinate representing reproductive value at conception *and, equivalently, marginal return*. The relevant marginal return on conceiving, say, a daughter is the daughter's reproductive value at conception minus its reproductive value just before conception, the latter being zero. *Uniquely at conception*, therefore, before which time current value is zero, Figure 14.1a also represents marginal return, the criterion by which selection acts on decision-making. The functions in Figure 14.1a therefore provide the answer to the allocation problem: if status is to the left of the crossover point, conceive a daughter; otherwise conceive a son (see also Williams 1979). For investment after conception the picture is different. Now, if Figure 14.1a is used to depict current value it will not also depict marginal return, and therefore fails as an illustration of the TW effect.

In empirical studies of the TW effect in postnatal investment the standard practice is to use current value as the criterion by which to test for the effect. In some cases, as we have seen, the RS of sons and daughters as a function of status is known (section 14.4.1), and follows the pattern of Figure 14.1a; at other times it is assumed to do so. Then high-status parents are predicted to favour sons, and low-status parents to favour daughters. As argued, however, this does not correctly predict how selection works. I suspect that the use of current value as the currency in postnatal investment studies has arisen by applying

TW logic, devised primarily for the primary sex ratio, directly to later investment decisions. Indeed Trivers and Willard themselves, at the end of their paper, predict postnatal investment in a way that is implicitly based on current value.

Whether using current value rather than marginal return as the currency for investment decisions leads to an incorrect prediction is an empirical question that would need to be examined case by case. For example, suckling a low-weight daughter rather than her plump twin brother (see the Figure 1 photograph in Hrdy 1990) will increase her chance of survival more than his. However, if her chances of marriage when mature are slim, while her brother can expect to marry polygynously, then her continued neglect, though tragic, is the adaptive maternal response. In this case we can assume that the brother has the higher current value *and* that marginal return will be enhanced more by suckling him than his sister. In contrast, consider the conclusion that, in the present-day US, involvement in a child's schooling shows a reverse TW effect, since *daughters* are increasingly favoured, relative to sons, as parents' educational level rises (even though, as a main effect, involvement is greater for sons than daughters, Freese & Powell 1999, section 14.4.1). Now, although the current value of the sons of highly educated parents may exceed that of their daughters, suppose that the same parents can improve their daughters' marriage prospects more through educational support than they can their sons'. In this case the reported failure of the TW hypothesis would be a success when measured by the correct index of marginal return. *To be more precise, it would be a success for a generalized TW hypothesis that predicted greater investment, as a function of status, in whichever sex gave the greater marginal return. I recommend that the TW hypothesis be reformulated in this way.*

This argument applies quite generally to investment decisions, not just to the TW effect. For example, the selected parental response to local factors (section 14.4.2) will not necessarily be to favour offspring who demonstrate more local resource enhancement or less local resource competition. Rather, it will be determined by marginal return, directed where it will have the greatest impact in *reducing* competition or *aug-*

menting enhancement. Data on the marginal returns of investment should now be sought.

14.8 | Prospects

Prospects for new work emerge from various analyses in this chapter. First, a number of potential Trivers-Willard mechanisms have been identified that have yet to be examined empirically: the proportions of X and Y sperm in ejaculates, coital frequency, embryonic mortality, family size and paternal age (sections 14.6.4 & 14.6.5.1). Second, I have outlined a method for predicting adaptations from proposed mechanisms and ancestral states, and this may be more widely applicable (section 14.6.5.1). Third, if the Trivers-Willard effect holds in populations where RS and status are negatively correlated, then women, rather than men, are predicted to have the greater variance in RS (section 14.7.2.2). Fourth, in tests of the Trivers-Willard effect many improvements are suggested in the measurement of status (section 14.7.2.3). Lastly, if investment in sons and daughters is under adaptive control it will be determined by the marginal return on the investment and not by the current reproductive value of the offspring (section 14.7.3). Measurement of the marginal return on parental investment should now be attempted.

14.9 | Coda

As this chapter shows, understanding variation in the human sex ratio requires contributions from medicine, physiology, evolutionary biology, psychology, anthropology, economics and demography. Human sex ratio research must be interdisciplinary if it is to be successful.

Acknowledgements

Stuart Laws has discussed human sex ratios with me a great deal over the years. Comments on an earlier version of the chapter by Monique Borgerhoff Mulder, Bruce Charlton, Andrew

Cockburn, Martin Daly and Ian Hardy led to important improvements. Martin Daly suggested to me the distinction between specific Trivers-Willard, and more general reproductive decision processes that might do the same job. Ian Croft, Chris Tolan-Smith and Anthony Downing helped me, respectively, with the social psychology and archaeology of status, and with path analysis. The Evolution and Behaviour Research Group at Newcastle provided a valuable platform for discussion, Olof Leimar discussed his work with me, and Stuart Laws and Ian Hardy provided references. I am most grateful to them all.

References

Abernethy V & Yip R (1990) Parent characteristics and sex differential infant mortality: the case in Tennessee. *Human Biology*, **62**, 279–290.

Anderson JL & Crawford CB (1993) Trivers-Willard rules for sex allocation: when do they maximise expected grandchildren in humans? *Human Nature*, **4**, 137–174.

Bereczkei T & Dunbar RIM (1997) Female-biased reproductive strategies in a Hungarian Gypsy population. *Proceedings of the Royal Society of London, series B*, **264**, 17–22.

Bernstein ME (1995) Genetic control of the secondary sex ratio. *Human Reproduction*, **10**, 2531–2533.

Bibbins PE Jr, Lipshultz LI, Ward JB Jr & Legator MS (1988) Fluorescent body distribution in spermatozoa in the male with exclusively female offspring. *Fertility and Sterility*, **49**, 670–675.

Boone JL III (1986) Parental investment and elite family structure in preindustrial states: a case study of late medieval-early modern Portuguese genealogies. *American Anthropologist*, **88**, 859–878.

Borgerhoff Mulder M (1994) On polygyny and sex ratio at birth: an evaluation of Whiting's study. *Current Anthropology*, **35**, 625–627.

Borgerhoff Mulder M (1998a) Brothers and sisters: how sibling interactions affect optimal parental allocations. *Human Nature*, **9**, 119–162.

Borgerhoff Mulder M (1998b) The demographic transition: are we close to an evolutionary explanation? *Trends in Ecology and Evolution*, **13**, 266–270.

Brown DE (1991) *Human Universals*. New York: McGraw-Hill.

Brown GR (2001) Sex-biased investment in nonhuman primates: can Trivers & Willard's theory be tested? *Animal Behaviour*, **61**, 683–694.

Buss D (1989) Sex differences in human mate preferences: evolutionary hypotheses tested in 37 cultures. *Behavioral and Brain Sciences*, **12**, 1–49.

Buss D (1992) Mate preference mechanisms: consequences for partner choice and intrasexual competition. In: JH Barkow, L Cosmides & J Tooby (eds) *The Adapted Mind: Evolutionary Psychology and the Generation of Culture*, pp 250–266. New York: Oxford University Press.

Buss DM (1999) *Evolutionary Psychology: The New Science of the Mind*. Boston, MA: Allyn and Bacon.

Chahnazarian A (1988) Determinants of the sex ratio at birth: review of recent literature. *Social Biology*, **35**, 214–235.

Charnov EL (1979a) Simultaneous hermaphroditism and sexual selection. *Proceedings of the National Academy of Sciences USA*, **76**, 2480–2484.

Charnov EL (1979b) The genetical evolution of patterns of sexuality: Darwinian fitness. *American Naturalist*, **113**, 465–480.

Clutton-Brock TH (1991) *The Evolution of Parental Care*. Princeton, NJ: Princeton University Press.

Clutton-Brock TH & Albon SD (1982) Parental investment in male and female offspring in mammals. In: King's College Sociobiology Group, Cambridge (eds) *Current Problems in Sociobiology*, pp 223–247. Cambridge: Cambridge University Press.

Constandse-Westermann TS & Newell RR (1989) Social and biological aspects of the Western European Mesolithic population structure: a comparison with the demography of North American Indians. In: C Bonsall (ed) *The Mesolithic in Europe*, pp 106–115. Edinburgh: John Donald.

Creasy MR (1977) The primary sex ratio of man. *Annals of Human Biology*, **4**, 390–392.

Cronk L (1989) Low socioeconomic status and female-biased parental investment: the Mukogodo example. *American Anthropologist*, **91**, 414–429.

Cronk L (1990) Comment on: Sieff DF, explaining biased sex ratios in human populations: a critique of recent studies. *Current Anthropology*, **31**, 35–36.

Cronk L (1991) Preferential parental investment in daughters over sons. *Human Nature*, **2**, 387–417.

Cronk L (2000) Female-biased parental investment and growth performance among the Mukogodo. In: L Cronk, N Chagnon & W Irons (eds) *Human Behavior* and Adaptation: An Anthropological Perspective. pp 203–221. New York: Aldine de Gruyter.

Cronk L, Chagnon N & Irons W (eds) (2000) Human Behavior and Adaptation: An Anthropological Perspective. New York: Aldine de Gruyter.

Cummins DD (1998) Social norms and other minds: the evolutionary roots of higher cognition. In: DD Cummins & C Allen (eds) The Evolution of Mind, pp 30–50. New York: Oxford University Press.

Daly M & Wilson M (1999) Human evolutionary psychology and animal behaviour. Animal Behaviour, 57, 509–519.

Daly M & Wilson M (2000) Reply to Smith et al. Animal Behaviour, 60, F27–F29:http://www.academicpress.com/anbehav/forum and http://www.idealibrary.com.

Davis JN & Todd PM (1999) Parental investment by simple decision rules. In: G Gigerenzer, PM Todd & The ABC Group (eds) Simple Heuristics that Make us Smart, pp 309–324. New York: Oxford University Press.

Dickemann M (1979) Female infanticide, reproductive strategies, and social stratification: a preliminary model. In: NA Chagnon & W Irons (eds) Evolutionary Biology and Human Social Behavior, pp 321–367. North Scituate, MA: Duxbury Press.

Dmowski WP, Gaynor L, Rao R, Lawrence M & Scommegna A (1979) Use of albumin gradients for X and Y sperm separation and clinical experience with male sex preselection. Fertility and Sterility, 31, 52–57.

Easterlin RA (1980) Birth and Fortune. London: Grant McIntyre.

Ellis BJ (1992) The evolution of sexual attraction: evaluative mechanisms in women. In: JH Barkow, L Cosmides & J Tooby (eds) The Adapted Mind: Evolutionary Psychology and the Generation of Culture, pp 267–288. New York: Oxford University Press.

Festa-Bianchet M (1996) Offspring sex ratio studies of mammals: does publication depend on the quality of the research or the direction of the results? Écoscience, 3, 42–44.

Fisher RA (1930) The Genetical Theory of Natural Selection. Oxford: Oxford University Press.

Fisher RA (1958) The Genetical Theory of Natural Selection, 2nd edn. New York: Dover Publications.

Foley R (1996) The adaptive legacy of human evolution: a search for the environment of evolutionary adaptedness. Evolutionary Anthropology, 4, 194–203.

Frank RH (1985) Choosing the Right Pond. Human Behavior and the Quest for Status. New York: Oxford University Press.

Frank SA (1987) Individual and population sex allocation patterns. Theoretical Population Biology, 31, 47–74.

Frank SA (1990) Sex allocation theory for birds and mammals. Annual Review of Ecology and Systematics, 21, 13–55.

Freedman DS & Thornton A (1982) Income and fertility: the elusive relationship. Demography, 19, 65–78.

Freese J & Powell B (1999) Sociobiology, status and parental investment in sons and daughters: testing the Trivers-Willard hypothesis. American Journal of Sociology, 106, 1704–1743.

Gangestad SW & Simpson JA (1990) Toward an evolutionary history of female sociosexual variation. Journal of Personality, 58, 69–96.

Gaulin SJC & Robbins CJ (1991) Trivers-Willard effect in contemporary North American society. American Journal of Physical Anthropology, 85, 61–69.

Graffelman J & Hoekstra RF (2000) A statistical analysis of the effect of warfare on the human secondary sex ratio. Human Biology, 72, 433–445.

Grant VJ (1990) Maternal personality and sex of infant. British Journal of Medical Psychology, 63, 261–266.

Grant VJ (1992) The measurement of dominance in pregnant women by use of the simple adjective test. Personality and Individual Differences, 13, 99–102.

Grant VJ (1994) Maternal dominance and the conception of sons. British Journal of Medical Psychology, 67, 343–351.

Grant VJ (1998) Maternal Personality, Evolution and the Sex Ratio. Do Mothers Control the Sex of the Infant? London: Routledge.

Hilsenrath RE, Buster JE, Swarup M, Carson SA & Bischoff FZ (1997) Effect of sexual abstinence on the proportion of X-bearing sperm as assessed by multicolor fluorescent in situ hybridization. Fertility and Sterility, 68, 510–513.

Hrdy SB (1987) Sex-biased parental investment among primates and other mammals: a critical evaluation of the Trivers-Willard hypothesis. In: RJ Gelles & JB Lancaster (eds) Child Abuse and Neglect: Biosocial Dimensions, pp 97–147. New York: Aldine de Gruyter.

Hrdy SB (1990) Sex bias in nature and in history: a late 1980s reexamination of the 'Biological Origins' argument. Yearbook of Physical Anthropology, 33, 25–37.

Hrdy SB & Judge DS (1993) Darwin and the puzzle of primogeniture: an essay on biases in parental investment after death. *Human Nature*, **4**, 1–45.

Irving J, Bittles A, Peverall J, Murch A & Matson P (1999) The ratio of X- and Y-bearing sperm in ejaculates of men with three or more children of the same sex. *Journal of Assisted Reproduction and Genetics*, **16**, 492–494.

Jacobsen R, Møller H & Mouritsen A (1999) Natural variation in the human sex ratio. *Human Reproduction*, **14**, 3120–3125.

James WH (1971) Cycle day of insemination, coital rate, and sex ratio. *The Lancet*, **i**, 112–114.

James WH (1974) Marital coital rates, spouses' ages, family size and social class. *Journal of Sex Research*, **10**, 205–218.

James WH (1980) Gonadotrophin and the human secondary sex ratio. *British Medical Journal*, **281**, 711–712.

James WH (1981) The honeymoon effect on marital coitus. *Journal of Sex Research*, **17**, 114–123.

James WH (1983) Decline in coital rates with spouses' ages and duration of marriage. *Journal of Biosocial Science*, **15**, 83–87.

James WH (1984) The sex ratios of black births. *Annals of Human Biology*, **11**, 39–44.

James WH (1985) The sex ratios of oriental births. *Annals of Human Biology*, **12**, 485–487.

James WH (1986) Hormonal control of sex ratio. *Journal of Theoretical Biology*, **118**, 427–441.

James WH (1987a) The human sex ratio. Part 1: a review of the literature. *Human Biology*, **59**, 721–752.

James WH (1987b) The human sex ratio. Part 2: a hypothesis and a program of research. *Human Biology*, **59**, 873–900.

James WH (1987c) Hormone levels of parents and sex ratios of offspring. *Journal of Theoretical Biology*, **129**, 139–140.

James WH (1990) The hypothesized hormonal control of human sex ratio at birth – an update. *Journal of Theoretical Biology*, **143**, 555–564.

James WH (1992) The hypothesized hormonal control of mammalian sex ratio at birth – a second update. *Journal of Theoretical Biology*, **155**, 121–128.

James WH (1993) Continuing confusion. *Nature*, **365**, 8.

James WH (1994) Comment on: changing sex ratios: the history of Havasupai fertility and its implications for human sex ratio variation. *Current Anthropology*, **35**, 268–270.

James WH (1995) What stabilizes the sex ratio? *Annals of Human Genetics*, **59**, 243–249.

James WH (1996) Evidence that mammalian sex ratios at birth are partially controlled by parental hormone levels at the time of conception. *Journal of Theoretical Biology*, **180**, 271–286.

James WH (1997a) Coital rates and sex ratios. *Human Reproduction*, **12**, 2083–2084.

James WH (1997b) Sex ratio, coital rates, hormones and time of fertilization within the cycle. *Annals of Human Biology*, **24**, 403–409.

James WH (1999) The status of the hypothesis that the human sex ratio at birth is associated with the cycle day of conception. *Human Reproduction*, **14**, 2177–2178.

James WH (2000) Secular movements in sex ratios of adults and of births in populations during the past half-century. *Human Reproduction*, **15**, 1178–1183.

James WH & Rostron J (1985) Parental age, parity and sex ratio in births in England and Wales, 1968–77. *Journal of Biosocial Science*, **17**, 47–56.

Joshi NV (2000) Conditions for the Trivers-Willard hypothesis to be valid: a minimal population-genetic model. *Journal of Genetics*, **79**, 9–15.

Kaplan HS & Lancaster JB (2000) The evolutionary economics and psychology of the demographic transition to low fertility. In: L Cronk, N Chagnon & W Irons (eds) *Human Behavior and Adaptation: An Anthropological Perspective*, pp 283–322. New York: Aldine de Gruyter.

Kaplan HS, Lancaster JB, Bock JA & Johnson SE (1995) Fertility and fitness among Albuquerque men: a competitive labour market theory. In: RIM Dunbar (ed) *Human Reproductive Decisions: Biological and Social Perspectives*, pp 96–136. Basingstoke: St. Martin's Press.

Katz DF (1991) Human cervical mucus: research update. *American Journal of Obstetrics and Gynecology*, **165**, 1984–1986.

Kelly RL (1995) *The Foraging Spectrum: Diversity in Hunter-Gatherer Lifeways*. Washington: Smithsonian Institution Press.

Kemper TD (1990) *Social Structure and Testosterone: Explorations of the Socio-bio-social Chain*. New Brunswick: Rutgers University Press.

Kenrick DT & Keefe RC (1992) Age preferences in mates reflect sex differences in human reproductive strategies. *Behavioral and Brain Sciences*, **15**, 75–133.

Krackow S (1995a) The developmental asynchrony hypothesis of sex ratio manipulation. *Journal of Theoretical Biology*, **176**, 273–280.

Krackow S (1995b) Potential mechanisms for sex ratio adjustment in mammals and birds. *Biological Reviews*, **70**, 225–241.

Laland KN, Kumm J & Feldman MW (1995) Gene-culture coevolutionary theory: a test case. *Current Anthropology*, **36**, 131–156.

Lazarus J & Inglis IR (1986) Shared and unshared parental investment, parent-offspring conflict and brood size. *Animal Behaviour*, **34**, 1791–1804.

Leimar O (1996) Life-history analysis of the Trivers and Willard sex-ratio problem. *Behavioral Ecology*, **7**, 316–325.

Low BS (2000) Sex, wealth and fertility: old rules, new environments. In: L Cronk, N Chagnon & W Irons (eds) *Human Behavior and Adaptation: An Anthropological Perspective*, pp 323–344. New York: Aldine de Gruyter.

Lummaa V, Merilä J & Kause A (1998) Adaptive sex ratio variation in pre-industrial human (*Homo sapiens*) populations? *Proceedings of the Royal Society of London, series B*, **265**, 563–568.

Mace R (1996) Biased parental investment and reproductive success in Gabbra pastoralists. *Behavioral Ecology and Sociobiology*, **38**, 75–81.

Mace R (2000) An adaptive model of human reproductive rate where wealth is inherited: why people have small families. In: L Cronk, N Chagnon & W Irons (eds) *Human Behavior and Adaptation: An Anthropological Perspective*, pp 261–281. New York: Aldine de Gruyter.

Mackey WC (1993) Relationships between the human sex ratio and the woman's microenvironment: four tests. *Human Nature*, **4**, 175–198.

Manning JT, Anderton R & Washington SM (1996) Women's waists and the sex ratio of their progeny: evolutionary aspects of the ideal female body shape. *Journal of Human Evolution*, **31**, 41–47.

Manning JT, Trivers RL, Singh D & Thornhill R (1999) The mystery of female beauty. *Nature*, **399**, 214–215.

Martin JF (1994) Changing sex ratios: the history of Havasupai fertility and its implications for human sex ratio variation. *Current Anthropology*, **35**, 255–280.

Martin JF (1995) Hormonal and behavioral determinants of the secondary sex ratio. *Social Biology*, **42**, 226–238.

Martin JF (1997a) Length of the follicular phase, time of insemination, coital rate and the sex of offspring. *Human Reproduction*, **12**, 611–616.

Martin JF (1997b) Coital rates and sex ratios. *Human Reproduction*, **12**, 2084–2085.

Maynard Smith J (1980) A new theory of sexual investment. *Behavioral Ecology and Sociobiology*, **7**, 247–251.

Maynard Smith J (1983) The economics of sex (review of E Charnov's *The Theory of Sex Allocation*). *Evolution*, **37**, 872–873.

Mealey L (1985) The relationship between cultural success and biological success: a case study of the Mormon religious hierarchy. *Ethology and Sociobiology*, **6**, 249–257.

Moore DH II & Gledhill BL (1988) How large should my study be so that I can detect an altered sex ratio? *Fertility and Sterility*, **50**, 21–25.

Mueller RO (1996) *Basic Principles of Structural Equation Modeling. An Introduction to LISREL and EQS*. New York: Springer-Verlag.

Myers JH (1978) Sex ratio adjustment under food stress: maximization of quality or numbers of offspring? *American Naturalist*, **112**, 381–388.

Nordborg M (1992) Female infanticide and human sex ratio evolution. *Journal of Theoretical Biology*, **158**, 195–198.

Oksanen L (1981) All-female litters as a reproductive strategy: defense and generalization of the Trivers-Willard hypothesis. *American Naturalist*, **117**, 109–111.

Pérusse D (1993) Cultural and reproductive success in industrial societies: testing the relationship at the proximate and ultimate levels. *Behavioral and Brain Sciences*, **16**, 267–322.

Potts M (1997) Sex and the birth rate: human biology, demographic change, and access to fertility-regulation methods. *Population and Development Review*, **23**, 1–39.

Potts M & Selman P (1979) *Society and Fertility*. Plymouth: Macdonald and Evans.

Rao KV & Demaris A (1995) Coital frequency among married and cohabiting couples in the United States. *Journal of Biosocial Science*, **27**, 135–150.

Renkonen KO (1970) Heterogeneity among first post-nuptial deliveries. *Annals of Human Genetics*, **33**, 319–321.

Roberts AM (1978) The origins of fluctuations in the human secondary sex ratio. *Journal of Biosocial Science*, **10**, 169–182.

Rosenheim JA, Nonacs P & Mangel M (1996) Sex ratios and multifaceted parental investment. *American Naturalist*, **148**, 501–535.

Rosenthal R (1991) *Meta-analytic Procedures for Social Research*. Newbury Park, CA: Sage.

Ruder A (1986) Paternal factors affect the human secondary sex ratio. *Human Biology*, **58**, 357–366.

Schwinger E, Ites J & Korte B (1976) Studies on frequency of Y chromatin in human sperm. *Human Genetics*, **34**, 265–270.

Sieff DF (1990) Explaining biased sex ratios in human populations: a critique of recent studies. *Current Anthropology*, **31**, 25–48.

Silk JB (1984) Local resource competition and the evolution of male-biased sex ratios. *Journal of Theoretical Biology*, **108**, 203–213.

Singh D & Zambarano RJ (1997) Offspring sex ratio in women with android body fat distribution. *Human Biology*, **69**, 545–556.

Smith EA (2000) Three styles in the evolutionary analysis of human behavior. In: L Cronk, N Chagnon & W Irons (eds) *Adaptation and Human Behavior: An Anthropological Perspective*, pp 27–46. New York: Aldine de Gruyter.

Smith EA, Borgerhoff Mulder M & Hill K (2000) Evolutionary analyses of human behaviour: a commentary on Daly & Wilson. *Animal Behaviour*, **60**, F21–F26: http://www.academicpress.com/anbehav/forum and http://www.idealibrary.com.

Smith HJ, Spears R & Hamstra IJ (1999) Social identity and the context of relative deprivation. In: N Ellemers, R Spears & B Doosje (eds) *Social Identity: Context, Commitment, Content*, pp 205–265. Oxford: Blackwells.

Smith MS, Kish BJ & Crawford CB (1987) Inheritance of wealth as human kin investment. *Ethology and Sociobiology*, **8**, 171–182.

Stinson S (1985) Sex differences in environmental sensitivity during growth and development. *Yearbook of Physical Anthropology*, **28**, 123–147.

Symons D (1989) A critique of Darwinian anthropology. *Ethology and Sociobiology*, **10**, 131–144.

Teitelbaum MS (1970) Factors affecting the sex ratio in large populations. *Journal of Biosocial Science, Supplement*, **2**, 61–71.

Teitelbaum MS (1972) Factors associated with the sex ratio in human populations. In: GA Harrison & AJ Boyce (eds) *The Structure of Human Populations*, pp 90–109. Oxford: Oxford University Press.

Teitelbaum MS & Mantel N (1971) Socio-economic factors and the sex ratio at birth. *Journal of Biosocial Science*, **3**, 23–41.

Todd PM & GF Miller (1999) From pride and prejudice to persuasion: satisficing in mate search. In: G Gigerenzer, PM Todd & The ABC Group (eds) *Simple Heuristics That Make Us Smart*, pp 287–308. New York: Oxford University Press.

Trivers RL (1974) Parent-offspring conflict. *American Zoologist*, **14**, 249–264.

Trivers R (1985) *Social Evolution*. Menlo Park, CA: Benjamin/Cummings.

Trivers RL & Willard DE (1973) Natural selection of parental ability to vary the sex ratio of offspring. *Science*, **179**, 90–92.

Turke PW (1990) Which humans behave adaptively, and why does it matter? *Ethology and Sociobiology*, **11**, 305–339.

Udry JR, Deven FR & Coleman SJ (1982) A cross-national comparison of the relative influence of male and female age on the frequency of marital intercourse. *Journal of Biosocial Science*, **14**, 1–6.

Van Schaik CP & Hrdy SB (1991) Intensity of local resource competition shapes the relationship between maternal rank and sex ratios at birth in cercopithecine primates. *American Naturalist*, **138**, 1555–1562.

Vining DR Jr (1986) Social versus reproductive success: the central theoretical problem of human sociobiology. *Behavioral and Brain Sciences*, **9**, 167–216.

Visaria PM (1967) Sex ratio at birth in territories with a relatively complete registration. *Eugenics Quarterly*, **14**, 132–142.

Voland E, Siegelkow E & Engel C (1991) Cost/benefit oriented parental investment by high status families. *Ethology and Sociobiology*, **12**, 105–118.

Voland E, Dunbar RIM, Engel C & Stephan P (1997) Population increase and sex-biased parental investment in humans: evidence from 18th- and 19th-century Germany. *Current Anthropology*, **38**, 129–135.

Waldron I (1983) Sex differences in human mortality: the role of genetic factors. *Social Science and Medicine*, **17**, 321–333.

Weinberg MS, Lottes IL & Gordon LE (1997) Social class background, sexual attitudes, and sexual behavior in a heterosexual undergraduate sample. *Archives of Sexual Behavior*, **26**, 625–642.

Wellings K, Field J, Johnson A & Wadsworth J (1994) *Sexual Behaviour in Britain: The National Survey of Sexual Attitudes and Lifestyles*. London: Penguin.

Wells JCK (2000) Natural selection and sex differences in morbidity and mortality in early life. *Journal of Theoretical Biology*, **202**, 65–76.

Werren JH & Charnov EL (1978) Facultative sex ratios and population dynamics. *Nature*, **272**, 349–350.

Whiting JWM (1993) The effect of polygyny on sex ratio at birth. *American Anthropologist*, **95**, 435–442.

WHO (1984) A prospective multicentre study of the ovulation method of natural family planning. IV. The outcome of pregnancy. *Fertility and Sterility*, **41**, 593–598.

Williams GC (1979) The question of adaptive sex ratio in outcrossed vertebrates. *Proceedings of the Royal Society of London, series B*, **205**, 567–580.

Wilson M & Daly M (1997) Life expectancy, economic inequality, homicide, and reproductive timing in Chicago neighbourhoods. *British Medical Journal*, **314**, 1271–1274.

Part 5

Sex ratios in plants and protozoa

Chapter 15

Sex ratios of malaria parasites and related protozoa

Andrew F. Read, Todd G. Smith, Sean Nee & Stuart A. West

15.1 | Summary

We review methods for studying the adaptive basis of sex allocation in the phylum Apicomplexa, a group of parasitic protozoa that includes the aetiological agents of malaria. It is our contention that analysis of apicomplexan sex ratios is not only interesting in its own right, but may actually provide insights into matters of clinical and epidemiological importance. We begin by justifying that position, and then summarize the natural history of these parasites and the sex ratio expectations that flow from that. Broadly speaking, these expectations are supported, but the evidence is scanty relative to that for many multicelled taxa. In the second half of the chapter, we give an overview of the theoretical and empirical methods available to take this work further. Much remains to be done: many key assumptions are currently little more than acts of faith.

15.2 | Introduction

Almost all work on the evolution of sex allocation is motivated by and tested on multicelled organisms. Yet the causative agents of some of the most serious diseases of humans and livestock have anisogamous sexual stages (Figure 15.1). These are all members of the protozoan phylum Apicomplexa, and include the malaria parasites (*Plasmodium* spp.). Species in other protozoan phyla can also have anisogamous sexual stages (e.g. some dinoflagellates, volvocidians and perhaps some foraminiferans; Lee *et al.* 1985) but we are unaware of any analysis of sex allocation in micro-organisms other than the Apicomplexa. An array of micro-organisms manipulate sex allocation in their hosts; the analysis of these extended phenotypes is discussed in Chapter 9.

There are at least four reasons for studying sex allocation in disease-causing protozoans. First, and not least, their sex ratios are often highly variable, both within and between hosts (Figure 15.2). Explaining the maintenance of diversity in a trait as closely related to fitness as is the sex ratio is of interest in its own right.

Second, the population structure of parasitic protozoa has proved hugely controversial, particularly in the context of *Plasmodium*, where population structures from effective panmixia to full clonality have been proposed (reviewed by Paul & Day 1998, Walliker *et al.* 1998). The population genetic structure of these parasites is likely to affect the evolution of drug resistance and virulence, and is also relevant to disease diagnosis and the development and assessment of chemo- and immuno-therapy (Tibayrenc *et al.* 1990, Herre 1993, Frank 1996, Hastings & Wedgewood-Oppenheim 1997, Mackinnon & Hastings 1998). The key issue is the frequency of self fertilization. Selfing rates can be measured directly and indirectly by population and molecular genetic methods. These approaches are expensive and the inferences drawn have frequently proved controversial. As we detail below, sex-allocation theory may provide a rapid and cheap

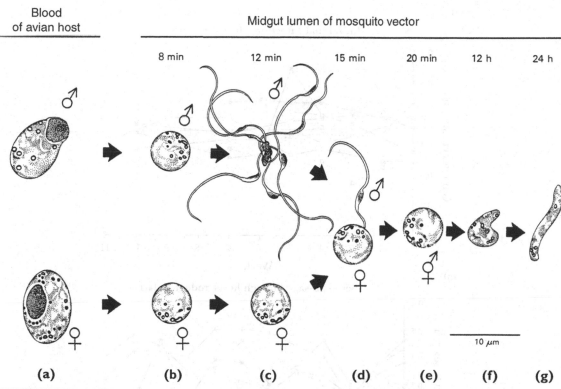

Blood of avian host

Midgut lumen of mosquito vector

8 min   12 min   15 min   20 min   12 h   24 h

(a)   (b)   (c)   (d)   (e)   (f)   (g)

10 μm

Fig 15.1 Malaria sex. Mosquito midgut stages of the avian malaria parasite *Plasmodium gallinaceum*. Times are post-bloodfeed and are approximate. (a) Male and female gametocytes in avian blood. (b) Male and female gametocytes after emergence from their host red cell in the midgut of the mosquito vector. The female gametocyte has now become the female gamete. (c) Formation and release of up to eight male gametes from the male gametocyte (exflagellation). (d) Fertilization. (e) Newly fertilized zygote. (f) Transformation into ookinete. (g) Mature ookinete, the stage which burrows through the midgut wall before the parasite develops into the oocyst bound to the haemolymph side of the midgut wall. Adapted with permission from Carter and Graves (1988).

method for inferring selfing rates in these populations (Read *et al.* 1992).

Third, much of what has been optimistically called 'Darwinian medicine' (Williams & Nesse 1991) involves the application of adaptationist arguments to infectious diseases, particularly in the context of disease severity (Ewald 1994, Williams & Nesse 1994). Optimality arguments typically assume population dynamic equilibria. These are not an obvious feature of many medically relevant diseases, where epidemic or unstable dynamics are both expected and frequently

observed (Anderson & May 1991). If infectious diseases do not successfully yield to sex-allocation theory, one of the most successful applications of adaptationist thinking, then there is little reason to think that adaptationist arguments concerning more complex phenotypes such as virulence will progress much beyond 'just so' stories.

More particularly, the evolution of both parasite sex ratio and virulence depends on the number of parasite clones per host, and they can be modelled in the same way (Frank 1992, 1996, 1998, Herre 1993, Pickering *et al.* 2000). A number of models have demonstrated that competition between unrelated parasite strains within a host generates selection in favour of higher virulence (reviewed by Read *et al.* in press). This occurs because increased competition is said to favour strains that more rapidly exploit their hosts and hence achieve greater relative transmission success than coinfecting strains, even at the expense of increased damage to hosts (virulence). Similarly, the presence of unrelated strains in the same host increases opportunities for outcrossing (section 15.4) and hence favours parasite strains with less female-biased sex ratios. Thus,

(a)

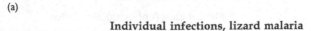

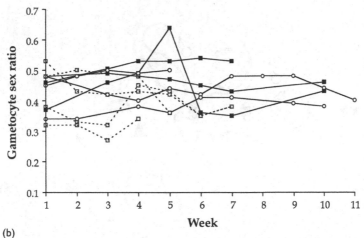

(b)

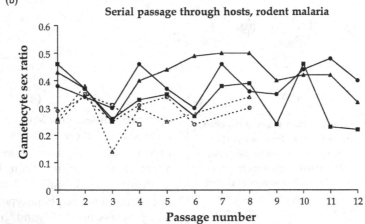

Fig 15.2 Patterns of sex ratio variation. (a) Within individual hosts through time. Data from ten naturally infected laboratory-maintained lizards sampled approximately weekly for up to 11 weeks. *Plasmodium mexicanum* in western fence lizards *Sceloporus occidentalis* (solid lines, squares), *P. giganteum* (dotted lines) and *P. agamae* (solid lines, circles) in rainbow lizards *Agama agama*. Redrawn from Schall (1989) with permission. (b) During serial passage through successive hosts. Three replicate lines of two clones (open and closed symbols) of *Plasmodium chabaudi* in laboratory mice. There was significant variation through time, and significant differences between clones but not between replicates within a clone. Redrawn from Taylor (1997) with permission. (c) & (d) Within populations of human malaria in Papua New Guinea ($n = 28$ infections) and Cameroon ($n = 54$ infections) respectively. Sex ratios are based on counts of ≥ 100 and ≥ 15 gametocytes per infection respectively. Redrawn from Read *et al.* (1992) and Robert *et al.* (1996) with permission.

(c)

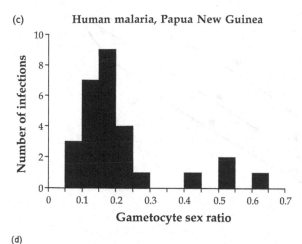

(d)

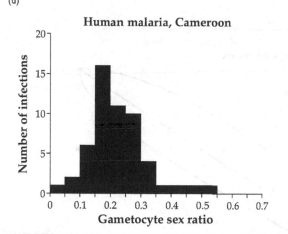

Fig 15.2 (cont.)

not only does analysis of sex ratios in these parasites allow testing of models mathematically analogous to virulence models, but optimal sex ratio and optimal virulence should be positively related (e.g. Pickering *et al.* 2000). It may therefore be possible to explain and predict virulence evolution using sex allocation.

Fourth, in light of the outstanding success of sex-allocation models across a range of biological systems, failure of an apparently appropriate model in a particular context probably points to flaws in our understanding of the basic natural history of the system. For instance, local mate competition theory (Hamilton 1967) successfully predicts sex ratios across populations of *Leucocytozoon* spp. in birds but it conspicuously fails to do so across populations of a related genus, *Haemoproteus*, also in birds (Figure 15.3; Read *et al.* 1995, Shutler *et al.* 1995). This immediately focuses

attention on details of breeding systems, vectors, or the epidemiology of *Haemoproteus* species which may not be, as assumed, simply analogous to species in better known taxa such as *Plasmodium* (Shutler & Read 1998). Formally, of course, the failure of apparently appropriate models might point to a failure of theory rather than misunderstandings of the natural history. As things stand, we see no reason to suppose that the study of apicomplexan sex ratios will require a radical expansion of sex-allocation theory, but we would be delighted to be wrong.

15.3 | Natural history

The Apicomplexa is a cosmopolitan group, all of which are parasites with sexual reproduction. Three subtaxa, all within the order Eucoccidiorida (class Sporozoasida, subclass Coccidiasina), are anisogamous (Figure 15.4). Most infamous are the haemospororins (suborder Haemospororina), which parasitize blood-feeding dipteran flies and the blood of various tetrapods. These include members of the genus *Plasmodium*, which are responsible for malaria in primates, birds and lizards, members of the genus *Leucocytozoon*, which infect birds, and members of the genus *Haemoproteus*, which infect birds, reptiles and amphibians (Atkinson & van Riper 1991). Also of substantial economic importance are the eimeriorins (suborder Eimeriorina), a diverse group that includes one- and two-host parasites of vertebrates and invertebrates. Important genera include *Cryptosporidium*, *Eimeria*, *Isospora*, *Sarcocystis*, *Neospora*, *Toxoplasma* and *Cyclospora*, members of which are pathogenic to immunocompromised humans or the causative agents of veterinary coccidiosis. Perhaps least well known are the adeleorins (suborder Adeleorina), one-host parasites of vertebrates and invertebrates or two-host parasites of blood-feeding invertebrates and the blood of vertebrates. Some *Hepatozoon* species cause often fatal disease in dogs.

In these three groups (haemospororins, eimeriorins and adeleorins), haploid infectious stages undergo a period of asexual proliferation and become multicelled stages called meronts, or *schizonts* (Figure 15.5). These rupture to produce

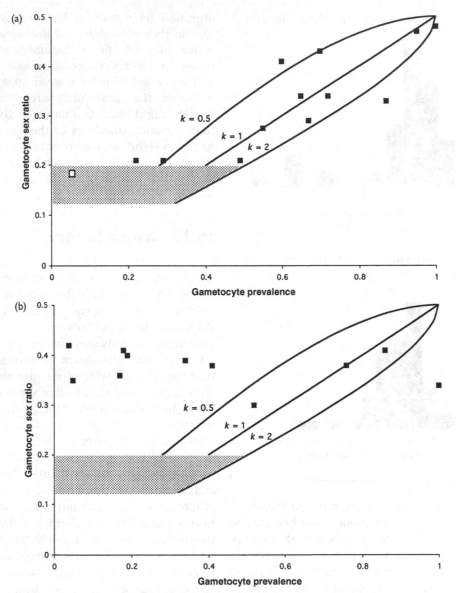

Fig 15.3 Relationship between gametocyte sex ratio and gametocyte prevalence across populations of blood parasites. The number of clones per host, and hence outcrossing rate and sex ratio, is expected to increase with increasing number of infectious hosts. (a) Filled squares represent populations of *Leucocytozoon* species; open square, *P. falciparum* in Papua New Guinea (Read *et al.* 1995). (b) Populations of *Haemoproteus* species (Shutler *et al.* 1995). In both, plotted points are means for different populations. Lines are theoretical expectation with various degrees of clonal aggregation, where *k* is a parameter of the negative binomial distribution which is inversely related to the degree of aggregation (see Read *et al.* 1995 for model details and derivation). A lower limit (horizontal part of the curve) is expected where there are just sufficient males to ensure fertilization of all the female gametes. This depends on the average number of viable gametes released per gametocyte, for which there is some uncertainty (shaded region). Adapted from Shutler and Read (1998) with permission.

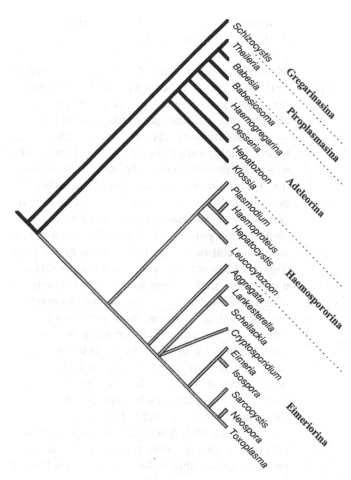

Fig 15.4 Phylogeny of representative genera of the phylum Apicomplexa. Genera in which syzygy occurs and in which 0.5 gametocyte sex ratios are expected are indicated by black branches. Genera in which gametocytes develop independently and gametes assort randomly, and in which biased sex ratios can be expected, are indicated by shaded branches. Clades are supported by character state changes detailed in Barta (1989), Barta *et al.* (1991), Carreno *et al.* (1997) and Smith *et al.* (2000b).

merozoites, some of which may continue to asexually replicate, or some or all of which transform into dioecious sexual stages, termed *gametocytes* for haemospororins and eimeriorins and gamonts for adeleorins. For the purposes of consistency, both across taxa and with existing protozoan sex ratio literature, we use the term gametocytes to cover both gamonts and gametocytes, and schizonts to cover meronts and schizonts.

Microgametocytes ('male') rupture to release a number of male *gametes*, whereas macrogametocytes ('female') give rise to just one female gamete. Male gametes fertilize the larger female gametes to form diploid zygotes, which undergo meiosis to restore the haploid state. The haploid products of this meiosis, encysted within the *oocyst*, initiate the period of asexual proliferation again. Throughout we take *sex ratio* to be the proportion of *gametocytes* that are male. The number of male gametes from a single

male gametocyte is reasonably species-specific and varies from a few to more than 1000. Thus, multiple mating by a single male gametocyte of a number of female gametocytes is possible.

There are a number of other relevant facts that we assume hold in all three of these groups but which have, for the most part, been formally demonstrated only in human or rodent *Plasmodium* species (for reviews see Carter & Graves 1988, Kemp *et al.* 1990, Walliker *et al.* 1998, Anderson *et al.* 2000). First, single haploid lineages can generate both male and female gametocytes which are self-compatible. Second, the sexual cycle is conventional, in that recombinants are generated when outcrossing occurs. Third, standard mendelian genetics occurs. Finally, mating is random with respect to genotype.

The production of gametocytes and fertilization occur by different means in a variety of host tissues in the different taxa. For instance, in the haemospororins (*Plasmodium*, *Leucocytozoon*, *Haemoproteus*), gametocytes are found in blood cells circulating in the peripheral blood of the vertebrate host (Figure 15.5a). When a vector takes a blood meal, gametocytes rupture in the midgut, releasing gametes. Male gametes are flagellated and swim in search of female gametes (Figure 15.1). Mating thus takes place among the gametes released from the gametocytes present in a blood meal (up to hundreds or even thousands). In contrast, the gametocytes of many

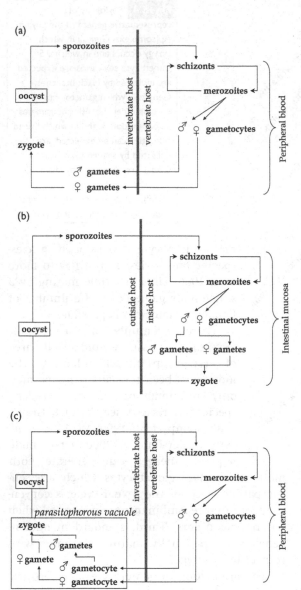

Fig 15.5 Apicomplexan life cycles. For simplicity, not all life stages are shown, and terminology is simplified for consistency (see text for further details).
(a) Haemospororina, e.g. *Plasmodium*, *Leucocytozoon* and *Haemoproteus* species. Mating takes place among gametes derived from gametocytes in a blood meal. (b) A one-host eimeriorin, e.g. *Eimeria* species. Mating takes place among gametes derived from gametocytes present in a localized area of the gastrointestinal tract. (c) A two-host adeleorin, e.g. *Hepatozoon* species. Mating takes place between gametes derived from male and female gametocytes, which have associated in a parasitophorous vacuole in a gut epithelial or fat body cell of an invertebrate (syzygy).

eimeriorins (e.g. *Eimeria*) are found in the cells of the intestinal muscosa (Figure 15.5b). Male gametocytes release a number of flagellated gametes which swim in search of a female gametocyte and fertilize her within her host cell. Thus, fertilization takes place among the gametes present in localized areas of the gastrointestinal tract.

The mating strategy of two-host adeleorins, which inhabit vertebrate blood (e.g. *Hepatozoon*), is somewhat different. Like the haemospororins, their gametocytes are found in vertebrate peripheral blood cells. But when these are ingested by a haematophagus invertebrate, the gametocytes pair up in a parasitophorous vacuole in a cell of the gut epithelium or fat body (Figure 15.5c). Mating takes place within this vacuole when male gametes are formed, one of which fuses with the female gamete to produce a zygote, while the excess male gametes die. This process of gametocytes associating within the parasitophorous vacuole in a host cell prior to gametogenesis and fertilization is known as *syzygy*. Thus in species where syzygy occurs, gametes from a single male gametocyte fertilize the gamete from a single female gametocyte. This contrasts with species without syzygy, where different gametes arising from a single male gametocyte are able to fertilize gametes from a number of female gametocytes.

15.4 | Theoretical expectations

Two features of this natural history are immediately striking in the context of sex-allocation theory. First, there is the potential for local mate competition (LMC, Hamilton 1967; see also Chapters 19 & 20). In the haemospororins and eimeriorins, gametes competing for matings will be those found in a single blood meal or localized area of the gut. Thus, matings will occur among the parasite genotypes present in the few microlitres of blood taken up by a vector or present in a few millimetres of gut tissue. This leads to the potential for LMC and selfing if there are low numbers of parasite genotypes present in those mating arenas. Not much is known about the genetic diversity in eimeriorin infections; in malaria infections, the number of clones per

(a)

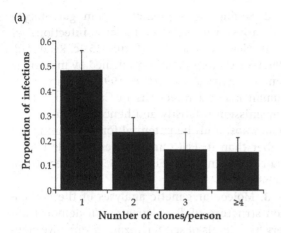

(b)

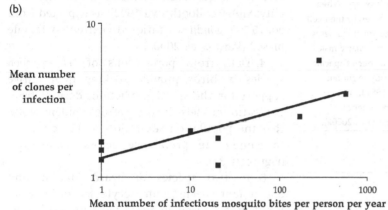

Mean number
of clones per
infection

Mean number of infectious mosquito bites per person per year

subject to the condition that there are sufficient male gametes to fertilize all female gametes in a blood meal (Figure 15.7; Read *et al.* 1992; see also Dye & Godfray 1993; Read *et al.* 1995; Pickering *et al.* 2000, West *et al.* 2000a). This relationship is the basis for our assertion that the population genetic structure of these parasites can be readily estimated from data on gametocyte sex ratio (Read *et al.* 1992). Below (section 15.6.4) we more precisely define *s*, and argue that eq. 15.1 is a general result that holds across a range of apicomplexan natural histories.

blood meal is variable but typically ranges from one to five, probably as a result of variation in transmission rates (Figure 15.6). (By 'clone' we specifically mean the asexually derived lineage arising from a haploid meiotic product.) Thus, at the limit, when all gametes in a mating pool are contributed by a single clone, selection should favour female-biased sex ratios, which maximize the number of zygotes that can be formed (i.e. the production of just enough males to fertilize all the females). Where gametes are contributed by many clones, sex ratios closer to 0.5 should be favoured because these will maximize the genetic representation of each clone in the zygote population.

This intuition can be formalized using Hamilton's standard LMC arguments to show that the optimal sex ratio, r^*, is related to the selfing rate *s* as

$$s = 1 - 2r^*, \tag{15.1}$$

Second, syzygy removes the factors that favour female-biased sex ratios under LMC. Even when only self fertilization is possible (as when, for example, only a single parasite genotype is present in a host) a sex ratio of 0.5 should be favoured by selection. This is because female-biased gametocyte sex ratios would not reduce competition among the male gametes from the single male gametocyte within a parasitophorous vacuole for access to the single female, nor would it increase the number of female gametes within a parasitophorous vacuole available for fertilization (Figure 15.5c). Thus, both factors favouring female-biased sex ratios under LMC, namely selfing and inbreeding (Taylor 1981), are missing where syzygy is present. This intuition is formalized by West *et al.* (2000a).

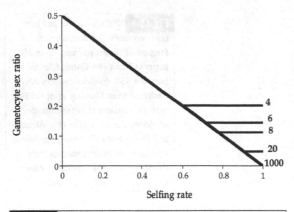

Fig 15.7 Predicted gametocyte sex ratios as a function of the selfing rate (eq. 15.1) for species without syzygy. When selfing rates are high, the sex ratio is constrained by the need to produce enough male gametes to fertilize female gametes. This constraint is determined by the number of viable male gametes released from a male gametocyte, a species-specific trait. In *Plasmodium* spp., this number is at most eight but probably between four and six (Read *et al.* 1992), although it may go as low as two (Schall 2000); in some eimeriorin species, there can be 20 or even 1000 (West *et al.* 2000a).

15.5 | Tests of theory

This is not the context for a thorough review of the relevant evidence. However, very briefly, we note the following.

1. Sex ratio in apicomplexan parasites that do not have syzygy are generally female biased; in species with syzygy, sex ratios are 0.5 (Figure 15.8).

2. Selfing rates predicted from gametocyte sex ratios from human malaria infections in Papua New Guinea (PNG) (Figure 15.2c; Read *et al.* 1992) were subsequently confirmed by molecular genetic analysis (Paul *et al.* 1995). Sex ratios in human malaria infections in Cameroon, where transmission intensity and hence the number of clones/host and the potential for outcrossing are higher than in PNG, are also less female biased than those in PNG (Figure 15.2d; Robert *et al.* 1996).

3. Molecular genetic analyses of the population structure of *Toxoplasma gondii* demonstrate very high levels of self fertilization (effective clonality; Sibley & Boothroyd 1992); as expected from eq. 15.1, *T. gondii* sex ratios are extremely female biased (West *et al.* 2000a).

4. Data from populations of *Leucocytozoon* species in birds provide striking quantitative support for the novel prediction, derived from sex ratio models incorporating epidemiology, that the population sex ratio should be related to gametocyte prevalence across populations (Figure 15.3a).

5. In two species of lizard malaria and one rodent species, gametocyte sex ratios were less female biased in hosts with more gametocytes (Taylor 1997, Pickering *et al.* 2000, Schall 2000). This is as expected if, as has been observed experimentally (Taylor *et al.* 1997), genetically diverse infections produce more gametocytes and if parasites are altering sex ratios in response to the presence of coinfecting genotypes.

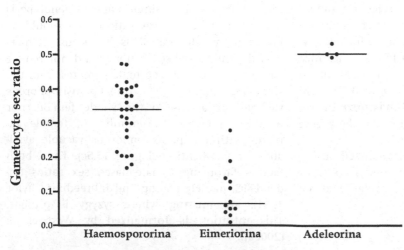

Fig 15.8 Population sex ratios from the three apicomplexan taxa with anisogamous gametes. In the haemospororins and eimeriorins, gametes from a single male gametocyte can fertilize gametes from a number of female gametocytes. The adeleorins have syzygy, where the gametes from a single male gametocyte can fertilize only a single female gamete. Horizontal lines are the medians for each taxon. Data from Read *et al.* (1995), Shutler *et al.* (1995), and West *et al.* (2000a).

Together, these lines of evidence generally support the theoretical arguments laid out above. However, the case is far from closed. In particular, sceptics might argue that some of these successes (e.g. 1–3 above) are little more than anecdotes (for instance, syzygy, probably the ancestral apicomplexan state, may have only been lost once (Figure 15.4; Barta 1989, West *et al.* 2000a) or involved predictions with rather large confidence intervals (62–100% selfing in PNG; Read *et al.* 1992). Data from populations of *Haemoproteus* species in birds most emphatically do not show the predicted relationship between sex ratio and gametocyte prevalence found with *Leucocytozoon* species (Figure 15.3b). Worse, male-biased sex ratios, which are never expected, have been reported in various species of *Haemoproteus* in lizards (Paperna & Landau 1991). In addition, there is a notable lack of support for any association between gametocyte densities and sex ratio in human malaria (Robert *et al.* 1996), other lizard malarias (Schall 1989), or in *Haemoproteus* species (Shutler *et al.* 1995). More generally, the substantial variation in sex ratio both within and across hosts (e.g. Figure 15.2) is still to be satisfactorily explained in terms of adaptive sex allocation.

Whether all this means that applications of sex-allocation theory to these parasites has so far been more than half right or more than half wrong is a matter for conjecture. What can be in no doubt is the need for more work.

15.6 | Basic methods

Methods for studying apicomplexan sex ratios, perhaps more than those in any other taxon, depend on advances in cell and molecular biology and are likely to evolve rapidly over the next few years, particularly as new tools become available from the malaria genome project. However, technical limitations are not the reason why studies of apicomplexan sex ratios are in their infancy. A greater hurdle is getting evolutionary biologists to grapple with new jargon and existing techniques (though it is probably still less than the hurdle of interesting parasitologists

in sex ratio evolution). In this section, we discuss methods generic to any study of apicomplexan sex ratios. In Section 15.7, we discuss approaches that are or could be used to address specific issues that have been of interest to date. Throughout, our discussion is focused on *Plasmodium* species unless stated otherwise, partly because we have most experience with this genus, and partly because the techniques for studying that genus, and particularly *P. falciparum*, far outstrip those available for other apicomplexans.

15.6.1 Measuring gametocyte sex ratios

Gametocyte sex ratios are typically estimated from stained blood or tissue samples (for details, see Schall 1989, Read *et al.* 1992, 1995, Shutler *et al.* 1995 and references therein for the haemospororins; for the other apicomplexans, see West *et al.* 2000a and Lee *et al.* 1985). We note the following points. First, sexing gametocytes requires more careful staining than is normally required for species identifications or for simply counting parasites or determining their stages. Second, sex ratios from thick (multiple cell layer) and thin (single cell layer) blood smears are highly correlated but sex ratios from thick smears are less female biased. The most likely explanation for this is that gametocytes in thick smears are not always viewed perpendicular to their long axis, making sexing more difficult; certainly, there are also more gametocytes that cannot be confidently sexed in thick smears (Read *et al.* 1992). For this reason, we recommend using thin smears where possible. Third, although sex differences in morphology could in principle result in some separation of the two sexes during the smearing process, sex ratios estimated from different regions of the same smear are highly repeatable (Read *et al.* 1992).

15.6.2 Field systems

Apicomplexan sex ratios have been studied in field populations of lizards (Schall 1989, 1996, Pickering *et al.* 2000), snakes, fish, cats, turtles (West *et al.* 2000a), birds (Read *et al.* 1995, Shutler *et al.* 1995, West *et al.* 2000a) and humans (Read *et al.* 1992, Robert *et al.* 1996). There is substantial potential for more descriptive field work in all these systems. With *Plasmodium* spp., it is possible (though logistically demanding) to look at transmission to vectors (e.g. Robert *et al.* 1996, Schall 1996); in other cases, the vectors involved in transmission have not been identified or are not easy to work with (e.g. *Leucocytozoon* spp., *Haemoproteus* spp.). Work with human malaria infections probably offers the greatest potential for field tests of theory, not just for fund-raising reasons, but also because the genetic and immunological markers currently and soon to become available are unlikely to be rivalled by anything available for the parasites of other hosts. However, field work with humans requires substantial collaboration with clinicians and malariologists, something for which evolutionary biologists are not noted. While experimental possibilities with humans are probably somewhat limited, it may be possible to manipulate parasite numbers and in-host relatedness in field populations of bird and lizard *Plasmodium* spp.

15.6.3 Biological models

There is also substantial potential for laboratory work with malaria sex ratios. *Plasmodium* species are readily cryopreserved and do not require mosquitoes for routine maintenance. Only *P. falciparum* can be reliably cultured *in vitro*. Doing so is not trivial (requiring sterile techniques operated in containment facilities plus reliable sources of safe, fresh human blood, and incubation at 37°C in a CO_2-rich atmosphere), but is routine in well set-up laboratories. Blood feeding of mosquitoes from cultures is routinely achieved by laboratories around the world but mosquito culture, particularly when working with human malarias, again requires substantial logistical support. In the only relevant study we know of, sex ratios *in vitro* and *in vivo* did not differ, though sample sizes were small ($n = 5$ culture adapted isolates; Read *et al.* 1992).

Work with animal models of malaria is technically easier (though, at least in the UK, the associated regulatory bureaucracy is substantially more burdensome). Various species of rodent malaria are maintained in laboratory mice worldwide (mainly *P. bergehi*, *P. vinckei* and *P. chabaudi*). While *Mus musculus* is not the natural host of any of these species, the use of laboratory mice allows the vast armoury of mouse genetics and reagents to be brought to bear. The maintenance of natural rodent hosts (e.g. the thicket rat, *Thamnomys rutilans*) was fairly routine in the 1970s but we know of no laboratory colonies operating now. Other laboratory models include *P. gallinaceum* in chickens, the natural host, and various primate malarias, some of which can be maintained in their natural host. Transmission to mosquitoes from rodents, chickens and simians is relatively routine. The biological properties of the various *Plasmodium* models are discussed by Collins (1998), Cox (1988) and McGhee (1988). With *P. gallinaceum* or various lizard malarias, it would be possible to couple laboratory work with field work on the same (natural) parasite–host combination (Schall 1996, R Paul pers. comm.), although the availablity of reagents, parasite lines and genetic markers is substantially less for those species than for rodent or human malaria species. Laboratory work with other haemospororins has been rare, largely because they lack blood-stage asexual replication so that a reliable source of their vectors, which are difficult to culture, is required.

15.6.4 Mathematical models

If we replace s, described as the 'selfing probability' by f, Wright's coefficient of inbreeding, then eq. 15.1 is still valid. Wright's coefficient has two interpretations that can be shown to be equivalent (Crow & Kimura 1970): (1) it is the probability that the uniting gametes are identical by descent, and (2) it is the correlation between the uniting gametes (here, the correlation between their sex ratio strategies). These can be taken as precise definitions of 'selfing probability'.

Claims that malaria populations can be considered clonal, despite obligate sexuality, have generated considerable interest in estimating Wright's coefficient of inbreeding, f, from oocyst data (reviewed by Paul & Day 1998, Walliker *et al.* 1998). Equation 15.1, with f instead of s, has been derived in the context of malaria by Dye and Godfray (1993). Their derivation makes numerous specific assumptions; for example, that different oocysts in a single mosquito are independent. This may have created the impression that eq. 15.1, while true in some particular ideal circumstances of a modelled world, may not be true in general. This may be one reason why efforts to estimate f have followed the expensive and laborious route of looking at oocysts instead of measuring sex ratios, which is much easier and cheaper.

However, as we show elsewhere (Nee *et al.* submitted), eq. 15.1 is true, in *general*. Regardless of how the parasite population is structured into nonrandom associations at the level of mosquitoes, houses or villages, and regardless of how mosquitoes behave, eq. 15.1 is the correct description of the relationship between sex ratio and selfing rate (or, more specifically, Wright's inbreeding coefficient).

15.7 | Applications

Here we discuss how these methods have been used to address questions that have been of interest to date: (1) is LMC involved, (2) are there conditional sex-allocation strategies, and (3) are key assumptions supported? Undoubtedly many of these approaches will also be useful in addressing a range of other issues. Our intention is to provide signposts to the relevant literature and a flavour of the possibilities.

15.7.1 Does the sex ratio conform to LMC?

Answering this is largely a matter of making inferences about parasite population genetic structure, particularly the frequency of self fertilization, and thus determining whether eq. 15.1 holds. Selfing rates have been estimated in three ways: by estimating the number of genotypes present in the mating arena, by genotyping zygotes and by analyses of population-wide linkage disequilibrium.

The number of genotypes likely to be present in a mating arena (e.g. the c. 2.5 µl of peripheral blood which forms a typical mosquito blood meal) can be estimated with what are now standard molecular genetic techniques, at least for *P. falciparum*, where sufficient polymorphic markers have been developed (reviewed by Arnot 1999).

However, all field studies to date have used polymerase chain reaction (PCR) amplification of samples of peripheral blood which contain numerous individual parasites. These studies are incapable of determining which of the clones present are actually producing gametocytes, though in principle they provide an upper boundary. But estimating even that is not straightforward. Obtaining unbiased estimates of the number of clones/host from this type of data is not trivial (many studies are in effect analysing a 'minimum maximum' number of clones present). Methods for obtaining an unbiased estimate of the population mean from these sort of data do exist (Hill & Babiker 1995), but have rarely been applied. In any case, there may be problems with the raw data: parasite samples contain an unknown number of clones at unknown frequencies. The probability of successfully detecting a clone by PCR is almost certainly affected by the concentration and ratios of competing template (e.g. Kyes *et al.* 1997), but precisely how is not well understood (Arnot 1999, Anderson *et al.* 2000).

Many of these problems should be overcome in the near future now that *in situ* PCR of individual gametocytes and reverse transcriptase PCR (RT-PCR) of polymorphic gametocyte genes is possible (Babiker *et al.* 1999, Ranford-Cartwright & Walliker 1999, Menegon *et al.* 2000). These should more readily provide information on the numbers of gametocytes likely to be found in blood meals, allowing within- and between-population analysis of gametocyte sex ratios.

The number of gametocyte-producing genotypes per host can be estimated by nonmolecular methods. The mean number of clones per host in a population should be a function of transmission intensity. This in turn should be a function of the proportion of hosts that are infectious. Thus outcrossing rates (and hence sex ratio) will be linked to the proportion of hosts that are gametocyte positive (Read *et al.* 1995). Data on sex ratios and gametocyte prevalences from blood parasite populations in birds have provided some of the best evidence that LMC shapes haemospororin sex ratios, and some of the best that it does not (Figure 15.3). While this approach cannot provide a means of analysing within-population variation in sex ratios, it has the huge advantage of being substantially easier and cheaper than approaches involving molecular genetics. Whether it involves more assumptions than molecular genetic approaches is currently a matter of conjecture. It is not yet obvious whether we have sufficient understanding of the biases involved in the PCR of samples of haploid parasites (Arnot 1999, Anderson *et al.* 2000).

A second way that selfing rates have been estimated is by molecular genetic analyses of the products of sex: the oocysts. These contain all the haploid products of a single fertilization. In principle this should provide the most direct assessment of mating patterns: heterozygote deficits are relatively easily related to selfing rates (Hill & Babiker 1995). Two populations of *P. falciparum* have been studied using this approach. In Tanzania, the inbreeding coefficient was estimated to be 0.33. This should accord with a gametocyte sex ratio of around 0.33 (from eq. 15.1), a prediction that has yet to be tested. In PNG, where transmission rates are much lower and there are fewer clones per host, a much higher inbreeding coefficient has been reported (0.92; Paul *et al.* 1995), consistent with what was predicted from gametocyte sex ratios in those populations (Read *et al.* 1992). But while PCR amplification of single oocysts is now a routine procedure, estimates of the inbreeding coefficient from the PNG oocyst data vary from 0.48 to 0.92 (!), depending on how one accounts for the existence of null alleles in the data (Anderson *et al.* 2000).

A third method for analysing the population genetic structure of apicomplexans is to analyse linkage disequilibria. Virulent strains of *Toxoplasma gondii*, for example, share identical multilocus genotypes distinct from moderately polymorphic nonvirulent strains (Sibley & Boothroyd 1992). Such observations are consistent with very frequent self fertilization and, as expected from eq. 15.1, *T. gondii* sex ratios are highly female biased (West *et al.* 2000a). A variety of molecular markers have been used to analyse linkage disequilibrium in blood-stage *P. falciparum* (Walliker *et al.* 1998). The problems involved in the interpretation of such data are summarized by Paul and Day (1998). So, as far as we can see, in all but extreme cases such as *T. gondii*, such data are unlikely to be useful as a way of studying LMC in these parasites because rates of outcrossing associated with even very substantial LMC can be sufficient to erode linkage disequilibria; moreover, other processes can promote linkage disequilibrium in outcrossing populations.

Finally, an altogether different tack can be used to investigate whether LMC ordinarily shapes apicomplexan sex ratios: find natural histories that violate the necessary assumptions, e.g. those with syzygy (Figures 15.5 and 15.8). We are unaware of other apicomplexan taxa where LMC should not be involved, but this may simply be ignorance. For instance, in any circumstances where gametes from many hosts are well mixed in a large mating arena, we would expect unbiased sex ratios.

15.7.2 Are there conditional sex ratio strategies?

Some of the most impressive evidence that adaptive sex-allocation theory actually works comes from the observation that the sex ratios of many metazoans change in response to environmental cues, often as predicted by theory (Charnov 1982, Godfray 1994, West *et al.* 2000b). Nothing is known of how sex is determined in apicomplexan parasites, except that segregating sex chromosomes cannot be involved: single haploid lineages can produce both male and female gametocytes (Smith *et al.* 2000a). Do apicomplexan parasites play conditional sex ratio strategies? There are various reasons why conditional strategies might be favoured by selection; for example, the number of clones (genetic relatedness) of parasites within hosts varies (e.g. Figure 15.6). This favours a conditional sex ratio strategy: when only one or two clones are in a host (high amount of selfing), an extremely female-biased sex ratio is favoured; with more clones per host (low amount of selfing), a more even sex ratio is favoured. However, such a conditional strategy could evolve only if parasites are able to detect the variation in clones per host, and then adjust their sex ratio accordingly. Perhaps such sophistication is beyond what are after all single-celled organisms. But the lesson from the last century is that these 'simple' organisms are sufficiently sophisticated to outwit biomedical science, and we can imagine various ways by which they might achieve kin recognition. For instance, the strength of antibody binding against clone-specific alleles may be a good indication of the presence of coinfecting nonrelatives. Other easily assayed surrogates could provide information about the number of clones in an infection (e.g. levels of anaemia, other host responses). Selection pressures other than LMC may also favour conditional sex ratios. The most obvious is the need to ensure that sufficient viable male gametes are present in the mating arena. Sex ratios should be closer to 0.5 in low-density infections (West *et al.* 2001) or as effective anti-male host responses develop (Paul *et al.* 1999).

If apicomplexan parasites are playing conditional sex ratio strategies, sex ratio distributions within a population may have a greater than binomial variance. In all *Plasmodium* populations so far examined, this has been so (Figure 15.2; Schall 1989, 2000, Read *et al.* 1992, Robert *et al.* 1996, Pickering *et al.* 2000). For instance, in human malaria infections in PNG, the median sex ratio was 0.16, but at least three of the 28 people sampled had infections with sex ratios significantly different from the population mean and indistinguishable from 0.5 (Figure 15.2c). Such a pattern would be consistent with a conditional sex ratio strategy, where the majority of parasite clones are in situations liable to generate very high selfing rates (e.g. single clone infections) and just a few are in a situation where substantial outcrossing is likely (multi-clone infections). Of course, greater than binomial sex ratio variation can also be generated even if sex ratios are genetically fixed (e.g. from fluctuating selection, which in this context could easily arise as a consequence of epidemic population dynamics). Ideally, what is required is the measurement of sex ratio from infections of known relatedness. This has not yet been analysed but is technically possible (section 15.7.1).

Laboratory data are inconsistent with sex ratio alterations in direct response to nonrelatives. In single clone infections, where LMC should be maximal, sex ratios of 0.5 have been recorded for *P. falciparum in vitro* and *P. chabaudi in vivo* (Trager *et al.* 1981, AF Read & MA Anwar unpublished observations). Moreover, sex ratios of two-clone

P. chabaudi infections were midway between those typical of the constituent clones in uniclonal infections (Taylor 1997). Nonetheless, the issue is far from closed. Can we generalize from observations of a few laboratory clones? Are parasites responding to surrogate measures of relatedness not appropriately manipulated in these laboratory experiments? Genetically diverse infections can have greater gametocyte densities than single clone infections (Taylor *et al.* 1997), and correlations between sex ratio and gametocyte density both across and, importantly, within infections have been found in some host–parasite systems (e.g. Taylor 1997, Pickering *et al.* 2000, Schall 2000). However, such correlations have not been found in other studies (Schall 1989, Shutler *et al.* 1995, Robert *et al.* 1996, Pickering *et al.* 2000). The other possibility, namely that conditional sex ratio strategies are played to ensure that sufficient males get into the mating arena, is in principle readily tested, but obtaining accurate sex ratios at very low gametocyte densities is a major technical challenge (section 15.7.3.3).

Even if these tests of theory are inconclusive, there is little doubt that sex ratios do vary more than expected by chance during the course of at least some infections (Figure 15.2; Schall 1989, Taylor 1997, Paul *et al.* 1999). Paul *et al.* (2000) have recently demonstated that sex ratios of *P. gallinaceum* in chickens and *P. vinckei* in mice are altered in response to levels of the hormone erythropoietin (Epo). The production of Epo is induced by anaemia and it in turn induces erythropoiesis, the release of immature red blood cells into the bloodstream. It is possible to conjure up adaptive – and testable – explanations for this (Paul *et al.* 2001), based around the idea that extra male gametocytes must be produced when conditions are unfavourable to fertilization. Nonetheless, the discovery that sex ratios are apparently being altered in response to environmental conditions makes it substantially more likely that these parasites can play conditional sex ratio games.

15.7.3 Tests of assumptions

Underpinning the application of sex-allocation theory to apicomplexans are various assumptions that have been tested to varying degrees. Several of these are of interest to malariologists as well as those of us interested in sex ratios for their own sake; all require further work.

15.7.3.1 Male and female gametocytes are equally costly to produce

To date, all models have assumed that this is so. It need not be: among haemospororins, for example, there is sexual size dimorphism of up to 15% (reviewed by Shutler & Read 1998). However, the size of mature gametocytes need not be related to the resource cost of producing them. At least in the case of *Plasmodium falciparum*, gametocytogenesis is triggered in the period of asexual replication within a red blood cell, with each of the resulting merozoites developing into a single gametocyte (Bruce *et al.* 1990). Thus, from the sex-allocation perspective, the key question is whether the number of gametocyte-committed merozoites emerging from a parent schizont is affected by their sex ratio. Determining this requires sexing and counting the progeny of single schizonts.

This has only recently been achieved (Smith *et al.* 2000a). Cultures of *P. falciparum* were overlaid on an attached monolayer of uninfected red blood cells. Merozoites emerging from each schizont then infected adjacent bound red cells to form a cluster of parasites called a plaque. After the cultures were washed off, an immunofluorescence assay with monoclonal antibodies specific to various stages and sexes of gametocytes was used to determine the sex of the parasites in each plaque. These experiments have shown that all the merozoites resulting from a single sexually committed schizont are of one sex, and, importantly, that the number of gametocytes produced per sexually committed parent schizont was similar for each sex, as expected if the sexes are equally costly to produce.

15.7.3.2 Gametocyte mortality is not sex biased

Sex-specific antibodies have been raised in rats and rabbits against *P. falciparum* (Guinet *et al.* 1996, Severini *et al.* 1999). This points to the possibility of sexual dimorphism in parasite antigens. If sex-specific antibodies are produced against gametocytes in natural hosts, and if they are functionally important, then males and females might be cleared by hosts at different rates. This would clearly have serious implications for interpreting observed sex ratios in terms of adaptive sex-allocation theory. Two lines of evidence suggest that gametocyte mortality is not sex biased. First, the sex ratios produced by a sample of cultured isolates of *P. falciparum* ($n = 5$) were indistinguishable from the sex ratios seen in a sample of people ($n = 7$) in the village from which the isolates came (Read *et al.* 1992). Second, using chloroquine treatment to prevent further gametocytogenesis, Smalley and Sinden (1977) found no evidence of sex differences in the gametocyte half-lives of *P. falciparum* in hospital patients ($n = 7$). These conclusions are based on perilously small sample sizes given the importance of the issue.

15.7.3.3 Sampling methods are unbiased

Sex ratios are reasonably easily assayed by light microscopy when gametocyte densities are relatively high. However, such densities can be very rare. For example, among over 4500 people sampled in PNG, only 28 had gametocyte densities sufficiently high that 100 or more gametocytes could be sexed. The estimate of the PNG population sex ratio is based on the data from these 28 (Figure 15.2c; Read *et al.* 1992). Are these people's infections representative? They formed just 10% of gametocyte-positive infections in the sample and just 1% of all detectable infections. It is unclear quite why most infections are not producing gametocytes, and why those with gametocytes have them at densities far below what is possible (Taylor & Read 1997). Whatever the explanation, the pattern is typical of malaria populations in humans and may cause problems for sex ratio studies. If a substantial proportion of the zygotes in a population come from infections where gametocytes are hard to sample by light microscopy, and certainly the majority of infectious hosts fall into this category, then we may not be looking at the sex ratio on which selection is predominantly acting. Sex ratios at low densities are also of interest in their own right: the need to ensure sufficient males to fertilize all the females may become the major source of selection on sex ratio (West et al. 2001). We note too that the conflicting picture for *Haemoproteus* and *Leucocytozoon* species is at its greatest in populations where gametocyte prevalence is low (Figure 15.3), the situation from which it is hardest to obtain further data. This makes it highly desirable to be able to obtain sex ratios from samples where gametocytes are rare. We are cautiously optimistic that advances in molecular genetics will soon make this possible. Using RT-PCR, it is already possible to detect low-density gametocytes in blood samples that are gametocyte-negative by light microscopy (Babiker et al. 1999). Using a combination of RT-PCR for sex-specific genes and quantitative PCR, it should be possible to obtain sex ratios.

15.7.3.4 Gametocyte sex ratio affects fitness

While it is hard to imagine that this assumption could not be true, there is little direct evidence. In an experimental cross between *P. falciparum* clones with different but nonetheless very female-biased sex ratios (<20% male), genetic representation in zygotes was greater for the least female-biased clone (Vaidya et al. 1993, 1995). When sex ratios in chicken malaria were artificially manipulated, zygote numbers were reduced, as expected if natural sex ratios maximize fertilization rates (Paul et al. 2000). While these observations are comforting, they are not necessarily compelling. In the cross, the least successful clone was barely producing any viable male gametocytes, and the methods used to manipulate the chicken malaria sex ratio may also affect other factors involved in fertilization success, such as transmission rate. More worrying is the unexpected finding from both lizard and human malaria infections (Schall 1996, Robert et al. 1996) that sex ratios closer to 0.5 generate more zygotes (oocysts) than the more female-biased sex ratios that dominate in nature. Such correlations, if they prove causal, would violate the assumptions of the standard LMC models. Several earlier studies have failed to find any association between sex ratio and measures of fertilization success (Boyd et al. 1935 cited in Carter & Graves 1988, Klein et al. 1986, Noden et al. 1994), although no data from any of those studies have been subject to a modern statistical analysis (*sensu* Chapter 3).

15.8 | Concluding remarks and future challenges

We hope this overview has demonstrated the necessity and potential for future work. It seems likely that apicomplexan sex ratios can be understood though the LMC lens, but the case is far from closed. Substantial work still needs to be done on some very basic issues. For instance, we estimate haemospororin sex ratios by drawing blood from relatively large blood vessels. Vectors typically feed on very small vessels. Are the sex ratios in the two the same? And in the absence of any understanding of sex determination in these parasites, are we really confident that sex ratios are under autosomal control? Cytoplasmic symbionts are common in *Plasmodium* species (Wang & Wang 1991), and, as with other eukaryotes, the male gamete contributes little to the zygote (Carter & Graves 1988). Do cytoplasmically transmitted symbionts contribute to observed sex ratios? The only relevant observation is that female-biased sex ratios are still seen in *Plasmodium falciparum* cultures maintained with gentamicin (Read et al. 1992). Symbionts not affected by this particular antibiotic could still be influential.

Direct field tests of the LMC view will frequently require molecular genetics. These studies are expensive and logistically difficult but, at least for human malarias, generating molecular genetic data is now relatively routine. In fact, a curious consequence of the 'have-machine-will-PCR' attitude in the malaria community in the 1990s is that we now have molecular genetic data on far more *Plasmodium* populations than we have sex ratio data (even if the biases in that glut of data are not fully understood; Arnot 1999, Anderson et al. 2000). It is both ironic and encouraging that the easiest data to obtain are the rarest.

Much of what is still required is relatively simple descriptive biology of the natural history of sex ratio in a wide range of Apicomplexa. How variable are sex ratios in infections and across individuals and populations? How does this relate to genetic diversity within infections and to other factors such as pathology, host immune status and gametocyte densities? Finally, we are only too aware of how much of the data and particularly the biological background is drawn from studies of human and laboratory *Plasmodium* species. Is extrapolation to other apicomplexans warranted?

Acknowledgements

Our empirical work is supported by the Biotechnology and Biological Sciences Research Council (UK). We are grateful to Steve Frank, Ric Paul, Joe Schall and Ken Wilson for comments, and to Ian Hardy for patience.

References

Anderson RM & May RM (1991) *Infectious Diseases of Humans. Dynamics and Control.* Oxford: Oxford University Press.

Anderson T, Paul R, Donnelly C & Day K (2000) Do malaria parasites mate non-randomly in the mosquito midgut? *Genetical Research,* **75,** 285–296.

Arnot D (1999) Clone multiplicity in *Plasmodium falciparum* infections in individuals exposed to variable levels of disease transmission. *Transactions of the Royal Society of Tropical Medicine and Hygiene,* **92,** 580–585.

Atkinson CT & van Riper C (1991) Pathogenicity and epizootiology of avian haematozoa: *Plasmodium, Leucocytozoon,* and *Haemoproteus.* In: JE Loye & M Zuk (eds) *Bird-Parasite Interactions. Ecology, Evolution and Behaviour,* pp 19–48. Oxford: Oxford University Press.

Babiker H, Abdel-Wahab A, Ahmed S, Suleiman S, Ranford-Cartwright L, Carter R & Walliker D (1999) Detection of low level *Plasmodium falciparum* gametocytes using reverse transcriptase polymerase chain reaction. *Molecular and Biochemical Parasitology,* **99,** 143–148.

Barta JR (1989) Phylogenetic analysis of the class Sporozoea (Phylum Apicomplexa Levine, 1970): evidence for the independent evolution of heteroxenous life cycles. *Journal of Parasitology,* **75,** 195–206.

Barta JR, Jenkins MC & Danforth HD (1991) Evolutionary relationships of avian Eimeria species among other apicomplexan protozoa: monophyly of the Apicomplexa is supported. *Molecular Biology and Evolution,* **8,** 345–355.

Boyd M, Stratman-Thomas W & Kitchen S (1935) On the relative susceptibility of Anopheles quadrimaculatus to *Plasmodium vivax* and *Plasmodium falciparum. American Journal of Tropical Medicine and Hygiene,* **15,** 485–493.

Bruce M, Alano P, Duthie S & Carter R (1990) Commitment of the malaria parasite *Plasmodium falciparum* to sexual and asexual development. *Parasitology,* **100,** 191–200.

Carreno RA, Kissinger JC, McCutchan TF & Barta JR (1997) Phylogenetic analysis of haemosporinid parasites (Apicomplexa: Haemosporina) and their coevolution with vectors and intermediate hosts. *Archiv für Protistenkunde,* **148,** 245–252.

Carter R & Graves PM (1988) Gametocytes. In: WH Wernsdorfer & I McGregor (eds) *Malaria. Principles and Practice of Malariology,* pp 253–305. Edinburgh: Churchill Livingstone.

Charnov EL (1982) *The Theory of Sex Allocation.* Princeton, NJ: Princeton University Press.

Collins W (1988) Major animal models in malaria research: simian. In: WH Wernsdorfer & I McGregor (eds) *Malaria. Principles and Practice of Malariology,* pp 1473–1501. Edinburgh: Churchill Livingstone.

Conway DJ, Greenwood BM & McBride JS (1991) The epidemiology of multiple-clone *Plasmodium falciparum* infections in Gambian patients. *Parasitology,* **103,** 1–6.

Cox (1988) Major animal models in malaria research: rodent. In: WH Wernsdorfer & I McGregor (eds) *Malaria. Principles and Practice of Malariology,* pp 1503–1543. Edinburgh: Churchill Livingstone.

Crow JF & Kimura M (1970) *An Introduction to Population Genetics Theory.* New York: Harper and Row.

Dye C & Godfray HCF [sic] (1993) On sex ratio and inbreeding in malaria parasite populations. *Journal of Theoretical Biology,* **161,** 131–134.

Ewald PW (1994) *Evolution of Infectious Diseases.* Oxford: Oxford University Press.

Frank SA (1992) A kin selection model for the evolution of virulence. *Proceedings of the Royal Society of London series B,* **250,** 195–197.

Frank SA (1996) Models of parasite virulence. *Quarterly Review of Biology,* **71,** 37–78.

Frank S (1998) *Foundations of Social Evolution.* Princeton, NJ: Princeton University Press.

Godfray HCJ (1994) *Parasitoids. Behavioral and*

Evolutionary Ecology. Princeton, NJ: Princeton University Press.

Guinet F, Dvorak J, Fujioka H, Keister D, Muratova O, Kaslow D, Aikawa M, Vaidya A & Wellems T (1996) A developmental defect in *Plasmodium falciparum* male gametogenesis. *Journal of Cell Biology*, **135**, 269–278.

Hamilton WD (1967) Extraordinary sex ratios. *Science*, **156**, 477–488.

Hastings IM & Wedgewood-Oppenheim B (1997) Sex, strains and virulence. *Parasitology Today*, **13**, 375–383.

Herre EA (1993) Population structure and the evolution of virulence in nematode parasites of fig wasps. *Science*, **259**, 1442–1445.

Hill WG & Babiker HA (1995) Estimation of numbers of malaria clones in blood samples. *Proceedings of the Royal Society of London, series B*, **262**, 249–257.

Kemp DJ, Cowman AF & Walliker D (1990) Genetic diversity in *Plasmodium falciparum*. *Advances in Parasitology*, **29**, 75–149.

Klein TA, Harrison B, Grove J, Dixon S & Andre R (1986) Correlation of survival rates of *Anopheles dirus* (Diptera, Culicidae) with different infection densities of *Plasmodium cynomolgi*. *Bulletin of the World Health Organization*, **64**, 901–907.

Kyes S, Harding R, Black G, Craig A, Peshu N, Newbold C & Marsh K (1997) Limited spatial clustering of individual *Plasmodium falciparum* alleles in field isolates from coastal Kenya. *American Journal of Tropical Medicine and Hygiene*, **57**, 205–215.

Lee J, Hunter S & Bovee E (eds) (1985) *Illustrated Guide to the Protozoa*. Lawrence, KA: Society of Protozoologists.

Mackinnon MJ & Hastings IM (1998) The evolution of multiple drug resistance in malaria parasites. *Transactions of the Royal Society of Tropical Medicine and Hygiene*, **92**, 188–195.

McGhee R (1988) Major animal models in malaria research: avian. In: WH Wernsdorfer & I McGregor (eds) *Malaria. Principles and Practice of Malariology*, pp 1545–1567. Edinburgh: Churchill Livingston.

Menegon M, Severini C, Sannella A, Paglia M, Abdel-Wahab A, Abdel-Muhsin A & Babiker H (2000) Genotyping of *Plasmodium falciparum* gametocytes by RT-PCR. *Molecular and Biochemical Parasitology*, **111**, 153–161.

Nee S, West SA & Read AF (2002) Inbreeding and parasite sex ratios. *Proceedings of the Royal Society of London, series B*, **269**, in press.

Noden B, Beadle P, Vaughan J, Pumpuni C, Kent M & Beier J (1994) *Plasmodium falciparum*: the population

structure of mature gametocyte cultures has little effect on their innate fertility. *Acta Tropica*, **58**, 13–19.

Paperna I & Landau I (1991) *Haemoproteus* (Haemosporidia) of lizards. *Bulletin of the Museum of Natural History Paris Series A*, **13**, 309–349.

Paul REL & Day KP (1998) Mating patterns of *Plasmodium falciparum*. *Parasitology Today*, **14**, 197–202.

Paul REL, Packer MJ, Walmsley M, Lagog M, Ranford-Cartwright LC, Paru R & Day KP (1995) Mating patterns in malaria parasite populations of Papua New Guinea. *Science*, **269**, 1709–1711.

Paul R, Raibaud A & Brey P (1999) Sex ratio adjustment in *Plasmodium gallinaceum*. *Parasitologica*, **41**, 153–158.

Paul R, Coulson T, Raibaud A & Brey P (2000) Sex determination in malaria parasites. *Science*, **287**, 128–131.

Paul R, Brey PT & Robert V (2002) *Plasmodium* sex determination and transmission to mosquitoes. *Trends in Parasitology*, **18**, 32–38.

Pickering J, Read AF, Guerrero S & West SA (2000) Sex ratio and virulence in two species of lizard malaria parasites. *Evolutionary Ecology Research*, **2**, 171–184.

Ranford-Cartwright L & Walliker D (1999) Intragenic recombination of *Plasmodium falciparum* identified by *in situ* polymerase chain reaction. *Molecular and Biochemical Parasitology*, **102**, 13–20.

Read AF, Narara A, Nee S, Keymer AE & Day KP (1992) Gametocyte sex ratios as indirect measures of outcrossing rates in malaria. *Parasitology*, **104**, 387–395.

Read AF, Anwar M, Shutler D & Nee S (1995) Sex allocation and population structure in malaria and related parasitic protozoa. *Proceedings of the Royal Society of London, series B*, **260**, 359–363.

Read AF, Mackinnon MJ, Anwar MA & Taylor LH (in press) Kin selection models as evolutionary explanations of malaria. In: U Dieckmann, JAJ Metz, MW Sabelis & K Sigmund (eds) *Virulence Management: The Adaptive Dynamics of Pathogen–Host Interactions*. Cambridge: Cambridge University Press.

Robert V, Read AF, Essong J, Tchuinkam T, Mulder B, Verhave J-P & Carnevale P (1996) Effect of gametocyte sex ratio on infectivity of *Plasmodium falciparum* to *Anopheles gambiae*. *Transactions of the Royal Society of Tropical Medicine and Hygiene*, **90**, 621–624.

Schall JJ (1989) The sex ratio of *Plasmodium* gametocytes. *Parasitology*, **98**, 343–350.

Schall JJ (1996) Malarial parasites of lizards: diversity and ecology. *Advances in Parasitology*, **37**, 255–333.

Schall JJ (2000) Transmission success of the malaria parasite *Plasmodium mexicanum* into its vector: role of gametocyte density and sex ratio. *Parasitology*, **121**, 575–580.

Severini C, Silvestrini F, Sannella A, Barca S, Gradoni L & Alano P (1999) The production of the osmiophilic body protein Pfg377 is associated with state of maturation and sex in *Plasmodium falciparum* gametocytes. *Molecular and Biochemical Parasitology*, **100**, 247–252.

Shutler D & Read AF (1998) Local mate competition, and extraordinary and ordinary blood parasite sex ratios. *Oikos*, **82**, 417–424.

Shutler D, Bennett GF & Mullie A (1995) Sex proportions of *Haemoproteus* blood parasites and local mate competition. *Proceedings of the National Academy of Sciences USA*, **92**, 6748–6752.

Sibley LD & Boothroyd JC (1992) Virulent strains of *Toxoplasma gondii* comprise a single clonal lineage. *Nature*, **359**, 82–85.

Smalley ME & Sinden RE (1977) *Plasmodium falciparum* gametocytes: their longevity and infectivity. *Parasitology*, **74**, 1–8.

Smith TG, Lourenco P, Carter R, Walliker D & Ranford-Cartwright L (2000a) Commitment to sexual differentiation in the human malaria parasite *Plasmodium falciparum*. *Parasitology*, **121**, 127–133.

Smith TG, Kim B, Hong H & Desser SS (2000b) Intraerythrocytic development of species of *Hepatozoon* infecting ranid frogs: evidence for convergence of life cycle characteristics among apicomplexans. *Journal of Parasitology*, **86**, 451–458.

Taylor LH (1997) *Epidemiological and Evolutionary Consequences of Mixed-Genotype Infections of Malaria Parasites*. PhD Thesis, University of Edinburgh.

Taylor LH & Read AF (1997) Why so few transmission stages? Reproductive restraint by malaria parasites. *Parasitology Today*, **13**, 135–140.

Taylor LH, Walliker D & Read AF (1997) Mixed-genotype infections of the rodent malaria *Plasmodium chabaudi* are more infectious to mosquitoes than single-genotype infections. *Parasitology*, **115**, 121–132.

Taylor PD (1981) Intra-sex and inter-sex sibling interactions as sex ratio determinants. *Nature*, **291**, 64–66.

Tibayrenc M, Kjellberg F & Ayala FJ (1990) A clonal theory of parasitic protozoa: the population structures of *Entamoeba, Giardia, Leishmania, Plasmodium, Trichomonas*, and *Trypanosoma* and their medical and taxonomic consequences. *Proceedings of the National Academy of Sciences USA*, **87**, 2414–2418.

Trager W, Tershakovec M, Lyandvert L, Stanley H, Lanners N & Gubert E (1981) Clones of the malaria parasite *Plasmodium falciparum* obtained by microscopic selection: their characterisation with regard to knobs, chloroquine sensitivity, and formation of gametocytes. *Proceedings of the National Academy of Sciences USA*, **78**, 6527–6530.

Vaidya AB, Morrisey J, Plowe CV, Kaslow DC & Wellems TE (1993) Unidirectional dominance of cytoplasmic inheritance in two genetic crosses of *Plasmodium falciparum*. *Molecular and Cell Biology*, **13**, 7349–7357.

Vaidya AB, Muratova O, Guinet F, Keister D, Wellems TE & Kaslow DC (1995) A genetic locus on *Plasmodium falciparum* chromosome 12 linked to a defect in mosquito infectivity and male gametocytogenesis. *Molecular and Biochemical Parasitology*, **69**, 65–71.

Walliker D, Babiker H & Ranford-Cartwright L (1998) The genetic structure of malaria parasite populations. In: I Sherman (ed) *Malaria: Parasite Biology, Pathogenesis and Protection*, pp 235–252. Washington DC: ASM Press.

Wang AL & Wang CC (1991) Viruses of parasitic protozoa. *Parasitology Today*, **7**, 76–80.

West S, Smith T & Read A (2000a) Sex allocation and population structure in apicomplexan (protozoa) parasites. *Proceedings of the Royal Society of London, series B*, **267**, 257–263.

West S, Herre E & Sheldon B (2000b) The benefits of allocating sex. *Science*, **290**, 288–290.

West SA, Smith TG, Nee S & Read AF (2001) Fertility insurance and the sex ratios of malaria and related haemospororin blood parasites. *Journal of Parasitology*, in press.

Williams GC & Nesse RM (1991) The dawn of darwinian medicine. *Quarterly Review of Biology*, **66**, 1–22.

Williams GC & Nesse RM (1994) *The Dawn of Darwinian Medicine*. New York: Time Books.

Chapter 16

Sex allocation in hermaphrodite plants

Peter G.L. Klinkhamer & Tom J. de Jong

16.1 | Summary

The flowers of hermaphrodite plants have both male and female parts. Hermaphrodite plants can change their allocation to both sexual functions in various ways, such as by changing the production ratios of pollen grains to ovules within flowers and of flowers to fruits. We discuss the problems involved in measuring sex allocation, trade-offs and fitness gain curves and present a simple model for the evolutionary stable allocation to fruits and flowers. The model provides an explanation for the low fruit-to-flower ratio found in many species and for the increasing allocation to female function with increasing selfing rate. Theoretical models predict that evolutionary stable sex allocation depends on plant size and this prediction is supported by literature data on monocarpic hermaphrodites and on monoecious species.

16.2 | Introduction

By far the most common mode of plant reproduction is through hermaphrodite flowers. Although such flowers serve both male and female functions, this does not mean that hermaphrodites are invariant in their sexual behaviour. Substantial variation in intraspecific sex allocation has been found and related to environmental conditions or plant size. A large body of theoretical literature is now accumulating that predicts how allocation to male and female reproduction should vary with a variety of factors such as pollination type, resource status, selfing rate, selective abortion, population structure, dispersal mechanisms, etc. Unfortunately empirical evidence lags far behind, mostly because the required measurements are notoriously difficult to collect and the methods full of pitfalls.

In this chapter we discuss why it is not possible to measure the exact allocation to male and female function in hermaphrodite plants and why it is more fruitful to ask questions about relevant trade-offs (section 16.3). Often the trade-off between allocation to sexual functions within a flower is considered in relation to optimal sex allocation. In many cases however, the trade-off between sexual functions at the level of the whole plant (i.e. the trade-off between flower and seed production) will be more relevant. We stress the importance of realizing what the relevant trade-offs are in relation to the problem studied (section 16.6) and discuss the problems involved in measuring trade-offs (section 16.4).

After defining the relevant trade-off, the next step towards modelling sex allocation is to describe the 'gain curves', i.e. the relationships between investments in the functions under consideration and fitness returns. Under the assumption of a trade-off between flower and seed production we present a simple model exploring optimal seed to flower ratios in plants (section 16.8). Finally, we discuss some of the predictions made by sex-allocation theory (sections 16.10 & 16.11).

16.3 | Measuring sex allocation

There are several problems involved in empirical measurements of plant sex allocation (Goldman & Willson 1986):

1. It is often unclear in what units sex allocation should be measured. Most studies use dry mass of reproductive structures but dry mass is not necessarily proportional to the factor that is limiting for the investment in male and female reproduction.

2. Male and female reproduction may be limited by different resources, such that they only partly draw on the same resource pool.

3. Male and female function may be separated in time so that female reproduction can be a function of the availability of resources both at the time of flowering and at the time of fruit maturation.

4. If allocation is measured in units of biomass or carbon it is often not clear what proportion of the total amount is photosynthesized by the seeds themselves.

5. Often investment in flowers (with pollen) is taken as the allocation to male function and investment in seeds is taken as the allocation to female function. This is reasonable if seed production is limited by resources rather than by the level of pollination, so that pollinator attraction mainly serves male function. However, if the seed set is limited by pollination it is not obvious how to attribute investment in the production of attractive structures such as petals or nectar to male and female function.

6. Even under the assumption of resource limitation for seed production, flowers that do not produce fruits or seeds may still contribute to female fitness because they increase the possibility for selective embryo abortion, leading to higher quality seeds.

Attempts to determine the exact sex allocation of individuals in hermaphrodite plants are unlikely to be successful. Fortunately, for most interesting questions concerning the evolution of mating systems and gender adjustment, it is not necessary to know the exact allocation to male and female functions. Instead of asking, 'what is the optimal allocation to male and female function?' we can formulate questions on the basis of the relevant relationships and trade-offs and ask, for example, 'what is the optimal allocation to anthers and ovules?' or 'what is the optimal allocation to flowers, seeds and attractive structures?' Most of the above-mentioned problems can be avoided by studying these trade-offs directly.

16.4 | Measuring trade-offs

Although trade-offs are central to most models of sex allocation, they have rarely been measured for plants (Charlesworth & Morgan 1991). A problem with measuring trade-offs in natural populations is that the overall resource pool differs strongly between individuals, even when corrected for size. When measuring trade-offs across individuals this often causes positive phenotypic correlations when negative ones are expected (van Noordwijk & de Jong 1986). Thus although negative phenotypic correlations do suggest trade-offs they are unlikely to be found (Charlesworth & Morgan 1991). One might also examine genetic correlations. In practice this involves a large amount of effort. The available genetic variation may be limited and environmental variation may be relatively large. Furthermore, genetic variation independent of the characters of interest may cause differences in resource acquisition, making some plants good at all traits (e.g. Koelewijn & Hunscheid 2000). Campbell (2000) discusses the few studies available: three of the four studies found positive genetic correlations between allocation to male versus female function. By far the most efficient way of demonstrating trade-offs is by manipulation experiments on individuals. Rademaker and Klinkhamer (1999) found no genetic correlation between flower and seed production, while both manipulation of the pollination level and flower-removal experiments demonstrated a highly significant increase in flower number with reduced seed production. In a qualitative way this trade-off between seed production and flower number is well known to most gardeners who practise the habit of removing old flowers to prevent seed production in order to prolong flowering. For other trade-offs, such as between anther and ovule or flower production, it is not clear how these can be measured with manipulation experiments and we probably have to rely on genetic correlations.

16.5 | Measuring gain curves

Once we have defined the relevant trade-off, the next step is to relate the investment in the components of the trade-off to fitness gains: such relationships are called 'gain curves'. Female fitness is determined by the number of seeds produced multiplied by a measure of seed quality. Similarly, male fitness is determined by the number of seeds sired multiplied by a measure of seed quality (e.g. survival until reproduction or the expected number of seeds produced or sired). In the model discussed in section 16.8 we have to describe male and female fitness as a function of flower and seed production, respectively. In other cases it might be anther mass and seed production. Again, discussions about the exact sex allocation to male or female function can often be avoided. Although perhaps trivial, this gives rise to much confusion. Researchers studying systems in which the seed set is limited by the level of pollination often take total fruit number or seed mass as a measure of female fitness and focus on how both seed production and male fitness depend on flower number. It is then often unclear what the trade-off under consideration is, and what the consequences of increased allocation to flower number are for the allocation to other relevant functions. In contrast, researchers studying systems in which seed production is limited by resources mostly take total seed mass as investment in female reproduction. They then focus on how seed dispersal or seedling survival depends on seed production to estimate female fitness, and on how male fitness is related to flower number. The trade-off then is between seed (or fruit) and flower production.

Gain curves are usually expressed as simple power functions. To calculate the exponent

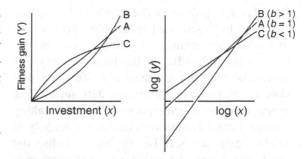

Fig 16.1 Gain curves. Gain curves are often assumed to be power functions of the level of investment in male or female reproduction. The exponent of the power function can be determined as the regression coefficient after log transformation of both dependent and independent variables. The left panel shows fitness gain (Y) as a power function of the investment, $Y = kX^b$. The right panel shows the linear relation $\log(Y) = b\log(X) + \log(k)$. Note that some authors (e.g. Devlin et al. 1992) only log transformed the Y axis. This will give decelerating curves even when gain curves are in fact linear.

of the gain curve it is often useful to take logarithms, because the exponent of the power function $Y = kX^b$ then becomes the regression coefficient $\log(Y) = b\log(X) + \log(k)$ (Figure 16.1). If $b > 1$ the gain curve is accelerating, if $b < 1$ the gain curve is decelerating and if $b = 1$ the gain curve is linear. The measurements of (the exponents of) the gain curves present severe problems in practice (Wilson et al. 1994). Many of these are basic problems with estimating fitness.

1. Ideally the success of the offspring produced by a mother or sired by a particular parent should be measured. Without molecular techniques it is impossible to assign a mother and a father to established seedlings. In many studies seed production itself is used as a measure of female fitness but this ignores the possible decelerating effects that can be caused by local resource competition. Male fitness is often estimated by indirect measures such as the number of pollinator visits or pollen removal. However, these parameters estimate the success during single steps of the total process that determines male fitness and they may be even negatively correlated in animal-pollinated plants (Klinkhamer et al. 1995).

2. In large populations the number of offspring that have to be screened with molecular markers to get reliable estimates of the gain curves may be too large to handle. Furthermore, unless many molecular markers are used it is often not possible to pinpoint a single candidate father. The practice of remaining possible fathers making an equal contribution to male fitness is likely to level-off gain curves. If the proportion of offspring sired that cannot be attributed to the father is equal for successful and unsuccessful fathers, the absolute number is highest for the successful father. Therefore, by attributing equal shares, the fitness of the unsuccessful fathers will be overestimated and that of the successful fathers underestimated. More sophisticated statistical methods for determining paternity are discussed by, for example, Smouse and Meagher (1994). It will not always be useful to engage in molecular methods because of the problems involved. Insight may sometimes be gained from experiments in which the separate steps that ultimately determine fitness are studied. For pollen dynamics, experiments can be performed in which fluorescent dye, for example, is used to estimate pollen transfer within and between plants (Rademaker & de Jong 1998). Similarly, if pollen availability is not limiting for seed production, insight in the shape of the female gain curve may be obtained by experiments in which (potted) plants with controlled seed numbers are arranged in such a way that seedlings can be attributed to mother plants without molecular methods.

3. For selfing organisms a measure of the level of inbreeding depression is needed. Since inbreeding effects may often occur late in the life cycle of the progeny, e.g. as a reduction of the siring success of the inbred offspring, this may involve a considerable quantity of work (Melser et al. 1999).

4. Seeds and pollen can disperse over large distances and both may depend strongly on local environmental conditions so that the variance within populations is extremely large. This will make it difficult to obtain any statistically reliable results.

5. Ideally, one should determine male and female fitness directly for plants that genetically differ in their allocation strategy (e.g. the fraction of resource allocated to flowers). However, the genetic variation within populations may be limited so that only part of the gain curve can be determined. Furthermore, this requires plants with similar budgets for reproduction, otherwise the interpretation of the data becomes difficult (section 16.6). Often it will prove extremely difficult to control for the plant's budget. An alternative can be to study gain curves indirectly. Perhaps the most powerful approach involves constructing an artificial, isolated population in which both genotype and biomass allocation can be manipulated (Campbell 2000). The relationship between flower number and male fitness can be determined in experiments in which flower number is manipulated (using plants of similar sizes, if necessary). Likewise the relationship between seed production and female fitness can be determined. The advantages of this method are that one is not restricted by the available genetic variation within a natural population and that one does not need plants with equal budgets.

6. Very often the shape of the gain curve is estimated by regressing fitness against, for example, flower number for a set of plants with different sizes. This method may give highly misleading results if the shape of the gain curve is influenced by plant size itself (Emms 1993, Klinkhamer et al. 1997). For instance, in wind-pollinated plants, larger individuals will produce more flowers and these will be more successful in dispersing pollen because it is released at a greater height. Sampling over individuals of different sizes may then lead to apparent accelerating gain curves while in fact the actual gain curves are linear or decelerating (Figure 16.2). In animal-pollinated plants direct effects of plant size on the shape of the gain curve are less likely because pollinators are expected to react to the number of flowers rather than to plant height. In such cases plants of different sizes may well be used to determine male gain curves on the basis of the relationship between flower number and pollen export. Of course this approach is based on the assumption (that should be tested) that variation in pollen production per flower is not related to plant size. Female

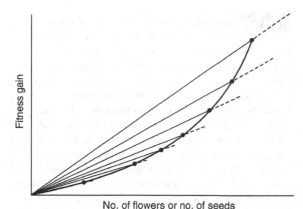

Fitness gain

No. of flowers or no. of seeds

Fig 16.2 Linear gain curves may appear nonlinear if different sized plants are used. In this example, we assume a positive correlation between plant size and flower or seed number, and that plant size directly affects the shape of the gain curve. In this case the slope of the linear gain curve increases with size. Straight lines indicate the gain curves for plants of different sizes, dots indicate sampling points. The curved thick line gives the resulting apparent 'gain curve' if different sized plants are used to determine the gain curve.

gain curves, however, may still be influenced by plant height if larger plants have better seed dispersal.

16.6 | The interaction between trade-offs and gain curves

The exact definitions of the gain curves depend on the trade-off under consideration. The importance of realizing what the appropriate trade-offs are is illustrated by the following example. In one of the very few attempts to measure gain curves in natural populations, Campbell (1998) tried to avoid the problem of using plants of different sizes by calculating the proportional investment in male and female reproduction on a per-flower basis for the semelparous scarlet gilia (*Ipomopsis agregata*). Proportional allocation to male function (r) was calculated as the dry mass of stamens divided by the summed dry masses of stamens, pistils and seeds. Female allocation was given by $1 - r$. Campbell used the number of seeds produced per flower as a fitness measure. Campbell concluded that the female gain curve was strongly accelerating.

This method is correct under the assumption that there is a trade-off only between anther mass and seeds produced per flower (which appears to be correct in Campbell's study system). The method produces, however, highly misleading results (i.e. apparently accelerating female gain curves even if they are in fact linear or decelerating) if seeds are (partly) traded-off against flowers, as will often be the case in semelparous and many other plant species (Figure 16.3). This can be shown mathematically: let j be the male allocation (anthers) of a flower and i the female allocation (pistils and seeds). The total budget for a plant equals T, so that the number of flowers equals $n = T/(i + j)$. Assume that the number of seeds per flower (Y) is linearly related to i: $Y = i/c$. The relative investment in female function $(1 - r)$ equals $i/(i + j)$. Then $Y = [(i + j)/c] \cdot (1 - r)$ or $Y = (T/nc) \cdot (1 - r)$. This only gives a linear relationship if the number of flowers (n) for a given budget (T) is constant and independent of the relative investment in female reproduction (Y); or, in other words, if there is only a trade-off between i and j ($i + j$ is constant). Alternatively, the allocation to anthers (j) may be constant and there may be a trade-off between the allocation to pistils and seeds (i) and the number of flowers (n). We then find an accelerating curve because n decreases with increasing $(1 - r)$.

16.7 | The shape of gain curves

Given the importance of measuring gain curves for our understanding of sex allocation in plants it is perhaps surprising that they are rarely estimated. Fortunately some predictions can be made based on our understanding of the ecological processes that determine the shape of the gain curves. Although some mechanisms have been proposed that may lead to accelerating gain curves, such as increased fruit dispersal with fruit number and increased pollen dispersal with flower number if pollinators are scarce, the prevailing opinion is currently that both male and female gain curves are linear or decelerating (de Jong & Klinkhamer 1994).

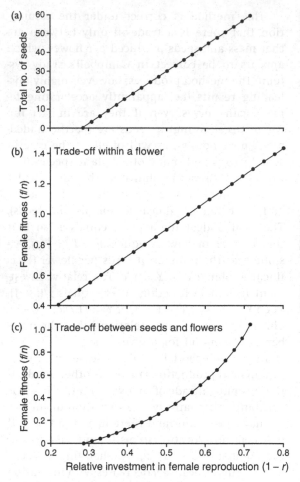

16.7.1 The male gain curve

This is assumed to be decelerating for two reasons. First, with increasing investment in pollen, a plant may saturate its neighbours (Nakamura *et al.* 1989) and in self-compatible species the fitness gains through selfing will be strongly decelerating. The effect of saturating the neighbours will only be important if neighbourhood sizes are very small. Charnov (1982) hypothesized that in wind-pollinated plants pollen is dispersed more homogeneously through the population than in animal-pollinated plants. Therefore, he expected the male gain curve to be more linear in wind-pollinated plants.

Second, in animal-pollinated species, pollinators move from one flower to another of the same plant. They thereby lose part of the pollen before they move to another plant and they will increase the level of selfing through pollination by neighbouring flowers of the same plant (geitonogamy) (e.g. de Jong *et al.* 1993). These effects are expected to increase with increasing flower number, making plants with large flower numbers relatively bad pollen dispersers (Hessing 1988, de Jong & Klinkhamer 1989, 1994, Harder & Barrett 1995a). The possible importance of geitonogamous selfing is demonstrated by higher selfing rates on plants with many flowers (e.g. Dudash 1991, Schoen & Lloyd 1992, Snow *et al.* 1995, Harder & Barrett 1995b, Vrieling *et al.* 1997). There is some support for the hypothesis that male fitness is a decelerating function of flower number (e.g. Hessing 1988 for *Geranium caespitosum*). Campbell (1998) found for *I. agregata* that pollen export was a linear function of flower number in one population but decreased after a certain flower number in another. Rademaker and de Jong (1998) studied pollen dynamics using fluorescent dye and concluded that, depending on the level of inbreeding depression assumed, the male gain curve was decelerating in two species and either linear or decelerating in one species.

Geitonogamy may also occur in wind-pollinated species. The architecture of the plant and physical factors such as wind speed or turbulence may determine the fraction of pollen that is dispersed (Freeman *et al.* 1997). However, the number of pollen grains caught by the plant's own flowers is unlikely to reduce pollen export significantly. Consequently, the selfing rate is expected to increase with the number of open flowers, but the number of pollen grains dispersed per flower will hardly be affected. Therefore, fitness gain is often assumed to be a more linear function of flower number (de Jong *et al.* 1999). On the basis of their data on *Picea alba* (white spruce), Schoen and Stewart (1986) assumed the male gain curve to be linear, but these data show much unexplained variance. Charnov (1987) found for wild rice that the relative allocation to male function was approximately 0.17 (in a population almost without selfing). He interpreted this as evidence for a decelerating male gain curve ($b < 1$) in combination with a linear female gain curve ($a = 1$, eq. 16.7).

16.7.2 The female gain curve

This is assumed to be decelerating because of increased local resource competition if seed or fruit dispersal is mostly within a restricted area around the mother plant (Lloyd 1984). However, when studying local resource competition in controlled experiments Rademaker and de Jong (1999) did not find any evidence for such a deceleration and assumed the female gain curves for two short-lived monocarpic perennials to be linear.

If the selfing rate depends on allocation to male reproduction, and inbreeding depression is high, even a small investment in male reproduction will strongly compromise female fitness. Greater investments in male reproduction will have diminishing negative effects on female fitness. As a result selfing in combination with high inbreeding depression may lead to accelerating female gain curves.

It is the ratio of the exponents of the male and female gain curves that is important for sex allocation, not the absolute values of these exponents (section 16.8.5). It should be clear from the above that there is currently too little empirical information to make definite statements about this ratio. However, the data of Rademaker and de Jong (1998, 1999) and Campbell (1998) strongly suggest that the male gain curve has a smaller exponent than the female gain curve in animal-pollinated plants.

16.8 | A basic model of seed-to-flower ratio in hermaphrodites

The problem that is perhaps the easiest to deal with is the optimal allocation to flowers and fruits. Through a combination of experiments that manipulate seed production and flower number, both the trade-off between fruit and flower production and the relationships between allocation to flowers and fruits and the fitness gained by it are easier to determine than, for instance, allocation to anthers and ovules.

The line of reasoning we use is that of most evolutionarily stable strategy (ESS) models on plant sex allocation (e.g. Charlesworth & Charlesworth 1981, Charnov 1982, Lloyd 1983, Morgan 1993). The basic assumption in almost all sex-allocation models is that there is a trade-off in resource allocation between male and female function. Hermaphrodites should invest in each sexual function to the point at which the marginal fitness returns on each function are equal. At that point the reduction in fitness caused by a small decrease in allocation to one function is exactly equal to the increase in fitness caused by the other function. Because the fitness of any sexual strategy depends on the gender distribution within the population, most studies adopt the ESS approach. ESS analyses attempt to identify a pattern of resource allocation such that if most individuals within the population adopt that strategy any rare individual with a slightly different allocation pattern would have a lower fitness. Here we use this approach and develop a simple model predicting the optimal allocation to flowers and fruits under the assumption that plants vary their sex allocation through a trade-off between these two.

16.8.1 Assumptions of the model

Seed production is assumed to be limited by the amount of resources allocated to female reproduction: siring success is limited by the access to ovules, which in turn is a function of the parents' allocation to flowers. Under this assumption flowers mainly serve a male function. This is a translation of Bateman's principle to plants and is known as the 'fleurs-du-male' or 'pollen donation hypothesis' (Queller 1997, Broyles & Wyatt 1990, 1997). Pollen and ovule production per flower are assumed to be fixed: seed number per fruit is constant and plants can produce one fruit per flower. We also assume self-incompatibility.

The model is a simplification, but gives the basic line of reasoning used in most allocation models. This basic model can be adjusted to incorporate further biological aspects (e.g. drawing a distinction between selfed and outcrossed seeds, or a different relationship between the investment in seeds and female fitness) and assumptions (e.g. a trade-off between anther weight and seed production within a flower, or assuming that flower and fruit production only partly depend on the same resources). We extend the model to allow self-fertilization in section 16.9.

16.8.2 Describing the trade-off

We start with a hermaphrodite species in which a fraction of the flowers produce fruits. The costs of producing f fruits are fc with c as the cost of producing one fruit. The costs of producing n flowers are nd, in which d is the cost of producing a flower (including pollen). If the plant produces n flowers and f fruits, $n - f$ fruits are aborted so that the costs of abortion equal $(n - f)z$ with z being the cost of an aborted fruit. If the total amount of resource allocated to reproduction equals T it follows that $T = fc + nd + (n - f)z$. From this n can be calculated as a function of f

$$n = (T - fc + fz)/(d + z). \tag{16.1}$$

16.8.3 Describing the gain curves

The next step is to relate the number of flowers and the number of fruits to male and female fitness, respectively. Mostly, in models a simple power function is used to relate n to male fitness, but more realistic expressions can easily be included if pollen dynamics are known. Here we assume that pollen export to other plants equals $E = q_1 n^b$, in which q_1 is a scaling factor and b is the exponent of the male gain curve. Furthermore, we assume that male fitness is linearly related to pollen export. In some cases this assumption may be violated; for example, when pollen grains from the same plant tend to stick together, resulting in sib-competition for a limited number of ovules. Female fitness $w_\female = q_2 f^a$, with q_2 again being a scaling factor and a being the exponent of the female gain curve.

16.8.4 Calculating the ESS

In a population of N resident plants that all adopt the strategy of producing f fruits and n flowers, the male fitness of a single rare mutant producing f_m fruits and n_m flowers can be calculated as its share in siring the successful offspring produced in the population. In total, the number of offspring produced equals: $Nw_\female = Nq_2 f^a$. The mutant's share in siring these offspring, under the assumption of equal opportunities for resident and mutant, equals $E_m/(EN + E_m)$. If the number of plants in the population (N) is large this reduces to E_m/EN so that its male fitness equals

$$w_{m\male} = (1/N)(E_m/E)Nq_2 f^a = (E_m/E)q_2 f^a. \tag{16.2}$$

Total fitness is given by: $w_m = w_{m\female} + w_{m\male}$, substitution of eq. 16.1 gives

$$w_m = [(T - f_m c + f_m z)/(T - fc + fz)]^b q_2 f^a + q_2 f_m{}^a. \tag{16.3}$$

The ESS allocation is found by maximizing w_m, that is finding the condition for which $dw_m/df_m = 0$; furthermore, in the ESS $f_m = f$. This gives the ESS number of fruits: $f^* = aT/(a + b)(c - z)$. Substitution in eq. 16.1 gives: $n^* = bT/(a + b)(d + z)$ so that the ESS ratio of fruits to flowers equals

$$f/n = a(d + z)/b(c - z). \tag{16.4}$$

Note that we would have found the same result if we had used the product theorem which states that at the ESS the product of male and female fitness is maximized. Although it is often used, it is better to set up a formal ESS reasoning because the product theorem is based on a mathematical coincidence and is only valid under restricted conditions, such as complete outbreeding (Charnov 1982).

16.8.5 Some conclusions

Several interesting conclusions can be drawn from this simple model (in various forms, similar conclusions can be found in, for example, Charlesworth & Charlesworth 1981, Charnov 1982, Lloyd 1984, Spalik 1991, Morgan 1993, de Jong et al. 1999):

1. From eq. 16.4 it can be concluded that the ESS allocation to fruits increases if the costs of fruit abortion (z) are high.
2. If the costs of fruit abortion are relatively small, eq. 16.4 reduces to

$$f/n = ad/bc. \tag{16.5}$$

Note that the absolute costs of fruits (c) and flowers (d) are not relevant, only the ratio is important. In other words, it only matters how flower production is reduced by the production of an additional seed. This trade-off can be determined experimentally by comparing the flower production of plants with different seed sets in, for example, experiments with different pollination levels or by the destruction of ovules. If the trade-off between f and n is given by a linear relationship with negative regression coefficient g, producing a fruit is $1/g$ times more costly than producing a flower. Equation 16.5 can then be rewritten as

$$f/n = g(a/b). \tag{16.6}$$

3. For given values of the exponents of the male gain curve (b) and the female gain curve (a), the optimal allocation does not depend on the scaling factors if, as we assumed, these are equal for all plants. This means that the optimal allocation does not depend on the fraction of seeds lost before germination, for example. If, however, this

fraction is high, this will bring the value of a closer to unity.
4. The ratio of allocation of resources to seed production (fc) and flower production (nd) equals: $fc/nd = a/b$. In other words, at the ESS the relative allocation is proportional to the exponents of the gain curves and does not depend on the costs of flowers and fruits. If the male and female gain curves have similar exponents, an equal proportion of resources should be invested in flower and fruit production. This is a special case of the more general result that the proportional allocation to male function (r) should be 0.5 without selfing and with equal exponents of the gain curves ($a = b$) (e.g. Chapters 1 & 17)

$$r = nd/(nd + fc) = b/(a + b). \tag{16.7}$$

5. Because the production of a fruit is usually more costly than the production of a flower (Ehrlen 1991) ($d < c$ or, put another way, $g < 1$) we would expect that at the ESS plants produce more flowers than would be necessary for the production of fruits, which means that in most plants we expect the fruit-to-flower ratio to be less than one.

16.9 | Effects of selfing on optimal fruit-to-flower ratios

It is well established empirically that the allocation to male function decreases with the selfing rate both among and within species (e.g. Cruden 1977, Charnov 1982, 1987, Schoen 1982, Lloyd 1984, Parker 1995). This observation has been explained using sex-allocation theory (e.g. Charlesworth & Charlesworth 1981, Charnov 1982). Following a similar line of reasoning we extend the model to plants that can self-fertilize. We need to distinguish between three ways in which an individual gains fitness: (1) through the number of outcrossed seeds produced, (2) through the number of selfed seeds produced (through these seeds the individual passes two sets of its haploid genome to its offspring and they may suffer from inbreeding depression), and (3) through its share in siring outcross seeds on other plants. It is important to know how selfing is brought about. In the case of geitonogamy (selfing by other flowers of the same individual) in animal-pollinated plants, selfing is a function of the number of simultaneously open flowers. The number of flowers determines how many flowers will be visited in a row on the same plant by a pollinator and thus determines pollen transfer within the plant. The selfing rate can also be a function of the number of pollen grains produced within a flower. Various functions may be incorporated but as a simple illustration we here assume a fixed selfing rate (S).

The number of selfed seeds equals fS and the number of outcrossed seeds $f(1 - S)$. Under the simplifying assumptions of linear gain curves, i.e. $a = b = 1$, and negligible costs of aborting fruits ($z = 0$), substitution in eq. 16.2 gives for the fitness of a mutant with a slightly different allocation to fruits

$$w_m = q_2 f_m(1 - S) + 2q_2 f_m S(1 - \delta)$$
$$+ [(T - f_m c)/(T - fc)]q_2 f(1 - S). \qquad (16.8)$$

The first term gives the fitness gained through the outcrossed seeds produced, the second term gives the fitness gained through selfed seeds produced (with δ denoting the inbreeding depression) and the third term gives the male fitness gained through seeds sired on other plants. Again the ESS can be found by setting $dw_m/df = 0$. This gives $f = (0.5T/c)[1 + S(1 - 2\delta)/(1 - S\delta)]$, with $n = (T - fc)/d$ this gives

$$f/n = (d/c)[1 + S(1 - 2\delta)]/(1 - S), \qquad (16.9)$$

and rearranging gives

$$r = n/(f + n) = (1 - S)/(2 - 2S\delta). \qquad (16.10)$$

This is a decreasing function of S. The model predicts that the allocation to male function decreases with increasing selfing rate. Similar results are found when we consider trade-offs between pollen and ovules or between pollen and seeds within a flower.

With geitonogamous selfing, the selfing rate is not fixed but instead is a function of flower number. It has been suggested that geitonogamous selfing does not affect the ESS allocation to flowers and seeds because with

geitonogamous selfing, in contrast to autogamous selfing, siring a selfed seed is as costly as siring an outcrossed seed (Lloyd 1987, Brunet 1992). de Jong et al. (1999) however showed this suggestion to be incorrect. They used the same approach as above with equations for the level of geitonogamous selfing depending on flower number, and showed that in this case selfing also reduces the optimal allocation to male function. The explanation is that by shifting the allocation to more seeds a mutant's male fitness loss is compensated for by the fact that it sires itself more of the extra seeds than the average individual in the population, because pollen dispersal is not homogeneous.

When testing the hypothesis that allocation to male function decreases with selfing rate, it should be noted that the arguments are based on ESS reasoning and that comparisons should be made between species and populations. Among the individuals within a population, positive correlations between male allocation and selfing can be expected because individuals that for any reason produce more pollen are likely to have more of their seeds self-fertilized (Damgaard & Loeschke 1994).

16.10 | Why do many plants have low fruit-to-flower ratios?

Even if pollen is not limiting seed production, many plants produce more flowers than is necessary for the production of seeds (e.g. Stephenson 1981, Sutherland 1986a,b, Klinkhamer & de Jong 1997). We have already explained (eq. 16.5 and 16.6) why this can be expected on the basis of simple sex-allocation arguments: if plants adjust allocation to male and female function, the fruit-to-flower ratio is smaller than unity because usually a fruit costs more resources than a flower. If on the other hand flowers are more costly than fruits the ESS fruit-to-flower ratio will be high and constrained by the simple fact that no fruits can be produced without flowers. The fruit-to-flower ratio is pushed even lower if increasing flower number has a positive effect on pollen dispersal per flower ($b > 1$). In the model we assumed that the number of pollen grains produced per flower is constant. Note, however, that if male fitness is a power function of allocation to pollen number within a flower, increasing the exponent leads to a higher ESS number of fruits per flower because it reduces optimal flower number (Morgan 1993).

Partial support for the explanation based on the difference between costs of fruits and flowers comes from the observation by Sutherland (1986b) that in self-compatible species there is an association between inexpensive fruits and higher fruit-to-flower ratios. However, such as association was absent for self-incompatible species.

If these were the only reasons for low fruit-to-flower ratios one would perhaps expect selection for the loss of female function in those flowers that do not produce seeds; in other words, we would expect many andro-monoecious species, while in fact this sexual system comprises only 1.7% of all plant species (Spalik 1991). Flowers that do not produce fruits may, however, still serve a female function through bet-hedging and selective embryo abortion.

Bet-hedging can be important if fruit and flower production are separated in time. Sutherland (1986b) suggested that plants produce excess ovules because when the decision about allocation of resources to flowers and ovules is made, plants do not 'know' how conditions will be for maturing fruits later on. If environmental conditions show large temporal variance, there will be unpredictable variation in the optimal allocation from one flowering period to another. If ovules and flowers are relatively cheap compared to fruits, it will pay to produce a number of flowers that will be 'too large' in most years, but this covers the possibility that in a year with unusually high pollinator and resource availability the number of ovules limits fruit or seed production (sometimes referred to as the 'ecological window hypothesis', Ehrlen 1991). Likewise, overproduction of flowers may provide a buffer against loss caused by damage to flowers or young fruits (sometimes referred to as the 'reserve ovary hypothesis,' Ehrlen 1991).

Despite the theoretical investigations of the bet-hedging hypothesis, there is only limited empirical evidence. A test would involve determining the causes of fruit production in a large number of years. For the short-lived monocarpic perennial *Cynoglossum officinale* (Boraginaceae), which produces a fixed number of four ovules per flower, average seed number per flower was close to one in five years of study (e.g. Klinkhamer & de Jong 1997). Among the thousands of plants studied, only one individual produced more than two seeds per flower. Microscopic observations revealed that on average at least 40% of all embryos are aborted (Melser 2001). In Sutherland's (1986b) survey hermaphrodite species had lower fruit-to-flower ratios than andro-monoecious and monoecious species, which in turn set fewer fruits than dioecious species. Morgan (1993) pointed out that this observation is hard to explain with the bet-hedging hypothesis because there is no reason to expect that greater environmental heterogeneity is associated with the hermaphrodite breeding system. In contrast, the observation is in line with the notion that low fruit-to-flower ratios in hermaphrodites result from the optimal division of resources over male and female reproduction.

Kozłowski and Stearns (1989) discussed bet-hedging and selective embryo abortion as hypotheses to explain the overproduction of ovules. Under the selective embryo abortion hypothesis one assumes that offspring vary in fitness and that the costs of producing ovules and aborting embryos are low. By providing a larger pool of flowers the best fruits can be selectively matured and plants can prevent the investment of resources in offspring with low fitness later in life (Stephenson 1981, Willson & Burley 1983). The hypothesis does not imply that the mother plant 'can tell' which fruits are best. Stearns (1992) suggested that embryos are in a competitive arena for resources. Under such competition a correlation between embryo abortion and the quality of offspring at later life-stages may exist. Several studies have shown that the selection of pollen sources leads to paternity percentages that deviate from those of the pollen originally applied (e.g. Marshall & Ellstrand 1988, Melser *et al.* 1997 and references therein). Helenurm and Schaal (1996) manipulated abortion rates and demonstrated how selfed seeds were preferentially aborted at high overall abortion rates. Deviations of the expected paternity percentages do not necessarily mean that offspring with the greatest fitness later in life are selected. However, in line with the selective abortion hypothesis, several studies report that selective seed abortion indeed increases offspring quality (Casper 1984, 1988, Stephenson & Winsor 1986). Too few empirical data are available at present, however, to evaluate the importance of selective embryo abortion for optimal sex allocation.

16.11 | Size-dependent sex allocation

A simple testable prediction of sex-allocation theory is that individuals should adjust sex allocation to their size (size-dependent sex allocation, SDS). This prediction is based on five arguments, two focusing on how reproductive investment influences mortality and three focusing on the shape of gain curves.

1. The first argument applies particularly to sequential hermaphrodites (perennial individuals that can change from male to female or vice versa depending on age, size or circumstances). For many species it is assumed that reproducing as a female is more costly than reproducing as a male and that mortality rates increase with increasing investment in reproduction. If, under poor environmental conditions, plants emphasizing female reproduction are more likely to die than plants emphasizing male reproduction, a genotype that grows in poor conditions, most likely to be small, should emphasize male reproduction. The same genotype that grows in rich conditions, most likely to be large, should emphasize female reproduction (Charnov & Dawson 1989, Iwasa 1991).

2. Day and Aarssen (1997) proposed the following explanation for increasing femaleness with plant size. Mortality rate per unit time diminishes strongly with plant size. Female function requires a longer reproductive time commitment relative to male function because of the time required for fruit production. When combined, these two factors promote the evolution of a positive correlation between femaleness and size. At present this hypothesis has not been tested.

3. Plant size can affect fitness returns directly, as in the case of wind pollination where dispersal efficiency increases with plant height (section 16.5). With such direct effects of plant size, fitness returns for a given absolute amount of resources invested differ for small and large plants, and plants should adjust gender to their size. With linear gain curves, the most likely outcome is an abrupt shift from male to female or vice versa at a certain size – sequential hermaphroditism (Klinkhamer et al. 1997). A large individual should be completely male if the slope of the male gain curve increases more than that of the female gain curve, while a small individual should be completely female. With nonlinear gain curves more complicated patterns of SDS will arise. Although complete sex change does occur in a number of species (reviewed in e.g. Freeman et al. 1980) it is relatively rare compared to the more gradually changing sex-allocation patterns found in simultaneous hermaphrodites.

4. If plant size has no direct effects, fitness returns for a given absolute amount of resources invested are equal for small and large plants. However, large plants will have a larger budget to invest and will, with the same proportion of resources allocated to male and female reproduction as a small plant, produce more flowers and seeds. Consequently, if one of the gain curves is nonlinear the ESS proportion of resources invested in male and female function depends on plant size. Allocation to female function should increase with increasing plant size if the male gain curve decelerates more quickly, whereas maleness should increase if the female gain curve decelerates more quickly (Klinkhamer et al. 1997).

5. If selfing is higher in large plants with many flowers, it will select for SDS (de Jong et al. 1999). As discussed in section 16.9, several authors have found that the selfing rate increases with plant size. With higher selfing rates plants should invest a smaller proportion of resources in male reproduction. The two factors combined lead to selection for increased femaleness with plant size.

Note that this argument applies both to animal- and wind-pollinated plants.

Given the above arguments one would expect SDS to be a common phenomenon. Three of the arguments presented (1, 2 & 5) point towards increasing femaleness with increasing plant size. However, given limited knowledge about the ratios of the gain curves' exponents, it is difficult to make an *a priori* prediction about the direction of SDS. We can test for the direction of SDS without an exact measure of sex allocation by scaling individuals relative to one another. In the perennial, insect-pollinated species *Asclepias syriaca*, large plants emphasized male fitness (Willson & Rathcke 1974). Increasing femaleness (measured as f/n) with plant size is, however, more common in animal-pollinated species. Data for monocarpic perennial species reviewed in Klinkhamer *et al.* (1997) showed that in 25/44 species femaleness increased significantly with size, while maleness increased in only one species. With regard to the allocation during the flowering phase, empirical evidence also suggests an increase in femaleness with plant size (e.g. Kudo 1993 and references in Klinkhamer *et al.* 1997).

Because in wind-pollinated species the male gain curve is assumed to be less steeply decelerating than in animal-pollinated species, and because plant height may promote pollen dispersal, increasing maleness with plant size should be more common in wind-pollinated species. Unfortunately this prediction cannot be tested with hermaphrodite species, because data on hermaphrodites include only a single wind-pollinated species. Bickel and Freeman (1993) studied floral sex ratios in monoecious plants. They found that in a total of 22 species femaleness increased with size in all eight animal-pollinated species while eight of the 14 wind-pollinated species showed increasing maleness with size.

16.12 | Conclusions

Measuring exact sex allocation in hermaphrodite plants is difficult or perhaps even impossible given the fact that flowers serve both male and female fitness in a variety of ways, even within single individuals. Fortunately, despite the fact that the issue is much debated, it is hardly ever interesting to do so. A much more fruitful approach is to carefully determine the trade-offs under consideration while checking if other plant characteristics are correlated with the components of this trade-off. Furthermore, the way gain curves should be determined also depends strongly on which trade-off related to male and female reproduction is considered. Trade-offs can often best be determined by manipulation experiments. Given the level of environmental variation, one should not be too eager to start with molecular methods in natural populations to determine the shape of the gain curves. Often measuring the relevant parameters experimentally for all the different steps that ultimately determine fitness may prove more useful. Models predict, and empirical evidence has shown, that hermaphrodites are not invariant in their sexual behaviour. However, many interesting hypotheses remain untested. Given the importance of the shape of gain curves for sex-allocation theory, priority should be given to determining gain curves in natural populations.

Acknowledgements

We thank Sarah Legge, Doug Taylor, Ian Hardy and an anonymous referee for comments on the manuscript.

References

Bickel AM & Freeman DC (1993) Effects of pollen vector and plant geometry on floral sex ratio in monoecious plants. *American Midland Naturalist*, **130**, 239–247.

Broyles SB & Wyatt R (1990) Paternity analysis in a natural population of *Asclepias exaltata*: multiple paternity functional gender, and the pollen donation hypothesis. *Evolution*, **44**, 1454–1468.

Broyles SB & Wyatt R (1997) The pollen donation hypothesis revisited: a response to Queller. *American Naturalist*, **149**, 595–599.

Brunet J (1992) Sex allocation in hermaphrodite plants. *Trends in Ecology and Evolution*, **7**, 79–84.

Campbell DR (1998) Variation in lifetime fitness in *Ipomopsis aggregata*: tests of sex allocation theory. *American Naturalist*, **152**, 338–353.

Campbell DR (2000) Experimental tests of sex-allocation theory in plants. *Trends in Ecology and Evolution*, **15**, 227–231.

Casper BB (1984) On the evolution of embryo abortion in the herbaceous perennial *Cryptantha flava*. *Evolution*, **38**, 1337–1349.

Casper BB (1988) Evidence for selective embryo abortion in *Cryptantha flava*. *American Naturalist*, **132**, 318–326.

Charlesworth D & Charlesworth B (1981) Allocation of resources to male and female function in hermaphrodites. *Biological Journal of the Linnean Society*, **15**, 57–74.

Charlesworth D & Morgan MT (1991) Allocation of resources to sex functions in flowering plants. *Philosophical Transactions Royal Society London, series B*, **332**, 91–102.

Charnov EL (1982) *The Theory of Sex Allocation*. Princeton, NJ: Princeton University Press.

Charnov EL (1987) On sex allocation and selfing in higher plants. *Evolutionary Ecology*, **1**, 30–36.

Charnov EL & Dawson EL (1989) Environmental sex determination with overlapping generations. *American Naturalist*, **134**, 806–816.

Cruden RW (1977) Pollen-ovule ratios: a conservative indicator of breeding systems in flowering plants. *Evolution*, **31**, 32–46.

Damgaard C & Loeschke V (1994) Genotypic variation for reproductive characters, and the influence of pollen-ovule ratio on selfing rate in rape seed (*Brassica napus*). *Journal Evolutionary Biology*, **7**, 599–607.

Day T & Aarssen LW (1997) A time commitment hypothesis for size-dependent gender. *Evolution*, **51**, 988–993.

de Jong TJ & Klinkhamer PGL (1989) Size dependency of sex allocation in plants. *Functional Ecology*, **3**, 201–206.

de Jong TJ & Klinkhamer PGL (1994) Plant size and reproductive success through female and male function. *Journal of Ecology*, **82**, 399–402.

de Jong TJ, Klinkhamer PGL & Rademaker MCJ (1999) How geitonogamous selfing affects sex allocation in hermaphrodite plants. *Journal of Evolutionary Biology*, **12**, 166–176.

Devlin B, Clegg J & Ellstrand NC (1992) The effect of flower production on male reproductive success in wild radish populations. *Evolution*, **46**, 1030–1042.

Dudash MR (1991) Plant size effects on male and female reproduction in hermaphrodite *Sabiata angularis* (Gentianaceae). *Ecology*, **72**, 1004–1012.

Ehrlen J (1991) Why do plants produce surplus flowers? A reserve-ovary model. *American Naturalist*, **138**, 918–933.

Emms SK (1993) On measuring fitness gain curves in plants. *Ecology*, **74**, 1750–1756.

Freeman DC, Harper KT & Charnov EL (1980) Sex change in plants: old and new observations and new hypotheses. *Oecologia*, **47**, 222–232.

Freeman DC, Lovett Doust J, El-Keblawy A, Miglia KJ & McArthur ED (1997) Sexual specialization and inbreeding avoidance in the evolution of dioecy. *Botanical Review*, **63**, 65–92.

Goldman DA & Willson MF (1986) Sex allocation in functionally hermaphroditic plants: a review and critique. *Botanical Review*, **52**, 157–194.

Harder LD & Barrett SCH (1995a) Mating costs of large floral displays in hermaphrodite plants. *Nature*, **373**, 512–515.

Harder LD & Barrett SCH (1995b) Pollen dispersal and mating patterns in animal-pollinated plants. In: DG Lloyd & SCH Barrett (eds) *Floral Biology, Studies on Floral Evolution in Animal-Pollinated Plants*, pp 140–190. New York: Chapman & Hall.

Helenurm K & Schaal BA (1996) Genetic load, nutrient limitation, and seed production in *Lupinus texensis* (Fabaceae). *American Journal of Botany*, **83**, 1585–1595.

Hessing MB (1988) Geitonogamous pollination and its consequences in *Geranium caespitosum*. *American Journal of Botany*, **75**, 1324–1333.

Iwasa Y (1991) Sex change evolution and cost of reproduction. *Behavioral Ecology*, **2**, 56–68.

Klinkhamer PGL & de Jong TJ (1997) Size-dependent allocation to male and female reproduction. In: FA Bazzaz & J Grace (eds) *Plant Resource Allocation*, pp 211–229. Physiological Ecology Series. San Diego: Academic Press.

Klinkhamer PGL, de Jong TJ & Metz H (1995) Why plants can be too attractive – a discussion of measures to estimate male fitness. *Journal of Ecology*, **82**, 191–194.

Klinkhamer PGL, de Jong TJ & Metz H (1997) Sex and size in cosexual plants. *Trends in Ecology and Evolution*, **12**, 260–265.

Koelewijn HP & Hunscheid MPH (2000) Intraspecific variation in sex allocation in hermaphrodite

Plantago coronopus (L.) *Journal of Evolutionary Biology*, **13**, 302–315.

Kozłowski J & Stearns SC (1989) Hypotheses for the production of excess zygotes: models for bet-hedging and selective abortion. *Evolution*, **43**, 1369–1377.

Kudo G (1993) Size-dependent resource allocation pattern and gender variation of *Anemone debilis* Fisch. *Plant Species Biology*, **8**, 29–34.

Lloyd DG (1983) Evolutionarily stable sex ratios and sex allocations. *Journal of Theoretical Biology*, **69**, 543–560.

Lloyd DG (1984) Gender allocations in outcrossing cosexual plants. In: J Dirzo & J Sarukhan (eds) *Perspectives on Plant Population Ecology*, pp 277–300. Sunderland MA: Sinauer.

Lloyd DG (1987) Allocations to pollen, seeds and pollination mechanisms in self-fertilizing plants. *Functional Ecology*, **1**, 83–89.

Marshall DL & Ellstrand NC (1988) Effective mate choice in wild radish: evidence for selective seed abortion and its mechanism. *American Naturalist*, **131**, 739–756.

Melser C (2001) Selective seed abortion and offspring quality. Ph.D. thesis, University of Leiden.

Melser C, Rademaker MCJ & Klinkhamer PGL (1997) Selection on pollen donors by *Echium vulgare* (Boraginaceae). *Sexual Plant Reproduction*, **10**, 305–312.

Melser C, Bijleveld A & Klinkhamer PGL (1999) Late acting inbreeding depression in both male and female function of *Echium vulgare* (Boraginaceae). *Heredity*, **83**, 162–170.

Morgan MT (1993) Fruit to flower ratios and trade-offs in size and number. *Evolutionary Ecology*, **7**, 219–232.

Nakamura RR, Stanton LS & Mazer SJ (1989) Effects of mate size and mate number on male reproductive success in plants. *Ecology*, **70**, 71–76.

Parker IM (1995) Reproductive allocation and the fitness consequences of selfing in two sympatric species of *Epilobium* (Onagraceae) with contrasting mating systems. *American Journal of Botany*, **82**, 1007–1016.

Queller D (1997) Pollen removal, paternity, and the male function of flowers. *American Naturalist*, **149**, 585–594.

Rademaker MCJ & de Jong TJ (1998) Effects of flower number on estimated pollen transfer in a natural population of three hermaphroditic species: an experiment with fluorescent dye. *Journal of Evolutionary Biology*, **11**, 623–641.

Rademaker MCJ & de Jong TJ (1999) The shape of the female gain curve for *Cynoglossum officinale* and *Echium vulgare*: quantifying seed dispersal and seedling survival in the field. *Plant Biology*, **1**, 451–356.

Rademaker MCJ & Klinkhamer PGL (1999) Size-dependent sex allocation in *Cynoglossum officinale* for different genotypes under uniform favourable conditions. *Plant Biology*, **1**, 108–114.

Schoen DJ (1982) Male reproductive effort and breeding system in a hermaphrodite plant. *Oecologia*, **53**, 255–257.

Schoen DJ & Lloyd DG (1992) Self- and cross-fertilization in plants. III Methods for studying modes and functional aspects of self-fertilization. *International Journal of Plant Science*, **153**, 381–393.

Schoen DJ & Stewart SC (1986) Variation in male reproductive investment and male reproductive success in white spruce. *Evolution*, **40**, 1109–1120.

Smouse PE & Meagher TR (1994) Genetic analysis of male reproductive contributions in *Chamaelirium luteum* (L) Gray (Liliaceae). *Genetics*, **136**, 313–322.

Snow AA, Spira TP, Simpson R & Klips RA (1995) The ecology of geitonogamous pollination. In: DG Lloyd & SCH Barrett (eds) *Floral Biology, Studies on floral evolution in animal-pollinated plants*, pp 191–216, New York: Chapman & Hall.

Spalik K (1991) On evolution of andromonoecy and 'overproduction' of flowers: a resource allocation model. *Biological Journal of the Linnean Society*, **42**, 325–336.

Stearns SC (1992) *The Evolution of Life Histories*. Oxford: Oxford University Press.

Stephenson AG (1981) Flower and fruit abortion: proximate causes and ultimate functions. *Annual Review of Ecology and Systematics*, **12**, 253–279.

Stephenson AG & Winsor JA (1986) *Lotus corniculatus* regulates offspring quality through selective fruit abortion. *Evolution*, **40**, 453–458.

Sutherland S (1986a) Floral sex ratios, fruit set and resource allocation in plants. *Ecology*, **67**, 991–1001.

Sutherland S (1986b) Patterns of fruit set: what controls fruit-flower ratios in plants? *Evolution*, **40**, 117–128.

van Noordwijk AJ & de Jong G (1986) Acquisition and allocation of resources: their influence on variation in life history tactics. *American Naturalist*, **128**, 342–350.

Vrieling K, Saumitou-Laprade P, Meelis E & Epplen JT (1997) Multilocus fingerprints in the plant *Cynoglossum officinale* L. and their use in the estimation of selfing. *Molecular Ecology*, **6**, 587–593.

Willson MF & Burley N (1983) *Mate Choice in Plants: Tactics, Mechanisms and Consequences*. Princeton, NJ: Princeton University Press.

Willson MF & Rathcke BJ (1974) Adaptive design of the floral display of *Asclepias syriaca* L. *American Midland Naturalist*, **92**, 47–57.

Wilson P, Thomson JD, Stanton ML & Rigney LP (1994) Beyond floral Batemania: gender biases in selection for pollination success. *American Naturalist*, **143**, 283–296.

Chapter 17

Sex ratios in dioecious plants

Tom J. de Jong & Peter G.L. Klinkhamer

17.1 | Summary

Some seeds of dioecious plants develop into male plants and others become females. Brothers and sisters can grow close together in the seed shadow of the maternal plant, which promotes sib-mating, and classical sex-allocation theory predicts a slight female bias among the seeds produced. We describe different ways of examining seed sex ratios and some of the pitfalls involved. The available direct (seed sex ratio) and indirect (proportions of male and female plants in the field) evidence suggests that the seed sex ratio is often close to 0.5, despite the fact that there is genetic variation in the seed sex ratio in some cases. The combination of significant sib-mating and an unbiased seed sex ratio is at odds with classical sex-allocation theory. Genetic conflict theory might provide new insights and should be a central theme in future research. The adult sex ratio can also become male or female biased due to sexually differential mortality, but this does not influence the seed sex ratio.

> Das Zahlenverhältnis [0.5] kann aber nur dann rein herauskommen, wenn eine ganze Reihe von Bedingungen erfüllt sind.
>
> The ratio [0.5] can, however, only emerge, when a whole range of conditions is satisfied.
>
> *(Correns 1928)*

17.2 | Introduction

Like most animals, but unlike the great majority of plant species, dioecious plants have separate male and female individuals. Both male and female organs develop in each of their flowers, in separate floral whorls, but the development of one type is halted before maturity (Grant *et al.* 1994), with the timing of the arrest differing between species. Taxonomically, dioecious plants form a mixed bag. For instance, 59 dioecious species are listed for Britain (Kay & Stevens 1986); there are completely dioecious families (e.g. Salicaceae) and species such as the creeping thistle (*Cirsium arvense*) with cosexual (male and female organs on the same individual) relatives. The sporadic occurrence of dioecy indicates that it evolved many times in the plant kingdom and also helps to explain the great variation in sex-determination systems among dioecious plants (section 17.3.1). Careful examination of all plants in a supposedly dioecious population often reveals a few cosexual individuals (Kay & Stevens 1986), which can indicate the ancestral state to be hermaphroditism (each flower with both male and female parts, Chapter 16) or monoecy (separate male and female flowers on the same plant).

Dioecy effectively bars self-pollination. The old idea that selection for outbreeding plays an important role in its evolution has recently been re-emphasized (de Jong *et al.* 1999) but opinions on the relative role of factors differ (Freeman *et al.* 1997). Dioecious species are overrepresented among trees and many depend on abiotic factors for their pollination (Renner & Rickleffs 1996). While the evolution of dioecy is an interesting subject for theoretical and empirical study in its own right, the focus of this chapter is on the sex ratio.

In dioecious plants some seeds grow into male flowering plants that produce pollen and others become females that produce seeds. It is interesting to ask whether parent plants can adjust the sex ratio in the seeds and, if so, whether this adjustment is consistent with theoretical predictions. If you were growing kiwi fruits (*Actinidia deliciosa*) for consumption, you would want to maximize fruit yield. You could do so by positioning several female plants around a central male plant. Such a female-biased sex ratio would also result in the highest seed output in natural populations. Although early papers on sex ratios assumed that natural populations maximize seed output in nature, such 'group selection' thinking is now considered flawed (Williams 1971). The current approach is to ask how selection acts on the sex ratio produced by individuals (or genes) within the population.

The path we take is well trodden by zoologists, but relatively untravelled by botanists studying dioecious plants. We begin by summarizing mechanisms of sex determination among plants and the possibilities of sex allocation control that these offer (section 17.3). Assuming adaptive control, we then illustrate the major predictions of sex-allocation theory (Maynard Smith 1978, Charnov 1982) in the terminology of plant ecologists (section 17.4). We then summarize methods for estimating seed sex ratios (section 17.5) and compare mean seed sex ratios with theoretical predictions (section 17.6). We briefly discuss genetic conflict, a novel extension of sex ratio theory, which examines the fate of sex ratio modifiers and restorers located outside the autosomes (section 17.7). Finally, we discuss population sex ratios of adult plants and how to unravel the factors that may lead to biased ratios (section 17.8).

17.3 | Sex ratio control

In this section we discuss whether plants potentially have control of sex allocation.

17.3.1 Sex-determination mechanisms

Plants have a variety of sex-determining mechanisms (reviewed by, e.g. Westergaard 1958, Irish & Nelson 1989, Chattopadhyay & Sharma 1991, Dellaporta & Calderon-Urrea 1993, Grant *et al.* 1994, Ainsworth *et al.* 1998, Grant 1999). Due to the disparate evolutionary origins of dioecy, sex-determining systems have also evolved independently, resulting in a diversity of mechanisms that produce similar results. For instance, in annual mercury (*Mercurialis annua*) sex is determined by two genes with multiple alleles segregating independently and two additional cytoplasmic factors (Durand & Durand 1991). Application of cytokinins in this species may induce males to form female flowers and to produce some seeds. In common sorrel (*Rumex acetosa*) the Y chromosome is inert and sex is determined by the ratio of X chromosomes to autosomes, which is similar to the system of sex determination in *Drosophila* flies (Ainsworth *et al.* 1998, Grant 1999, Chapter 8). Cytoplasmic factors can also play a role, selecting for the production of more daughters (section 17.7), and theoretically it is even possible that cytoplasmic factors take over from the nuclear genes as a female determining factor (Taylor 1990).

Heterochromatic sex-determination systems (XX-XY, Chapter 7), in which males are the heterochromatic sex, are also found in plants. Direct evidence for this comes from the few species with sex chromosomes. In other species sex chromosomes cannot be distinguished under the microscope but males may still be heterozygous at a single sex-determining locus. Testolin *et al.* (1995) found that some male kiwi (*Actinidia deliciosa*) plants were 'inconstant' and produced a few hermaphrodite flowers. When selfed, 75% of the seeds from these inconstant males developed into a male plant and 25% into a female, demonstrating that males are the heterozygous sex. It also shows that YY males are viable as long as the differentiation of the male-determining chromosome is limited. In the commercial production of *Asparagus* seed, YY males are crossed with females (XX) to produce an all-male (XY) seed crop. Although generally the male is the heterozygous sex, this is not always the case. The wild strawberry (*Fragaria vesca*), the shrubby cinquefoil (*Potentilla fruticosa*), *Cotula* and spanish catchfly (*Silene otites*, Richards 1997) are some exceptions in which the female is heterozygous at the sex-determining locus.

17.3.2 Mechanisms of sex ratio adjustment

While some genes are responsible for sex determination, other genes can be responsible for modifying seed sex ratios. Most directly, the ratio of X- to Y-bearing pollen may differ from 1:1 because of selective abortion of one of the pollen types. This phenomenon, known as meiotic drive, has been most studied in animals (Chapter 9). Under meiotic drive the father's genotype is the most influential on the seed sex ratio. If meiosis does lead to half of the pollen bearing Y and half bearing X, the mother needs to discriminate to adjust the sex ratio. The strength of the discrimination mechanism can again be under genetic control. Many studies have indicated the importance of the prezygotic phase. Pollen tube growth is then supported by maternal tissue. The support and growth may depend on the haploid genotype, X or Y, of the pollen. This mechanism cannot work in species in which the female is the heterochromatic (or heterozygous) sex, so that all pollen is of type X. Alternatively, the mother plant discriminates after fertilization by selective abortion of embryos. However, discarding embryos is only favourable if resources become available to produce more offspring: a higher fitness is gained if, say, five male and five female seeds are produced, than if three male and five female seeds are produced, even if the latter matches the optimal sex ratio. For this reason one expects females to discriminate as early as possible.

Pollination intensity can affect the sex ratio of the seeds formed. Environmental effects (especially pollination intensity) on the sex ratio are referred to as 'certation'. From the observation that more daughters are produced with abundant pollination (see below), Correns (1928) suggested that competition between pollen grains is more intense under abundant pollination and that X-bearing pollen outcompete Y-bearing pollen, leading to female-biased sex ratios. Under sparse pollination, there will be sufficient ovules for all pollen grains, leading to a sex ratio of 0.5. While the suggested mechanism can result in unbiased or female-biased sex ratios, male bias is not expected. The relatively poor performance of Y-bearing pollen could be due to accumulation of mutations in the nonhomologous region of the Y chromosome. Such mutations cannot be easily repaired by recombination, as there is no crossing over (genes on the Y chromosome are passed on unaltered from father to son).

Working on *Silene latifolia*, Correns (1928, p 93) found that with abundant pollination the sex ratio of seed batches of three females was extremely female biased (0.076, 0.085 & 0.086) whereas with the sparse pollination it was less female biased (0.178, 0.151 & 0.077). The sex ratios with sparse pollination are much lower than the expected 0.5, suggesting that additional factors (meiotic drive, discrimination by the mother) play a role. Certation has also been reported to occur in *Rumex acetosa* (Correns 1928, Rychlewski & Zarzycki 1975) and heart sorrel (*R. hastatulus*, Conn & Blum 1981). However, certation and the mechanism behind it remain controversial (Richards 1997). Other researchers found no effects of pollination intensity (Carroll & Mulcahy 1993) or position (Purrington 1993) on the sex ratio in *Silene latifolia*. Taylor *et al.* (1999) recently reported results similar to those of Correns (1928), although the effect of pollination intensity was relatively small compared to other effects on the sex ratio, especially the identity of the male used in the cross. Taylor *et al.* also found that the effect varies among parents, which may explain the different results obtained by studies using genetically different *Silene*. Even if effects of pollen density on seed sex ratio can be demonstrated in the laboratory, there is still no evidence that such effects operate in nature. If they do, the plant can use information about the number of males present and adjust the sex ratio of its seeds accordingly.

17.3.3 Sex ratio variation

Heterochromatic sex determination in birds and mammals (Chapter 7) has been considered to impose constraints on sex-allocation control and lead to sex ratios (proportion males = males/(males+females)) close to 0.5 (Bull & Charnov 1988, Chapters 7 & 13). Similarly, restricted variation in plants can be tested for by assessing the genetic variation in the sex ratio of seed batches. Seeds from several mothers should be kept separate and grown to maturity under uniform conditions. Unfortunately in many studies seeds have been mixed, which yielded only an estimate of the overall seed sex ratio of the population. Several studies have, however, assessed seed batches separately. In the lily, *Chamaelirium luteum*, the sex ratios of seed batches from ten different parents had no significant variation and were close to 0.5 (Meagher 1981). In *Silene latifolia*, however, sex ratio differed greatly between families, from 0.169 to 0.823 (DR Taylor 1994). In crosses between *Salix viminalis* clones, carried out by Alström-Rapaport *et al.* (1997), sex ratio varied between 0.02 and 0.82 (Table 17.1). In these experiments male and female clones came from different locations in Sweden and one male clone came from The Netherlands. In the fruitfly *Drosophila simulans* such interpopulation crosses have revealed sex ratio modifiers and suppressors (Merçot *et al.* 1995, Atlan *et al.* 1997,

Chapter 9) that would have gone unnoticed in intrapopulation crosses. It would therefore be interesting to have additional data on the sex ratio of crosses using parents from a single *Salix* population.

TJ de Jong collected seeds of *Salix repens* from naturally pollinated female plants in the field. Seeds were germinated and 50 seedlings were planted under fertile garden conditions in the spring of 1999. In the spring of 2000 not all 50 plants had flowered and the experiment is still continuing so results reported here are preliminary (Table 17.2). The sex ratio appears to differ significantly between mothers, with five mothers producing sex ratios close to 0.5 and two mothers producing only daughters. Similarly, reanalysis of data on *Rumex acetosa* (Table 1B in Rychlewski & Zarzycki 1975) shows family-level sex ratio heterogeneity (Table 17.2). Field data estimate the overall adult sex ratio to be female biased (0.43, Rychlewski & Zarzycki 1975).

One hundred and ninety-five individuals of the common nettle (*Urtica dioica*) were surveyed in dunes in The Netherlands (TJ de Jong, unpublished). The overall sex ratio (0.48) was very close to 0.5. From 480 seeds taken from eight motherplants and sown in the greenhouse, 472 plants could be raised to maturity. Of these, 234 were male (sex ratio = 0.495), which is again close to 0.5. However, considering the 60 seeds from each motherplant separately reveals large and significant deviations from sex ratio equality (Table 17.2).

These data show that there is genetic variation of the sex ratio and hence potential for evolutionary change. The genetic background seems to differ between species. Quantitative variation exists in *S. latifolia*, *U. dioica* and *R. acetosa* while the situation in *S. repens* suggests a single gene, located in the cytoplasm or on the X chromosome, that distorts the sex ratio and which can be restored. The extent to which these conclusions can be generalized is unknown. The evidence in this section, however, strongly suggests that at least some plants have the capacity to control the sex allocation and the sex ratio of their offspring. We now discuss the sex-allocation decisions that evolutionary considerations lead us to expect

Table 17.1 Sex ratio (proportion males) of controlled crosses in *Salix viminalis*. For all crosses the number of observations ranges between 79 and 124. M = mother, F = father. Data from Alström-Rapaport *et al.* (1997)

	M1	M2	M3	M4
F5	0.276	–	0.559	0.483
F6	0.020	0.689	0.289	0.823
F7	0.348	–	–	0.414
F8	0.421	0.265	0.525	0.470

Table 17.2 Variation in seed sex ratio in seed samples from different motherplants (M)

Species	Motherplant	Sample size	Sex ratio	Analysis	Reference
Rumex acetosa	M11	24	0.083	$\chi^2 = 28.55, 5$ df, $P < 0.05$	adapted from Rychlewski & Zarzycki (1975)
	M12	19	0.210		
	M13(a+b)	44	0.432		
	M14	24	0.125		
	M15(a+b)	35	0.057		
	M18	19	0		
Urtica dioica	M1	58	0.534	$\chi^2 = 32.31, 7$ df, $P < 0.05$	TJ de Jong, unpublished
	M2	60	0.333		
	M3	55	0.618		
	M4	60	0.483		
	M5	60	0.683		
	M6	60	0.616		
	M7	59	0.389		
	M8	60	0.317		
Salix repens	M1	35	0.486	$\chi^2 = 52.60, 6$ df, $P < 0.05$	TJ de Jong, unpublished
	M2	48	0.417		
	M3	45	0.444		
	M4	38	0		
	M5	47	0.383		
	M6	43	0.348		
	M7	49	0		

plants to make, given that they have sex ratio control.

17.4 | Theory for adaptive sex allocation

We introduce sex ratio theory by formulating a model for dioecious plants. In this model we assume that plants are annuals and that each individual has a resource (T) available for making seeds and is free to allocate a certain fraction (r) to 'sons' and the remainder of the resource ($1 - r$) to 'daughters'. Male and female seeds may be of different size, so we distinguish between the costs of producing a male (C_m) and a female (C_f). Sons and daughters have probability s_m and s_f, respectively, of surviving until flowering. The

variable to be optimised is r. Although most dioecious plants are not annuals and individuals differ in how much resource is available for reproduction, these assumptions allow us to illustrate the basic theory in a comprehensible way, e.g. by avoiding the complication of overlapping generations. It is well known that the predictions of sex-allocation theory hold under a broader range of conditions (Charnov 1982).

Male and female siblings often grow relatively close together in the 'seed shadow' of the mother plant. In animal-pollinated species pollination will often be local, for instance because bumblebees move between neighbouring plants. Wind pollination may distribute pollen more homogeneously through the population, but still neighbouring plants will be disproportionately pollinated (Proctor *et al.* 1996). Thus, plant populations have mating substructure, not population-wide random mating (panmixis, e.g. Ingvarsson & Giles 1999). A convenient way of dealing with nonrandom mating is to assume that some of the pollen is distributed locally, leading to sib-mating, and the rest is dispersed randomly through the larger population (Maynard Smith 1978, Uyenoyama & Bengtsson 1982). The local pollination implies that a fraction (S) of the daughters will be fertilized by their brothers, as opposed to the Fisherian assumption (Bull & Charnov 1988, Chapter 1) that all pollen is dispersed randomly through the population ($S = 0$). Note that Maynard Smith's (1978) model is a simplification: in practice, plants interact most with their neighbours and pollen dispersal will decline rapidly with distance. Models that explicitly incorporate spatial scale have yet to be developed for plants. In this respect the dispersal-mating-dispersal model of PD Taylor (1994) seems relevant. In this model males disperse first and after mating the fertilized females disperse. This is similar to the situation in plants with pollen and seed dispersal: a nuclear gene in the mother is dispersed once with the seed while a nuclear gene in the father is dispersed twice, first with the pollen and then with the seed.

If r is the dominant strategy in a large population, then each motherplant will have $T r s_m / C_m$ sons and $T(1 - r) s_f / C_f$ daughters that survive until flowering. The reproductive success of each male will then depend on the ratio of females to males in the population. A male's expected success will equal 1 if there is an unbiased sex ratio (proportion males = 0.5). Male success will be doubled if each male has two females to fertilize (sex ratio = 0.33) and halved if two males have to compete for a single female (sex ratio = 0.66). At flowering, the ratio of females/males in a population in which all plants allocate r to sons is $(1 - r)(s_f / C_f)(r)(s_m / C_m)$. We denote this ratio as P. A fraction S of the ovules is used in sib-mating, so the competition between males is only for the fraction $1 - S$ of the ovules. The expected success of a single surviving male plant then equals the product of $(1 - S)$ and P. For the $T r s_m / C_m$ sons that each mother produces a fraction s_m survives until flowering and their contribution to fitness is $T r (s_m / C_m)(1 - S)P$. As inbreeding depression

is common in plants, we assume that the offspring of mating between sisters and brothers perform less well, by a factor $1 - \delta$, than the offspring from fertilization with unrelated pollen. δ is defined as the performance of a selfed seed divided by the performance of an outcrossed seed. This could be due to higher mortality, or to slower growth resulting in lower seed production.

Now suppose an autosomal gene in the mother can modify the amount of resource allocated to sons, r (sex ratio modifiers on the Y chromosome are discussed in section 17.7). How many copies of this gene are then transmitted to the grandchildren of a parent plant? The gene is transmitted along three routes. First, the gene may pass through daughters that cross with unrelated males. Second, two copies of the gene are transmitted in sib-matings. Third, it may pass through sons that mate with unrelated females. Fitness (w) of a common plant with phenotype r is then

$$w = T(1 - r)(1 - S)(s_f / C_f) + 2T(1 - r)S(1 - \delta)$$
$$\times (s_f / C_f) + Tr(s_m / C_m)(1 - S)P. \tag{17.1}$$

By substituting the equation for the ratio of females to males (P) and rearranging we find

$$w = (T s_f / C_f)(2 - 2S\delta)(1 - r). \tag{17.2}$$

Next we calculate the fitness of a mutant that allocates r_m of her resources to sons and $1 - r_m$ to daughters. When she is rare her fitness is

$$w_m = (T s_f / C_f)[(1 - r_m)(1 - S) + 2(1 - r_m)S(1 - \delta)$$
$$+ (r_m / r)(1 - r)(1 - S)]. \tag{17.3}$$

The equation is identical to that used previously by many authors to study optimal allocation to the sexual functions in hermaphrodites (e.g. Charlesworth & Charlesworth 1978, de Jong et al. 1999, Chapter 16). For hermaphrodites, S denotes the selfing rate of plants and $1 - \delta$ denotes inbreeding depression. As the inbreeding coefficient is 1/4 in the case of selfing in hermaphrodites and in the case of sib-mating in dioecious plants, the two measures of inbreeding depression are very similar.

We can explore different values of r and r_m, in order to find the evolutionarily stable

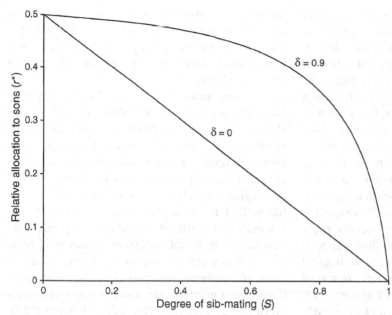

Fig 17.1 Predicted relative allocation to sons as a function of the fraction sib-mating with no inbreeding depression ($\delta = 0$) and high inbreeding depression ($\delta = 0.9$). It is assumed that sib-mating is independent of the relative allocation of the parent.

and $(1 - r)/C_f$ daughters. Therefore, when $r = 0.5$ the sex ratio is $C_f/(C_m + C_f)$, which is only 0.5 if both sexes are equally costly. This result is well known in animal ecology (Hamilton 1967, Trivers & Hare 1976). Taylor (1996) tested it in *Silene latifolia* and found no weight difference between male and female seeds.

value of r, which cannot be invaded by another strategy. When the population adopts this value, no mutant can attain a higher fitness and the population cannot be invaded by a mutant with a different value of r. In general, the easiest way to do this is to use a spreadsheet program on a computer. In this case (S is constant and independent of r) there is also an analytical solution; to find it we use standard methods (differentiating w_m with respect to m, setting the derivative equal to zero and finally setting $r_m = r$, Parker 1984). The evolutionarily stable allocation to sons (r^*) is

$$r^* = (1 - S)/(2 - 2S\delta). \tag{17.4}$$

The same equation (with $\delta = 0$) was derived by Maynard Smith (1978), Uyenoyama and Bengtsson (1982) and Werren and Hatcher (2000). Several conclusions can be drawn from eq. 17.4.

1. Mothers should optimize the amount of resource they allocate to male and female seeds. Without sib-mating ($S = 0$), the maternal plant should allocate 50% of its resources to each sex. If male and female seeds are equally costly to produce (expected if male and female seeds have similar weights), then the expected numerical sex ratio is 0.5. If seed weight is sexually dimorphic, the plant should produce more seeds of the cheaper sex. With a given relative investment in sons (r), r/C_m sons are produced

2. With sib-mating a female bias is expected (e.g. Hamilton 1967, Hardy 1994) and inbreeding depression has the effect of pushing optimal allocation back to 0.5 (eq. 17.4, Figure 17.1). For low levels of sib-mating, expected values of r^* lie between 0.4 and 0.5. It is convenient that eq. 17.3 is identical to that for hermaphrodite plants (analysed in detail by, for example, Charlesworth & Charlesworth 1978, 1981, de Jong et al. 1999, Chapter 16). In the simplest case, the one considered here, the selfing rate (S) is assumed constant and all plants export the same proportion of pollen to the outcross pollen pool, independent of their selfing rate. This means that there is no pollen discounting. Pollen discounting means that a higher selfing rate implies a cost in terms of pollen used or lost otherwise, so that genotypes with high selfing contribute less pollen to the outcross pollen pool. For hermaphrodites the selfing rate may well increase with the relative allocation to male function (r) (de Jong et al. 1999). For dioecious plants the equivalent is that sib-mating increases with the proportion of sons produced. In such cases there is no analytical solution for r^* but it is still easy to compute w_m numerically so that r^* can be calculated. The effect of inbreeding depression on r^* then depends on how S increases with r. If the shape of the curve is very flat, inbreeding depression continues to push r^* back to 0.5

as in Figure 17.1. If the curve is steep, a higher level of inbreeding depression may result in an even lower r^* than without inbreeding depression (Charlesworth & Charlesworth 1981, de Jong et al. 1999). There are currently no empirical data on the shape of the relationship between the sex ratio and the degree of sib-mating. High inbreeding depression could either push the optimal sex ratio back to 0.5 or push it towards a greater female bias, compared to optima in the absence of inbreeding depression. The important result is that, unless for some reason sib-mating is extremely rare, female bias in the seeds is expected in dioecious plants. This should especially apply to short-lived plants with poor pollen and seed dispersal, while for long-lived trees with good pollen and seed dispersal the appropriateness of the prediction is less clear. These ideas are poorly diffused among plant ecologists, and many articles still refer to sex ratios of 0.5 as 'what theory predicts'.

3. There may well be differences between the survival of male and female plants. Herbivores often distinguish between the sexes (Mutikainen et al. 1994, Ågren et al. 1999) and the investment in reproduction may also differ, the females generally investing a greater proportion of their resources than the males (Delph 1999). In addition, male plants may start reproducing earlier in life and/or flower more frequently (Delph 1999). However, these factors will not affect sex ratio optima. Sex ratio optima may be affected by the developmental mortality of males under extreme sib-mating (Nagelkerke & Hardy 1994), but dioecious plants are unlikely to experience these conditions and mortality is thus unlikely to influence the seed sex ratio. The possibility of sexually differential developmental mortality also means that adult sex ratios do not necessarily reflect seed sex ratios.

17.5 | Methods for estimating sex allocation and the seed sex ratio

If we are to test sex-allocation theory, we need to be able to estimate crucial variables, such as resource allocation to males and females. Since maternal investment in offspring is made at the seed stage, obvious questions are, 'do male and female seeds differ in weight?' and 'what is the sex ratio of seeds?'.

In some plant species, adults are sexually size-dimorphic (Lloyd & Webb 1977); for instance, male Asparagus plants are taller and more slender then their female counterparts. However, differences between the sexes are less pronounced than in many animals and sexual size-dimorphism in plant seeds seems generally unlikely. Taylor (1996), for instance, found no sex-related weight difference in white campion (Silene latifolia) seeds. In spinach (Spinacia oleracea), however, males were overrepresented among larger seeds and underrepresented among smaller seeds (Freeman et al. 1994). Germination of large seeds (81%) was greater than that of small seeds (75%), so seed size had a pronounced effect on survival until flowering (the time at which sex could be assessed) and calculation of the relative allocation of resource to sons (r) is thus complex. A further complication is that the sex of spinach seeds is partly determined by hormones and can be altered by external conditions (Chailakhyan & Khrianin 1987). The exposure of cloned plants to extreme conditions can establish whether the environmental conditions affect sex expression, as found for spinach (Freeman et al. 1994) and hemp (Cannabis sativa, Chailakhyan & Khrianin 1987).

Even in species with stable sex expression (i.e. a seed's sex is unaffected by its environment) calculating the seed sex ratio is complicated if seeds cannot be sexed prior to or immediately after germination (see below). For instance, some seeds may remain dormant while others germinate readily; if one sex tended to germinate earlier, only including the readily germinating seeds in samples would lead to estimation errors. Thus, all seeds should be screened, which includes breaking dormancy. Similarly, the sexually differential survival of seeds until the time that sex is assessed (usually at flowering) complicates the estimation of seed sex ratios and preflowering mortality should thus be minimized (see also Fiala's 1979 comment on avian developmental mortality and sex ratio). If this is not possible one could assign juvenile plants to one of two treatments (good versus poor conditions); a lack of effect on the adult sex ratio would suggest that preflowering mortality is independent of sex. Finally, data should be collected until all plants in a sample have flowered because males and females may differ in the timing of flowering (Testolin *et al.* 1995, Delph 1999).

While the established method of estimating seed sex ratios is to grow plants under ideal greenhouse conditions, this method requires a large amount of greenhouse space and is impractical for shrubs and trees with long juvenile periods. There are two alternatives with the advantage that juvenile (vegetative) plants can be sexed, so that changes in sex ratio in a cohort can be followed through time (Taylor 1996).

First, young seedlings of some species can be screened for sex chromosomes under the microscope. Despite considerable effort in this field in the first half of the twentieth century, few species have well-established heteromorphic sex chromosomes (Chattopadhyay & Sharma 1991); the list includes hemp, hop (*Humulus lupulus*), several *Rumex* species, *Silene latifolia* and red campion (*S. dioica*).

Second, molecular markers for sexuality can now be developed. The usual method of bulked segregant analysis is to take plants from a single cross. This minimizes variation when compared to sampling males and females randomly in the population. Then females and males (usually about ten individuals) are pooled to find markers unique to one sex. Next it should be checked if the putative markers are also diagnostic outside the cross. Taylor (1996) used random amplified polymorphic DNAs (RAPDs) successfully to sex mature and developing seeds of *Silene latifolia*, a species with sex chromosomes. In the same species Zhang *et al.* (1998) extended the RAPD method to the more reliable sequence characterized amplified region markers. With RAPDs, Alström-Rapaport *et al.* (1998) found a single band (out of 1080 examined) difference between the sexes in a cross of *Salix viminalis*. This marker did not predict sex correctly outside the specific cross. Raemon Buttner *et al.* (1998) developed an amplified fragment-length polymorphism (AFLP) probe that distinguishes between male and female *Asparagus* plants.

Molecular genetic methods may still be problematic to use with species without sex chromosomes. In these, sex may be determined by a single locus and much work will be involved in its identification. Undoubtedly, advances in molecular methods will lead to more bands and variation to score, which will facilitate the pinpointing of sex-determining genes on a genetic map. Apart from their usefulness in estimating the sex ratio, these markers are bound to become important in agriculture. Often the breeder has a greater need for plants of a particular sex. For instance, until recently papayas could only be sexed at an age of six to eight months, and typically produce 50% sons. Using a microsatellite probe, breeders can now discriminate between male and female papaya seedlings and the surplus males can be discarded, preventing much resource wastage (Parasnis *et al.* 1999).

17.6 | Analysis of mean seed sex ratios

Simple germination tests and assessment of sex ratios of plants grown in ideal greenhouse conditions (section 17.5) suggest that the mean sex ratio in seed samples pooled from many plants is often close to 0.5. Examples are sheep's sorrel (*Rumex acetosella*), *R. acetosa* (Putwain & Harper 1972), *Chamaelirium luteum* (Meagher 1981), white bryony (*Bryonia dioica*, Correns 1928), *Urtica dioica* (Table 17.2), *Actinidia* (Testolin *et al.* 1995), *Silene otites* (Soldaat *et al.* 1997) and so on. It has been shown that the average seed sex ratio is less than 0.5 in at least three species. First, a small but consistent female bias is present in *Silene* species (Prentice 1984), notably *S. latifolia* (sex ratio = ca. 0.46, DR Taylor 1994, 1996). Second, in 6/13 crosses in *Salix viminalis*, offspring had significantly female-biased sex ratios and 2/13 showed a significant male bias; in total, a fraction of 0.425 of the seeds grew into male plants (Alström-Rapaport *et al.* 1997). Female bias has also been reported to occur in populations of other *Salix* species (summarized in Alström-Rapaport 1997). Third, in *Rumex hastatulus*, Smith (1963) found that 0.43 of seeds grew into male plants and Conn and Blum (1981) estimated the sex ratio as 0.40 after abundant pollination.

With sex ratios between 0.4 and 0.5, large numbers of observations are needed in order to detect significant differences (see also Chapter 3). For instance, Putwain and Harper (1972) pooled the seeds of several *Rumex acetosella* parents and sowed these in six replicate samples, generating 81 males and 101 females. A sex ratio of 0.445 males is well within the range discussed above, but is far from statistically significant with this sample size (χ^2 test, $P = 0.20$, after pooling the sample, 1 *df*). Clearly larger sample sizes are necessary to increase the power of the test, i.e. to minimize the probability of accepting a false null-hypothesis (a type II error; in this case the null hypothesis, H_0, is that the underlying sex ratio is 0.5). Sokal and Rohlf (1995, pp 157–169) explain how to perform power analysis. It involves estimating the type II error (β) as the area of the probability distribution under the alternative hypothesis (H_1) that lies within the acceptance region of H_0. It is helpful to remember that the number of males in a sample of n observations follows a binomial distribution with standard deviation $(npq)^{1/2}$, in which p and q are the expected frequencies of males and females respectively ($p = q = 0.5$). With $n = 100$ the power $(1 - \beta)$ of the test (H_0:sex ratio $= 0.5$, H_1:sex ratio $= 0.45$) is only 0.16. Even with $n = 400$ the probability of finding a significant deviation from a sex ratio of 0.5 if H_1 is true is only 0.5. Therefore, large sample sizes of several hundred individuals are needed when studying small deviations from a sex ratio of 0.5. Consequently, some authors may have concluded too readily that sex ratios estimated from small samples do not deviate from the 'expected' 0.5.

More data are clearly needed before we can firmly conclude that the majority of plant species conform to seed sex ratio equality (0.5). Currently available data are biased towards short-lived plants from temperate regions; much less is known about tropical trees. If further work shows biased seed sex ratios to be more common than currently thought, it would be of considerable interest to examine if the seed sex ratio is associated with the life-history of the plant and specifically with the intensity of sib-mating.

17.7 | New theory: genetic conflict and the sex ratio

Classic sex-allocation theory is compatible with genetics based on nuclear inheritance of autosomal genes and maternal control over the sex ratio of the offspring. Theory predicts that, with sib-mating, female-biased sex ratios in the seeds should be common in plants. If the mean seed sex ratio is nearly always close to 0.5 (which is still uncertain, sections 17.6 & 17.8), what can explain this discrepancy between predictions and observations? One possibility is that plants have not evolved adaptive control of sex allocation, and sex ratios of 0.5 simply reflect a constraint of sex-determination mechanisms, but this does not seem likely (section 17.3.3).

A different possibility is that genes that affect the sex ratio can be selected for or against, depending on where they are located on the genome. Sex ratio genes can be located on the autosomes, on the X chromosome, on the Y chromosome or in the cytoplasm. As soon as sex-determining genes evolve there will be conflict between sex ratio modifiers located on X and Y, each modifier being selected for if it promotes its own transmission at the cost of the other (e.g. Hamilton 1967, Werren & Beukeboom 1998, Werren & Hatcher 2000, Chapter 9). Conflict between nuclear genes and cytoplasmic genes is also expected. In most cosexual plant species, cytoplasmic genes are transferred through seeds and not through pollen (conifers are an exception). In dioecious plants, a cytoplasmic gene can be transferred from a mother to a son, but is then at a dead end. Consequently, selection favours cytoplasmic genes that increase sex ratio bias towards daughters. The conflict between cytoplasmic and nuclear genes in cosexual plants has received considerable attention in relation to gynodioecy, the occurrence of female and cosexual individuals within the same population (Samitou-Laprade *et al.* 1994), but genetic conflict is a relatively novel subject in relation to sex ratio in dioecious plants.

Seed sex ratios have been most studied in *Silene latifolia*, a species with sex chromosomes (references in Kay & Stevens 1986, DR Taylor

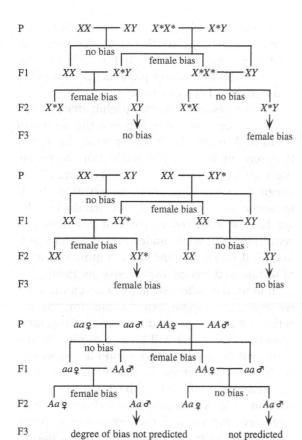

Fig 17.2 Reciprocal crossing scheme. DR Taylor (1994) used this to determine whether sex ratio modifiers are inherited through X chromosomes, Y chromosomes or autosomes. For all patterns of inheritance, it is assumed that the sex ratio bias is expressed in the male parent.

1994, 1996). Reciprocal crosses (as in Figure 17.2) can show whether offspring sex ratio variation is due mostly to the maternal genotype, the paternal genotype, or an interaction. DR Taylor's (1994) studies show that the paternal genotype has the largest effect, a result consistent with the older literature (Correns 1928). DR Taylor (1994) further devised a crossing scheme to examine whether the sex-ratio-modifying genes were located on the X chromosome, Y chromosome or on autosomes (Figure 17.2). For instance, if bias is due to a gene on the Y chromosome, then the offspring of a son should also be biased, while those of a daughter should not. DR Taylor (1994) demonstrated Y-linked modifiers and an interaction with the maternal genotype. He referred to the 'restorer hypothesis', which states that seed sex ratio evolution in *Silene* is an interactive

process with X-linked and cytoplasmatic genes pulling towards a female bias and Y-linked genes pulling towards a male bias. We can add that, with significant sib-mating, autosomal genes will also prefer a female bias. Y-linked genes will then be outnumbered, but given the usually small deviations from seed sex ratios of 0.5, even in *S. latifolia*, they are apparently quite successful in this genetic battle.

Correns (1928) reported that some plants produce progeny containing only sons or only daughters, which suggests a lack of constraints on *S. latifolia* sex ratio and also that conflict resolution is not just a question of restoring a female bias to 0.5. Table 17.2 also lists some genotypes with male-biased sex ratios. The 'restorer hypothesis' needs to be tested in other species, especially those previously considered uninteresting because they produced sex ratios of 0.5. It also seems important to document the seed sex ratios of motherplants (half- or full-sib families) and to document the scale over which driving and restoring genes are distributed through the population. The sex ratio resulting from opposing genetic forces may differ when measured over a spatial scale of several kilometres (Taylor 1999). High variation in intra- and interpopulation crosses (as in *Drosophila*) would support the 'restorer hypothesis'. Low variation would be more in line with stabilizing selection on a sex ratio of 0.5 and/or meiotic constraints.

17.8 | Population sex ratios of adult plants

Many articles on field sex ratios of flowering plants open with a statement such as, 'Despite the theoretical prediction that sex ratios should be 1:1 in natural populations, there have been many reports of biased sex ratio in populations'. By now it should be clear that theory predicts the sex ratio of seeds, not of adult plants, and the predictions are generally not affected by differences in mortality between the sexes.

Lewis (1942) claimed that sex ratios are often female biased and this view persists in the literature. His data refer mostly to the sex ratio of adult plants, not that of seeds. Moreover, his list of eight out of nine species with a female bias requires scrutiny. The citation of the German literature (Correns 1928) is incomplete. For instance, *Asparagus officinalis*, with a sex ratio of 0.53 (Correns 1928), is missing. While Correns (1928) reports a field sex ratio of 0.514 for *Mercurialis annua*, Lewis gives 0.467 (citing another study). In two species listed (*Spinacia oleracea* and *Cannabis sativa*) sex expression is not stable and is affected by the environment. Finally some sex ratios may not be representative of the whole species. For instance, *Humulus lupulus* is reported to be extremely female biased (0.098, Lewis 1942), but a brief survey of 39 plants in a natural population in the dunes of Meijendel (near The Hague, The Netherlands) estimated the sex ratio as 0.615 (TJ de Jong, unpublished results). Other recent studies do not support the claim of a general female bias either. Lloyd and Webb (1977, Webb & Lloyd 1980) found that long-lived plants typically have male-biased ratios. A review of data on 44 dioecious species for which information was available on the reproductive effort of male and female plants found that in 25/44 species the adult sex ratio was male biased, 13 species had unbiased ratios and six species had female bias (Delph 1999). Within the flora of the British Isles, Kay and Stevens (1986) found strongly biased sex ratios to be much rarer than sex ratios close to 0.5. If we treat for simplicity (Kay and Stevens do not give enough detail for each species separate) all *Salix* as one species and all *Populus* as one species, 11 species do not deviate significantly from 0.5, six are female biased and four are male biased. Rottenberg (1998) documents a sex ratio of 0.5 in 33/41 species examined in Israel: six species showed a consistent female-biased ratio, six showed a consistent male bias and no simple explanation could be provided for these differences.

Usually, in dioecious plants genetic variation between individuals is so large that, with some acquired skills, genotypes can be discriminated from each other. However, Rottenberg (1998) found this to be difficult in *Salix* species, which showed strong vegetative spread, with some stands consisting of a single individual. In small populations chance effects will then have a large effect on the local sex ratio (Proctor *et al.* 1996).

Both Correns (1928) and Rottenberg (1998) emphasized that populations need to be visited at least twice as, depending on the time of year, one sex will be more conspicuous than the other and may be missed in counts.

Differences in life-history (Delph 1999) or herbivory (Ågren *et al.* 1999) between the sexes can result in diverging habitats, especially in species that rely on vegetative reproduction. Some authors even referred to 'niche-partitioning' between the sexes (Cox 1981). An extreme case is butterbur (*Petasites hybridus*); in Britain females are largely restricted to parts of northern and central Britain, while males have a broader geographical range and are fairly common in most of Britain and Ireland (Kay & Stevens 1986).

If mortality differs strongly between the sexes we can derive some expectations for the sex ratio in different populations. Under favourable conditions mortality will be at a minimum and sexes will be present in the ratio as present in seeds, while under stress conditions the more vulnerable sex will suffer the higher mortality (section 17.5). As female plants frequently invest a greater proportion of their resources in reproduction and over a longer time period (Charnov 1982, Delph 1999), we expect the sex ratio to increase with environmental stress. Figure 17.3 illustrates that under sufficiently stressful conditions only males survive, leading to all-male populations, while under favourable conditions a sex ratio of 0.5 is expected. *Rumex acetosa* does not, however, conform to this expectation: females grow more vigorously and outcompete males if the population becomes older (Putwain & Harper 1972).

The measurement of stress is problematic. Perhaps it is useful to define stress as the mean size of the median male and female plant in a habitat. At least this measure is independent of the sex ratio. Verdu and Garcia-Fayos (1998) did not find a male-biased sex ratio in stressful habitats, in contrast to expectations under sexually differential mortality. Also Korpelainen (1991) found spatial variation in the sex ratio, but could not correlate this with any of the environmental factors measured. However, in some cases, the mechanism behind changes in the sex ratio has been

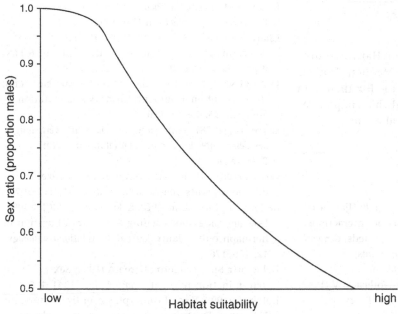

Fig 17.3 The influence of habitat suitability on adult sex ratio. The population sex ratio may differ from the sex ratio in seeds due to sexually differential mortality, especially in long-lived plants with vegetative reproduction. If the two sexes are differently affected, then under favourable conditions a sex ratio of 0.50 is expected, while under stressful conditions a male bias is predicted. A possible way of scaling the X-axis is taking the average size of the median male and female plant in a habitat.

elucidated by experimental work. We give two examples (see also Delph 1999 for a review).

First, Meagher (1981) carefully studied the demography of *Chamaelirium luteum* and found higher mortality rates of females after reproduction and that males tended to flower at an earlier age. The demographic data were combined into a life-history projection model. Starting with a seed sex ratio of 0.5, the model correctly predicts that after a large number of generations the adult sex ratio will increase to 0.63–0.77, depending on the demographic parameters of the specific population (Meagher 1981).

Second, Escarré and Houssard (1991) sowed seeds of *Rumex acetosella* in experimental plots that they then followed over a number of years. They counted only male and female shoots and did not discriminate between individual plants. Their 'ramet' sex ratio therefore has two components: the proportion of males (the relevant parameter in the evolutionary models discussed above) and the size (number of flowering stems) per individual (which reflects differences in growth between males and females). In the first year the ramet sex ratio was 0.5, but in following years the plots became dominated by males. Houssard *et al.* (1994) showed that females allocated more resources to reproduction than males and consequently had less to devote

to clonal growth. As the plots became more closed, making seedling establishment more difficult, the males took over. Interestingly, in the congener *R. acetosa* females grow more vigorously and therefore ramet sex ratios in older populations become female biased (Putwain & Harper 1972, Korpelainen 1991).

17.9 | Conclusions

We advocate that plant ecologists give greater attention to the seed sex ratio: it cannot be assumed that seed sex ratios are always unbiased as, with sib-mating, classic sex-allocation theory predicts female-biased sex ratios. The available data suggest that seed sex ratios are quite variable and not constrained to remain close to 0.5. This is an interesting field for further study, especially of the species that apparently defy theory and combine sib-mating with unbiased seed sex ratios. Even if mean seed sex ratios are unbiased, genetic variation for the ratio may exist and it seems worthwhile to document the spatial heterogeneity in the sex ratio. Further genetic studies are required to elucidate whether the dynamic sex ratio in *Silene latifolia*, with sex ratio modifiers on the Y chromosome, is generally representative of dioecious plant sex ratios. Although the study of dioecious plant sex ratios has a long and rich history, many questions regarding their evolution are still unanswered.

Acknowledgements

We thank Leo Beukeboom, Ian Hardy, Andrew Lack, Sarah Legge, Ed van der Meijden, Douglas Taylor and an anonymous referee for their comments, which greatly improved this chapter. We thank Evert Meelis for statistical advice.

References

Ågren J, Danell J, Elmquist T, Ericson L & Hjältén J (1999) Sexual dimorphism and biotic interactions. In: MA Geber, TE Dawson & LF Delph (eds) *Gender and Sexual Dimorphism in Flowering Plants*, pp 217–246. Berlin: Springer-Verlag.

Ainsworth C, Parker J & Buchanan-Wollaston V (1998) Sex determination in plants. *Current Topics in Developmental Biology*, 38, 167–223.

Alström-Rapaport C (1997) On the sex determination and evolution of mating systems in plants. *Acta Universitatis Agriculturae Suecia Agraria*, 66.

Alström-Rapaport C, Lascoux M & Gullberg U (1997) Sex determination and sex ratio in the dioecious shrub *Salix viminalis*. *Theoretical and Applied Genetics*, 94, 493–497.

Alström-Rapaport C, Lascoux M, Wang YC, Roberts G & Tuskan GA (1998) Identification of a RAPD marker linked to sex determination in the basket willow (*Salix viminalis* L.). *Journal of Heredity*, 89, 44–49.

Atlan A, Merçot H, Landre C & Montchamp-Moreau C (1997) The sex-ratio trait in *Drosophila simulans*: geographical distribution of distorters and restorers. *Evoluion*, 51, 1886–1895.

Bull JJ & Charnov EL (1988) How fundamental are Fisherian sex ratios? *Oxford Surveys in Evolutionary Biology*, 5, 96–135.

Carroll SB & Mulcahy DL (1993) Progeny sex ratios in dioecious *Silene latifolia* (Caryophyllaceae). *American Journal of Botany*, 80, 551–556.

Chailakhyan MK & Khrianin VN (1987) *Sexuality in Plants and its Hormonal Regulation*. Berlin: Springer-Verlag.

Charlesworth B & Charlesworth D (1978) A model for the evolution of dioecy and gynodioecy. *American Naturalist*, 112, 975–997.

Charlesworth D & Charlesworth B (1981) Allocation of resources to male and female functions in hermaphrodites. *Biological Journal of the Linnean Society*, 19, 57–74.

Charnov EL (1982) *The Theory of Sex Allocation*. Princeton, NJ: Princeton University Press.

Chattopadhyay D & Sharma AK (1991) Sex determination in dioecious species of plants. *Feddes Repertorium*, 102, 29–55.

Conn JS & Blum U (1981) Sex ratio of *Rumex hastatulus*: the effect of environmental factors and certation. *Evolution*, 35, 1108–1116.

Correns C (1928) *Bestimmung, Vererbung und Verteilung des Geschlechtes bei den höheren Pflanzen*. Berlin: Borntraeger.

Cox PA (1981) Niche partitioning between sexes of dioecious plants. *American Naturalist*, 117, 295–307.

de Jong TJ, Klinkhamer PGL & Rademaker MCJ (1999) How geitonogamous selfing affects sex allocation in hermaphrodite plants. *Journal of Evolutionary Biology*, 12, 166–176.

Dellaporta SL & Calderon-Urrea A (1993) Sex determination in plants. *Plant Cell*, 5, 1241–1251.

Delph LF (1999) Sexual dimorphism in live history. In: MA Geber, TE Dawson & LF Delph (eds) *Gender and Sexual Dimorphism in Flowering Plants*, pp 149–174. Berlin: Springer-Verlag.

Durand B & Durand R (1991) Sex determination and reproductive organ differentiation in *Mercurialis*. *Plant Science*, 80, 49–65.

Escarré J & Houssard C (1991) Changes in sex ratio in experimental populations of *Rumex acetosella*. *Journal of Ecology*, 79, 379–388.

Fiala KL (1979) On estimating the primary sex ratio from incomplete data. *American Naturalist*, 113, 442–445.

Freeman DC, Wachocki BA, Stender MJ, Goldschlag DE & Michaels HJ (1994) Seed size and sex ratio in spinach: application of the Trivers-Willard hypothesis to plants. *Ecoscience*, 1, 54–63.

Freeman DC, Lovett Doust J, El-Keblawy A, Miglia KJ & McArthur ED (1997) Sexual specialization and inbreeding avoidance in the evolution of dioecy. *Botanical Review*, 63, 65–92.

Grant S (1999) Genetics of gender dimorphism in higher plants. In: MA Geber, TE Dawson & LF Delph (eds) *Gender and Sexual Dimorphism in Flowering Plants*, pp 247–274. Berlin: Springer-Verlag.

Grant S, Houben A, Vyskot B, Siroky J, Pan W-H, Macas J & Saedler H (1994) Genetics of sex determination in flowering plants. *Developmental Genetics*, 15, 214–230.

Hamilton WD (1967) Extraordinary sex ratios. *Science*, 156, 477–488.

Hardy ICW (1994) Sex ratio and mating structure in the parasitoid Hymenoptera. *Oikos*, 69, 3–20.

Houssard C, Thompson JD & Escarré J (1994) Do sex-related differences in response to environmental variation influence the sex-ratio in the dioecious *Rumex acetosella*? *Oikos*, **70**, 80–90.

Ingvarsson PK & Giles BE (1999) Kin-structured colonization and small-scale genetic differentiation in *Silene dioica*. *Evolution*, **53**, 605–611.

Irish EE & Nelson T (1989) Sex determination in monoecious and dioecious plants. *Plant Cell*, **1**, 737–744.

Kay QON & Stevens DP (1986) The frequency, distribution and reproductive biology of dioecious species in the native flora of Britain and Ireland. *Botanical Journal of the Linnean Society*, **92**, 39–64.

Korpelainen H (1991) Sex ratio variation and spatial segregation of the sexes in populations of *Rumex acetosa* and *R. acetosella* (Polygonaceae). *Plant Systematics and Evolution*, **174**, 183–195.

Lewis D (1942) On the evolution of sex in flowering plants. *Biological Review*, **17**, 46–76.

Lloyd DG & Webb CJ (1977) Secondary sex characters in plants. *Botanical Review*, **43**, 177–216.

Maynard Smith J (1978) *The Evolution of Sex*. Cambridge: Cambridge University Press.

Meagher TR (1981) Population biology of *Chamaelirium luteum*, a dioecious lily. II. Mechanism governing sex ratios. *Evolution*, **35**, 557–567.

Merçot H, Atlan A, Jacques M & Montchamp-Moreau C (1995) Sex-ratio distortion in *Drosophila simulans*: co-occurrence of drive and suppressors of drive. *Journal of Evolutionary Biology*, **8**, 283–300.

Mutikainen P, Walls M & Ojala A (1994) Sexual differences in responses to simulated herbivory in *Urtica dioica*. *Oikos*, **69**, 397–404.

Nagelkerke CJ & Hardy ICW (1994) The influence of developmental mortality on optimal sex allocation under local mate competition. *Behavorial Ecology*, **5**, 401–411.

Parasnis AS, Ramakrishna W, Chowdari KV, Gupta VS & Ranjekar PK (1999) Microsatellite (GATA)(n) reveals sex-specific differences in Papaya. *Theoretical and Applied Genetics*, **99**, 1047–1052.

Parker GA (1984) Evolutionarily stable strategies. In: JR Krebs & ND Davies (eds) *Behavioural Ecology*, pp 30–61. Cambridge: Cambridge University Press.

Prentice HC (1984) The sex ratio in a dioecious endemic plant *Silene diclinis*. *Genetica*, **64**, 129–133.

Proctor M, Yeo P & Lack A (1996) *The Natural History of Pollination*. London: HarperCollins.

Purrington CB (1993) Parental effects on progeny sex ratio, emergence, and flowering in *Silene latifolia* (Caryophyllaceae). *Journal of Ecology*, **81**, 807–811.

Putwain PD & Harper JL (1972) Studies in the dynamics of plant populations. V. Mechanisms governing the sex ratio in *Rumex acetosa* and *R. acetosella*. *Journal of Ecology*, **60**, 113–130.

Raemon Buttner SM, Schondelmaier J & Jung C (1998) AFLP markers tightly linked to the sex locus in *Asparagus officinalis* L. *Molecular Breeding*, **4**, 91–98.

Renner SS & Ricklefs RE (1996) Dioecy and its correlates in the flowering plants. *American Journal of Botany*, **82**, 596–606.

Richards AJ (1997) *Plant Breeding Systems*. London: Chapman & Hall.

Rottenberg A (1998) Sex ratio and gender stability in the dioecious plants of Israel. *Botanical Journal of the Linnean Society*, **128**, 137–148.

Rychlewski J & Zarzycki K (1975) Sex ratio in seeds of *Rumex acetosa* L. as a result of sparse and abundant pollination. *Acta Biologica Cracoviensia, Series Botanica*, **18**, 101–114.

Samitou-Laprade P, Cuguen J & Vernet P (1994) Cytoplasmic male sterility in plants: molecular evidence and the nucleocytoplasnic conflict. *Trends in Ecology and Evolution*, **9**, 431–435.

Smith BW (1963) The mechanism of sex determination in *Rumex hastatulus*. *Genetics*, **48**, 1265–1288.

Sokal RR & Rohlf FJ (1995) *Biometry*, 3rd edn. New York: Freeman.

Soldaat LL, Vetter B & Klotz S (1997) Sex ratio in populations of *Silene otites* in relation to vegetation cover, population size and fungal infection. *Journal of Vegetation Science*, **8**, 697–702.

Taylor DR (1990) Evolutionary consequences of cytoplasmatic sex ratio distorters. *Evolutionary Ecology*, **4**, 235–248.

Taylor DR (1994) The genetic basis of sex ratio in *Silene alba* (= *S. latifolia*). *Genetics*, **136**, 641–651.

Taylor DR (1996) Parental expenditure and offspring sex ratios in the dioecious plant *Silene alba* (= *S. latifolia*). *American Naturalist*, **147**, 870–879.

Taylor DR (1999) Genetics of sex ratio variation among natural populations of a dioecious plant. *Evolution*, **53**, 55–62.

Taylor DR, Saur MJ & Adams E (1999) Pollen performance and sex-ratio evolution in a dioecious plant. *Evolution*, **53**, 1028–1036.

Taylor PD (1994) Sex ratio in a stepping stone population with sex-specific dispersal. *Theoretical Population Biology*, **45**, 203–218.

Testolin R, Cipriani G & Costa G (1995) Sex segregation ratio and gender expression in the genus Actinidia. *Sexual Plant Reproduction*, **8**, 129–132.

Trivers RL & Hare H (1976) Haplodiploidy and the evolution of the social insects. *Science*, **191**, 249–263.

Uyenoyama MK & Bengtsson BO (1982) Towards a genetic theory for the evolution of the sex ratio. III. Parental and sibling contol of brood investment ratio under partial sib-mating. *Theoretical Population Biology*, **22**, 43–68.

Verdu M & Garcia-Fayos P (1998) Female biased sex ratios in *Pistacia lentiscus* L. (Anacardiaceae). *Plant Ecology*, **135**, 95–101.

Webb CJ & Lloyd DG (1980) Sex ratios in New Zealand apioid Umbelliferae. *New Zealand Journal of Botany*, **18**, 121–126.

Werren JH & Beukeboom LW (1998) Sex determination, sex ratios, and genetic conflict. *Annual Review of Ecology and Systematics*, **29**, 233–261.

Werren JH & Hatcher MJ (2000) Maternal-zygotic gene conflict over sex determination: effects of inbreeding. *Genetics*, **155**, 1469–1486.

Westergaard M (1958) The mechanism of sex determination in flowering plants. *Advances in Genetics*, **9**, 217–281.

Williams GC (1971) *Group Selection*. Chicago: Aldine.

Zhang YH, Di Stillo VS, Rehman F, Avery A, Mulcahy D & Kesseli R (1998) Y chromosome specific markers and the evolution of dioecy in the genus *Silene*. *Genome*, **41**, 141–147.

Part 6

Applications of sex ratios

Chapter 18

Operational sex ratios and mating competition

Charlotta Kvarnemo & Ingrid Ahnesjö

18.1 | Summary

This chapter deals with the operational sex ratio (OSR) and its importance for understanding mating competition, which is a key component of sexual selection. We focus on OSR as an empirical measurement with important applications in sexual selection, but we also pay considerable attention to the question of how to estimate sexual differences in potential reproductive rates (PRR). The sexual difference in PRR and the adult (or qualified) sex ratio are the most important factors influencing the OSR, and thus the pattern of sexual selection in a population. We illustrate our points using examples from a wide range of taxa. In particular, we investigate how environmental factors, through their effects on PRR and OSR, often add a dynamic to mating competition, which sometimes results in different sex roles and varying intensities being displayed in different populations or at different times within a breeding season. Finally, we consider some examples of contrasting patterns, how OSR relates to mate choice and prospects for further research.

18.2 | Sexual selection

Males of many animals have evolved conspicuous traits that seem to reduce their survival. Darwin (1871) proposed the theory of sexual selection to explain the evolution of such traits. Sexual selection arises through competition over mates or matings, and assumes that individuals with a certain trait, whether a red tail or a specific behaviour, will have an advantage when competing for matings. Such competition between same-sex individuals includes both intrasexual selection (through aggression, dominance and displays among members of the same sex) and intersexual selection (through competition by attractiveness to be chosen by members of the other sex) (Andersson 1994, Andersson & Iwasa 1996). What is important in relation to mating dynamics is not simply the overall adult sex ratio in a breeding population, but the ratio of males to females that are ready to mate, which is termed the operational sex ratio (OSR, Box 18.1) (Emlen 1976, Emlen & Oring 1977). If there are many sexually active males that are prepared to mate most of the time, whereas females are ready to mate only for brief periods, it is clear that males will have to compete intensely among themselves for mating opportunities. Whatever the mechanisms of mating competition, the OSR will greatly influence its intensity and hence also the intensity of sexual selection (Emlen & Oring 1977). The OSR is thus a very powerful tool to assess this intensity. Furthermore, one of the more important factors influencing OSR is the sexual difference in the potential reproductive rate (PRR, Box 18.2), along with the adult or qualified sex ratio (Figure 18.1; Ahnesjö et al. 2001).

Box 18.1 | The operational sex ratio (OSR)

OSR is the ratio of males to females ready to mate in a population at a given moment (Emlen 1976, Emlen & Oring 1977), and can be estimated as a snap-shot count. An alternative approach was provided by Clutton-Brock and Parker (1992), and further developed by Parker and Simmons (1996). They divided the reproductive cycle (T) primarily between being ready to mate ('time in', S_m for males and S_f for females) and not being ready to mate ('time out', G_m and G_f; see Box 18.2) such that $T = G + S$. OSR is then calculated as S_m/S_f. Note that S_m and S_f are the sums of 'time in' for each sex in a representative sample of the population over one reproductive cycle. The sex with the longer summed 'time in' is expected to have more time available for mating and thus to be the predominant competitor for mates.

A male/female ratio of OSR may range between 0 and infinity, being 1 when unbiased. Similar biases in either direction give very different deviations from equality, e.g. 4 males/2 females = 2, whereas 2 males/4 females = 0.5. To avoid such skewed distributions when analysing OSR data, values of OSR (x) are preferably transformed by $x/x+1$ into a relative ratio (males/(males+females)) of the number of the individuals being ready to mate, or 'time in'. This relative OSR will then range from 0 when only females are ready to mate, through 0.5 when unbiased, to 1 when only males are ready to mate (as used in Enders 1993, Vincent et al. 1994, Kvarnemo 1996, Jirotkul 1999).

Box 18.2 | Potential reproductive rate (PRR)

For each sex PRR is the population's *mean* value of the individuals' maximum reproductive rates (Clutton-Brock & Parker 1992, Kvarnemo & Ahnesjö 1996, Parker & Simmons 1996), measured when mating partners are freely available, while other constraints typical for the population and the time of the breeding season (food and nest site availability, body size distribution, etc.) remain (Kvarnemo & Ahnesjö 1996). Thus, the sexual difference in PRR has to be estimated experimentally for representative samples of the population as the mean number of offspring produced per unit time and sex.

Parker and Simmons (1996) mathematically expressed PRR as $1/G_m$ for males and $1/G_f$ for females, where G is the mean value of 'time out' for each sex (i.e. the time spent during a reproductive cycle unavailable for mate acquisition as a result of reproduction; for instance, parental care, gamete or resource replenishment). Though, according to Parker and Simmons (1996) two factors of further importance are the adult sex ratio M (males/females) and the 'collateral investment'. A 'collateral investment' occurs when a reproductive event of an individual of one sex involves more than one member of the opposite sex, such that 'f females typically spread their clutches between m successful males'. If a reproductive cycle is defined as $T = fG_f + S_f$, then males are expected to compete if $T(M - 1) > mG_m - fG_f$. OSR (S_m/S_f) is then calculated from $S_m = MT - mG_m$ and $S_f = T - fG_f$. Taking resource competition into account, we suggest that M should be replaced by a 'qualified sex ratio' (Q; Ahnesjö et al. 2001; Figure 18.1).

It is very important to examine the roles of the sexes in animals with a wide variety of mating patterns in order to gain a better understanding of the dynamics of sexual selection and the multitude of reproductive behaviours among animals. In this chapter we illustrate our points with a wide range of examples, yet with a bias in favour of two species of fish that are our own study animals: the sand goby (*Pomatoschistus minutus*, Gobiidae) and the broad-nosed pipefish (*Syngnathus typhle*, Syngnathidae). In most animals, males compete for matings, while females are selective in their choice of mates, as in the sand goby (Kvarnemo *et al.* 1995, Kvarnemo & Forsgren 2000). Some animals, however, are sex role reversed, such as the broad-nosed pipefish, where females compete for matings (Berglund 1991, Vincent *et al.* 1994, Berglund & Rosenqvist 2001a) and males are more choosy (Berglund & Rosenqvist 1993, 2001b). However, sex roles often vary over time within populations or between, such that the sexes behave differently depending on the circumstances. In such cases, the concepts of OSR (Box 18.1) and PRR (Box 18.2) provide us with important tools to explain and understand the variation of reproductive behaviour and mating patterns, and how the operation of sexual selection varies in time and space.

18.3 | The operational sex ratio (OSR)

18.3.1 Background
A landmark paper by Emlen and Oring (1977) initiated the use of the term OSR and provided a general framework for understanding animal breeding systems. Emlen (1976) had made nocturnal behavioural observations on a marked population of bullfrogs (*Rana catesbeiana*) in order to understand the process of their mate selection. Bullfrogs have a long mating season, in which males arrive at breeding ponds before the females. Males congregate in choruses and remain sexually active for much of the season. In contrast, individual females remain sexually active for just one night. Consequently, 'this produces a strong male bias in the *operational sex ratio* (defined as the ratio of potentially receptive males

to receptive females at any time). The result is intense sexual selection and strong male-male competition for the few available females.' (Emlen 1976).

18.3.2 OSR as a tool
The OSR is a central concept in explaining variation in sex roles and intensity in mating competition (Clutton-Brock & Parker 1992, Andersson 1994, Andersson & Iwasa 1996, Reynolds 1996). In recent years, it has become prominent in empirical studies (reviewed in Kvarnemo & Ahnesjö 1996) and received considerable theoretical attention (Clutton-Brock & Vincent 1991, Clutton-Brock & Parker 1992, Parker & Simmons 1996, Kokko & Monaghan 2001). A bias in OSR can predict which sex competes for access to mates, and the intensity of the competition. The more biased the OSR, the more intense the competition, and the stronger the sexual selection on the sex in excess, which will be the predominant competitor for access to mating partners (Emlen & Oring 1977, Kvarnemo & Ahnesjö 1996). OSR has proved successful in predicting both the relative competitiveness of the two sexes and variations in the intensity of mating competition within a sex role, whether conventional or reversed (Clutton-Brock & Parker 1992, Kvarnemo & Ahnesjö 1996).

Among the many processes resulting in sexual selection (Andersson & Iwasa 1996), it is important to realize that it is only mating competition that is predicted by OSR. Sexual selection may, for instance, operate prior to the actual mating competition, when individuals compete for the resources necessary to become ready to mate (Figure 18.1). Such resource competition, e.g. for nest sites, is not predicted by the OSR. Yet, any limitation in resources will often profoundly influence the OSR and the pattern of sexual selection through its influence on the ratio of males to females that are 'qualified to mate' (Ahnesjö *et al.* 2001). For instance, the number of nest-holding individuals may be reduced considerably by a scarcity of nest sites and thus bias the OSR, as in the blenny *Salaria pavo* (section 18.5, Almada *et al.* 1995, Oliveira *et al.* 1999). Similarly, in a sand goby population with a pronounced nest-site shortage, male mating success

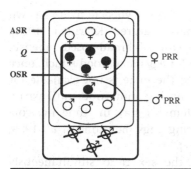

Fig 18.1 Schematic representation. Imagine a nest-site-limited population of fish with paternal care. About half of the males get a nest site, and, since each male cares for one female clutch at a time, some females will not find a nest-holding male to spawn with. All adult males and females are part of the adult sex ratio (ASR or M in Parker & Simmons 1996, Box 18.2), but only the ones that are ready to mate (filled symbols) make up the operational sex ratio (OSR). Open symbols represent individuals that are not ready to mate (having 'time out'): recently spawned females are unavailable as mates until new eggs have matured; males with filled nests are unavailable until hatching; and males without nest sites (crossed open symbols) are not ready to mate at all. When predicting OSR based on potential reproductive rates (PRR), we therefore suggest an important distinction, namely that only the individuals of each sex that can have 'time in' or 'time out' are considered to be qualified. Thus, individuals lacking a resource necessary for reproduction, such as a nest site, do not qualify as mates and are therefore not part of what we call the 'qualified sex ratio' (Q) (Ahnesjö et al. 2001). In this scenario, there is female–female competition for nest-holding males. This can only be successfully predicted by OSR if we redefine M to equal Q, whereas when using the original definition of M as ASR such competition remains unpredicted.

is primarily determined by intrasexual competition over nest sites and only marginally by female choice. However, in a population with nest sites in excess, it was found that any male could acquire a nest site and, consequently, mating success to a larger extent was determined by female mate choice (Forsgren *et al.* 1996). Moreover, in an experimental field study of the common goby (*Pomatoschistus microps*) in two nearby bays, one bay had been provided with a large number of additional nest sites. Common gobies are normally considered to have conventional sex roles, but in this study females interacted competitively with each other more often in the area with a natural shortage of nest sites than in the manipulated area. Furthermore,

when nest sites were scarce, females courted the males more often than males courted females, while the reverse was true in the manipulated area (Åsa Borg, Elisabet Forsgren and Carin Magnhagen, unpublished data). In these examples, the sexual difference in PRR is unaltered whereas the change in mating competition is caused by nest-site availability, influencing the sex ratio among individuals 'qualified to mate' (Figure 18.1).

Moreover, in the case of multiple mating, sexual selection may occur after mating through sperm competition (e.g. Birkhead & Møller 1998), which is not addressed by OSR either. A strongly male-biased OSR, however, often coincides with an enhanced risk of sperm competition (e.g. Jirotkul 1999, but see section 18.5.3). Finally, although the bias in OSR relates to the intensity in mating competition, the costs of competitive behaviour may decrease the intensity at strongly biased OSRs, as suggested to occur in the European lobster (Debuse *et al.* 1999).

Despite these limitations, the following examples illustrate the usefulness of the concept of OSR, where the degree of bias correlates with the intensity of mating competition, as in the bullfrog described above (Emlen 1976). In the common spider mite (*Tetranychus urticae*), male–male competition in the form of antagonistic male interactions increases under increasingly male-biased OSR, and larger male size is favoured accordingly (Enders 1993). Numerous studies of orthopterans show that the direction of mating competition relates to the bias in OSR (Gwynne 1983, 1990, Gwynne & Simmons 1990, Simmons & Bailey 1990). In a field study of the sex role reversed broad-nosed pipefish, the proportions of interactions (indicating competition) within each sex were found to relate to the degree of OSR bias (Vincent *et al.* 1994). In the Majorcan midwife toad (*Alytes muletensis*), which is also sex role reversed, female–female competition was found to predominate under the whole range of OSRs, but more so under a female bias. In contrast, male–male competition also occurred, but only when there was a male-biased OSR (Bush & Bell 1997). Similarly, in the sand goby, not only do males compete more frequently under male bias than female bias, but females increase their competitive interactions under female bias, even though males are the keener competitors (Kvarnemo et al. 1995). Importantly, this illustrates that although the bias in OSR predicts which sex is the principal competitor, both sexes may simultaneously compete for mates.

18.3.3 Measuring OSR

In all measurements of OSR, it is important to define carefully the specific time span and population in focus (Box 18.1). In some animals, the major changes in OSR occur at an annual level. For example, in the adder (Vipera berus), OSR is determined mainly by the adult sex ratio and competition between males is stronger in years with more male-biased sex ratios (Madsen & Shine 1993). In other animals, the adult sex ratio is of minor importance, while momentary differences in the distribution of the sexes are central, as in lekking ruff (Philomachus pugnax, Höglund et al. 1993).

Important mating competition often occurs shortly before the actual reproduction has begun, as individuals are preparing to mate. The OSR can then be estimated in snap-shot censuses, counting the number of males and females that are ready to mate, or by comparing their 'times in' (Box 18.1). However, once reproduction has started, counting the numbers of males and females ready to mate will not reveal whether individuals have the capacity to mate several times during a reproductive cycle, whereas a time-based calculation of OSR will. Consequently, snap-shot OSRs would have to be performed repeatedly, and therefore time-based estimates are generally preferable.

It may be difficult to assess whether or not an individual is prepared to mate. If the non-mating status (e.g. brooding individuals) is easier to determine, an alternative approach is suggested in the model by Clutton-Brock and Parker (1992), namely to calculate the adult sex ratio and then exclude all individuals that are not ready to mate (e.g. Vincent et al. 1994), or to calculate the average time fraction of the reproductive cycle when each sex is not ready to mate (Boxes 18.1 & 18.2, section 18.4). In the sand goby, for example, a male guarding a nest full of eggs is not ready to mate until the eggs have hatched, and

neither can a recently spawned female respawn until new eggs have matured (Kvarnemo 1994). Similarly, in the case of broad-nosed pipefish, where males care for eggs in a brood pouch, once the pouch is filled the male is not available for further matings, having 'time out' until parturition, whereas females stay in 'time in', continuously producing eggs (Berglund et al. 1989, Vincent et al. 1994).

A change in the sex ratio simultaneously changes the within-sex density (i.e. the density of males, irrespective of the number of females present, and vice versa), and sometimes it is this gender density rather than the OSR that influences mating competition. However, male fighting in the natterjack toad (Bufo calamita) is influenced by the OSR, but neither male nor female density per se has any significant effect (Tejedo 1988). Similarly, in the sand goby, OSR has a clear effect on intrasexual interactions among males and females, whereas density has no influence (Kvarnemo et al. 1995). Yet, in other cases, density is also an important factor (e.g. Alonso-Pimentel & Papaj 1996, Otronen 1996). Thus, it should be remembered that density may act as a confounding variable.

18.4 | Potential reproductive rate (PRR)

18.4.1 PRR as a tool

The OSR is strongly influenced by any sexual difference in potential reproductive rate, PRR (Box 18.2), such that more individuals of the sex with the potential to reproduce at a higher rate will be ready to mate at any one time (Clutton-Brock & Parker 1992, Parker & Simmons 1996). This has been demonstrated, for instance, in giant water bugs (Kraus 1989, Kruse 1990), bushcrickets (Gwynne 1990, Simmons 1995), lobsters (Debuse et al. 1999), pipefishes (Berglund et al. 1989, Vincent et al. 1994, Ahnesjö 1995), seahorses (Masonjones & Lewis 2000), gobies (Kvarnemo 1994, 1996, Swensson 1997) and frogs (Pröhl & Hödl 1999).

The concept of PRR had been used in earlier studies (Berglund et al. 1989, Kraus 1989, Gwynne 1990), but the phrase was not coined until 1991, when Clutton-Brock and Vincent introduced it as 'the maximum number of independent offspring that parents can produce per unit time'. Later, PRR was mathematically expressed as the inverse of 'time out' for males and females, respectively (Clutton-Brock & Parker 1992) (Box 18.2). Often an empirically more practical solution is to count the number of offspring produced per reproductive cycle when given free access to mates, which directly includes the effects of any 'collateral investment' (Box 18.2, Ahnesjö et al. 2001). We wish to point out that the definition of PRR has developed over the years (Clutton-Brock & Vincent 1991, Clutton-Brock & Parker 1992, Kvarnemo & Ahnesjö 1996, Parker & Simmons 1996, Box 18.2). In particular Clutton-Brock and Vincent's (1991) definition aimed to identify which sex may be the keenest competitor for access to mates among species providing paternal care. To calculate the PRR of a sex they therefore searched the literature for the greatest value ever recorded of one individual's reproductive rate. Thus, their estimate depended solely on the upper value of a range and ignored the rest of the distribution; therefore, it is likely to be misleading (Kvarnemo & Ahnesjö 1996, Parker & Simmons 1996) when used to predict and understand the dynamics of mating competition in a population. Also, the reason for the general bias towards species with paternal care in empirical studies of PRR and OSR is mainly 'historical', as Clutton-Brock and Vincent (1991) focused on that group. It is important to emphasize that the concepts of OSR and PRR are highly useful and relevant to any kind of breeding system for investigating mating competition (Andersson 1994, Reynolds 1996).

The difference in PRR between males and females is an important source of variation in OSR (Berglund et al. 1989, Clutton-Brock & Vincent 1991, Clutton-Brock & Parker 1992). In fact, if the adult (or qualified) sex ratio is unbiased, then the sexual difference in PRR will be the main determinant of OSR and of the sex that is the principal competitor for mates (Clutton-Brock & Parker 1992, Parker & Simmons 1996). In sexually reproducing animals, each offspring has one mother and one father. Therefore, when the adult sex ratio is unbiased, the realized reproductive rates of

males and females will on average be equal. In contrast, the PRR may differ between the sexes (Clutton-Brock & Vincent 1991). This difference will only be apparent when individuals are experimentally provided with an unlimited access to mates.

18.4.2 Measuring PRR

'Time out' is that part of the reproductive cycle devoted to reproduction, such as parental care, egg production or ejaculate replenishment. PRR is the number of offspring that potentially could be produced during this 'time out'. Alternatively, if the unit of one female clutch (or male brood) is used, PRR equals 1/'time out' (Clutton-Brock & Parker 1992). When estimating the PRR for males and females in a population, the calculations should always be based on all the adult individuals of each sex that are 'qualified to mate' (section 18.7, Figure 18.1, Ahnesjö et al. 2001), or representative samples thereof. PRR can only be measured by manipulating the access of mating partners, which is particularly important for the limited sex. For each sex, PRR is *the population mean value of the rate of reproduction each individual achieves when not constrained by mate availability, while other natural limitations remain in operation*. Such natural limitations are characteristic for each population and may include factors such as temperature, food, sizes of nest sites, as well as body size and age distributions of the individuals within the population. Any changes in these factors often influence the PRR of males and females differently, with a consequent influence on OSR and sexual selection (section 18.5.2).

When one sex is able to simultaneously raise multiple clutches produced by the other sex, this will not be reflected in measures of PRR as 1/'time out', unless multiplied by the number of offspring or clutches that are brooded simultaneously (Berglund et al. 1989, Kraus 1989, Kvarnemo 1994) or dealt with as a separate term (collateral investment, Parker & Simmons 1996, Box 18.2). For example, in a pollen-feeding Australian bush-cricket (*Kawanaphila nartee*), the male transfers a protein-rich spermatophore to the female as a nuptial gift when mating. If food is limited, females often mate with multiple males to gain extra nutrients, before depositing a clutch of eggs.

Any two males mating with the same female will end up spending a full 'time out' for the same clutch, essentially halving their PRR. Recent theoretical elaborations have included situations like these, where a single reproductive event involves a 'collateral investment', i.e. the 'times out' of more than one individual of the other sex (Parker & Simmons 1996, Simmons & Parker 1996, Box 18.2). Therefore, the procedure of giving free access to mates allows the determination of how long 'time out' lasts and if individuals of one sex have the capacity to care for multiple clutches or need multiple mates to care for one clutch during 'time out'. One can, however, expect that the potential and realized reproductive rates in the slower, limiting, sex will coincide.

For each particular animal species, one has to find the relevant unit to measure PRR. For some animals, it is preferable to use the time spent 'out' after a reproductive event, particularly when both sexes reproduce in discrete clutches with distinguishable time spans of 'in' and 'out'. For example, female sand gobies always spawn their complete clutch in one male's nest, and the males can care for multiple clutches (two or more female clutches) at one time in a brood (Kvarnemo 1994; Jones et al. 2001). Here the unit is a female clutch and the 'collateral investments' are $f = 2$ and $m = 1$ (Box 18.2). For other animals, however, where both sexes remain 'in' after one or more matings, the number of offspring rather than the number of clutches or matings may determine when an individual enters 'time out'. Here a PRR based on 'the number of offspring produced per unit time' is preferable. In the case of broad-nosed pipefish, in which females continuously produce eggs, a female clutch is difficult to define. A female transfers some of her eggs in each mating, often only partially filling the male's brood pouch, which in turn will rapidly mate with additional females until his pouch is filled (on average three females per brood, Jones et al. 1999). Therefore, when both sexes copulate with several mates, the rate of offspring production per male brooding period ('time out') is a straightforward estimate of PRR (Berglund et al. 1989, Berglund & Rosenqvist 1990, Ahnesjö 1995), combining the effects of one

clutch divided by 'time out' $(1/G)$ and the 'collateral investment'.

18.5 | The dynamics of OSR and PRR

Although typical sex roles can be attributed to most animal species, recent studies have produced numerous examples of species showing predominant male–male competition for mates in some circumstances and predominant female–female competition (i.e. the definition of sex role reversal, Vincent et al. 1992) in others. This may result in shifts in sex roles within populations over time or different sex roles in different populations. For example, a field study of the bushcricket K. nartee showed that, depending on the type of flowering plants and hence the abundance of pollen and nectar, OSR differed between two nearby sites, and within another site it shifted from being female to male biased over just a few days (Gwynne et al. 1998). Another example of differing sex roles within a species comes from a blenniid fish (Salaria pavo), in which males provide parental care in nests and, in most populations, court females. However, in a population with a severe nest site shortage, the OSR was biased towards females because only a small proportion of the males were able to acquire nests, but all females produced eggs (Almada et al. 1995, Oliveira et al. 1999, cf. Figure 18.1). Although males competed for nest sites, females competed for nest-holding males, and thus females were the sex predominantly competitive for mates. Hence, varying environmental factors, including the monopolizable resources required for being ready to mate, may change the OSR in a dynamic fashion, leading to shifts in sex roles, and consequent changes in intensities in mating competition and sexual selection.

Changes in OSR may gradually alter the intensity of competition for mates, without shifting the sex roles. Such variation in OSR may occur within populations, usually over the course of a breeding season. In a population of broad-nosed pipefish, OSR was found to vary over the season, gradually becoming more and more female biased as males became pregnant and so

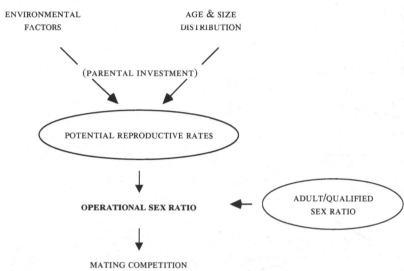

ENVIRONMENTAL FACTORS

AGE & SIZE DISTRIBUTION

(PARENTAL INVESTMENT)

POTENTIAL REPRODUCTIVE RATES

OPERATIONAL SEX RATIO

ADULT/QUALIFIED SEX RATIO

MATING COMPETITION

Fig 18.2 A summary of the principal factors influencing operational sex ratio. OSR may be influenced through the adult/ qualified sex ratio or through sex differences in the potential reproductive rates. This figure should be seen as an illustration of how we view these direct and indirect influences on OSR. (For comparable flow charts see, e.g., Clutton-Brock & Parker 1992, Andersson 1994, Reynolds 1996.)

unavailable for matings. As predicted, the proportion of female–female encounters, indicating competition among females, increased with the degree of female bias in OSR (Vincent *et al.* 1994). In the sex role reversed shorebird Wilson's phalarope (*Phalaropus tricolor*), the intensity of female–female competition weakened as OSR became less female biased with the arrival of more males at the breeding grounds (Colwell & Oring 1988). Finally, in the smooth newt (*Triturus vulgaris*) males competed for access to females during the greater part of the breeding season. However, early in the season a physiological constraint in spermatophore production caused sperm to be a limiting resource, resulting in female competition for mates (Waights 1996).

18.5.1 Factors influencing OSR

We have summarized much of what is covered in this section in Figure 18.2. Together with sex differences in PRR, the adult sex ratio is considered a major determinant of the OSR (Berglund *et al.* 1989, Kraus 1989, Clutton-Brock & Parker 1992, Madsen & Shine 1993, Parker & Simmons 1996). The adult sex ratio is obviously influenced by the primary sex ratio, and also by sex differences in time of emergence, age at maturation, reproductive life span, mortality rate and migration pattern (e.g. Kynard 1978, Reynolds *et al.* 1986, Björklund 1991, Acharya 1995, Maxwell 1998). In this context, a cactophilic fruitfly (*Drosophila pachea*) provides an example of how a

sexual difference in age at maturity can affect OSR. Males, which produce giant sperm, need four times as many days as females to reach sexual maturity, and consequently OSR is usually female biased (Pitnick 1993). Also, migration schedules in several birds differ between the sexes, creating seasonal changes in OSR, as in phalaropes (Reynolds *et al.* 1986, Colwell & Oring 1988). Finally, a female-biased OSR in the butterfly *Euphydryas editha* is brought about by males running a larger risk than females of being killed in spider webs since males sometimes mistake dead butterflies in webs for newly hatched virgin females (Moore 1987).

18.5.2 Factors influencing PRR

Factors that influence the sexual difference in PRR will consequently influence a population's OSR (Berglund *et al.* 1989, Gwynne 1990, Clutton-Brock & Parker 1992, Kvarnemo 1996, Parker & Simmons 1996; Boxes 18.1 & 18.2). Such factors may include food and temperature, parental investment, as well as the age and body size distribution of the population (Figure 18.2). Many of the environmental factors will affect PRR and thus OSR dynamically, as illustrated in the following examples.

18.5.2.1 Environmental factors

Temperature variations can affect the intensity of sexual selection in many ectotherms. In some species of giant water bugs, water temperature has been shown to have a considerable effect on the sexual difference in PRR and OSR, because

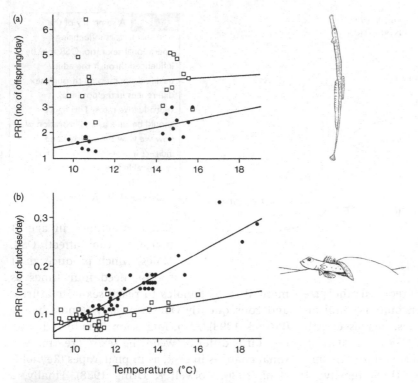

Fig 18.3 Variation in ambient temperature may considerably affect the sexual difference in potential reproductive rates (PRR) and the operational sex ratio (OSR). As found (a) in the sex role reversed broad-nosed pipefish, *Syngnathus typhle* (Ahnesjö 1995), and (b) in the sand goby, *Pomatoschistus minutus*, with conventional sex roles (Kvarnemo 1994), both species show paternal care. Characteristically, an increased temperature increases the PRR of males more than that of females. OSR becomes biased towards the sex with the higher PRR, which is predicted to compete more intensely for matings than the other sex, and with increasing intensity the greater the discrepancy in PRR. Accordingly, female broad-nosed pipefish compete more intensely in colder water (Vincent *et al.* 1994), as do male sand gobies in warmer water (Kvarnemo 1996). Filled circles, males; open boxes, females. Reprinted from Kvarnemo & Ahnesjö (1996) with permission from Elsevier Science.

temperature influences male brooding time, egg-synthesizing time and clutch size in females (Kraus 1989, Ichikawa 1993). Other examples are the facultatively sex role reversed Majorcan midwife toad (*Alytes muletensis*, Bush 1993), the sex role reversed broad-nosed pipefish (Ahnesjö 1995) and the sand goby with conventional sex roles (Kvarnemo 1994). In these species an increased

ambient temperature affects the sexual difference in PRR by increasing the reproductive rates of males more than that of females (Figure 18.3). As temperature changes, the sexual difference in PRR and consequently OSR and intensities in mating competition will change over the breeding season as well as between seasons in these species.

Food availability is another environmental factor that can affect sexual selection. As mentioned above, bushcricket (*K. nartee*) males produce an energetically expensive spermatophore as a nuptial gift to the female. During the early breeding season, the pollen-poor flowers of kangaroo paws provide a limited food source. The OSR is then female biased and females are the more competitive sex, because the males' spermatophore production rate is less than the rate at which females are willing to mate. This is reinforced by females 'foraging for spermatophores' by repeated matings. In contrast, later in the season when the pollen-rich grass trees burst into flower, males start competing for matings since they then rapidly produce new spermatophores. Consequently, more males than females become ready to mate, and thus the OSR changes to being male biased (Gwynne 1990, Gwynne & Simmons 1990, Simmons & Bailey 1990, Simmons 1992, 1995). Food availability has also been found to determine the mating pattern in red-winged blackbirds (*Agelaius phoeniceus*). Male parental care is less important to offspring survival when food is abundant than when it is scarce, allowing males to reduce their relative parental investment at food-rich sites, and also reducing the cost to a female of mating with an already mated male. Consequently, males can invest less in feeding the offspring, but more in additional matings, resulting in a more male-biased OSR and increased mating competition, as well as variance in mating success among males when food is abundant (Whittingham & Robertson 1994).

As a rule, several environmental factors will influence OSR simultaneously. For example, in the sand goby, the PRR of males is more strongly affected by water temperature than is the PRR of females (Kvarnemo 1994, Figure 18.3), whereas food availability limits the PRR of females but not of males (Kvarnemo 1997). Consequently, when the water is warm, the OSR becomes male biased and males compete more strongly for matings (Kvarnemo 1996). This pattern would also be expected when food is scarce (Kvarnemo 1997). In contrast, the PRR of females does not exceed that of males, not even in a situation with cold water and food in excess (Kvarnemo 1994). This suggests that females are unlikely to become the more competitive sex unless additional factors affect the OSR. Such factors may be the availability of nest sites limiting the reproductive success of males (Forsgren *et al.* 1996), or a strongly female-biased adult sex ratio. In an experimental study manipulating the adult sex ratio, females competed more with each other under female than male bias (Kvarnemo *et al.* 1995). Thus, many factors can influence the sexual differences in PRR, further affecting the OSR and patterns of competition for mates. Consequently, the inevitable variation in such environmental factors may result in shifting directions and intensities of mating competition, which thus will influence sexual selection.

18.5.2.2 Parental investment

Traditionally, sexual selection is thought to be governed by relative parental investment, which is 'any investment by the parent in an individual offspring that increases the offspring's chance of surviving (and hence reproductive success) at the cost of the parent's ability to invest in other offspring' (Trivers 1972). In theory, the sex with the lower parental investment will be the sex towards which OSR is biased. However, parental investment is empirically inaccessible, as it has proven extremely difficult to measure the cost in terms of future offspring (but see Simmons 1992). Moreover, the two sexes may pay the cost in different currencies, such as time or energy, making sexual comparisons difficult (Knapton 1984). Parental expenditure (i.e. resources spent on the production of offspring regardless of the fitness costs, Evans 1990, Clutton-Brock 1991) has often been used to approximate parental investment (Clutton-Brock & Parker 1992) but it does by no means replace it. Yet, the energy expenditure pattern often correlates to the PRR of a sex (Clutton-Brock & Parker 1992). For example, the low parental expenditure that is typical for males of most animal species requires less time, and therefore males generally have the potential to reproduce at a higher rate than females. However, there are examples of a decoupling between parental energy expenditures and PRR, in which case the energy expenditure of a sex is unrelated to the time requirements, and hence PRR. This has been demonstrated for the butterfly *Pieris napi* (Wiklund *et al.* 1998) and the broadnosed pipefish (Ahnesjö 1995). Still, a difference between the sexes in parental investment (Trivers 1972) should lead to a difference in the frequency with which males and females can afford to engage in reproductive events. Therefore, as has recently been suggested (Simmons 1995, Parker & Simmons 1996), relative parental investment may be assessed as relative PRR, if the fitness cost of reproduction is viewed as the parent's 'time out', which in turn relates directly to the OSR.

18.5.2.3 Age and size distributions

The PRR is often positively related to body size and/or age, and differently so between the sexes

(Berglund & Rosenqvist 1990, Ahnesjö 1995). Thus, as a consequence of a sexual difference in PRR, OSR will be sensitive to the size and age distributions in a population. Variation of this kind within and among populations has not yet been well studied.

18.5.3 Contrasting patterns

In contrast to studies showing a positive correlation between the sex bias in OSR and mating competition, there are also studies demonstrating opposite patterns. In Richardson's ground squirrel, male-male combats were most intense when most females were in oestrus and the OSR least male biased (Michener & McLean 1996). Similarly, in water striders sexual selection on males is most intense under a female-biased OSR, because females then resist male mating attempts more vigorously than under a male-biased OSR (reviewed in Rowe *et al.* 1994, Arnqvist 1997). The male trait under selection varies between water strider species, but includes the size of abdominal claspers, body size and leg size. In general, female water striders are reluctant to mate, because most matings are costly in terms of increased risk of predation and energy expenditure, as well as reduced foraging success and mobility. However, dislodging a male is also costly, and it does not pay to resist when male abundance is high (Watson *et al.* 1998). Accordingly, females are less reluctant to mate at a male-biased OSR, thereby avoiding being harassed by other males. Alternatively, the decrease in number of dislodgements may be due to postcopulatory mate guarding by the male (Parker 1970, Thornhill & Alcock 1983, Clark 1988). A male-biased OSR will increase the risk of sperm competition, and to counter this risk water striders (Clark 1988, Vepsäläinen & Savolainen 1995), other insects (e.g. McLain 1989, Telford & Dangerfield 1990, Alonso-Pimentel & Papaj 1996), and crustaceans (reviewed in Jormalainen 1998) may show prolonged mate guarding when experiencing a male-biased OSR. In the water strider *Gerris lacustris*, prolonged mate guarding at a male-biased OSR seems to be the result both of females resisting less and of males being more persistent (Vepsäläinen & Savolainen 1995). Thus, sperm competition may primarily influence sexual selection on males under a male bias, whereas female reluctance to mate may be relatively more important for sexual selection on males under a female-biased OSR.

Sperm competition is an important mechanism of sexual selection when females store sperm and have multiple mates, and a male-biased OSR often increases the risk and intensity of sperm competition. However, in the bushcricket *K. nartee*, the greatest risk of sperm competition coincides with a female-biased OSR. This is generated by a higher female remating rate, which occurs at food shortage, as the female try to acquire more of the resources offered by the males at mating. In contrast, males have a greatly reduced remating rate, resulting in a strongly female-biased OSR (Simmons & Gwynne 1991). Also, larger females have a greater mating success when competing for calling males (Gwynne & Simmons 1990). Thus, the intensity of sperm competition covaries with female size. This has been shown to have consequences for male ejaculate expenditure: when experiencing a female-biased OSR, males reduce their expenditure when copulating with large females (Simmons & Kvarnemo 1997), in accordance with theoretical expectations (Parker *et al.* 1996).

18.6 | OSR influences mate choice

OSR and the degree of choosiness are related, as the sex in short supply may have many potential mates to choose among and can therefore be selective without losing mating opportunities. In accordance, the limiting sex has been demonstrated to be the more choosy sex; for example, the broad-nosed pipefish (Berglund 1994) and the field cricket, *Gryllus pennsylvanicus* (Souroukis & Murray 1995).

Choosiness has often been viewed as a 'sex role' opposite to competition, since this follows the most common pattern, namely that males compete and females are choosy. However, often both sexes may be choosy although the less competitive sex is more selective than the other (Berglund & Rosenqvist 1993, Berglund 1994, Grant *et al.* 1995). Alternatively, the less competitive sex may be unselective and still mate with fitter mates following the outcome of a contest competition (e.g. Cox & LeBoeuf 1977, but see Qvarnström & Forsgren 1998). Other studies have shown that the competitive sex may also be the choosiest sex (Summers 1992, Owens et al. 1994). A large variance in mate quality may promote both high selectivity (Real 1990) and intense competition over those high-quality mates (Summers 1992, Owens & Thompson 1994). In fact, when individuals vary substantially in quality only a subset of members of that sex may be competed for as mates (Summers 1992, Owens et al. 1994). Then, the OSR will no longer be a sufficient predictor of mating competition (Parker 1983, Clutton-Brock & Parker 1992, Owens & Thompson 1994). Furthermore, especially when there is mutual mate choice, an individual of good quality may have better prospects of being accepted as a mate by the opposite sex than an individual of poorer quality, resulting in an assortative mating pattern (Parker 1983, Johnstone 1997).

When the relative influence of PRR and variance in mate quality on mate choice have been modelled, differences between the sexes in PRR have been found to have a greater impact than variance in mate quality on choosiness (Johnstone et al. 1996). In this model, high variance in mate quality increased the potential benefits of choosiness, while the costs were expressed as reduced PRR due to delayed matings. Thus, when alternative mates are hard to find, e.g. due to a low PRR of the opposite sex, the costs outweigh potential benefits and it pays to mate indiscriminately (Johnstone et al. 1996; NB their term 'processing time' is equivalent to 'time out', JD Reynolds and RA Johnstone, personal communication).

The relative influence of PRR and variance in mate quality on mate choice have also been studied empirically, with results varying between studies. In the biparental cichlid St. Peter's fish (Sarotherodon galilaeus), female choosiness is determined by the bias in OSR (influenced by PRR and the adult sex ratio), but not by the variance in mate quality (Balshine-Earn 1996). Similarly, in an experimental study of a bushcricket, Requena verticalis, PRR but not variance in female quality influenced the level of choosiness among males (Kvarnemo & Simmons 1998). In another

bushcricket, K. nartee, male mate choice was influenced by both OSR and variance in mate quality, as males and females engaged in sexual interactions sooner under a male than a female bias, but, importantly, males were then more likely to reject the female when there was high variance in female quality (Kvarnemo & Simmons 1999). Finally, in the case of the sand goby, in which PRR and variance in mate quality for both sexes were manipulated separately, the fact that females were more choosy than males was best explained by PRR. However, within the choosy sex (females), variation in mate quality had the strongest influence on their choosiness (Kvarnemo & Forsgren 2000).

18.7 | Future developments

When there is competition for the resources that are prerequisites for being able to mate (for instance, a territory or a nest site), not all individuals in the adult sex ratio will qualify to mate (Figure 18.1, Ahnesjö et al. 2001), which will clearly influence the OSR. Such effects must be taken into account in order to avoid estimates of mating competition being confounded by resource competition.

Another issue that requires attention arises when sneaking, as an alternative mating strategy, is prevalent. Sneaking usually occurs among males that are otherwise excluded from the mating pool because of strong male–male competition. Consequently, sneakers influence the variance in reproductive success, and thus the opportunity for sexual selection, through sperm competition. Generally, when the mating pattern is polygamous, one can predict that sneaking will decrease the variance in reproductive success among males, compared to a situation without sneaking where only a fraction of males would gain reproductive success (Jones et al. 2001). In contrast, for socially monogamous birds with extrapair fertilizations, the opposite prediction has been made, namely that the variation in male reproductive success increases with the degree of extrapair paternity (Møller and Birkhead 1994). However, further complexity can be found when satellite males sneak matings and at the same time attract more mates to their host's territory (Widemo & Owens 1995). Another consequence of sneaking may be that more time is required to replenish sperm, decreasing male PRR (Simmons & Parker 1996). Future work is needed to investigate the effect of sneakers on OSR and mating competition, theoretically as well as empirically.

Finally, we should also bear in mind that estimates of OSR are likely to be influenced by whether matings are evenly distributed or more clumped among partners, in time as well as in space: details of such influences have yet to be investigated.

18.8 | Conclusion

The operational sex ratio, OSR, can successfully be used to explain and predict which sex is the predominant competitor for mates, to examine cases where, under different circumstances, males or females act as principal competitors, and to trace how the intensity in mating competition changes within the competing sex. Major factors determining the OSR in a population are the adult sex ratio, or the qualified sex ratio, and the sexual difference in PRR, estimated as the potential number of offspring produced per unit time (Figure 18.2; Ahnesjö et al. 2001).

In this chapter, we have focused on two methods of estimating the OSR in a population (Box 18.1): to directly count the number of males and females that are ready to mate or, preferably, to predict OSR by measuring the PRR of each sex (Box 18.2). To measure PRR, both sexes have to be experimentally provided with an unlimited access to mating partners. Then, PRR can be measured as the potential number of offspring produced per unit time or as the inverse 'time out' for a clutch. The sexual difference in PRR is likely to be influenced simultaneously by several factors of varying impact; for example, food availability, temperature, size and age distributions. To be able to estimate male and female PRR for a certain population one needs to assess these factors. Thus, using the PRR approach, we achieve more than just a simple prediction of which sex roles are likely to prevail. We will also become aware of the constraints that determine male and female PRR and thus OSR, deepening our understanding of the processes that influence the direction and intensity of mating competition, and hence sexual selection, in a population.

Acknowledgements

We thank the Swedish Natural Sciences Research Council and Uppsala University for funding, and are most grateful to Göran Arnqvist, Anders Berglund, James Cook, Ian Hardy, Staffan Ulfstrand, Christer Wiklund and two anonymous referees for useful comments.

References

Acharya L (1995) Sex-biased predation on moths by insectivorous bats. *Animal Behaviour*, 49, 1461–1468.

Ahnesjö I (1995) Temperature affects male and female potential reproductive rates differently in the sex-role reversed pipefish *Syngnathus typhle*. *Behavioral Ecology*, 6, 229–233.

Ahnesjö I, Kvarnemo C & Merilaita S (2001) Using potential reproductive rates to predict mating competition among individuals qualified to mate. *Behavioral Ecology*, 12, 397–401.

Almada VC, Gonçalves EJ, Oliveira RF & Santos AJ (1995) Courting females: ecological constraints affect sex roles in a natural population of the blenniid fish *Salaria pavo*. *Animal Behaviour*, 49, 1125–1127.

Alonso-Pimentel H & Papaj DR (1996) Operational sex ratio versus gender density as determinants of copulation duration in the walnut fly, *Rhagoletis juglandis* (Diptera: Tephritidae). *Behavioral Ecology and Sociobiology*, 39, 171–180.

Andersson MB (1994) *Sexual Selection*. Princeton, NJ: Princeton University Press.

Andersson M & Iwasa Y (1996) Sexual selection. *Trends in Ecology and Evolution*, 11, 53–58.

Arnqvist G (1997) The evolution of water strider mating systems: causes and consequences of sexual conflicts. In: JC Choe & BJ Crespi (eds) *The Evolution of Mating Systems in Insects and Arachnids*, pp 146–163. Cambridge: Cambridge University Press.

Balshine-Earn S (1996) Reproductive rates, operational sex ratios and mate choice in St. Peter's fish. *Behavioral Ecology and Sociobiology*, 39, 107–116.

Berglund A (1991) Egg competition in a sex-role reversed pipefish: subdominant females trade reproduction for growth. *Evolution*, **45**, 770–774.

Berglund A (1994) The operational sex ratio influences choosiness in a pipefish. *Behavioral Ecology*, **5**, 254–258.

Berglund A & Rosenqvist G (1990) Mate limitation of female reproductive success in a pipefish: effects of body size differences. *Behavioral Ecology and Sociobiology*, **27**, 129–133.

Berglund A & Rosenqvist G (1993) Selective males and ardent females in pipefishes. *Behavioral Ecology and Sociobiology*, **32**, 331–336.

Berglund A & Rosenqvist G (2001a) Male pipefish prefer dominant over attractive females. *Behavioral Ecology*, **12**, 402–406.

Berglund A & Rosenqvist G (2001b) Male pipefish prefer ornamented females. *Animal Behaviour*, **61**, 345–350.

Berglund A, Rosenqvist G & Svensson I (1989) Reproductive success of females limited by males in two pipefish species. *American Naturalist*, **133**, 506–516.

Birkhead TR & Møller AP (eds) (1998) *Sperm Competition and Sexual Selection*. Cambridge: Academic Press.

Björklund M (1991) Coming of age in fringillid birds: heterochrony in the ontogeny of secondary sexual characters. *Journal of Evolutionary Biology*, **4**, 83–92.

Bush SL (1993) *Courtship and Male Parental Care in the Mallorcan Midwife Toad*, Alytes muletensis. Ph.D. thesis. Norwich: University of East Anglia.

Bush SL & Bell DI (1997) Courtship and female competition in the Majorcan midwife toad, *Alytes muletensis*. *Ethology*, **103**, 292–303.

Clark SJ (1988) The effects of operational sex ratio and food deprivation on copulation duration in the water strider (*Gerris remigis* Say). *Behavioral Ecology and Sociobiology*, **23**, 317–322.

Clutton-Brock TH (1991) *The Evolution of Parental Care*. Princeton, NJ: Princeton University Press.

Clutton-Brock TH & Parker GA (1992) Potential reproductive rates and the operation of sexual selection. *The Quarterly Review of Biology*, **67**, 437–456.

Clutton-Brock TH & Vincent ACJ (1991) Sexual selection and the potential reproductive rate of males and females. *Nature*, **351**, 58–60.

Colwell MA & Oring LW (1988) Sex ratios and intrasexual competition for mates in a sex-role reversed shorebird, Wilson's phalarope (*Phalaropus tricolor*). *Behavioral Ecology and Sociobiology*, **22**, 165–173.

Cox CR & LeBoeuf BJ (1977) Female incitation of male competition: a mechanism in sexual selection. *American Naturalist*, **111**, 317–335.

Darwin C (1871) *The Descent of Man, and Selection in Relation to Sex*. London: Murray.

Debuse VJ, Addison JT & Reynolds JD (1999) The effects of sex ratio on sexual competition in the European lobster. *Animal Behaviour*, **58**, 973–981.

Emlen ST (1976) Lek organization and mating strategies in the bullfrog. *Behavioral Ecology and Sociobiology*, **1**, 283–313.

Emlen ST & Oring LW (1977) Ecology, sexual selection, and the evolution of mating systems. *Science*, **197**, 215–223.

Enders MM (1993) The effect of male size and operational sex ratio on male mating success in the common spider mite, *Tetranychus urticae* Kock (Acari: Tetranychidae). *Animal Behaviour*, **46**, 835–846.

Evans RM (1990) The relationship between parental input and investment. *Animal Behaviour*, **39**, 797–798.

Forsgren E, Kvarnemo C & Lindström K (1996) Modes of sexual selection determined by resource abundance in two sand goby populations. *Evolution*, **50**, 646–654.

Grant JWA, Casey PC, Bryant MJ & Shahsavarani A (1995) Mate choice by Japanese medaka (Pisces, Oryziidae). *Animal Behaviour*, **50**, 1425–1428.

Gwynne DT (1983) Male nutritional investment and the evolution of sexual differences in the Tettigoniidae and other Orthoptera. In: DT Gwynne & GK Morris (eds) *Orthopteran Mating Systems: Sexual Competition in a Diverse Group of Insects*, pp 337–366. Boulder: Westview Press.

Gwynne DT (1990) Testing parental investment and the control of sexual selection in katydids: the operational sex ratio. *American Naturalist*, **136**, 474–484.

Gwynne DT & Simmons LW (1990) Experimental reversal of courtship roles in an insect. *Nature*, **346**, 172–174.

Gwynne DT, Bailey WJ & Annells A (1998) The sex in short supply for matings varies over small scales in a katydid (*Kawanaphila nartee*, Orthoptera: Tettigoniidae). *Behavioral Ecology and Sociobiology*, **42**, 157–162.

Höglund J, Montgomerie R & Widemo F (1993) Costs and consequences of variation in the size of ruff leks. *Behavioral Ecology and Sociobiology*, **32**, 31–39.

Ichikawa N (1993) Biased operational sex ratio causes the female giant water bug *Lethocerus deyrollei* to destroy egg masses. *Journal of Ethology*, **11**, 151–152.

Jirotkul M (1999) Operational sex ratio influences female preference and male-male competition in guppies. *Animal Behaviour*, **58**, 287–294.

Johnstone RA (1997) The tactics of mutual mate choice and competitive search. *Behavioral Ecology and Sociobiology*, **40**, 51–59.

Johnstone RA, Reynolds JD & Deutsch JC (1996) Mutual mate choice and sex differences in choosiness. *Evolution*, **50**, 1382–1391.

Jones AG, Rosenqvist G, Berglund A & Avise JC (1999) The genetic mating system of a sex-role reversed pipefish (*Syngnathus typhle*): a molecular inquiry. *Behavioral Ecology Sociology*, **46**, 357–365.

Jones AG, Walker DE, Kvarnemo C, Lindström K & Avise JC (2001). How cuckoldry can decrease the opportunity for sexual selection: data and theory from a genetic parentage analysis of the sand goby, *Pomatoschistus minutus*. *Proceedings of the National Academy of Sciences USA*, **98**, 9151–9156.

Jormalainen V (1998) Precopulatory mate guarding in crustaceans: male competitive strategy and intersexual conflict. *Quarterly Review of Biology*, **73**, 275–304.

Knapton RW (1984) Parental investment: the problem of currency. *Canadian Journal of Zoology*, **62**, 2673–2674.

Kokko H & Monaghan P (2001) Predicting the direction of sexual selection. *Ecology Letters*, **4**, 159–165.

Kraus WF (1989) Is male back space limiting? An investigation into the demography of the giant water bug *Abedus indentatus* (Heteroptera: Belostomatidae). *Journal of Insect Behaviour*, **2**, 623–648.

Kruse KC (1990) Male backspace availability in the giant waterbug (*Belostoma flumineum* Say). *Behavioral Ecology and Sociobiology*, **26**, 281–289.

Kvarnemo C (1994) Temperature differentially affects male and female reproductive rates in the sand goby: consequences for operational sex ratio. *Proceedings of the Royal Society of London, series B*, **256**, 151–156.

Kvarnemo C (1996) Temperature affects operational sex ratio and intensity of male-male competition: an experimental study of sand gobies, *Pomatoschistus minutus*. *Behavioral Ecology*, **7**, 208–212.

Kvarnemo C (1997) Food affects the potential reproductive rates of sand goby females but not of males. *Behavioral Ecology*, **8**, 605–611.

Kvarnemo C & Ahnesjö I (1996) The dynamics of operational sex ratios and competition for mates. *Trends in Ecology and Evolution*, **11**, 404–408.

Kvarnemo C & Forsgren E (2000) The influence of potential reproductive rates and variation in mate quality on male and female choosiness in the sand goby. *Behavioral Ecology and Sociobiology*, **48**, 378–384.

Kvarnemo C & Simmons LW (1998) Male potential reproductive rate influences mate choice in a bushcricket. *Animal Behaviour*, **55**, 1499–1506.

Kvarnemo C & Simmons LW (1999) Variance in female quality affects male mate choice in a bushcricket. *Behavioral Ecology and Sociobiology*, **45**, 245–252.

Kvarnemo C, Forsgren E & Magnhagen C (1995) Effects of sex ratio on intra- and intersexual behaviour in sand gobies. *Animal Behaviour*, **50**, 1455–1461.

Kynard BE (1978) Breeding behaviour of a lacustrine population of three-spined sticklebacks (*Gasterosteus aculeatus* L.). *Behaviour*, **67**, 178–207.

Madsen T & Shine R (1993) Temporal variability in sexual selection acting on reproductive tactics and body size in male snakes. *American Naturalist*, **141**, 167–171.

Masonjones HD & Lewis SM (2000) Differences in potential reproductive rates of male and female seahorses related to courtship roles. *Animal Behaviour*, **59**, 11–20.

Maxwell MR (1998) Seasonal adult sex ratio shift in the praying mantid *Iris oratoria* (Mantodea: Mantidae). *Environmental Entomology*, **27**, 318–323.

McLain DK (1989) Prolonged copulation as a post-insemination guarding tactic in a natural population of the ragwort seed bug. *Animal Behaviour*, **38**, 659–664.

Michener GR & McLean IG (1996) Reproductive behaviour and operational sex ratio in Rickardson's ground squirrels. *Animal Behaviour*, **52**, 743–758.

Møller AP & Birkhead TR (1994) The evolution of plumage brightness in birds is related to extrapair paternity. *Evolution* **48**, 1089–1100.

Moore SD (1987) Male-biased mortality in the butterfly *Euphydryas editha*: A novel cost of mate acquisition. *American Naturalist*, **130**, 306–309.

Oliveira RF, Almada VC, Forsgren E & Gonçalves EJ (1999) Temporal variation in male traits, nesting aggregations and mating success in the peacock blenny. *Journal of Fish Biology*, **54**, 499–512.

Otronen M (1996) Effects of seasonal variation in operational sex ratio and population density on the mating success of different sized and aged males in the yellow dung fly, *Scathophaga stercoraria*. *Ethology, Ecology and Evolution*, **8**, 399–411.

Owens IPF & Thompson DBA (1994) Sex differences, sex ratios and sex roles. *Proceedings of the Royal Society of London, series B*, **258**, 93–99.

Owens IPF, Burke T & Thompson DBA (1994) Extraordinary sex roles in the Eurasian dotterel: female mating arenas, female-female competition, and female mate choice. *American Naturalist*, **144**, 76–100.

Parker GA (1970) Sperm competition and its evolutionary consequences in the insects. *Biological Reviews*, **45**, 525–567.

Parker GA (1983) Mate quality and mating decisions. In: P Bateson (ed) *Mate Choice*, pp 141–66. Cambridge: Cambridge University Press.

Parker GA & Simmons LW (1996) Parental investment and the control of sexual selection: predicting the direction of sexual competition. *Proceedings of the Royal Society of London, series B*, **263**, 315–321.

Parker GA, Ball MA, Stockley P & Gage MJG (1996) Sperm competition games: individual assessment of sperm competition intensity by group spawners. *Proceedings of the Royal Society London, series B*, **263**, 1291–1297.

Pitnick S (1993) Operational sex ratios and sperm limitation in populations of *Drosophila pachea*. *Behavioral Ecology and Sociobiology*, **33**, 383–391.

Pröhl H & Hödl W (1999) Parental investment, potential reproductive rates, and mating system in the strawberry dart-poison frog, *Dendrobates pumilio*. *Behavioral Ecology and Sociobiology*, **46**, 215–220.

Qvarnström A & Forsgren E (1998) Should females prefer dominant males? *Trends in Ecology and Evolution*, **13**, 498–501.

Real L (1990) Search theory and mate choice. I. Models of single-sex discrimination. *American Naturalist*, **136**, 376–404.

Reynolds JD (1996) Animal breeding systems. *Trends in Ecology and Evolution*, **11**, 68–72.

Reynolds JD, Colwell MA & Cooke F (1986) Sexual selection and spring arrival times of red-necked and Wilson's phalaropes. *Behavioural Ecology and Sociobiology*, **18**, 303–310.

Rowe L, Arnqvist G, Sih A & Krupa JJ (1994) Sexual conflict and the evolutionary ecology of mating patterns: water striders as a model system. *Trends in Ecology and Evolution*, **9**, 289–293.

Simmons LW (1992) Quantification of role reversal in relative parental investment in a bushcricket. *Nature*, **358**, 61–63.

Simmons LW (1995) Relative parental expenditure, potential reproductive rates, and the control of sexual selection in katydids. *American Naturalist*, **145**, 797–808.

Simmons LW & Bailey WJ (1990) Resource influenced sex roles of zaprochiline tettigoniids (Orthoptera: Tettigoniidae). *Evolution*, **44**, 1853–1868.

Simmons LW & Gwynne DT (1991) The refractory period of female katydids (Orthoptera: Tettigoniidae): sexual conflict over the remating interval? *Behavioral Ecology*, **2**, 276–282.

Simmons LW & Kvarnemo C (1997) Ejaculate expenditure by male bushcrickets decreases with sperm competition intensity. *Proceedings of the Royal Society of London, series B*, **264**, 1203–1208.

Simmons LW & Parker GA (1996) Parental investment and the control of sexual selection: can sperm competition affect the direction of sexual competition? *Proceedings of the Royal Society of London, series B*, **263**, 515–519.

Souroukis K & Murray A-M (1995) Female mating behavior in the field cricket, *Gryllus pennsylvanicus* (Orthoptera: Gryllidae) at different operational sex ratios. *Journal of Insect Behavior*, **8**, 269–279.

Summers K (1992) Dart-poison frogs and the control of sexual selection. *Ethology*, **91**, 89–107.

Swensson RO (1997) Sex-role reversal in the tidewater goby, *Eucyclogobius newberryi*. *Environmental Biology of Fishes*, **50**, 27–40.

Tejedo M (1988) Fighting for females in the toad *Bufo calamita* is affected by the operational sex ratio. *Animal Behaviour*, **36**, 1765–1769.

Telford SR & Dangerfield JM (1990) Manipulation of the sex ratio and duration of copulation in the tropical millipede *Alloporus uncinatus*: a test of the copulatory mate guarding hypothesis. *Animal Behaviour*, **40**, 984–985.

Thornhill R & Alcock J (1983) *The Evolution of Insect Mating Systems*. Cambridge, MA: Harvard University Press.

Trivers RL (1972) Parental investment and sexual selection. In: BG Campbell (ed) *Sexual Selection and the Descent of Man, 1871–1971*, pp 136–179. Chicago, IL: Aldine.

Vepsäläinen K & Savolainen R (1995) Operational sex ratios and mating conflict between the sexes in the water strider *Gerris lacustris*. *American Naturalist*, **146**, 869–880.

Vincent A, Ahnesjö I, Berglund A & Rosenqvist G (1992) Pipefishes and seahorses: are they all sex role reversed? *Trends in Ecology and Evolution*, **7**, 237–241.

Vincent A, Ahnesjö I & Berglund A (1994) Operational sex ratios and behavioural sex differences in a pipefish population. *Behavioral Ecology and Sociobiology*, **34**, 435–442.

Waights V (1996) Female sexual interference in the smooth newt, *Triturus vulgaris vulgaris*. *Ethology*, **102**, 736–747.

Watson PJ, Arnqvist G & Stallman RR (1998) Sexual conflict and the energetic costs of mating and mate choice in water striders. *American Naturalist*, **151**, 46–58.

Whittingham LA & Robertson RJ (1994) Food availability, parental care and male mating success in red-winged blackbirds (*Agelaius phoeniceus*). *Journal of Animal Ecology*, **63**, 139–150.

Widemo F & Owens IPF (1995) Male mating skew and the evolution of lekking. *Nature*, **373**, 148–151.

Wiklund C, Kaitala A & Wedell N (1998) Decoupling of reproductive rates and parental expenditure in a polyandrous butterfly. *Behavioral Ecology*, **9**, 20–25.

Chapter 19

Using sex ratios: the past and the future

Steven Hecht Orzack

19.1 | Summary

Our understanding of the evolution of sex ratios has advanced substantially in recent decades, in part due to the important work of Hamilton (1967) on 'extraordinary sex ratios'. However, important aspects of the biology have largely remained unstudied, including the mating dynamics and structure of natural populations, the role of the individual in producing observed sex ratios, and the nature of sex ratio control. Also little studied is how tests of sex ratio models should be structured so as to provide maximum insight. Perhaps ignoring these facets of the biology has aided progress in the past, but these gaps in the study of sex ratios are now blindspots, which hinder understanding. Further progress in the evolutionary analysis of sex ratios requires their elimination. This can be accomplished only by direct investigations of mating dynamics, population structure and the behaviour of individuals.

19.2 | Introduction

Every great scientific theory is a partial lie about nature. This is not a claim about scientific fraud, as the issue is not one of honesty in the usual sense. Instead, it is a claim about how theories succeed (Cartwright 1984). A theory can succeed by explaining facts correctly. It can also succeed by making facts appear irrelevant or unimportant, sometimes correctly but sometimes incorrectly. Such facts sometimes resurface, but sometimes they are never seen again (Lewontin 1991). To this extent, the lie arises because nature is not described, as much as it is created from some facts while others are eliminated and left unexplained.

Such a lie may arise more from the organization of science than from direct intention. It is a large shared body of data that makes sustaining a lie more difficult; ideally, data should be a defining basis for an intellectual community. Although this ideal has never been met exactly, intellectual communities are increasingly distant from it, as they are increasingly organized around theoretical claims, and less around common empirical knowledge. In evolutionary biology, scientists are now trained more in particular skills, say sequencing or algebra, and less in the biology of organisms. This is an ideal circumstance for sustaining lies about nature.

It is this perspective on scientific theories that proves useful in assessing WD Hamilton's 1967 paper on 'extraordinary sex ratios'. This *is* one of the great papers in evolutionary biology. Beyond establishing sex ratio distorters and population subdivision as important topics in the study of sex ratio evolution, it foreshadows work on other important topics in this area including the effects of resource quality and of the relatedness of individuals (Trivers & Willard 1973, Trivers & Hare 1976, Charnov 1979, 1982, Charnov *et al.* 1981); it is also a canonical game-theoretical analysis of an evolutionary question.

However, one legacy of Hamilton's paper is scientific lies of the kind described above. Some are good, some are bad; our resulting understanding of sex ratio evolution is clear in some directions but very unclear in others because of blindspots in our field of vision. My goal here is to provide some guide for future analyses so as to eliminate these blindspots. I first discuss Hamilton's claim (sections 19.2.1–19.2.3), and then discuss the study of population structure (section 19.3), recent empirical work on sex ratios (section 19.4), the importance of studying individuals (section 19.5) and the importance of studying sex ratio control (section 19.6). In section 19.7, I consider claims about the general importance of sex ratio studies in evolutionary biology and discuss the testing of sex ratio models. Finally, in section 19.8 I consider the future of sex ratio studies.

19.2.1 Extraordinary sex ratios

One of the central aspects of Hamilton's paper is an explanation of the female-biased sex ratios found in many arthropod species. This presupposes that there is a well-defined phenomenon in need of explanation. Is this true? To answer this question one must determine if the sex ratios have a female bias and, if so, how it is spatially and temporally expressed (Clausen 1939). For example, a female bias could be expressed only between populations. This would not be consistent with Hamilton's claim that the bias is an evolutionary response to local competition for mates in a structured population (LMC). Hamilton assumed that populations are subdivided such that a finite number of females oviposit together, leaving offspring that mate amongst themselves. Males do not disperse, while mated females depart and distribute themselves randomly into new groups and the cycle is repeated. Under these circumstances, it is evolutionarily advantageous for a female to control her reproduction so as to produce a female-biased sex ratio. What do we know about the validity of Hamilton's idealization? As discussed below, although it appears to be correct for some species, for most we lack sufficient data to meaningfully judge its relevance to nature.

Nonetheless, most theoreticians have assumed that local interactions underlie the evolution of female-biased sex ratios and that females can precisely control their sex ratios. The result is an impressive body of theory, including models predicting a female's sex ratios as determined by information she has about her offsprings' resource availability and mating opportunities (Orzack 1993 and Hardy 1994 contain discussions of theory prior to 1994; newer work includes Nagelkerke & Hardy 1994, Nagelkerke 1996, Greeff 1997, Proulx 2000). These models predict the behaviour of an individual, as they use the Darwinian assumption that natural selection discriminates among individuals. In addition, most of these models determine trait optima.

Hamilton's idealization has also been used by many empiricists (e.g. Wrensch & Ebbert 1993 and Chapter 20). One consequence is an impressive body of experiments and observations, including many studies that compare the predictions of optimality models with data, usually by assessing whether a predicted trend is observed.

What is a reasonable perspective on these endeavours? They have led to a substantial increase in our understanding of how sex ratios probably evolved in a number of well-studied species (Godfray & Werren 1996). These are facts explained. At the same time, however, important facts have been eliminated by the activities of both theoreticians and empiricists. The concern is that Hamilton's idealization may come to determine our facts about nature, instead of having facts determine the truth of the idealization. There is good reason to think that it can continue to be fruitful as a conceptual organizer and basis for model construction; however, this heuristic role should not be assumed to be an explanatory role. Other idealizations should be explored, as has been done to some extent (e.g. Uyenoyama & Bengtsson 1982, Nunney & Luck 1988). Recall that theory construction is itself an important accomplishment. What has been problematic is that theoreticians have often stated or implied that sex ratio theory helps explain a well-defined phenomenon (perhaps to counter the attitude that theory is not 'real' biology). Theory also tends to generate further theory with similar or identical assumptions. The result is a tendency for Hamilton's idealization to be viewed as true, not because of evidence, but simply because it is plausible.

The activities of empiricists have also contributed to this tendency. Sex ratio models are usually tested by comparing theoretical predictions with data. But what data? One usually has laboratory data, which are possibly not relevant to nature, *or* field data, which are usually uncontrolled. Accordingly, drawing conclusions involves making plausible inferences, as in many areas of evolutionary biology. This is perfectly appropriate except when, as in the case of the generation of theory, the accumulation of inferences blurs the distinction between what is known definitively and what is not. I now discuss examples of this blurring.

19.2.2 The existence of local mate competition

An important fact to be acknowledged is that the occurrence of LMC is not documented for most species whose sex ratios have been explained by it. Female-biased sex ratios have been recorded for many species of arthropods, especially among the Hymenoptera. Are such excesses spatially and temporally localized? They are in some species (Hamilton 1979, Herre 1985, 1987, Herre *et al.* 1997, but see Greeff & Ferguson 1999). However, most female-biased sex ratios in Hymenoptera are known only from field collections, which are possibly subject to a number of collection biases, such as differential habitat use and the differential survival of males and females (Kirkendall 1993 discusses such biases in studies of bark beetles). While suggestive, most field collections do not provide compelling evidence about the ecological and evolutionary context in which sex ratios are expressed. We know little about the mating behaviour and population structure of most species with female-biased sex ratios (Hardy 1994). How did the opposite impression become widespread? Consider two of the most important attempts to integrate sex ratio theory and data: Herre's (1985, 1987) papers on the female-biased sex ratios of some Panamanian fig wasp species.

For the species studied, there is good evidence that LMC occurs. However, many subsequent authors, lacking such evidence, have claimed that sex ratio trends in other species have evolved in response to LMC because they parallel those described by Herre. Such claims lend momentum to Hamilton's general claim about the localized structure of populations, despite the absence of general data. Why is such an inference inappropriate? First, general claims need general evidence. Second, female-biased sex ratios can evolve in populations with structures distinct from that assumed by Hamilton (Uyenoyama & Bengtsson 1982, Godfray 1994) and some appear to have done so (Hardy & Godfray 1990, Antolin & Strand 1992, Guertin *et al.* 1996, Ode *et al.* 1997).

19.2.3 Hamilton's evidence

The absence of general evidence for LMC is highlighted an appraisal of Table 1 of Hamilton's paper (1967, p 482), which contains examples of 'Insects and mites *having* usual sibmating combined with arrhenotoky and spanandry [emphasis added].' This is a claim about evidence. I have examined all of the papers cited by Hamilton. In every paper there is a lack of evidence or inconsistency in the evidence for one or more important aspects of the relevant biology. I discuss some of these gaps below; space considerations do not allow a complete enumeration here but a list is available upon request.

In every paper, there is either no claim about a species' 'usual site of mating' or there is an absence of data for such a claim. The absence of a claim is exemplified by the entry for the wasp *Nasonia vitripennis*: Hamilton cited Graham-Smith (1919) and Moursi (1946) in his notes 36 and 47, but neither paper provides any claim or data about mating behaviour (*N. vitripennis* is described as *N. brevicornis* by Graham-Smith and *Mormoniella vitripennis* by Moursi). The absence of data for a claim is illustrated by Cooper's (1937) paper on the mite *Pediculopsis graminumis* (see note 65): this is one of the best papers in regard to its documentation of a claim about mating and in reporting sample sizes. Yet, these consist only of a statement (p 42) that, 'copulation may be seen to take place while the mites are still within their mother's body' and (p 43) 'In only four instances has copulation been observed in these cultures outside the maternal body.' Not one paper having a claim about the 'usual site of mating' presents the total number of observations. Jackson's (1966) article on the aquatic wasp, *Caraphractus cinctus*, is typical. Jackson claimed (p 25) that, 'Mating may occur in the water, on the surface film or on emergent plants. It has never been observed within the host egg. It occurs most commonly under water, for the male usually emerges first, waits upon the host egg for the females to come out and then mates with one female after another.' There are no data presented, nor is there further mention of sibmating.

The lack, or inconsistency, of evidence with respect to the sex ratio is underscored by the fact that only one paper in Hamilton's Table 1 (Entwistle 1964) has any statistical analysis of sex ratio data (and this lacks information on sample size). For most, there is no information about the area and time over which the sample was obtained, the numbers of broods scored, the brood sizes involved, or whether sex ratio data are based on the offspring of individual females. To this extent, these papers do not provide *evidence* about sex ratios and mating. This is not a comment about the truth of the claims in these papers; many may be correct. Older standards for evolutionary claims are partially responsible for the lack of evidence in Table 1. It is reasonable to have a different expectation for the science in Enock's (1898) paper (note 54) on the wasp *Prestwichia aquatica*, as compared to the expectation for our papers. Nonetheless, this attitude goes only so far, as some 'long-ago' authors distinguished between what we know and what we infer and were aware of the need to reconcile conflicting accounts. For example, Henriksen (1922) (note 54) mentioned Enock's claim (p 153) that, 'In all I have examined at least a dozen eggs containing parasites, and in each there were one or two pairs *in copula inside the egg-shell(!!)*' and then stated (p 28), 'This does not agree with my observations which do, on the other hand, agree with those of Heymons. In the great material which I had at my disposal. . . . I never saw any copulation take place in the host egg, on the contrary

all imagines will lie quite motionless until at last one lying next to the shell will gnaw a hole in it and escape. . .' Henrikson concluded 'Enock and Rimsky-Korsakov as well as Heymons and I are quite sure of having observed exactly, but it passes my understanding how so widely differing observations can be made.' Hamilton recognized this discrepancy (note 54) but this species is listed as having a usual site of mating 'in the host'.

Even if one regards verbal claims about sex ratios and mating structure as compelling, evidence for the existence of LMC in some of these species is ambiguous. As Hamilton emphasized (p 482), the evolution of a female-biased sex ratio in response to LMC depends upon females being more vagile than males. Yet, in his paper on the wasp *Melittobia chalybii* (note 51) Buckell (1928) stated (p 19) that, 'The females, although fully-winged, were never seen to fly, and could not be induced to do so.' Rimsky-Korsakov (1916) and Henriksen (1922) described short-winged and long-winged females in *P. aquatica*. Hence, for neither species is it clear whether females and males differ in their dispersal from their natal site. (It is also unclear how any particular wing morphology affects dispersal in *P. aquatica* since adults live under water.)

Another ambiguity involves the wasp *Monodontomerus mandibularis*. Hamilton listed the 'usual site of mating' as 'in host cell' but cited Rau (1947) (note 46) who stated (p 223), 'It seems improbable that mating occurs in the dark cell before [individuals] emerge. . . .'

Finally, there is conflicting evidence in some of these papers for the existence of arrhenotoky. Busck (1917) (note 62) described arrhenotoky and thelytoky in one collection of the wasp *Goniozus emigratus* (*Perisierola emigrata*), while Willard (1927) found no thelytoky in another collection. Similarly, Henriksen (1922, p 29) described conflicting evidence for arrhenotoky and thelytoky in *Prestwichia aquatica*.

These gaps relate to important aspects of the biology under investigation. Accordingly, I regard the claims about sex ratios and mating behaviours in Hamilton's Table 1 as inadequately substantiated; they are best viewed as plausible hypotheses in need of testing. At present they do not support Hamilton's claim that the female-biased sex ratios in question *are* associated with 'usual sibmating'.

19.3 | How and why has population structure been neglected?

The gap between the content of these papers and Hamilton's use of them probably reflects an eagerness on his part to support an imaginative model, especially at a time when mathematics was less accepted in biology than it is today. How did Hamilton's claim gain acceptance? There are at least three reasons. First, although early discussions such as Hartl (1971) refer to it only as 'plausible', many later discussions have taken Hamilton's claim as proven, either for the papers he cited (Antolin 1993, 1999, Godfray & Cook 1997, Godfray & Shimada 1999) or more generally (e.g. Charnov 1982, Sober & Wilson 1998). Many studies (mostly of Hymenoptera) have appeared since 1967 but virtually none provides data on population structure and mating dynamics (see Hardy *et al.* in press).

Second, important textbooks have presented Hamilton's claim as a fact. For example, Futuyma writes (1998, p 614), 'In many species, mating occurs not randomly among members of a large population, but within small groups descended from one or a few founders. After one or a few generations, progeny emerge into the population at large, then colonize patches of habitat and repeat the cycle. In many species of parasitoid wasps, for example, the progeny of one or a few females emerge from a single host and almost immediately mate with each other; the daughters then disperse in search of new hosts.' This is a description of Hamilton's idealization (see also Ridley 1996, p 309). Even if one considers only parasitoid wasps, there are no data to substantiate this as a general claim.

Finally, verbal arguments have a long tradition in behavioural ecology. Their familiarity should not compensate for their weakness. Of course, it is wrong to say that such arguments are never acceptable, but the acceptability of a verbal claim should decline as the importance of the inference increases. Population subdivision is essential to the evolutionary workings of LMC; to

this extent, *evidence* for sibmating and population subdivision is essential to a claim that female-biased sex ratios have evolved due to LMC. Verbal claims cannot usually count as evidence since they are inherently privileged, whereas data can be analysed by others. At best, perhaps a compilation of the very best verbal evidence would be compelling. To qualify, any particular study would need well-delineated claims about female-biased sex ratios and local mating (much better than those described above). Additionally, a meaningful assemblage of verbal claims would be substantially larger than Hamilton's and be corrected for a possible lack of independence among species. The creation of such an assemblage has long been possible, especially since the appearance of Peck's (1963) catalogue of the Nearctic Chalcidoidea, which contains thousands of citations. This, along with Noyes (1998), has much of the necessary information.

While surely all claims about Hamilton's hypothesis are benignly intended, their consequence is not benign, as a plausible inference about the evolution of female-biased sex ratios has become a 'fact' about nature. Population structure thereby has become a blindspot, so much so that many well-informed evolutionary biologists are surprised to learn that so little is definitively known about it (Godfray & Cook 1997).

19.4 | What have we learned since Hamilton 1967?

Given the ambiguity of Hamilton's evidence, it is important to understand what we have learned from subsequent research about population structure and female-biased sex ratios. Much important work has concerned pollinating and non-pollinating fig wasps (Hamilton 1979, Frank 1985, Herre 1985, 1987, 1989, Herre *et al.* 1997, 2001, Chapter 20). The reason is that females oviposit into or inside of a fig and their offspring eclose in its interior; to this extent, the fig 'creates' population subdivision and it is likely that LMC occurs in many fig wasp species. Note the dependency of this inference on the specialized ecology of the fig; there are few other species with female-biased sex ratios and an ecology similarly specialized enough to infer the existence of LMC.

Yet, the inference that LMC occurs in many species of fig wasp needs important clarification. For example, the presence of dead females of pollinating species inside figs allowed Herre (1985) to test his important LMC model, which predicts the optimal set of sex ratios that a female should produce when encountering variable numbers of other ovipositing females (see also Frank 1985). The dead females (or 'foundresses') are assumed to have oviposited and then died. To my knowledge, only this study along with Frank (1985), Orzack et al. (1991) and Molbo and Parker (1996) have tested this kind of conditional sex ratio model in Hymenoptera (Orzack et al. 1991 provide the only data on sex ratios produced by individuals within groups; see also Roeder 1992 and Fig. 13.2 in Herre et al. 1997). In Herre's (1985) analysis, the model predictions are qualitatively accurate, but quantitatively inaccurate. But Herre's estimates of foundress numbers could be inaccurate if a given female oviposits in more than one fig (as has been reported to occur in other species by Okamato & Tashiro 1981, Gibernau et al. 1996) and if she uses past as well as present encounters with other females when deciding on a sex ratio in a given fig. Herre (1996) regarded this as unlikely (as does F Kjellberg, pers. comm.), but we lack data. In any case, it is quite possible that this phenomenon has been generally overlooked among pollinating fig wasps.

The need for detailed behavioural analyses is reinforced by observations that males of some species leave the fig after emergence (Grandi 1929, Greeff & Ferguson 1999, S Compton, pers. com., F Kjellberg, pers. comm.). Accordingly, the claim that mating occurs only in the natal fig may be incorrect. Males leaving figs may mate with females from other figs; this is unknown at present, but unstudied as well. Such mating could substantially change evolutionary dynamics. The point is that our view of fig wasps as ideal examples of LMC needs investigation, not canonization (see also Kathuria et al. 1999, Herre et al. 2001).

In population genetics it is generally accepted that claims about mating and population subdivision based upon inferences from ecology or morphology are not acceptable, as they have not always been sustained by behavioural or genetic data (Taylor et al. 1984, Leibherr 1988; see Kirkendall 1993, Orzack 1993, Bossart & Prowell 1998 and Peterson & Denno 1998 for other examples and discussion). This fact weakens even the best analyses with such claims (e.g. West & Herre 1998a).

The weakness of morphological inferences is illustrated by the case of the wasps *Nasonia giraulti* and *N. vitripennis*. The female-biased sex ratios commonly produced by female *N. vitripennis* have often been regarded as resulting from LMC (e.g. Werren 1980, 1983, Orzack 1986, 1990, Orzack & Parker 1986, 1990, Orzack et al. 1991), since males have vestigial wings and females do not. King and Skinner (1991) reported the paradox that *N. giraulti*, in which both sexes have apparently functional wings (implying less-subdivided populations and less female-biased sex ratios), has *more* female-biased average sex ratios than does *N. vitripennis*. Drapeau and Werren (1999) reported that *N. giraulti* has a higher frequency of mating within the host than does *N. vitripennis*; perhaps this difference resolves the paradox. The lesson is that we need more detailed behavioural, ecological and genetic studies of population structure; this need is underscored by the important papers of Nadel and Luck (1992), Fauvergue et al. (1999) and Hardy et al. (1999), which indicate that strict local mating does not appear to occur in several parasitoids. Some important progress has been made with genetic analyses. Kazmer and Luck (1991) and Antolin (1999) studied population structure in *Trichogramma* species with electrophoretic analysis of soluble enzyme loci. Their results imply that there is a mix of sib and local mating in the populations studied. Molbo and Parker (1996) presented a similar analysis of data from *N. vitripennis*. Taken together, these studies offer both partial support for the existence of Hamilton's idealization *and* an indication that surprises await us in the study of population structure. I am currently undertaking such a study of *N. vitripennis* and *N. giraulti*, using denaturing liquid chromatography to analyse a widespread base polymorphism in the coding sequence of a LIM protein (see Dawid *et al.* 1998 for a discussion of these proteins). All of these studies of population structure will help to eliminate this blindspot in our vision.

19.5 | What we do know and should know about individuals

Another blindspot is that we know so little about the sex ratio behaviours of individuals. Almost all present analyses concern sex ratios produced by groups of individuals. It is telling to enumerate the few studies out of hundreds that have analysed the sex ratio produced by an individual within a group of individuals; to my knowledge, the only such studies are Werren (1980), Orzack and Parker (1986), Orzack *et al.* (1991), Orzack and Gladstone (1994) and Flanagan *et al.* (1998). Observations on sex ratios produced by groups are important, but they cannot underwrite a claim for the optimality of a trait (Orzack & Sober 1994a,b), which has been a common use for them. An optimality model, by definition, concerns a trait of an individual, even if the trait's fitness is affected by the frequencies of other individuals, as for sex ratio traits. Beyond the need to match the nature of data with the nature of predictions, attention to the sex ratio traits of individuals will provide a much greater understanding of sex ratio evolution. Such data *are* the domain of optimality models. Some have succeeded at accurately predicting the traits of individuals (e.g. Brockmann & Dawkins 1979, Brockmann *et al.* 1979). In this sense, it is not asking too much of such a model to test it with data on individuals; this is no misunderstanding of behavioural ecology (*contra.* Godfray 1994, p 182). Instead, it is a matter of taking model assumptions and structure seriously. Optimality models deserve this, given their importance (see also Vet 1995; the importance of attention to individuals is well recognized in related areas of analysis, e.g. Caswell & John 1992).

One consequence of the inattention to individuals is that important studies, such as Herre (1987), are limited in significant ways. This study demonstrates the importance of selective context. Herre observed that the magnitude of the discrepancy between sex ratio data and model predictions for a given foundress number generally increases as the frequency of occurrence of the foundress number decreases. He claimed that this occurs because a lower frequency means less opportunity for natural selection to optimize the sex ratio. But his analysis involves the average sex ratios produced by groups of individuals, which yield no information as to the variability among individuals, which is the best measure of the strength of selection on a trait. If Herre is correct, this variability should decrease as the frequency of occurrence of foundress number increases. West and Herre (1998b) demonstrated such a relationship for single-foundress broods; they lack data for other group sizes. If such a relationship is not generally observed, it casts doubt on Herre's claim about the importance of selective context, although the importance of his study would remain.

Tools for the analyses of sex ratios produced by individuals (or isofemale strains, which can often be regarded as proxies for individuals) include direct observation and manipulation of oviposition and the use of genetic markers to distinguish between the offspring of different females. Species differences may also sometimes be used to track the offspring of individuals (see Orzack 1993 for discussion). In this context, 'Necessity is the mother of invention'.

Many who claim that it is asking too much of optimality models to apply to individuals or to be quantitatively accurate appear to have the attitude that nature is too 'complex' to be fully understood. It is not clear that evolutionary phenomena are any more 'complex' than the phenomena that are accurately predicted by models in other natural sciences or that it is impossible to create explanatory models of individual behaviour (see Orzack & Sober 1993 for discussion).

What do optimality models predict about individuals? An optimum implies monomorphism or near-monomorphism. Recall what underlies the evolution of such a trait. The performance of the optimal trait is such that the carrier has a greater level of fitness than the carriers of other traits. All other things being equal, this implies that other traits are eliminated from the population or cannot enter as rare mutants. The consequence is little or no genetic variation for the trait.

This (near-)monomorphism does not mean the absence of variation. A snapshot of individuals with the optimal trait could reveal differences among them if the trait is a mixture of subtraits, as in the case of sex ratio optima having a mixture of two sexes. What is not consistent with the evolution of the optimum is the presence of many variants of the trait, with all or most being common. Many regard (near-)monomorphism as extremely unlikely since, 'all traits are genetically variable'. This perspective overlooks some important ambiguities in our understanding of genetic variation in nature. Many traits in evolutionarily defined populations do have substantial heritabilities (Mousseau & Roff 1987). However, many studies of quantitative genetic variation involve an amalgamation of geographically disparate strains (many analyses of *Drosophila* species) or they involve organisms of little relevance to nature (corn and laboratory strains of mice). Significant heritabilities stemming from an analysis of geographically disparate strains are consistent with monomorphism *or* polymorphism within local populations. Whether the variation underlying these results occurs within populations remains to be seen. Accordingly, (near-)monomorphism of sex ratio traits is a plausible empirical outcome. The capricious nature of expectations is easy to see. If one regards a sex ratio trait to be a quantitative trait 'like any other', it 'should' be variable since most quantitative traits exhibit genetic variation. Alternatively, if one regards a particular sex ratio trait to be 'like any other' sex ratio trait, it 'should not' be variable since many such traits do not exhibit genetic variation (e.g. Falconer 1954, Toro & Charlesworth 1982). It is imperative that any expectation as to how the disposition of genetic variation 'must be' does not take precedence over data.

19.6 | What do we know about the control of sex ratios?

Hamilton's idealization has reinforced a tradition of interpreting female-biased sex ratios in Hymenoptera as being due to the control of egg fertilization by individual females. The haplo-diploidy of many of these species implies that the control of sperm is the same as the control of sex ratio. In addition, sex ratio behaviours are thought to be determined by 'regular' genetic loci, those that obey Mendelian rules of inheritance and accordingly obey standard evolutionary dynamics. Although these assumptions have been demonstrated to occur in some species of Hymenoptera, there is no general evidence that they are true for most species with female-biased sex ratios (see also Cornell 1988, Godfray & Cook 1997).

In fact, recent work suggests that 'irregular' genetic influences on sex ratio evolution may occur. Much important work concerns PSR, a B chromosome in *Nasonia vitripennis* (Werren 1991, Beukeboom & Werren 1992, Chapter 9), which results in male-biased sex ratios. PSR appears to be rare in this species; whether similar chromosomes affect other species is unknown.

Another possible 'irregular' influence on sex ratios relates to the mixed and labile sexuality long known in the Hymenoptera. In particular, there are reports of parthenogenetic females coexisting with sexual females, of strains in which both kinds of females occur, and of *thelytokous females that produce rare males* (Keeler 1929, Flanders 1945, 1965, Tardieux & Rabasse 1988, Aeschlimann 1990, Fucheng & Zhang 1991, Luck *et al.* 1993, Weinstein & Austin 1996, Belshaw *et al.* 1999); this can occur without fertilization (Chen *et al.* 1992, Stary 1999). Recent work suggests an association between micro-organisms and these phenomena. Much work was stimulated by the report by Stouthamer *et al.* (1990) that parthenogenetic strains of several *Trichogramma* species can be rendered sexual by treatment with antibiotics or high temperatures (see also Chen *et al.* 1992, Stouthamer & Luck 1993, Stouthamer & Kazmer 1994). In some cases parthenogenesis is associated with endosymbiotic bacteria in the genus *Wolbachia* (Pijls *et al.* 1996, Plantard *et al.* 1998, Pintureau *et al.* 1999). The point is that many female-biased sex ratios may be partially the result of temporal and/or spatial variation in the mode of reproduction. Standard evolutionary considerations (the cost of meiosis argument) make it clear that an infectious agent causing an unmated female to produce daughters and occasional sons could readily spread in a panmictic population; no population subdivision and local interactions would be needed.

The focus on the fertilization of eggs as the determinant of the sex ratio has created another blindspot: hymenopteran males are thought to have no influence on the sex ratios produced by their mates. Yet, it is plausible that this occurs even if the female has proximate control of fertilization. Male ejaculatory fluid is known to affect sperm usage in *Drosophila melanogaster* and polymorphism in some component proteins is associated with different sperm-usage patterns (Clark *et al.* 1995). Ejaculate components could affect sex ratios by affecting female oviposition and fertilization behaviours (Hawkes 1992) and there is some circumstantial evidence for this (Legner 1988, 1989a,b). (Male genotype could also affect sex ratios by influencing the proportions of different kinds of sperm, Wilkes & Lee 1965.) Genetic influences of males on their mates is a situation in which an individual's genotype is expressed in another individual; the locus and the locus of expression are not coincident, as they are for most traits. Such influences on sex ratios have received very little attention; this is partially due to the blindspot created by the idea that females 'must' control the sex ratios they produce.

19.7 | What is the importance of studies of sex ratio evolution?

There are claims that the study of sex ratios (and of sex allocation) provides 'powerful' evolutionary insights and is one of the most 'successful' areas of evolutionary biology (e.g. Godfray 1994, p 151, Hamilton 1996, p 132, Hardy & Mayhew 1998, p 431). What could such a claim mean? Its basis is the belief that a sex ratio is a 'real' trait, since it is just a count of males and females. Second, since this count also relates directly to evolutionary fitness, one can construct an evolutionary explanation whose only important causal component is natural selection. In practice, this usually means that one constructs an optimality model.

This approach has tremendous power because it describes the possible; it is important to have a well-motivated claim about what an organism *should* do if natural selection is the only important force affecting its evolution (Orzack & Sober 1994a,b). The study of sex ratios is also successful because it has led to the creation of theory that reveals the common aspects of distinct sex-allocation problems (Charnov 1982, Queller 1984), although other areas have such a unifying framework (e.g. Roff 1992, Orzack 1997). These are significant accomplishments; however, most claims about power and success relate to our ability to provide explanations for sex ratio data. The essence of these claims is that model predictions match the observations in most, if not nearly all, studies.

By this criterion, I do not believe that the study of sex ratios is more successful than many other areas of ecology and evolution. One reason is how most model testing has proceeded. Usually it involves only determining whether the model successfully predicts a qualitative trend in the data. If so, it is published as an explanation; if data and model are deemed to conflict, the model often undergoes 'behind the scenes' revision until a satisfactory match is attained, which is published. (Many proponents of optimality models make the claim that optimality is not even under test when predictions are compared with data, e.g. Parker & Maynard Smith 1990.) The result is that only studies providing qualitative support for optimality models are usually published. This tendency makes claims about the success of sex ratio theory somewhat circular; failures are not available for counting (this problem of selective reporting is not unique: Simberloff 1983 discusses another possible example).

Another reason for scepticism as to the special success of sex ratio studies is that even the qualitative matches between sex ratio theory and data are mixed in their quality; one can find contradictory conclusions about the nature of qualitative fit (Orzack 1990, 1993). Of course, these can be found elsewhere in evolutionary biology. But this similarity is the point; the study of sex ratios often has much explanatory power, but no more than in many other areas of study. It is a significant accomplishment that the approach I've outlined is successful in particular instances.

However, claims that the study of sex ratios is nearly at the point of 'dotting i's and crossing t's' are an exaggeration (see also Chapter 20).

One aspect of the study of sex ratios in need of greater scrutiny is the use of competing hypotheses. In the study of community ecology, one can compare the explanatory power of a neutral model and a non-neutral model or compare alternative non-neutral models (e.g. Strong *et al.* 1979, Schluter & Grant 1982); in the study of population dynamics, one can compare the explanatory power of density-dependent and density-independent models (e.g. Stiling 1988, Hassell *et al.* 1989) and in the study of molecular population genetics, one can compare the explanatory power of genetic drift and natural selection (e.g. Kimura 1983, Gillespie 1991). But, almost without exception, most tests of sex ratio models do not involve alternative adaptive hypotheses, much less nonadaptive ones. In this sense, testing is not comparative. The use of competing hypotheses can strengthen qualitative and quantitative tests of models because it forces one to more precisely define the predictions of each model.

An example will illustrate what this claim means. Several authors have noted that the qualitative predictions of LMC models and host quality models can overlap (Waage 1986, King 1992, Orzack 1993). As more foundresses contribute offspring to a population, both types of models predict an increased proportion of males. LMC models do so because there is an increased opportunity for outcrossing, while resource quality models do so because males are assumed to be less harmed by reduced resource levels (Charnov *et al.* 1981, van den Assem *et al.* 1989). The evolution of a sex ratio response to host quality can occur with or without population subdivision (reviewed in Godfray 1994, Hardy 1994).

Is it likely then that many sex ratio trends not attributed to LMC are an evolved response to variable resource quality? Naively, it would be hard *not* to make this conclusion. For example, Wylie (1973) described increased proportions of males when females of *Nasonia vitripennis* oviposited into hosts previously parasitized by conspecific females. This shift has been construed as an evolved response to the increased opportunity for outcrossing (Suzuki & Iwasa 1980, Werren 1980, Orzack & Parker 1986, 1990). Yet, Wylie observed the same increase when females oviposited into hosts previously parasitized by females of *other* species. Such a shift should not occur if the sex ratio behaviour is an evolved response to LMC since, of course, individuals of different species do not mate each other. Accordingly, females either could not manifest a species-specific response or they responded to a change in resource quality. Most sex ratio studies have not controlled for changing resource quality. But even this sort of control in the laboratory (as in Orzack *et al.* 1991) only provides circumstantial evidence that female-biased sex ratios in the field evolved due to LMC.

One can decide between the two hypotheses by testing a qualitative or quantitative prediction that only one makes. For example, only LMC models predict a difference between the sex ratios produced by females in hosts they previously parasitized as compared to those produced in hosts other females parasitized. Another distinction is that as foundress number increases the change in optimal sex ratio is gradual under LMC, but abrupt under resource quality.

The need for care about testing hypotheses is more basic than one might imagine. Many studies have no explicit criteria for the acceptance or rejection of *one* hypothesis. An optimality model predicting, say, an association between two variables is often tested by determining whether they are significantly correlated. Should any significantly nonzero correlation be a basis for accepting the model as explanatory? Nonoptimal trait values have lower fitnesses by definition. Accordingly, the issue is whether trait values that are statistically consistent with the optimal prediction should be taken as *evolutionarily* consistent with it. The fitness surfaces around sex ratio optima can be nearly flat or sharply peaked (Orzack 1990, Orzack *et al.* 1991, West & Herre 1998b); as a result, there is no general statement one can make about 'acceptable' deviations from predictions, other than that the evolutionary cost of differing from the optimum will generally increase as population size decreases. For foundress group sizes smaller than ten, often thought to underlie the evolution of female-biased sex ratios, sex ratios that differ from the optimum by a small percentage can have fitness levels at least a few percent lower than the optimum; this is a large selection coefficient. The shape of the fitness surface can enter into a statistical assessment of an optimality model's predictive power. One way is to weight trait values by their absolute fitnesses or by their relative fitnesses (Orzack *et al.* 1991). Another possibility is to score as supportive of the hypothesis only those observed trait values having predicted fitnesses within a given standard percentage of the optimal fitness, and to score as nonsupportive all other observed trait values. The proportion of the two types could then be compared statistically with, say, a χ^2 or exact test; a significant excess of supportive trait values would support the model. Although any deviation from the optimum implies evolutionary instability, if such a procedure reveals trait values tightly clustered around predictions it would be compelling support for the model.

Finally, statistical testing of sex ratio models and analysis of sex ratio data have often not accounted for the interdependency of variables that can arise when one analyses ratios.

Imagine, for example, that one statistically assesses the relationship between sex ratio and brood size, as when testing an optimality model predicting that sex ratio is increasingly biased as relative clutch size gets larger (Werren 1980, Flanagan *et al.* 1998) or when assessing the order in which males and females are produced (King 1993). In either case, one calculates the correlation between the sex ratio and the relative or absolute clutch size. If sex ratio is calculated as the proportion of, say, males, the number of males is present in both the numerator and denominator of the proportion and in the clutch size; the result is a 'spurious' nonzero correlation even if the underlying variables such as the number of males and females are independent of one another (see Kenney 1982, Jackson & Somers 1991 for further background). This has been discussed in the context of the testing of sex ratio models (Orzack 1986) but it is still common to see studies that ignore the issue, although it is relatively straightforward to account for a spurious correlation in testing (Chayes 1971, Pendleton *et al.* 1983). It is quite possible that many sex ratio analyses have incorrectly rejected a null hypothesis of no association because of a spurious correlation.

19.8 | What is the future of the unknown in sex ratio studies?

There is much more to be said about sex ratio evolution than I have said here. As outlined above, important aspects of population structure, of sex ratio expression and of underlying sexual biology are in need of careful study and elaboration. Similar concerns could be raised about our understanding of the comparative fitness performance of males and females (see Kazmer and Luck 1995) and about how individuals process information on the environment (i.e. how the spatial distribution of resources maps onto the distribution of oviposition sites). We are at an important point in the development of our understanding of sex ratio evolution; it is time to celebrate our successes, but also time to comfortably acknowledge our failures and to rescue important facts from being forgotten. We can only benefit from such actions, as they will help eliminate blindspots in our field of vision.

Acknowledgements

I thank Ian Hardy for his support, patience and comments; Allen Herre, Dave Parker and Jack Werren, for being stimulating colleagues; and Jaco Greeff, Finn Kjellberg, Peter Mayhew, Dave Parker and Stuart West for comments. I thank Christa Modschiedler of the University of Chicago and Mary Sears of Harvard University for excellent library assistance. This research was supported by National Science Foundation awards DEB-9407965 and SES-9906997.

References

Aeschlimann JP (1990) Simultaneous occurrence of thelytoky and bisexuality in Hymenopteran species, and its implications for the biological control of pests. *Entomophaga*, **35**, 3–5.

Antolin MF (1993) Genetics of biased sex ratios in subdivided populations: models, assumptions, and evidence. In: D Futuyma & J Antonovics (eds) *Oxford Surveys in Evolutionary Biology*, pp 239–281. Oxford: Oxford University Press.

Antolin MF (1999) A genetic perspective on mating systems and sex ratios of parasitoid wasps. *Researches on Population Ecology*, **41**, 29–37.

Antolin MF & Strand MR (1992) Mating system of *Bracon hebetor* (Say) (Hymenoptera:Braconidae). *Ecological Entomology*, **17**, 1–7.

Belshaw R, Quicke DLJ, Volkl W & Godfray HCJ (1999) Molecular markers indicate rare sex in a predominantly asexual parasitoid wasp. *Evolution*, **53**, 1189–1199.

Beukeboom LW & Werren JH (1992) Population-genetics of a parasitic chromosome – experimental-analysis of psr in subdivided populations. *Evolution*, **46**, 1257–1268.

Bossart JL & Prowell DP (1998) Genetic estimates of population structure and gene flow: limitations, lessons and new directions. *Trends in Ecology and Evolution*, **13**, 202–206.

Brockmann HJ & Dawkins R (1979) Joint nesting in a digger wasp as an evolutionarily stable preadaptation to social life. *Behaviour*, **71**, 203–245.

Brockmann HJ, Grafen A & Dawkins R (1979) Evolutionarily stable nesting strategy in a digger wasp. *Journal of Theoretical Biology*, **77**, 473–496.

Buckell ER (1928) Notes on the life history and habits of *Melittobia chalybii* Ashmead (Chalcidoidea, Elachertidae). *Pan-Pacific Entomologist*, **5**, 14–22.

Busck A (1917) Notes on *Perisierola emigrata* Rohwer, a parasite of the pink boll worm (Hymenoptera, Bethyliidae). *Insectutor Inscitiae Menstruus*, **5**, 3–5.

Cartwright N (1984) *How the Laws of Physics Lie*. Oxford: Oxford University Press.

Caswell HA & John AM (1992) From the individual to the population in demographic models. In: DL Angelis & LJ Gross (eds) *Individual-based Models and Approaches in Ecology*, pp 36–61. New York: Chapman & Hall.

Charnov EL (1979) The genetical evolution of patterns of sexuality: Darwinian fitness. *American Naturalist*, **113**, 465–480.

Charnov EL (1982) *The Theory of Sex Allocation*. Princeton, NJ: Princeton University Press.

Charnov EL, Los-den Hartogh RL, Jones WT & van den Assem J (1981) Sex ratio evolution in a variable environment. *Nature*, **289**, 27–33.

Chayes F (1971) *Ratio Correlation*. Chicago, IL: University of Chicago Press.

Chen BH, Kfir R & Chen CN (1992) The thelytokous *Trichogramma chilonis* in Taiwan. *Entomologia Experimentalis et Applicata*, **65**, 187–194.

Clark AG, Aguadé M, Prout T, Harshman LG & Langley CH (1995) Variation in sperm displacement and its association with accessory gland protein loci in *Drosophila melanogaster*. *Genetics*, **139**, 189–201.

Clausen CP (1939) The effect of host size upon the sex ratio of hymenopterous parasites and its relation to methods of rearing and colonization. *Journal of the New York Entomological Society*, **47**, 1–9.

Cooper KW (1937) Reproductive behaviour and haploid parthenogenesis in the grassmite, *Pedichlopsis graminum* (Rent.) (Acarina, Tarsonemidae). *Proceedings of the National Academy of Sciences USA*, **23**, 41–44.

Cornell HV (1988) Solitary and gregarious brooding, sex ratios and the incidence of thelytoky in the parasitic Hymenoptera. *American Midland Naturalist*, **119**, 63–70.

Dawid IB, Breen JJ & Toyama R (1998) LIM domains: multiple roles as adapters and functional modifiers in protein interactions. *Trends in Genetics*, **14**, 156–162.

Drapeau MD & Werren JH (1999) Differences in mating behaviour and sex ratio between three sibling species of *Nasonia*. *Evolutionary Ecology Research*, **1**, 223–234.

Enock F (1898) Notes on *Prestwichia aquatica*, Lubbock. *Entomologist's Monthly Magazine*, **34**, 152–153.

Entwistle PF (1964) Inbreeding and arrhenotoky in the ambrosia beetle *Xyleborus compactus* (Eichh.) (Coleoptera: Scolytidae). *Proceedings of the Royal Entomological Society of London, series B*, **39**, 83–88.

Falconer DS (1954) Selection for sex ratio in mice and Drosophila. *American Naturalist*, **88**, 385–397.

Fauvergue X, Fleury F, Lemaitre C & Allemand R (1999) Parasitoid mating structures when hosts are patchily distributed: field and laboratory experiments with *Leptopilina boulardi* and *L. heterotoma*. *Oikos*, **86**, 344–356.

Flanagan KE, West SA & Godfray HCJ (1998) Local mate competition, variable fecundity and information use in a parasitoid. *Animal Behaviour*, **56**, 191–198.

Flanders SE (1945) The bisexuality of uniparental Hymenoptera, a function of the environment. *American Naturalist*, **79**, 122–141.

Flanders SE (1965) On the sexuality and sex ratios of Hymenopterous populations. *American Naturalist*, **99**, 489–494.

Frank SA (1985) Hierarchical selection theory and sex ratios. II. On applying the theory, and a test with fig wasps. *Evolution*, **39**, 949–964.

Fucheng W & Zhang S (1991) *Trichogramma pintoi* (Hym.: Trichogrammatidae): deuterotoky, laboratory multiplication and field releases. In: E Wajnberg, SB Vinson (eds) *Trichogramma and other egg parasitoids*, pp 155–157. Paris: Institut National de la Recherche Agronomique.

Futuyma D (1998) *Evolutionary Biology*. Sunderland, MA: Sinauer Associates.

Gibernau M, Hossaert-McKey M, Anstett M & Kjellberg F (1996) Consequences of protecting flowers in a fig: a one-way trip for pollinators? *Journal of Biogeography*, **23**, 425–432.

Gillespie JH (1991) *The Causes of Molecular Evolution*. Oxford: Oxford University Press.

Godfray HCJ (1994) *Parasitoids: Behavioral and Evolutionary Ecology*. Princeton, NJ: Princeton University Press.

Godfray HCJ & Cook JM (1997) Mating systems of parasitoid wasps. In: J Choe & B Crespi (eds) *Social Competition and Cooperation in Insects and Arachnids: I. Evolution of Mating Systems*, pp 211–225. Cambridge: Cambridge University Press.

Godfray HCJ & Shimada M (1999) Parasitoids as model organisms for ecologists. *Researches on Population Ecology*, **41**, 3–10.

Godfray HCJ & Werren J (1996) Recent developments in sex ratio studies. *Trends in Ecology and Evolution*, **11**, 59–63.

Graham-Smith GS (1919) Further observations on the habits and parasites of common flies. *Parasitology*, **11**, 347–384.

Grandi G (1929) Studio Morphologico e biologico della *Blastophaga psenes* (L.). *Bollettino del Laboratorio di Entomologia del Royale Instituto Superiore Agrario di Bologna*, **2**, 1–147.

Greeff JM (1997) Offspring allocation in externally ovipositing fig wasps with varying clutch size and sex ratio. *Behavioral Ecology*, **8**, 500–505.

Greeff JM & Ferguson JWH (1999) Mating ecology of the nonpollinating fig wasps of *Ficus ingens*. *Animal Behaviour*, **57**, 215–222.

Guertin DS, Ode PJ, Strand MR & Antolin MF (1996) Host-searching and mating in an outbreeding parasitoid wasp. *Ecological Entomology*, **21**, 27–33.

Hamilton WD (1967) Extraordinary sex ratios. *Science*, **156**, 477–488.

Hamilton WD (1979) Wingless and fighting males in fig wasps and other insects. In: MS Blum & NA Blum (eds) *Sexual Selection and Reproductive Competition in Insects*, pp 167–220. New York: Academic Press.

Hamilton WD (1996) *Narrow Roads of Gene Land*, volume 1: *Social Behaviour*. Oxford: WH Freeman.

Hardy ICW (1994) Sex ratio and mating structure in the parasitoid Hymenoptera. *Oikos*, **69**, 3–20.

Hardy ICW & Godfray HCJ (1990) Estimating the frequency of constrained sex allocation in field populations of Hymenoptera. *Behaviour*, **114**, 137–147.

Hardy ICW & Mayhew P (1998) Partial local mating and the sex ratio: indirect comparative evidence. *Trends in Ecology and Evolution*, **13**, 431–432.

Hardy ICW, Pedersen JB, Sejr MK & Linderoth UH (1999) Local mating, dispersal and sex ratio in a gregarious parasitoid wasp. *Ethology*, **105**, 57–72.

Hardy ICW, Ode PJ & Siva-Jothy M (2002) Mating systems. In: M Jervis (ed) *Insects as Natural Enemies: Practical Approaches to their Study and Evaluation*, 2nd edn. Dordrecht: Kluwer Academic Publishers, in press.

Hartl DL (1971) Some aspects of natural selection in arrhenotokous populations. *American Zoologist*, **11**, 309–325.

Hassell MP, Latto J & May RM (1989) Seeing the woods for the trees: detecting density dependence from existing life-table studies. *Journal of Animal Ecology*, **58**, 883–892.

Hawkes PG (1992) Sex ratio stability and male-female conflict over sex ratio control in hymenopteran parasitoids. *South African Journal of Science*, **88**, 423–430.

Henriksen KJ (1922) Notes upon some aquatic Hymenoptera (*Anagrus brocheri* Schulz, *Prestwichia aquatica* Lubb., *Agriotypus armatus* Walk.). *Annales de Biologie Lacustre*, **11**, 19–37.

Herre EA (1985) Sex ratio adjustment in fig wasps. *Science*, **288**, 896–898.

Herre EA (1987) Optimality, plasticity and selective regime in fig wasp sex ratios. *Nature*, **329**, 627–629.

Herre EA (1989) Coevolution of reproductive characteristics in 12 species of new world figs and their pollinator wasps. *Experientia*, **45**, 637–647.

Herre EA (1996) An overview of studies on a community of Panamanian figs. *Journal of Biogeography*, **23**, 593–607.

Herre EA, West SA, Cook JM, Compton SG & Kjellberg F (1997) Fig wasps: pollinators and parasites, sex ratio adjustment and male polymorphism, population structure and its consequences. In: J Choe & B Crespi (eds) *Social Competition and Cooperation in Insects and Arachnids: I. Evolution of*

Mating Systems, pp 226–239. Cambridge: Cambridge University Press.

Herre EA, Machado CA & West SA (2001) Selective regime and fig wasp sex ratios: towards sorting rigor from pseudo-rigor in tests of adaptationism. In: S Orzack & E Sober (eds) *Adaptationism and Optimality*, pp 191–218. Cambridge: Cambridge University Press.

Jackson DA & Somers KM (1991) The spectre of 'spurious' correlations. *Oecologia*, **86**, 147–151.

Jackson DJ (1966) Observations on the biology of *Caraphractus cinctus* Walker (Hymenoptera: mymaridae), a parasitoid of the eggs of Dytiscidae (Coleoptera). *Transactions of the Royal Entomological Society of London*, **118**, 23–49.

Kathuria P, Greeff JM, Compton SG & Ganeshaiah KN (1999) What fig wasp sex ratios may or may not tell us about sex allocation strategies. *Oikos*, **87**, 520–530.

Kazmer DJ & Luck RF (1991) The genetic-mating structure of natural and agricultural populations of *Trichogramma*. In: E Wajnberg & SB Vinson (eds) *Trichogramma and other Egg Parasitoids*, pp 25–28. Paris: Institut National de la Recherche Agronomique.

Kazmer DJ & Luck RF (1995) Field tests of the size-fitness hypothesis in the egg parasitoid *Trichogramma pretiosum*. *Ecology*, **76**, 412–425.

Keeler CE (1929) Thelytoky in *Scleroderma immigrans*. *Psyche*, **36**, 41–44.

Kenney BC (1982) Beware of spurious self-correlations! *Water Resources Research*, **18**, 1041–1048.

Kimura M (1983) *The Neutral Theory of Molecular Evolution*. Cambridge: Cambridge University Press.

King BH (1992) Sex-ratios of the wasp *Nasonia vitripennis* from self- versus conspecifically-parasitized hosts: local mate competition versus host quality models. *Journal of Evolutionary Biology*, **5**, 445–455.

King BH (1993) Sequence of offspring sex production in the parasitoid wasp, *Nasonia vitripennis*, in response to unparasitized versus parasitized hosts. *Animal Behavior*, **45**, 1236–1238.

King BH & Skinner SW (1991) Sex ratio in a new species of *Nasonia* with fully-winged males. *Evolution*, **45**, 225–228.

Kirkendall LR (1993) Ecology and evolution of biased sex ratios in bark and ambrosia beetles. In: D Wrensch & M Ebbert (eds) *Evolution and Diversity of Sex Ratio in Insects and Mites*, pp 235–345. New York: Chapman & Hall.

Legner EF (1988) *Muscidifurax raptorellus* (Hymenoptera: Pteromalidae) females exhibit postmating oviposition behavior typical of the male genome. *Annals of the Entomological Society of America*, **81**, 522–527.

Legner EF (1989a) Paternal influences in males of *Muscidifurax raptorellus* (Hymenoptera: Pteromalidae). *Entomophaga*, **34**, 307–320.

Legner EF (1989b) Wary genes and accretive inheritance in Hymenoptera. *Annals of the Entomological Society of America*, **82**, 245–249.

Leibherr JK (1988) Gene flow in ground beetles (Coleoptera: Carabidae) of differing habitat preference and flight-wing development. *Evolution*, **42**, 129–137.

Lewontin RC (1991) Facts and factitious in natural sciences. *Critical Inquiry*, **18**, 140–153.

Luck RF, Stouthamer R & Nunney LP (1993) Sex determination and sex ratio patterns in parasitic Hymenoptera. In: D Wrensch & M Ebbert (eds) *Evolution and Diversity of Sex Ratio in Insects and Mites*, pp 442–476. New York: Chapman & Hall.

Molbo DE & Parker ED Jr (1996) Mating structure and sex ratio variation in a natural population of *Nasonia vitripennis*. *Proceedings of the Royal Society of London, Series B*, **263**, 1703–1709.

Moursi AA (1946) The effect of temperature on the sex ratio of parasitic hymenoptera. *Bulletin de la Societé Fouad Ier d'Entomologie*, **33**, 21–37.

Mousseau TA & Roff DA (1987) Natural selection and the heritability of fitness components. *Heredity*, **59**, 181–197.

Nadel H & Luck RF (1992) Dispersal and mating structure of a parasitoid with a female-biased sex ratio: implications for theory. *Evolutionary Ecology*, **6**, 270–278.

Nagelkerke CJ (1996) Discrete clutch sizes, local mate competition, and the evolution of precise sex allocation. *Theoretical Population Biology*, **49**, 314–343.

Nagelkerke CJ & Hardy ICW (1994) The influence of developmental mortality on optimal sex allocation under local mate competition. *Behavioral Ecology*, **5**, 401–411.

Noyes JS (1998) *Catalogue of the Chalcidoidea of the World*. Amsterdam: Expert Center for Taxonomic Identification, University of Amsterdam.

Nunney L & Luck RF (1988) Factors influencing the optimum sex ratio in a structured population. *Theoretical Population Biology*, **33**, 1–30.

Ode PJ, Antolin MF & Strand MR (1997) Constrained oviposition and female-biased sex allocation in a parasitic wasp. *Oecologia*, **109**, 547–555.

Okamoto M & Tashiro M (1981) Mechanism of pollen transfer and pollination in *Ficus erecta* by *Blastophaga nipponica*. *Bulletin of the Osaka Museum of Natural History*, **34**, 7–16.

Orzack SH (1986) Sex-ratio control in a parasitic wasp, *Nasonia vitripennis*. II. Experimental analysis of an optimal sex-ratio model. *Evolution*, **40**, 341–356.

Orzack SH (1990) The comparative biology of second sex ratio evolution within a natural population of a parasitic wasp, *Nasonia vitripennis*. *Genetics*, **124**, 385–396.

Orzack SH (1993) Sex ratio evolution in parasitic wasps. In: D Wrensch & M Ebbert (eds) *Evolution and Diversity of Sex Ratio in Insects and Mites*, pp 477–511. New York: Chapman & Hall.

Orzack SH (1997) Life history evolution and extinction. In: S Tuljapurkar & H Caswell (eds) *Structured Population Models in Marine, Terrestrial, and Freshwater Systems*, pp 273–302. New York: Chapman & Hall.

Orzack SH & Gladstone J (1994) Quantitative genetics of sex ratio traits in the parasitic wasp, *Nasonia vitripennis*. *Genetics*, **137**, 211–220.

Orzack SH & Parker ED Jr (1986) Sex-ratio control in a parasitic wasp, *Nasonia vitripennis*. I. Genetic variation in facultative sex-ratio adjustment. *Evolution*, **40**, 331–340.

Orzack SH & Parker ED Jr (1990) Genetic variation for sex ratio traits within a natural population of a parasitic wasp, *Nasonia vitripennis*. *Genetics*, **124**, 373–384.

Orzack SH & Sober E (1993) A critical assessment of Levins's strategy of model building in population biology (1966). *Quarterly Review of Biology*, **68**, 533–546.

Orzack SH & Sober E (1994a) Optimality models and the test of adaptationism. *American Naturalist*, **143**, 361–380.

Orzack SH & Sober E (1994b) How (not) to test an optimality model. *Trends in Ecology and Evolution*, **9**, 265–267.

Orzack SH, Parker ED Jr & Gladstone J (1991) The comparative biology of genetic variation for conditional sex ratio adjustment in a parasitic wasp, *Nasonia vitripennis*. *Genetics*, **127**, 583–599.

Parker GA & Maynard Smith J (1990) Optimality theory in evolutionary biology. *Nature*, **348**, 27–33.

Peck O (1963) Catalogue of the Nearctic Chalcidoidea (Insecta: Hymenoptera). *Canadian Entomologist*, *Supplement*, **30**, 1–1092.

Pendleton BF, Newman I & Marshall RS (1983) A Monte Carlo approach to correlation spuriousness and ratio variables. *Journal of Statistical Computing and Simulation*, **18**, 93–124.

Peterson MA & Denno RF (1998) Life-history strategies and the genetic structure of phytophagous insect populations. In: S Mopper & SY Strauss (eds) *Genetic Structure and Local Adaptation in Natural Insect Populations*, pp 263–322. New York: Chapman & Hall.

Pijls JWAM, van Steenbergen HJ & van Alphen JJM (1996) Asexuality cured: the relations and differences between sexual and asexual *Apoanagyrus diversicornis*. *Heredity*, **76**, 506–513.

Pintureau B, Chapelle L & Delobel B (1999) Effects of repeated thermic and antibiotic treatments on a *Trichogramma* (Hym., Trichogrammatidae) symbiont. *Zeitschrift für Angewandte Entomologie*, **123**, 473–483.

Plantard O, Rasplus JY, Mondor G, Le Clainche I, & Solignac M (1998) *Wolbachia*-induced thelytoky in the rose gallwasp *Diplolepis spinosissimae* (Giraud) (Hymenoptera Cynipidae), and its consequences on the genetic structure of its host. *Proceedings of the Royal Society of London, series B*, **265**, 1075–1080.

Proulx SR (2000) The ESS under spatial variation with applications to sex allocation. *Theoretical Population Biology*, **58**, 33–47.

Queller DC (1984) Pollen-ovule ratios and hermaphrodite sexual allocation strategies. *Evolution*, **38**, 1148–1151.

Rau P (1947) Bionomics of *Monodontomerus mandibularis* Gahan, with notes on other Chalcids of the same genus. *Annals of the Entomological Society of America*, **40**, 221–226.

Ridley M (1996) *Evolution*. Cambridge, MA: Blackwell Science, Inc.

Rimsky-Korsakov M (1916) Observations biologiques sur les Hymenopteres aquatiques. *Russkoe Entomologicheskoe Obozrenie*, **16**, 209–225.

Roeder CM (1992) Sex ratio response of the two-spotted spider mite (*Tetranychus urticae* Koch) to changes in density under local mate competition. *Canadian Journal of Zoology*, **70**, 1965–1967.

Roff DA (1992) *The Evolution of Life Histories: Theory and Analysis*. New York: Chapman & Hall.

Schluter D & Grant PR (1982) The distribution of *Geospiza difficilis* in relation to *G. fuliginosa* in the Galapagos Islands: tests of three hypotheses. *Evolution*, **36**, 1213–1226.

Simberloff D (1983) Sizes of coexisting species. In: DJ Futuyma & M Slatkin (eds) *Coevolution*, pp 404–430. Sunderland, MA: Sinauer Associates, Inc.

Sober E & Wilson DS (1998) *Unto Others*. Cambridge, MA: Harvard University Press.

Stary P (1999) Biology and distribution of microbe-associated thelytokous populations of aphid parasitoids (Hym., Braconidae, Aphidiinae). *Journal of Applied Entomology*, **123**, 231–235.

Stiling P (1988) Density-dependent processes and key factors in insect populations. *Journal of Animal Ecology*, **57**, 581–593.

Stouthamer R & Kazmer DJ (1994) Cytogenetics of microbe-associated parthenogenesis and its consequences for gene flow in *Trichogramma* wasps. *Heredity*, **73**, 317–327.

Stouthamer R & Luck RF (1993) Influence of microbe-associated parthenogenesis on the fecundity of *Trichogramma-deion* and *T-pretiosum*. *Entomologia Experimentalis et Applicata*, **67**, 183–192.

Stouthamer R, Luck RF & Hamilton WD (1990) Antibiotics cause parthenogenetic *Trichogramma* (Hymenoptera/Trichogrammatidae) to revert to sex. *Proceedings of the National Academy of Sciences USA*, **87**, 2424–2427.

Strong DR Jr, Szyska LA & Simberloff DS (1979) Tests of community-wide character displacement against null hypotheses. *Evolution*, **33**, 897–913.

Suzuki Y & Iwasa Y (1980) A sex ratio theory of gregarious parasitoids. *Researches on Population Ecology*, **22**, 366–382.

Tardieux I & Rabasse JM (1988) Induction of a thelytokous reproduction in the *Aphidius colemani* (Hym., Aphidiidae) complex. *Journal of Applied Entomology*, **106**, 58–61.

Taylor CE, Powell JR, Kekic V, Andjelkovic M & Burla H (1984) Dispersal rates of species of the *Drosophila obscura* group: implications for population structure. *Evolution*, **38**, 1397–1401.

Toro MA & Charlesworth B (1982) An attempt to detect genetic variation in sex ratio in *Drosophila melanogaster*. *Heredity*, **49**, 199–209.

Trivers RL & Hare H (1976) Haplodiploidy and the evolution of the social insects. *Science*, **191**, 249–263.

Trivers RL & Willard DE (1973) Natural selection of parental ability to vary the sex ratio of offspring. *Science*, **179**, 90–92.

Uyenoyama MK & Bengtsson BO (1982) Towards a genetic theory for the evolution of the sex ratio III. Parental and sibling control of brood investment ratio under partial sib-mating. *Theoretical Population Biology*, **22**, 43–68.

van den Assem J, van Iersel JJA & Los-den Hartogh RL (1989) Is being large more important for female than for male parasitic wasps? *Behaviour*, **108**, 160–195.

Vet LEM (1995) Parasitoid foraging: the importance of variation in individual behavior for population dynamics. In: RB Floyd & AW Sheppard (eds) *Frontiers and Applications of Population Ecology*, pp 1–17. Melbourne: CSIRO.

Waage JK (1986) Family planning in parasitoids: adaptive patterns of progeny and sex allocation. In: JK Waage & D Greathead (eds) *Insect Parasitoids*, pp 63–95. London: Academic Press.

Weinstein P & Austin AD (1996) Thelytoky in *Taeniogonalos venatoria* Riek (Hymenoptera: Trigonalyidae), with notes on its distribution and first description of males. *Australian Journal of Entomology*, **35**, 81–84.

Werren JH (1980) Sex ratio adaptations to local mate competition in a parasitic wasp. *Science*, **208**, 1157–1159.

Werren JH (1983) Sex ratio evolution under local mate competition in a parasitic wasp. *Evolution*, **37**, 116–124.

Werren JH (1991) The paternal-sex-ratio chromosome of *Nasonia*. *American Naturalist*, **137**, 392–402.

West SA & Herre EA (1998a) Partial local mate competition and the sex ratio: a study on non-pollinating fig wasps. *Journal of Evolutionary Biology*, **11**, 531–548.

West SA & Herre EA (1998b) Stabilizing selection and variance in fig wasp sex ratios. *Evolution*, **52**, 475–485.

Wilkes A & Lee PR (1965) The ultrastructure of dimorphic spermatozoa in the hymenopteran *Dahlbominus fuscipennis*. *Canadian Journal of Genetics and Cytology*, **7**, 609–619.

Willard HF (1927) Parasites of the pink bollworm in Hawaii. *United States Department of Agriculture Technical Bulletin*, **19**, 1–15.

Wrensch DL & Ebbert MA (1993) *Evolution and Diversity of Sex Ratio in Insects and Mites*. New York: Chapman & Hall.

Wylie HG (1973) Control of egg fertilization by *Nasonia vitripennis* (Hymenoptera: pteromalidae) when laying on parasitized house fly pupae. *Canadian Entomologist*, **105**, 709–718.

Chapter 20

Using sex ratios: why bother?

Stuart A. West & Edward Allen Herre

20.1 | Summary

Many see research into sex allocation as the jewel in the crown of evolutionary ecology. There is a very rich experimental literature providing qualitative, and in some cases quantitative, support for the predictions of numerous theoretical models. Consequently, it might be argued that future work will primarily be concerned with dotting i's and crossing t's. Given that there are still so many relatively untamed areas in evolutionary biology, we should therefore ask – why bother with more sex-allocation studies? Our aim in this chapter is to address this question (why?), complementing the more methodological (how?) parts of this book. We argue that sex allocation is an excellent model trait for examining general questions in evolutionary biology.

20.2 | The usefulness of sex allocation

The strength of sex-allocation research arises for both theoretical and empirical reasons. Sex allocation has a direct and potentially large influence on fitness, and the relevant trade-offs are easy to quantify. Consequently, optimality models are able to make clear theoretical predictions in many specified cases. Empirically, sex allocation can be a relatively easy trait to measure. This is especially true in cases where males and females are equally costly to produce, and so we can con-

cern ourselves simply with the sex ratio (defined as proportion males, i.e. males/(males+females)). In this case, all we must do is count the number of male and female offspring that are produced. Taken together, these theoretical and empirical considerations mean that in many cases we can make clear, and often quantitative, predictions for an easily measurable trait.

Sex-allocation research is useful for at least three broader reasons. First, work on sex allocation can have applied uses, such as in biological control, whilst Chapter 15 illustrates how sex allocation can be used to estimate the population structure of clinically important parasites. Second, the pattern of sex allocation has fundamental consequences for numerous other fields of research, such as the evolution of other life-history traits and population dynamics. An example in this book is given by Chapter 18, which discusses how sex ratio may influence the form that sexual selection takes.

Our chapter is concerned with one other use of sex allocation. We argue that sex allocation is an excellent model trait (or tool) with which to study general questions. The main part of our chapter is split into three sections. Section 20.3 briefly summarizes relevant theoretical predictions, section 20.4 provides some examples of how sex-allocation research has been used to examine general questions. Specifically, we examine some of the insights that sex-allocation research has provided into the levels at which selection acts (genetic element, individual, kin). Then, section 20.5 discusses several areas in which we believe future research into sex allocation provides excellent opportunities to address general questions. We suggest how sex-allocation research can be used to: (1) examine how individuals process information about their environment; (2) test whether and how animals recognize kin; (3) indirectly estimate important characteristics of natural populations, such as the selfing rate or factors limiting reproductive success, and (4) determine the importance of different factors that may lead to phenotypic and genetic variation in traits.

20.3 | Background theory

Before discussing how sex-allocation research can be used as a tool for addressing general questions, we briefly summarize relevant predictions of theory. In particular, we explain: (1) Fisher's (1930) principle of equal investment in the sexes and (2) the evolution of biased sex allocation in structured populations (local mate competition; Hamilton 1967).

20.3.1 Fisher's theory of equal investment in the sexes

Fisher (1930) pointed out that, all else being equal, natural selection favours equal investment in the two sexes. Consider the case where males and females are equally costly to produce. If there were an excess of males, they would on average obtain less than one mate, and so the fitness of females would be greater, favouring parents that produced a relative excess of female offspring. In contrast, if there were an excess of females, males would on average obtain more than one mate, and so the fitness of males would be greater, favouring parents that produced a relative excess of male offspring. Consequently, the fitness of males and females is only equal when equal numbers of the two sexes are produced (a sex ratio of 0.5). If males and females are not equally costly to produce then the argument is phrased in terms of investment, and the evolutionarily stable strategy (ESS) is to invest equally in male and female offspring.

Fisher's principle clearly shows the frequency-dependent nature of selection on the sex ratio,

and it provides a null model (equal investment in the sexes) which is the foundation block upon which most areas of sex-allocation research are built. There has been relatively little empirical work on Fisher's principle itself (Basolo 1994, Carvalho et al. 1998): much of the most productive research has investigated what happens when its explicit and implicit assumptions are violated (Bull & Charnov 1988, Frank 1990).

Fisher's principle can apply widely to organisms where equal investment is made in male and female offspring (e.g. equal egg size and no parental care) and mating occurs in large, effectively panmictic, populations. However, Fisher's principle may not hold if parents are able to provide different amounts of resources to male and female offspring. If one sex gains more from added investment, then selection favours: (1) investing more resources in each individual of that sex (Trivers & Willard 1973, Charnov 1979, Charnov et al. 1981); (2) a numerical sex ratio usually biased towards the other sex (Bull 1981, Charnov 1982, Frank & Swingland 1988, Charnov & Bull 1989) and (3) an overall resource allocation ratio that may be biased towards either sex, but generally towards the sex that gains more from added investment (Frank 1987, 1995, 1998, Frank & Swingland 1988). It is frequently underappreciated that equal overall investment in the sexes is only favoured when the shape of the relationships between fitness and investment are the same (identical or both linear) for male and female offspring (Frank 1990, 1998).

20.3.2 Local mate competition

Several different mechanisms can lead to the evolution of biased sex allocation. However, the branch of sex-allocation research that has been the most productive, and is the most useful for several areas in this chapter, is the study of female-biased sex allocation in species with structured populations, frequently termed local mate competition (LMC; Hamilton 1967). LMC is one of a general class of models in which interactions between relatives can lead to biased sex allocation (Taylor 1981).

Hamilton (1967) was the first to demonstrate that when the offspring of one or a few mothers mate amongst themselves in their natal patch,

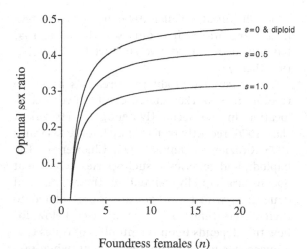

Foundress females (n)

Fig 20.1 The unbeatable evolutionarily stable (ESS) sex ratio under conditions of local mate competition. The ESS sex ratio (r) is plotted against the number of foundresses (n) for diploid species (eq. 20.1), and for haplodiploid species with the proportion of sibmated females (s) equal to 0, 0.5 or 1.0 (eq. 20.2).

before daughters disperse, a female-biased sex ratio is favoured by natural selection. Hamilton predicted that, assuming sons and daughters are equally costly to produce, the unbeatable (ESS) sex ratio (proportion of males) on a patch (r') would be given by:

$$r' = \frac{n - 1}{2n}, \qquad (20.1)$$

where n is the number of foundress females that contribute offspring to the patch (Figure 20.1). Sex ratios thus decline from 0.5 for large n to 0 when $n = 1$, the latter interpreted as meaning that the female should produce the minimum number of males required to fertilize all her daughters.

The reasons for biased sex allocation under LMC have been the subject of considerable debate. However, it is now generally agreed that LMC can be equally well explained by at least two approaches (Godfray 1994, Frank 1998). These two approaches are both correct (give the same answers), representing different notational conventions and tools of reasoning, for the same evolutionary logic. Taylor (1981) showed, using an approach based on selection at the individual level, that the female bias arises because it reduces competition among brothers for mates,

and because it increases the number of mates for each of the female's sons. In contrast, Frank (1986, 1998, see also Taylor & Bulmer 1980, Colwell 1981) has shown that the bias can be explained with an approach that emphasizes selection within and between groups. Producing a female-biased sex ratio decreases fitness relative to other members of the same group (patch), but increases the overall productivity of the patch.

In haplodiploid species an additional factor biases the sex ratio. Inbreeding causes mothers to be more related to their daughters than their sons, and so a slightly more female-biased sex ratio is favoured (Frank 1985, Herre 1985). Specifically, the unbeatable (ESS) sex ratio becomes:

$$r' = \frac{(n - 1)(2 - s)}{n(4 - s)}, \qquad (20.2)$$

where s is the proportion of sibmated females (Frank 1985, Herre 1985, Werren 1987, Figure 20.1). Several features of LMC make it particularly useful. First, Hamilton's original model has been extended in numerous ways, providing a very strong theoretical background for studies in this area (reviewed by Godfray 1994, Hardy 1994). In particular, we know which parameters can have the most influence on quantitative predictions, and so what is and is not likely to be important. Second, it is one of the branches of sex-allocation research in which we are able to make the clearest (quantitative) predictions about how sex allocation should vary under different field or laboratory conditions. Third, in many species studied the investment in male and female offspring appears to be equal (e.g. an egg), and so we must merely determine the numerical sex ratio.

20.4 | Sex allocation and levels of selection

In this section we briefly review two examples of how sex-allocation research has been used to address general evolutionary questions. Specifically, we show how sex-allocation research has provided some of the clearest demonstrations of how selection at different levels (genetic element, individual, kin) interact (Leigh *et al.* 1985).

20.4.1 Intragenomic conflict and selfish genetic elements

Theory suggests that a genetic element can be selected to increase its own transmission, even if this carries a fitness cost to the individual that carries it (Hurst 1993). Some of the clearest examples of selfish genetic elements are the sex ratio distorters that have been discovered in a wide range of organisms (Chapter 9). For example, the parasitoid wasp *Nasonia vitripennis* harbours a nonessential (supernumerary) or B chromosome that has been termed PSR (paternal sex ratio). Like other Hymenoptera, *N. vitripennis* is haplodiploid: females develop from fertilized eggs, and males from unfertilized eggs (see Chapter 8). Conflict arises because PSR is only present in and passed on to male offspring. In order to maximize its own transmission, PSR causes improper condensation and eventual loss of the paternal chromosomes, except for itself (Chapter 9). Consequently, diploid eggs, which would normally develop into females, are converted into haploid eggs that turn into males infected with PSR. This is effective in >99% of fertilized eggs, leading to the production of broods containing only males by females that mated with a PSR-carrying male. In each generation, PSR ensures its own transmission at the cost of the rest of the genome of males that carry it: you cannot get much more selfish than that.

20.4.2 Kin selection

Hamilton (1964) showed that individuals can be selected to increase their own (inclusive) fitness by helping relatives reproduce and hence gain indirect fitness (kin selection). Kin selection theory provides a theoretical basis for understanding the evolution of social behaviour, and a framework for predicting reproductive characteristics in social organisms. The same theoretical principles underlie kin selection, social evolution and sex allocation (Frank 1998). Consequently, sex allocation has been central to our understanding of kin selection, and has allowed the clearest empirical tests of theoretical predictions (Frank 1998, Chapuisat & Keller 1999). Indeed, with few, if any, exceptions, it is only sex allocation that has provided quantitative support for kin selection theory.

A particularly productive area has been the resolution of worker–queen conflict over sex allocation in the social Hymenoptera (Trivers & Hare 1976, recently reviewed by Bourke & Franks 1995, Crozier & Pamilo 1996, Chapter 4). The haplodiploid genetics of such species means that queens are equally related to their sons and daughters, whilst workers are more related to their sisters than their brothers (see below for how this depends upon the number of times that a queen has mated). This means that, when producing reproductives, queens should favour an equal investment in the sexes, whilst workers should favour an investment biased towards females (Trivers & Hare 1976).

Intraspecific variation in queen mating frequencies provides a powerful opportunity to study the resolution of conflict over sex allocation because the extent of conflict depends upon how many times a queen has mated (Boomsma & Grafen 1990, 1991). A queen is always equally related to both sons and daughters, irrespective of how many times she has mated (assuming no inbreeding). However, the differential relatedness of a worker to females and males produced in the colony decreases as queen mating frequency increases. Theory therefore predicts that workers should rear mainly or only females in colonies with relatedness asymmetries above the population average, and mainly or only males in colonies with relatedness asymmetries below the average (Boomsma & Grafen 1990, 1991).

Empirical work has shown that the sex ratio of adults produced correlates with relatedness asymmetry as predicted by theory (Sundstrom 1994), and that these result from worker manipulation (Sundstrom *et al.* 1996). These results provide clear support for the theoretical concepts of social evolution (other examples are reviewed by Chapuisat & Keller 1999). In addition, they suggest that workers have a mechanism for accurately determining queen mating frequency, possibly by assessing genetic relatedness to other workers (section 20.5.2).

20.5 | Future possibilities

In this section we discuss some further areas where studies of sex allocation provide excellent opportunities for addressing general evolutionary and ecological questions.

20.5.1 How do individuals process relevant information about their environment?

This is a general problem that has implications for many optimality models (Parker & Maynard Smith 1990), because the optimal behaviour can depend upon a number of factors that the organism must assess. Indeed, the lack of a mechanistic basis to most theoretical models has been a major criticism of the optimality approach (Krebs & Kacelnik 1991). In this section we argue the usefulness of sex-allocation research for investigating this issue.

The advantage of sex allocation for investigating information processing is that theoretical predictions have been shown to depend very clearly upon the quality of information that individuals have about their environment (Nunney & Luck 1988, Stubblefield & Seger 1990, Taylor & Crespi 1994, Greeff 1997). Usually two extreme cases are modelled, with individuals having either 'complete knowledge' (i.e. perfect knowledge of all relevant parameters) or 'self knowledge' only (i.e. no information about what other individuals are doing). For example, Stubblefield and Seger (1990) considered an LMC model in which the different females on a patch lay different numbers of eggs. The general prediction is that females which lay larger clutches should produce a more female-biased sex ratio. However, the exact form of the relationship between sex ratio (or number of males produced) and number of eggs laid, both within and across patches, depends upon whether individuals have self or complete knowledge.

A start at testing some of these models has been made by Flanagan et al. (1998). They examined the sex ratio produced by females of the parasitoid N. vitripennis under LMC conditions. Female size was manipulated in order to alter fecundity, and therefore the number of eggs that a female lays. The results gave a qualitative fit to the complete knowledge model of Stubblefield and Seger (1990). The lack of a perfect fit may be explained by the fact that females appeared to use the body size of other females ovipositing on a patch to estimate how many eggs they would lay, rather than being able to actually count the eggs. Body size is a reasonable predictor of how many eggs a female will lay (Flanagan et al. 1998). The use of such a 'rule of thumb' behaviour is not unusual in cases where the complete information cannot be processed (e.g. Davies 1992).

Note that most theoretical models implicitly assume complete knowledge. This has two important consequences. First, empirical and theoretical research on the consequences of how individuals process information about the environment could be extended to numerous fields of sex allocation where facultative strategies are predicted. For example, in section 20.5.2 we discuss how sex allocation can be used to determine the actual cues that an organism uses to recognize kin. Other possibilities include how individuals assess foundress number under conditions of LMC (Strand 1988, Godfray 1994, West & Herre 1998a), resource availability (Charnov et al. 1981, Frank 1995) or population perturbations (Werren & Taylor 1984, West & Godfray 1997). Second, deviations from supposed optimal strategies predicted by any model can arise because the model assumes complete knowledge, while the organisms considered do not possess this. For example, Orzack and Parker (1990) suggest that female N. vitripennis may not be able to determine the exact number of eggs that had been laid in a previously parasitized host, a crucial parameter influencing the extent of LMC (Werren 1980). In the extreme case, a complete lack of knowledge can lead to individuals not being able to alter facultatively their behaviour in response to variable conditions, and just pursuing the optimal 'average' strategy (Nunney & Luck 1988). This point has a number of general implications and we return to it when discussing the cause of individual variation in traits (section 20.5.4) and limits to the application of optimality models (section 20.6).

20.5.2 Do individuals recognize kin?

The ability of individuals to recognize kin has implications for many areas in evolutionary biology, such as mate choice and the evolution of cooperative behaviour (reviewed by Charlesworth & Charlesworth 1987, Frank 1998). Despite its importance, identifying the cues that organisms use to recognize their kin remains controversial (Grafen 1990). In some cases individuals may use genetic cues (e.g. Sundstrom et al. 1996), whilst in other cases environmental cues are used to determine indirectly which individuals are *likely* to be close relatives (e.g. Ode et al. 1995).

Sex-allocation theory provides simple and clear ways to test whether animals recognize kin, and to investigate the cues involved. In some cases, the optimal sex-allocation strategy depends upon relatedness between interacting organisms, and consequently whether and how individuals can recognize kin. We have already discussed how the patterns of sex allocation in social insect colonies show that workers can assess their relatedness to nest mates (section 20.4.2). However, this method is only applicable to social haplodiploid species.

LMC theory allows us to test the ability of solitary haplodiploid species to recognize kin. In haplodiploid species, all else being equal, individuals that are mated to a sibling are predicted to produce more female-biased sex ratios than outbred individuals (Greeff 1996, eq. 20.2, Figure 20.1). This prediction could be tested with experiments on any haplodiploid species that has female-biased sex ratios due to LMC (e.g. the parasitoid N. vitripennis). All that would be required is to measure and compare the sex ratios produced by individuals that were either inbred or outbred, when on a patch with a certain number of other females. This could be done by measuring the average sex ratios of groups of individuals that were all inbred or outbred (e.g. Werren 1983), or by using an eye colour mutant (or genetic markers) to follow the behaviour of individuals in a group (e.g. Orzack et al. 1991, Molbo & Parker 1996, Flanagan et al. 1998). Alternatively, this prediction could be tested on natural populations with the use of genetic markers to determine if individuals were inbred or sibmated (Molbo & Parker

1996). If such differences did occur, manipulation experiments could then be used to test whether kin are recognized by some environmental cue (e.g. developing in the same host, Ode et al. 1995) or direct genetic recognition.

LMC models could also be used to test whether parasitic micro-organisms are able to recognize kin, in terms of determining with how many genotypes (clones) they are sharing a host. If they can do this, then species subject to LMC would be able to adjust their sex ratio facultatively in response to the likelihood of outcrossing: less female-biased sex ratios should be produced in hosts that have been infected with more genotypes and where outcrossing is more likely (Chapter 15). There is some evidence from rodent and lizard malaria parasites that such facultative sex allocation does occur (Taylor 1997, Pickering et al. 2000). Moreover, these data suggest that an indirect environmental cue (infection levels) and not direct genetic recognition is the mechanism involved. However, there are instances where the pattern is not shown, and further work is required to resolve this issue (Chapter 15, West et al. 2001).

20.5.3 What can sex allocation tell us about a species or its ecology?

Traditionally, work on sex allocation has centred around constructing theoretical models and collecting data to test those models. More recently, it has been realized that in cases where we believe the major assumptions of models to apply, this idea can be turned around, and sex allocation used as a relatively easy (and cheap) way to gain an indirect estimate of something about an organism and its ecology (West et al. 2000a). The importance of sufficient biological knowledge to appropriately carry out such work cannot be understated. In cases where theory fails, attention is focused on the biological attributes of the species involved, which may lead to previously unrecognized features of the natural history which violated the assumptions of sex-allocation models (e.g. Shutler & Read 1998). This is an area with enormous scope for future research, and we discuss three examples where there has already been progress.

20.5.3.1 Estimating selfing rates

Understanding the population structure of parasitic micro-organism populations is important for a number of clinical and epidemiological reasons such as the evolution of virulence or drug resistance. However, direct genetic estimates of population structure can be expensive and time consuming to obtain. Read *et al.* (1992) pointed out that LMC theory offers a relatively cheap and easy way to indirectly estimate an important component of parasite population structure, the selfing rate (for a more detailed discussion see Chapter 15). Basically, if the sex ratio observed in a natural population is assumed to reflect the level of LMC then eq. 1 can be rearranged to predict the selfing rate (s; assuming $s = 1/n$) from an observed sex ratio (r). Specifically, $s = 1 - 2r$ (Read *et al.* 1992). Importantly, in cases where there are indirect sex ratio and direct genetic estimates of the population structure, they are in quantitative agreement, supporting the use of this methodology (Read *et al.* 1992, Paul *et al.* 1995, West *et al.* 2000b).

So far these methods have been most used with parasitic protozoa, where estimates of the selfing rate have clinical importance. However, similar methodology could be applied to a variety of taxa, such as estimating relative selfing rates across hermaphroditic plants (Cruden 1977).

20.5.3.2 How are parasites distributed across hosts?

Another important aspect of parasite population structure is the distribution between hosts. Parasites may be distributed across their hosts in one of three possible ways: (1) randomly distributed between hosts (Poisson distribution); (2) aggregated at certain hosts (overdispersed; negative binomial distribution) and (3) spread evenly among hosts (underdispersed).

Sex allocation is an indirect method for estimating the form of aggregation in species subject to LMC, when mating occurs between individuals from a host (i.e. the host represents the patch in Hamilton's (1967) classic model). If the proportion of hosts that are infected is known, then the average number of strains in an infected host can be estimated for any form of distribution (Read *et al.* 1995, West & Herre 1998a). Given this,

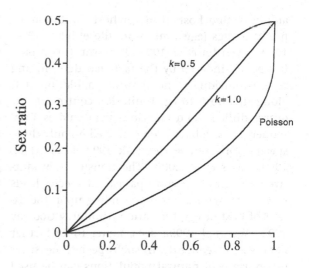

Fig 20.2 The predicted relationship between sex ratio and proportion of hosts parasitized. The different lines represent cases when parasites are distributed randomly between hosts (Poisson) or aggregate at certain fruit (k is the aggregation parameter of the negative binomial distribution – low k describe high degrees of aggregation).

the predicted relationship between proportion of hosts infected and the sex ratio can be calculated for any type of distribution (Figure 20.2). These predicted relationships can then be compared to data from one or more species in order to determine the form of clumping that the sex ratio data suggest.

Sex ratio data from several protozoan blood parasite species in the genera *Leucocytozoon* and *Plasmodium* suggest that strains are aggregated (Read *et al.* 1995). Although data from these genera are lacking (Read *et al.* 1995), aggregated distributions are extremely common in other parasites (Anderson & May 1991). Similarly, sex ratio data from several nonpollinating fig wasp species in the genera *Idarnes* suggest that ovipositing females were aggregated (West & Herre 1998a). This suggestion has been supported by other data (West *et al.* 1996, JM Cook, unpublished data).

20.5.3.3 What limits reproductive success?

There has recently been considerable theoretical debate over the extent to which the reproductive success of parasitoids is limited by the ability to find hosts (time or host limitation) or produce eggs (egg limitation) (Rosenheim 1996, 1999, Sevenster *et al.* 1998). However, this debate has been hindered by the fact that determining where individuals and natural populations fall along this host- to egg-limitation continuum requires difficult and laborious field studies. Consequently, data have been collected on only three species (Driessen & Hemerik 1992, Ellers *et al.* 1998, Casas *et al.* 2000). The unusual reproductive strategies of certain parasitoid species leads to the optimal sex ratio depending upon the extent of host or egg limitation (Hunter & Godfray 1995, West *et al.* 1999a, see Chapter 10 for further details). Consequently, in such species, measures of sex ratio in natural populations can be used as an easy way to indirectly estimate the extent of host or egg limitation (West & Rivero 2000). The relative ease with which these data can be collected means that we can obtain estimates for a large number of species and/or multiple estimates from the same species.

Currently, suitable sex ratio data from field populations are available for eight parasitoid species (Donaldson & Walter 1991, Hunter 1993, West *et al.* 1999a). Overall, the data suggest that: (1) the extent of host or egg limitation in a species varies between site collected and time of year; (2) the overall species means are at an intermediate position on the egg- to host-limitation continuum, with a bias towards host limitation (West & Rivero 2000).

20.5.4 What is the cause of individual variation in traits?

One of the fundamental aims of evolutionary biology is to understand and predict the distributions of phenotypic traits. In this section we concentrate on two areas where work on sex allocation is able to provide particular insights: (1) the evolution of facultative or conditional strategies (phenotypic plasticity) and (2) explaining heritable genetic variation.

20.5.4.1 Phenotypic plasticity

Selection can favour individuals who facultatively adjust their phenotype in response to the conditions in which they find themselves (Stearns 1992). Sex allocation is an excellent trait for examining general points about how phenotypic

plasticity evolves because: (1) relevant environmental variation (e.g. foundress number or host quality) can be easily quantified and/or manipulated; (2) the fitness consequences of different phenotypes (e.g. offspring sex ratio) in different environments can be easily calculated and (3) the phenotype expressed in different environments can be easily measured. These points have led to sex allocation providing some of the most striking and clear examples of adaptive phenotypic plasticity (e.g. Werren 1980, Charnov *et al.* 1981, Herre 1985, 1987, Sundstrom 1994, Komdeur *et al.* 1997). For example, under conditions of LMC, females are expected to adjust their sex ratio in response to a number of factors such as the number of females ovipositing in a patch (section 20.3) and the relative fecundity of the different females (section 20.5.1). More generally, these data provide some of the best evidence for adaptationist theory (West *et al.* 2000a, Herre *et al.* 2001).

Here we discuss how work on sex allocation can be used to investigate the constraints on the evolution of phenotypic plasticity. This is clearly related to the more general question of the limits to natural selection and adaptation. We would not expect natural selection to produce an organism that responds perfectly to any situation it might encounter, let alone situations it never encounters. Limits will be placed upon selection by trade-offs, physical constraints or by the continuous input of deleterious mutations (Stearns 1992, Kawecki 1994, Whitlock 1996, Reboud & Bell 1997). Given this, how precisely should an organism be adapted? This is extremely hard to study because it requires a detailed knowledge of how variation in a trait affects fitness under all the situations that are encountered, as well as how frequently different situations occur (Seger & Stubblefield 1996). To our knowledge, studies of sex allocation in fig-pollinating wasps (Herre 1987, Herre *et al.* 1997, 2001) are the only ones that have been able to address this question in natural populations.

Female fig-pollinating wasps adjust the sex ratio of their offspring in response to the number of foundress females that lay eggs in a patch, as predicted by eq. 20.2 (Herre 1985, 1987, Herre *et al.* 1997, 2001). However, in cases where the fit of data to theory was not perfect, the deviations from the optimal sex ratio were not random.

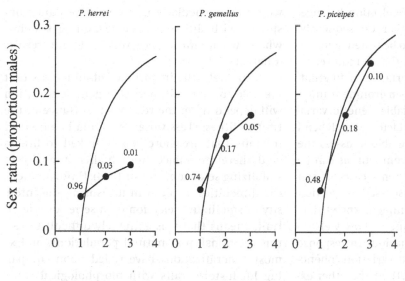

Number of females laying eggs in a fruit

Fig 20.3 Precision of adaptation and the sex ratios of fig-pollinating wasps. The observed sex ratios (circles) and theoretical optima (curved lines) when different numbers of females lay eggs in a fruit, for three fig pollinating wasp species in the genus *Pegoscapus*. The numbers show the relative frequency with which different numbers of females lay eggs in a fruit in nature. In these and other species, the observed sex ratios are closest to theoretical predictions in situations that were encountered most frequently. Also, the observed shifts in sex ratio are greatest in species where the number of mothers laying eggs in a fruit are more variable.

Specifically, across species, the mean sex ratios produced for a given number of foundresses were closest to theoretical predictions in the situations (number of foundresses entering a fruit) that they encountered most frequently (Figure 20.3). Females showed a greater ability to alter their brood sex ratios in response to variable foundress number (phenotypic plasticity) in species where foundress number showed greater variation (more variable selective regime).

There is enormous scope for future research in this area, with both natural and laboratory populations. For example, in natural populations such as fig wasps, given the distribution of foundress numbers that occur in nature and the sex ratios produced, to what extent does fitness deviate from the maximum possible? Another possible line of research would be to carry out artificial selection experiments in the laboratory, manipulating how foundress number varies. Such laboratory experiments would be important for at least two reasons. First, the field data are correlational and so other factors may have feasibly contributed to the observed patterns (Kathuria *et al.* 1999). Second, experimental manipulations would allow more detailed questions to be posed, such as about how quickly phenotypic plasticity evolves or fades away. Similar approaches could be taken when investigating the consequences of the selective regime for sex-allocation problems other than LMC, such as

the resources available for reproduction (Charnov *et al.* 1981). Another related issue is that the extent and form of phenotypic plasticity are predicted to depend enormously upon how individuals process information about their environment (section 20.5.1).

20.5.4.2 Heritable genetic variation

Simple optimality models generally predict a single value for a trait. However, even when measured in the same environment, traits usually show phenotypic variation, and this variation can be genetic (Orzack *et al.* 1991). Explaining such variation remains a major challenge, and potential explanations could involve any of a number of factors acting at several levels (Barton & Turelli 1989). Numerous theoretical quantitative genetic models have been developed to examine the factors that can maintain genetic and therefore phenotypic variation in a trait (e.g. Lande 1975, Turelli 1984, Barton & Turelli 1989). For example, deleterious mutations may maintain genetic and phenotypic variation at mutation–selection balance, or fluctuating selection and genotype-by-environment interactions (different genotypes have the highest fitness in different environments) may prevent a single phenotype from spreading to fixation. Sex allocation provides a trait with which the fitness consequences of variation can be determined more easily than with other traits used frequently to study mutation–selection balance, such as *Drosophila* bristle number. Consequently, it allows general predictions to be tested, and emphasizes the theoretical work that is required.

We turn first to the importance of fluctuating selection and genotype-by-environment interactions in maintaining heritable genetic variation. The importance of these factors will depend upon how well individuals are able to assess the relevant variation in the environment, as can be shown by considering the extreme cases of self and complete knowledge (discussed in section 20.5.1). If individuals have complete knowledge, then the question becomes one of how well individuals can adjust their behaviour in response to the relevant environmental variation (phenotypic plasticity, section 20.5.4.1). At the other extreme, if individuals cannot assess variation (self knowledge) then they can only adopt an average strategy for their genotype (Nunney & Luck 1988, Stubblefield & Seger 1990) and genotype-by-environment interactions can occur, helping to maintain genetic variation in a trait (Barton & Turelli 1989). Realistically, we might expect many if not most situations to fall between these two extremes, with individuals having some, but not perfect, knowledge about relevant environmental variation (e.g. Flanagan *et al.* 1998). Theoretical work is required to determine the consequences of this for maintaining variation: most relevant quantitative genetic models implicitly assume self knowledge.

We turn next to the importance of mutation–selection balance in maintaining heritable genetic variation. In contrast to work on many other life-history traits, studies of sex allocation have rarely considered mutation–selection balance. Consequently, before we can progress further in this area, empirical work is required to estimate a number of parameters such as the amount of genetic variation in sex allocation produced each generation by mutation (Houle *et al.* 1996), the heritability of sex allocation (Parker & Orzack 1985, Varandas *et al.* 1997), the level of genetic variation for sex allocation in natural populations (Orzack & Parker 1990, Orzack *et al.* 1991, Orzack & Gladstone 1994), and how quickly sex allocation can evolve (Parker & Orzack 1985, Carvalho *et al.* 1998, Conover & van Voorhees 1990, Basolo 1994). It

would be particularly useful to have data from species with different sex-determining systems, where we might expect very different results (Chapters 7–9).

Ultimately, the importance of all factors that are able to maintain heritable genetic variation will depend upon the relative importance of the trait to fitness. Less variance should be observed in traits that are more closely linked to fitness (and therefore subject to a higher intensity of stabilizing selection, Turelli 1984). Unfortunately, the difficulties involved in measuring the intensity of stabilizing selection on a series of related traits means that it is extremely difficult to test this relationship in natural populations, and so most generalizations have relied upon comparing life-history traits with morphological traits (Merila & Sheldon 1999). Sex allocation has allowed this prediction to be tested, by examining a case in which phenotypic variation (which is likely to be linked with genotypic variation) could be related to the predicted intensity of stabilizing selection (West & Herre 1998b).

The offspring sex ratios produced under extreme LMC, when only one female oviposits in a patch (single foundress sex ratios), are subject to stabilizing selection: too many males reduces the total number of dispersing females, and too few males will result in unmated females (Green *et al.* 1982, West & Herre 1998b). West and Herre (1998b) showed theoretically that the intensity of stabilizing selection on single-foundress sex ratios is correlated with how frequently a species produces single-foundress broods in nature. Specifically, the intensity of stabilizing selection will be greater in species that encounter single-foundress broods more frequently, both because the trait is expressed more often and because fitness shows a greater sensitivity to variation (narrower fitness profile) when that trait is expressed. This prediction was supported by data from 16 species of Panamanian pollinating fig wasps, where the phenotypic variance in single-foundress sex ratios was negatively correlated with the frequency with which that species encounters single-foundress broods in nature (Figure 20.4, West & Herre 1998b).

As with the fig wasp studies on phenotypic plasticity described in section 20.5.4.1, experimental work is required to back up and extend

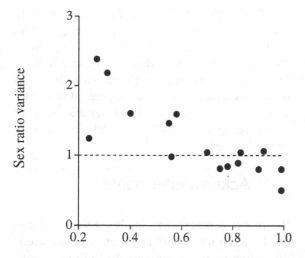

Fig 20.4 Sex ratio variance and single-foundress broods. The relationship, across 16 Panamanian fig-pollinating wasp species, between variance in single-foundress sex ratios (observed variance divided by that expected given a binomial distribution, see West & Herre, 1998b) and the proportion of single-foundress broods encountered in nature by that species. The dashed line represents binomial variation in the sex ratio.

this work on stabilizing selection. For example, the frequency with which females encounter single-foundress broods (selective regime) could be experimentally manipulated, and the consequences for the observed phenotypic variation determined. Further experiments could determine to what extent phenotypic variation represents heritable genetic variation, and assess the relative roles of the factors that may help to maintain (mutation, fluctuating selection) or reduce (removal of deleterious mutations, canalization) this variation (Barton & Turelli 1989, Stearns & Kawecki 1994).

20.6 | Conclusions

Sex allocation is possibly the most quantitatively well verified area in evolutionary biology. This has been achieved because clear predictions can be made for a trait that can be measured relatively easily. Our aim has been to emphasize that this allows sex allocation to be used for addressing a number of other general questions, of both pure and applied importance.

Sex-allocation is only one aspect of an organism's reproductive strategy. Other aspects, such as clutch size and mate choice/selection, also offer good opportunities for the study of general problems although, in our opinion, to a lesser extent than sex allocation. It is no coincidence that these have also been extremely productive areas of research (Seger & Stubblefield 1996). As with sex allocation, the advantage of these other traits is that reproduction is tightly linked to fitness and so it is possible to examine the consequences of variation in behaviour relatively simply. One field of research into reproductive strategies that does not benefit from the same advantages is the question of why organisms produce sexually, a topic that remains the subject of considerable debate (Hurst & Peck 1996, West *et al.* 1999b). The crucial difference here is that the fitness benefits of sex can take multiple generations to arise, and so are much harder to study.

The enormous body of research on sex allocation, and especially LMC, provides a case study of the use and limits of optimality (ESS) models when applied to real organisms (Herre *et al.* 2001). The aim of optimality models is to help us understand traits in terms of the selective forces and constraints that act upon them (Maynard Smith 1982, Parker & Maynard Smith 1990). Generally, if the predictions of the model match biological observations then we can hope to have identified the underlying and important biological assumptions. However, the theoretical and empirical advantages of sex allocation that we have emphasized in this chapter mean it is one of the few areas where we aspire to identify the important assumptions to a degree that a reasonably quantitative fit of data to theory can be expected (Seger & Stubblefield 1996). Consequently, sex-allocation studies have been taken to a more detailed level, enabling us to explore when and why there is a lack of fit between data and theory. The most obvious take-home messages from studies of sex allocation are: (1) more complex models can give different results, and even predict adaptive variation; (2) estimating the parameters of optimality models is as important as testing their predictions; (3) the most efficient progress can be made when theory and data go hand in hand, each stimulating the other, and gradually building our knowledge of a trait;

and (4) results should be considered within the context of their selective regime (see Herre *et al.* 2001, for a more detailed discussion).

There is increasing evidence for adaptive patterns of sex allocation in response to many factors in numerous taxa. Examples in this book include studies of parasitic protozoa (Chapter 15), plants (Chapters 16 & 17), invertebrates (Chapters 10–12) and vertebrates (Chapter 13). However, our chapter has been biased towards certain branches of sex-allocation research, both conceptually (primarily LMC) and empirically (particularly haplodiploid invertebrates). This is because the conceptual areas that are most useful for addressing general questions are those such as LMC, in which clear and dramatic shifts in sex allocation are predicted, and have been observed. The most useful study organisms for using sex allocation to address more general questions are those which have the greatest control over their sex allocation (allowing natural selection to produce adaptive sex-allocation patterns) and in which sex allocation is easily measured (e.g. by measuring the sex ratio), such as haplodiploid invertebrates.

The shifts in sex allocation of other taxa, such as mammals (Clutton-Brock & Iason 1986, Kruuk *et al.* 1999) and birds (Ellegren *et al.* 1996), are often more subtle, which can decrease their usefulness as a tool for addressing more general questions. It is conventionally assumed that chromosomal sex determination acts as a strong constraint on adaptive sex allocation in these species. However, some dramatic patterns of sex allocation have been observed in these taxa (e.g. Komdeur *et al.* 1997, Sheldon *et al.* 1999), suggesting that this may not be the only explanation (West *et al.* 2000a). Alternative possibilities include: (1) the adult lifespans in these taxa are relatively long compared with their juvenile development time, and so a greater number of factors can influence the optimal sex ratio (e.g. the added complications that can arise with overlapping generations (Werren & Taylor 1984, West & Godfray 1997), or from multiple types of interactions between relatives (Taylor 1981)); (2) differences in the reliability with which different taxa can process the relevant information (e.g. host quality or foundress number in parasitoids, versus mate quality or social rank in birds and mammals; West *et al.* 2000a). Comparative analyses of sex ratio skews from a wide range of taxa could be used to test these alternative hypotheses (West *et al.* 2000a). Nonetheless, we hope that the increasing attention being given to a wide variety of taxa, as illustrated by the chapters of this book, will lead to further examples of sex-allocation research being used as a tool for addressing more general issues.

Acknowledgements

We thank Ashleigh Griffin, Ian Hardy, Andrew Read, Ana Rivero and Ben Sheldon for useful discussion and/or comments on the manuscript. Funding was provided by the BBSRC, NERC, the Leverhulme Trust and STRI.

References

Anderson RM & May RM (1991) *Infectious Diseases of Humans. Dynamics and Control.* Oxford: Oxford University Press.

Barton NH & Turelli M (1989) Evolutionary quantitative genetics: how little do we know? *Annual Review of Genetics*, **23**, 337–370.

Basolo AL (1994) The dynamics of Fisherian sex-ratio evolution: theoretical and experimental investigations. *American Naturalist*, **144**, 473–490.

Boomsma JJ & Grafen A (1990) Intraspecific variation in ant sex ratios and the Trivers-Hare hypothesis. *Evolution*, **44**, 1026–1034.

Boomsma JJ & Grafen A (1991) Colony-level sex ratio selection in the eusocial Hymenoptera. *Journal of Evolutionary Biology*, **3**, 383–407.

Bourke AFG & Franks NR (1995) *Social Evolution in Ants.* Princeton, NJ: Princeton University Press.

Bull JJ (1981) Sex ratio when fitness varies. *Heredity*, **46**, 9–26.

Bull JJ & Charnov EL (1988) How fundamental are Fisherian sex ratios? In: PH Harvey & L Partridge (eds) *Oxford Surveys in Evolutionary Biology*, pp 96–135. Oxford: Oxford University Press.

Carvalho AB, Sampaio MC, Varandas FR & Klackzo LB (1998) An experimental demonstration of Fisher's principle: evolution of sexual proportions by natural selection. *Genetics*, **148**, 719–731.

Casas J, Nisbet RM, Swarbrick S & Murdoch WM (2000) Eggload dynamics and oviposition rate in a wild population of a parasitic wasp. *Journal of Animal Ecology*, **69**, 185–193.

Chapuisat M & Keller L (1999) Testing kin selection with sex allocation data in eusocial Hymenoptera. *Heredity*, **82**, 473–478.

Charlesworth D & Charlesworth B (1987) Inbreeding depression and its evolutionary consequences. *Annual Review of Ecology and Systematics*, **18**, 237–268.

Charnov EL (1979) The genetical evolution of patterns of sexuality: Darwinian fitness. *American Naturalist*, **113**, 465–480.

Charnov EL (1982) *The Theory of Sex Allocation*. Princeton, NJ: Princeton University Press.

Charnov EL & Bull JJ (1989) Non-Fisherian sex ratios with sex change and environmental sex determination. *Nature*, **338**, 148–150.

Charnov EL, Los-den Hartogh RL, Jones WT & van den Assem J (1981) Sex ratio evolution in a variable environment. *Nature*, **289**, 27–33.

Clutton-Brock TH & Iason GR (1986) Sex ratio variation in mammals. *Quarterly Review of Biology*, **61**, 339–374.

Colwell RK (1981) Group selection is implicated in the evolution of female-biased sex ratios. *Nature*, **290**, 401–404.

Conover DO & van Voorhees DA (1990) Evolution of a balanced sex ratio by frequency dependent selection in a fish. *Science*, **250**, 1556–1558.

Crozier RH & Pamilo P (1996) *Evolution of Social Insect Colonies: Sex Allocation and Kin Selection*. Oxford: Oxford University Press.

Cruden RW (1977) Pollen-ovule ratios: a conservative indicator of breeding systems in flowering plants. *Evolution*, **31**, 32–46.

Davies NB (1992) *Dunnock Behaviour and Social Evolution*. Oxford: Oxford University Press.

Donaldson JS & Walter GH (1991) Host population structure affects the field sex ratios of the heteronomous hyperparasitoid, *Coccophagus atratus*. *Ecological Entomology*, **16**, 25–33.

Driessen G & Hemerik L (1992) The time and egg budget of *Leptopilina clavipes*, a parasitoid of larval *Drosophila*. *Ecological Entomology*, **17**, 17–27.

Ellegren H, Gustafsson L & Sheldon BC (1996) Sex ratio adjustment in relation to paternal attractiveness in a wild bird population. *Proceedings of the National Academy of Science USA*, **93**, 11 723–11 728.

Ellers J, van Alphen JJM & Sevenster JG (1998) A field study of size-fitness relationships in the parasitoid *Asobara tabida*. *Journal of Animal Ecology*, **67**, 318–324.

Fisher RA (1930) *The Genetical Theory of Natural Selection*. Oxford: Clarendon Press.

Flanagan KE, West SA & Godfray HCJ (1998) Local mate competition, variable fecundity, and information utilization in a parasitoid. *Animal Behaviour*, **56**, 191–198.

Frank SA (1985) Hierarchical selection theory and sex ratios. II. On applying the theory, and a test with fig wasps. *Evolution*, **39**, 949–964.

Frank SA (1986) Hierarchical selection theory and sex ratios. I. General solutions for structured populations. *Theoretical Population Biology*, **29**, 312–342.

Frank SA (1987) Individual and population sex allocation patterns. *Theoretical Population Biology*, **31**, 47–74.

Frank SA (1990) Sex allocation theory for birds and mammals. *Annual Review of Ecology and Systematics*, **21**, 13–55.

Frank SA (1995) Sex allocation in solitary bees and wasps. *American Naturalist*, **146**, 316–323.

Frank SA (1998) *Foundations of Social Evolution*. Princeton, NJ: Princeton University Press.

Frank SA & Swingland IR (1988) Sex ratio under conditional sex expression. *Journal of Theoretical Biology*, **135**, 415–418.

Godfray HCJ (1994) *Parasitoids. Behavioral and Evolutionary Ecology*. Princeton, NJ: Princeton University Press.

Godfray HCJ & Werren JH (1996) Recent developments in sex ratio studies. *Trends in Ecology and Evolution*, **11**, 59–63.

Grafen A (1990) Do animals really recognise kin? *Animal Behaviour*, **39**, 42–54.

Greeff JM (1996) Alternative mating strategies, partial sibmating and split sex ratios in haplodiploid species. *Journal of Evolutionary Biology*, **9**, 855–869.

Greeff JM (1997) Offspring sex allocation in externally ovipositing fig wasps with varying clutch size and sex ratio. *Behavioral Ecology*, **8**, 500–505.

Green RE, Gordh G & Hawkins B (1982) Precise sex ratios in highly inbred parasitic wasps. *American Naturalist*, **120**, 653–665.

Hamilton WD (1964) The genetical evolution of social behaviour, I & II. *Journal of Theoretical Biology*, **7**, 1–52.

Hamilton WD (1967) Extraordinary sex ratios. *Science*, **156**, 477–488.

Hardy ICW (1994) Sex ratio and mating structure in the parasitoid Hymenoptera. *Oikos*, **69**, 3–20.

Herre EA (1985) Sex ratio adjustment in fig wasps. *Science*, **228**, 896–898.

Herre EA (1987) Optimality, plasticity and selective regime in fig wasp sex ratios. *Nature*, **329**, 627–629.

Herre EA, West SA, Cook JM, Compton SG & Kjellberg F (1997) Fig wasp mating systems: pollinators and parasites, sex ratio adjustment and male polymorphism, population structure and its consequences. In: J Choe & B Crespi (eds) *Social Competition and Cooperation in Insects and Arachnids*. Volume 1: *The Evolution of Mating Systems*, pp 226–239. Princeton, NJ: Princeton University Press.

Herre EA, Machado CA & West SA (2001) Selective regime and fig wasp sex ratios: towards sorting rigor from pseudo-rigor in tests of adaptation. In: S Orzack & E Sober (eds), *Adaptionism and Optimality*, pp 191–218. Cambridge: Cambridge University Press.

Houle D, Moriwaka B & Lynch M (1996) Comparing mutational variabilities. *Genetics*, 143, 1467–1483.

Hunter MS (1993) Sex allocation in a field population of an autoparasitoid. *Oecologia*, 93, 421–428.

Hunter MS & Godfray HCJ (1995) Ecological determinants of sex allocation in an autoparasitoid wasp. *Journal of Animal Ecology*, 64, 95–106.

Hurst LD (1993) The incidences, mechanisms and evolution of cytoplasmic sex ratio distorters in animals. *Biological Reviews*, 68, 121–193.

Hurst LD & Peck JR (1996) Recent advances in understanding of the evolution and maintenance of sex. *Trends in Ecology and Evolution*, 11, 46–53.

Kathuria P, Greeff JM, Compton SG & Ganeshaiah KN (1999) What fig wasp sex ratios do and do not tell us about sex allocation strategies. *Oikos*, 87, 520–530.

Kawecki TJ (1994) Accumulation of deleterious mutations and the evolutionary cost of being a generalist. *American Naturalist*, 144, 833–838.

Komdeur J, Daan S, Tinbergen J & Mateman C (1997) Extreme modification of sex ratio of the Seychelles warbler's eggs. *Nature*, 385, 522–525.

Krebs JR & Kacelnik A (1991) Decision-making. In: JR Krebs & NB Davies (eds) *Behavioural Ecology, an Evolutionary Approach*, pp 105–136. Oxford: Blackwell Scientific.

Kruuk LEB, Clutton-Brock TH, Albon SD, Pemberton JM & Guinness FE (1999) Population density affects sex ratio variation in red deer. *Nature*, 399, 459–461.

Lande R (1975) The maintenance of genetic variability by mutation in a polygenic character with linked loci. *Genetical Research*, 26, 221–236.

Leigh EG, Herre EA & Fischer EA (1985) Sex allocation in animals. *Experientia*, 41, 1265–1276.

Maynard Smith J (1982) *Evolution and the Theory of Games*. Cambridge: Cambridge University Press.

Merila J & Sheldon BC (1999) Genetic architecture of fitness and nonfitness traits: empirical patterns and development of ideas. *Heredity*, 83, 103–109.

Molbo D & Parker ED (1996) Mating structure and sex ratio variation in a natural population of *Nasonia vitripennis*. *Proceedings of the Royal Society London, series B*, 263, 1703–1709.

Nunney L & Luck RF (1988) Factors influencing the optimum sex ratio in structured populations. *Journal of Theoretical Biology*, 33, 1–30.

Ode PJ, Antolin MF & Strand MR (1995) Brood-mate avoidance in the parasitic wasp *Bracon hebetor* Say. *Animal Behaviour*, 49, 1239–1248.

Orzack SH & Gladstone J (1994) Quantitative genetics of sex ratio traits in the parasitic wasp, *Nasonia vitripennis*. *Genetics*, 137, 211–220.

Orzack SH & Parker ED (1990) Genetic variation for sex ratio traits within a natural population of a parsitic wasp. *Genetics*, 124, 373–384.

Orzack SH, Parker ED & Gladstone J (1991) The comparative biology of genetic variation for conditional sex ratio behaviour in a parasitic wasp, *Nasonia vitripennis*. *Genetics*, 127, 583–599.

Parker ED & Orzack SH (1985) Genetic variation for the sex ratio in *Nasonia vitripennis*. *Genetics*, 110, 93–105.

Parker GA & Maynard Smith J (1990) Optimality theory in evolutionary biology. *Nature*, 348, 27–33.

Paul REL, Packer MJ, Walmsley M, Lagog M, Ranford-Cartwright LC, Paru R & Day KP (1995) Mating patterns in malaria parasite populations of Papua New Guinea. *Science*, 269, 1709–1711.

Pickering J, Read AF, Guerrero S & West SA (2000) Sex ratio and virulence in two species of lizard malaria parasites. *Evolutionary Ecology Research*, 2, 171–184.

Read AF, Narara A, Nee S, Keymer AE & Day KP (1992) Gametocyte sex ratios as indirect measures of outcrossing rates in malaria. *Parasitology*, 104, 387–395.

Read AF, Anwar M, Shutler D & Nee S (1995) Sex allocation and population structure in malaria and related parasitic protozoa. *Proceedings of the Royal Society London, series B*, 260, 359–363.

Reboud X & Bell G (1997) Experimental evolution in *Chlamydomonas*. III. Evolution of specialist and generalist types in environments that vary in time and space. *Heredity*, 78, 507–514.

Rosenheim JA (1996) An evolutionary argument for egg limitation. *Evolution*, 50, 2089–2094.

Rosenheim JA (1999) The relative contribution of time and eggs to the cost of reproduction. *Evolution*, 53, 376–385.

Seger J & Stubblefield JW (1996) Optimization and adaptation. In: MR Rose & GV Lauder (eds) *Adaptation*, pp 93–123. San Diego CA: Academic Press.

Sevenster JG, Ellers J & Driessen G (1998) An evolutionary argument for time limitation. *Evolution*, **52**, 1241–1244.

Sheldon BC, Andersson S, Griffith SC, Ornborg J & Sendecka J (1999) Ultraviolet colour variation influences blue tit sex ratios. *Nature*, **402**, 874–877.

Shutler D & Read AF (1998) Extraordinary and ordinary blood parasite sex ratios. *Oikos*, **82**, 417–424.

Stearns SC (1992) *Evolution of Life Histories*. Oxford: Oxford University Press.

Stearns SC & Kawecki TJ (1994) Fitness sensitivity and the canalization of life-history traits. *Evolution*, **48**, 1438–1450.

Strand MR (1988) Variable sex ratio strategy of *Telenomus heliothidis* (Hymenoptera: Scelionidae): adaptation to host and conspecific density. *Oecologia*, **77**, 219–224.

Stubblefield JW & Seger J (1990) Local mate competition with variable fecundity: dependence of offspring sex ratios on information utilization and mode of male production. *Behavioural Ecology*, **1**, 68–80.

Sundstrom L (1994) Sex ratio bias, relatedness asymmetry and queen mating frequency in ants. *Nature*, **367**, 266–268.

Sundstrom L, Chapuisat M & Keller L (1996) Conditional manipulation of sex ratios by ant workers: a test of kin selection theory. *Science*, **274**, 993–995.

Taylor LH (1997) *Epidemiological and evolutionary consequences of mixed-genotype infections of malaria parasites*. Ph.D. thesis, University of Edinburgh.

Taylor PD (1981) Intra-sex and inter-sex sibling interactions as sex determinants. *Nature*, **291**, 64–66.

Taylor PD & Bulmer MG (1980) Local mate competition and the sex ratio. *Journal of Theoretical Biology*, **86**, 409–419.

Taylor PD & Crespi BJ (1994) Evolutionary stable strategy sex ratios when correlates of relatedness can be assessed. *American Naturalist*, **143**, 297–316.

Trivers RL & Hare H (1976) Haplodiploidy and the evolution of the social insects. *Science*, **191**, 249–263.

Trivers RL & Willard DE (1973) Natural selection of parental ability to vary the sex ratio of offspring. *Science*, **179**, 90–92.

Turelli M (1984) Heritable genetic variation via mutation-selection balance: Lerch's zeta meets the abdominal bristle. *Theoretical Population Biology*, **25**, 138–193.

Varandas FR, Sampaio MC & Carvalho AB (1997) Heritability of sexual proportion in experimental *sex ratio* populations of *Drosophila mediopunctata*. *Heredity*, **79**, 104–112.

Werren JH (1980) Sex ratio adaptations to local mate competition in a parasitic wasp. *Science*, **208**, 1157–1159.

Werren JH (1983) Sex ratio evolution under local mate competition in a parasitic wasp. *Evolution*, **37**, 116–124.

Werren JH (1987) Labile sex ratios in wasps and bees. *Bioscience*, **37**, 498–506.

Werren JH & Taylor PD (1984) The effects of population recruitment on sex ratio selection. *American Naturalist*, **124**, 143–148.

West SA & Godfray HCJ (1997) Sex ratio strategies after perturbation of the stable age distribution. *Journal of Theoretical Biology*, **186**, 213–221.

West SA & Herre EA (1998a) Partial local mate competition and the sex ratio: a study on non-pollinating fig wasps. *Journal of Evolutionary Biology*, **11**, 531–548.

West SA & Herre EA (1998b) Stabilizing selection and variance in fig wasp sex ratios. *Evolution*, **52**, 475–485.

West SA & Rivero A (2000) Using sex ratios to estimate what limits reproduction in parasitoids. *Ecology Letters*, **3**, 294–299.

West SA, Herre EA, Windsor DW & Green PRS (1996) The ecology and evolution of the New World non-pollinating fig wasp communities. *Journal of Biogeography*, **23**, 447–458.

West SA, Flanagan KE & Godfray HCJ (1999a) Sex allocation and clutch size in parasitoid wasps that produce single sex broods. *Animal Behaviour*, **57**, 265–275.

West SA, Lively CM & Read AF (1999b) A pluralist approach to the evolution of sex and recombination. *Journal of Evolutionary Biology*, **12**, 1003–1012.

West SA, Herre EA & Sheldon BC (2000a) The benefits of allocating sex. *Science*, **290**, 288–290.

West SA, Smith TG & Read AF (2000b) Sex allocation and population structure in Apicomplexan (Protozoa) parasites. *Proceedings of the Royal Society London, series B*, **267**, 257–263.

West SA, Reece SE & Read AF (2001) Evolution of gametocyte sex ratios in malaria and related apicomplexan (protozoan) parasites. *Trends in Parasitology*, **17**, 525–531.

Whitlock MC (1996) The red queen beats the jack-of-all-trades: the limitations on the evolution of phenotypic plasticity and niche breadth. *American Naturalist*, **148**, S65–S77.

Index

Printed in the United States
By Bookmasters